Distribution of Zeros of Entire Functions

Revised Edition

Translations of
MATHEMATICAL
MONOGRAPHS

Volume 5

Distribution of Zeros of Entire Functions

Revised Edition

B. Ja. Levin

American Mathematical Society
Providence, Rhode Island

РАСПРЕДЕЛЕНИЕ КОРНЕЙ ЦЕЛЫХ ФУНКЦИЙ

Б. Я. ЛЕВИН

Государственное Издательство
Технико-Теоретической Литературы

Москва 1956

Translated from the Russian by R. P. Boas, J. M. Danskin, F. M. Goodspeed, J. Korevaar, A. L. Shields, H. P. Thielman

Publication aided by grant NSF-GN 57
from the
NATIONAL SCIENCE FOUNDATION

Library of Congress Cataloging in Publication Data

Levin, Boris I͡Akovlevich.
Distribution of zeros of entire functions.

(Translations of mathematical monographs ; v. 5)
Translation of Raspredelenie korneĭ t͡selykh funkt͡siĭ.
Bibliography: p.
1. Functions, Entire. I. Title. II. Series.
QA353.E5L4813 1980 515.9'8 80-36891
ISBN 0-8218-4505-5

Second printing, first edition, 1972

Printed in the United States of America.

♾ The paper used in this book is acid-free and falls within the guidelines established to ensure permanence and durability.
Visit the AMS home page at URL: http://www.ams.org/

10 9 8 7 6 5 4 3 04 03 02 01 00

TABLE OF CONTENTS

PREFACE TO THE SECOND EDITION

In this edition I have corrected some inaccuracies and misprints, of which there were a rather large number in the first edition. Since the appearance of the first edition many papers have been published on the theory of functions of completely regular growth and related topics. However, I do not have the time to expound these new results, which would fill an additional volume.

I have tried to make up for this shortcoming by stating new results in a number of places, without proofs, when they are connected with the topics under discussion. I have also added an appendix, "Further development of the theory of functions of completely regular growth," in which I present a survey of recent results. In addition, I have cited papers in which this material is presented in more detail, with proofs.

B. Ja. Levin
1978

PREFACE TO THE FIRST EDITION

One of the most important problems in the theory of entire functions is the problem of the connection between the growth of an entire function and the distribution of its zeros. Many other problems in fields close to complex function theory lead to this problem.

The connection between the growth of an entire function and the distribution of its zeros was investigated in the classical works of Borel, Hadamard, Lindelöf and other authors at the end of the last and the beginning of the present century.

More precise characterizations of the growth and of the distribution of the zeros of an entire function led to the possibility of establishing more precise relations between them. At the same time analogous results were obtained for a wider class of functions, holomorphic in an angle.

Especially precise relations were obtained for a special class of functions which it is natural to call the functions of completely regular growth. In this book the theory of functions of completely regular growth is applied systematically to the investigation of various questions in the theory of entire functions.

Chapters II and III are devoted to the theory of functions of completely regular growth.

Then in Chapters IV, V, and VI this theory is applied to various questions of completeness and uniqueness, interpolation, distribution of the roots of exponential sums, properties of entire functions bounded on the real axis, and so forth.

In the succeeding chapters we study a range of questions connected with carrying over to entire functions certain properties of polynomials. This includes the generalization of the Hermite-Biehler theorem, begun in the works of N. G. Čebotarev and developed in the works of other Soviet mathematicians (Chapter VII), a presentation of the results of Laguerre-Pólya and Schur on algebraic properties of entire functions (Chapter VIII) and a far-reaching generalization of the inequalities of A. A. and V. A. Markov and of S. N. Bernšteĭn (Chapter IX).

In these chapters the machinery of the theory of functions of completely regular growth does not play a decisive role; however, it simplifies the investigation and shortens some of the proofs.

Since I was unable to find a suitable reference where the general theory of entire functions is presented in the generality and in the form that I require, I felt it necessary to expand the introduction I had originally planned into a rather extensive chapter devoted to the general theory of entire functions.

Chapter V can be read independently of Chapter IV, and VII, VIII and IX can be read independently of Chapters IV and VI. At the end of the book there are several supplements in which applications of the results and methods presented in the book are given to related fields.

The range of questions treated in this book was determined by the interests of the author and the direction of his own investigations.

At the end of the book there is a list of the most important theorems and concepts. By using this list the reader can quickly recall not only the definitions of the concepts, but also the basic facts relating to them. The classical concepts and theorems are not included in this list.

I wish to thank M. G. Kreĭn, who persuaded me to write this book, and who read it in manuscript and made a number of important comments. I also wish to thank N. I. Ahiezer, who read the manuscript as it was being written and made a number of suggestions that I followed. I am also very thankful to N. S. Landkoff who read the entire manuscript very carefully and made a number of comments enabling me to improve the book.

CHAPTER I

THE GENERAL THEORY OF THE GROWTH OF ENTIRE FUNCTIONS

In this chapter we present the elements of the general theory of growth of entire functions. An *entire function* is a function of a complex variable holomorphic in the entire plane and consequently represented by an everywhere convergent power series

$$f(z) = a_0 + a_1 z + a_2 z^2 + \dots + a_n z^n + \dots \qquad (1.00)$$

These functions form a natural generalization of the polynomials, and are close to polynomials in their properties. The theorem of Weierstrass on the expansion of entire functions into infinite products provided the basic apparatus for the investigation of the properties of entire functions and was the starting point for their classification. At approximately the same time as this work of Weierstrass, Laguerre studied the connection between entire functions and polynomials, and introduced the important concept of the genus of an entire function.

The classical investigations of Borel, Hadamard and Lindelöf dealt with the connection between the growth of an entire function and the distribution of its zeros. The rate of growth of a polynomial as the independent variable goes to infinity is determined, of course, by its degree. On the other hand, the number of roots of a polynomial is equal to its degree. Thus, the more roots a polynomial has, the greater is its growth. This connection between the set of zeros of the function and its growth can be generalized to arbitrary entire functions. The content of most of the classical theorems of the theory of entire functions consists in establishing relations between the distribution of the roots of an entire function and its asymptotic behavior as $z \to \infty$.

To measure the growth of an entire function and the density of its zeros, a special growth scale is introduced.

1. The Growth Scale

To characterize the growth of an entire function $f(z)$ we introduce the function

$$M_f(r) = \max_{|z|=r} |f(z)|$$

(sometimes we shall simply write $M(r)$). It is easily seen that $M_f(r)$ is continuous. In fact, let $r_1 < r_2$ and $|f(r_2e^{i\theta_0})| = M(r_2)$. Then

$$0 < M(r_2) - M(r_1) \leq |f(r_2e^{i\theta_0})| - |f(r_1e^{i\theta_0})| < \varepsilon$$

for $r_2 - r_1 < \delta_\varepsilon$.

It follows from the maximum modulus principle that as r increases $M_f(r)$ grows monotonically. The rate of growth of the function $M_f(r)$ is an important characteristic of the behavior of the entire function. Let us show that for an entire function that is not a polynomial $M_f(r)$ grows faster than any positive power of r.

THEOREM 1. *If there exists a positive integer n such that*

$$\varliminf_{r\to\infty} \frac{M_f(r)}{r^n} < \infty,$$ [1]

then $f(z)$ is a polynomial of degree at most n.

PROOF. If

$$f(z) = a_0 + a_1z + \ldots + a_nz^n + a_{n+1}z^{n+1} + \ldots$$

and

$$P_n(z) = a_0 + a_1z + \ldots + a_nz^n,$$

then the function

$$\varphi(z) = [f(z) - P_n(z)]\, z^{-n-1}$$

is entire and tends uniformly to zero on some sequence of circles $|z| = r_n$ $(r_n \to \infty)$. It follows from the maximum principle that $\varphi(z) \equiv 0$, that is,

$$f(z) \equiv P_n(z).$$

Thus, in order to estimate the growth of transcendental entire functions, one must choose comparison functions that grow more rapidly then powers of r. We choose comparison functions of the form

$$e^{r^k},$$

where $k > 0$.

An entire function $f(z)$ is said to be a *function of finite order* if there exists a positive constant k such that the inequality

$$M_f(r) < e^{r^k}$$

is valid for all sufficiently large values of r $(r > r_0(k))$.

The greatest lower bound of such numbers k is called the *order* of the entire function $f(z)$.

---

[1] Here, as usual,

$$\varlimsup_{r\to\infty} \varphi(r) = \lim_{r\to\infty} \left(\sup_{t \geq r} \varphi(t) \right) \quad \text{and} \quad \varliminf_{r\to\infty} \varphi(r) = \lim_{r\to\infty} \left(\inf_{t \geq r} \varphi(t) \right).$$

It follows from this definition that if ρ is the order of the entire function $f(z)$, and if ε is an arbitrary positive number, then

$$e^{r^{\rho-\varepsilon}} < M_f(r) < e^{r^{\rho+\varepsilon}}, \qquad (1.01)$$

where the inequality on the right is satisfied for all sufficiently large values of r, and the inequality on the left holds for some sequence $\{r_n\}$ of values of r, tending to infinity. It is easy to verify that condition (1.01) is equivalent to the equation

$$\rho = \overline{\lim_{r\to\infty}} \frac{\ln \ln M_f(r)}{\ln r}, \qquad (1.02)$$

which could therefore be taken as the definition of the order of the function.

An inequality that holds for *all sufficiently large* values of r will be called an *asymptotic inequality*.

For functions of a given order a more precise characterization of the growth is given by the type of the function. By the *type* σ *of an entire function* $f(z)$ *of order* ρ we mean the greatest lower bound of positive numbers A for which asymptotically

$$M_f(r) < e^{Ar^\rho}.$$

Just as with the definition of order it is easy to verify that the type σ of a function $f(z)$ of order ρ is given by the equation

$$\sigma = \overline{\lim_{r\to\infty}} \frac{\ln M_f(r)}{r^\rho}.$$

If $\sigma = 0$, the function $f(z)$ is said to be of *minimal* type, if $0 < \sigma < \infty$ of *normal* type, and if $\sigma = \infty$ of *maximal* type.

An example of an entire function of order n and type σ is the function

$$e^{\sigma z^n}$$

for integral n. It is easy to verify also that $\sin z$ is an entire function of order one and normal type $\sigma = 1$. The function $\cos \sqrt{z}$ has order $\rho = \frac{1}{2}$ and type $\sigma = 1$. The function

$$e^{e^z}$$

is an example of an entire function of infinite order.

In the following section we construct examples of entire functions of arbitrary order and type.

We shall say that the function $f_2(z)$ is of larger growth than the function $f_1(z)$ if the order of $f_2(z)$ is greater than the order of $f_1(z)$, or if the orders are equal and the type of $f_2(z)$ is larger than the type of $f_1(z)$.

It is easy to see that the order of the sum of two functions is not greater than the larger of the orders of the two summands, and if the orders of the summands and of the sum are all equal, then the type of the sum is not greater than the larger of the types of the two summands. In addition, if one of the two functions is of larger growth than the other, then the sum has the same order and type as the function of larger growth.

2. The Connection Between the Growth of an Entire Function and the Rate of Decrease of its Taylor Coefficients

Let the entire function $f(z)$ have the power series

$$f(z) = c_0 + c_1 z + c_2 z^2 + \dots + c_n z^n + \dots . \tag{1.03}$$

The radius of convergence is infinite and therefore

$$\lim_{n \to \infty} \sqrt[n]{|c_n|} = 0. \tag{1.04}$$

The following theorem enables one to determine the order and type of an entire function by the rate of decrease of its Taylor coefficients.

THEOREM 2. *The order and type of an entire function are expressed in terms of its Taylor coefficients by the following equations*

$$\rho = \overline{\lim_{n \to \infty}} \frac{n \ln n}{\ln \frac{1}{|c_n|}}, \tag{1.05}$$

$$(\sigma e \rho)^{\frac{1}{\rho}} = \overline{\lim_{n \to \infty}} \left(n^{\frac{1}{\rho}} \sqrt[n]{|c_n|} \right). \tag{1.06}$$

PROOF. By a well-known inequality for the coefficients of a power series

$$|c_n| \leqslant \frac{M_f(r)}{r^n}. \tag{1.07}$$

If $f(z)$ is of finite order, then asymptotically

$$M_f(r) < e^{Ar^k}, \qquad |c_n| < e^{Ar^k} r^{-n}. \tag{1.08}$$

Employing the usual method for finding extrema it is easy to see that the function on the right side of this inequality takes its largest value in the range $r > 0$ for

$$r = \left(\frac{n}{Ak}\right)^{\frac{1}{k}}$$

and therefore asymptotically

$$|c_n| < \left(\frac{eAk}{n}\right)^{\frac{n}{k}}. \tag{1.09}$$

Conversely, assume that (1.09) holds for all indices n greater than $n_0(k, A)$, and let us estimate $M_f(r)$. For $n > m_r = [2^k eAkr^k]$ and all sufficiently large values of r we have by (1.09)

$$|c_n z^n| < 2^{-n}$$

and therefore

$$|f(z)| < \sum_{n=0}^{m_r} |c_n| r^n + 2^{-m_r}.$$

Introducing the notation

$$\mu(r) = \max_n |c_n| r^n,$$

we have

$$M_f(r) \leqslant (1 + 2^k eAkr^k)\mu(r) + 2^{-m_r}. \tag{1.10}$$

If $f(z)$ is not a polynomial, then $M_f(r)$, and therefore by (1.10) also $\mu(r)$, grows faster than any power of r, and therefore the index of the largest term in the series (1.03) increases without bound as r grows. It follows from (1.09) that asymptotically

$$\mu(r) \leqslant \max_n \left(\frac{eAk}{n}\right)^{\frac{n}{k}} r^n.$$

The maximum of the right side is attained for

$$n = Akr^k$$

and therefore asymptotically

$$\mu(r) \leqslant e^{Ar^k}.$$

From (1.10) we see that asymptotically

$$M_f(r) < (2 + 2^k eAkr^k)\, e^{Ar^k}. \tag{1.11}$$

Thus (1.09) follows from (1.08), and (1.11) follows from (1.09).

This shows that the order ρ of an entire function $f(z)$ is equal to the greatest lower bound of numbers k for which the asymptotic inequality (1.09) holds, and the type is equal to the greatest lower bound of numbers A for which the asymptotic inequality (1.09) holds with $k = \rho$. From this both assertions of the theorem follow at once.

We note an interesting corollary that follows at once from formulas (1.07) and (1.10). The maximal term $\mu(r)$ in the power series expansion of an entire function of finite order satisfies the asymptotic inequalities

$$\mu(r) \leqslant M(r) < r^{\rho+\varepsilon}\mu(r),$$

where ρ is the order of the function $f(z)$ and ε is an arbitrary positive number.

With the aid of Theorem 2 one can easily construct entire functions of arbitrary order and type. To this end consider the entire function

$$f(z) = \sum_{n=0}^{\infty} \frac{(A^\alpha z)^n}{\Gamma(\alpha n + 1)},$$

where $A > 0$ and $\alpha > 0$. By Stirling's formula

$$\Gamma(\alpha n + 1) = \left(\frac{\alpha n}{e}\right)^{\alpha n} \sqrt{2\pi\alpha n}\left[1 + O\left(\frac{1}{n}\right)\right]$$

we have[2] from (1.05) and (1.06)

$$\rho = \frac{1}{\alpha} \text{ and } \sigma = A.$$

Functions of maximal and minimal type can also be constructed. For example, for

$$c_n = \left(\frac{\ln n}{n}\right)^{\frac{n}{\rho}} \qquad (n = 1,\ 2,\ 3,\ \ldots)$$

the series (1.03) represents a function of order ρ and maximal type, while for

$$c_n = \left(\frac{1}{n \ln n}\right)^{\frac{n}{\rho}} \qquad (n = 2,\ 3,\ \ldots)$$

it represents a function of order ρ and minimal type. For $c_n = e^{-n^2}$ we obtain an entire function of order zero. All these assertions follow at once from (1.05) and (1.06). Similarly one can construct entire functions of infinite order.

The limit superiors in formulas (1.05) and (1.06) are unchanged if c_n is replaced by $(n + 1)c_n$. Consequently, order and type are unchanged by differentiation.

3. The Expansion of Entire Functions in Infinite Products

It is well known that every polynomial can be written as the product of linear factors. The analogue of this assertion for entire functions is the theorem of Weierstrass on the representation of entire functions by infinite products. This representation is the basis for the investigation of the basic question of the theory of entire functions—the question of the relation between the growth of the entire function and the distribution of its zeros in the complex plane.

Let $a_1, a_2, \ldots, a_n, \ldots$ be a sequence of complex numbers, none of which is zero, with the point at infinity as the only limit point. We shall construct an entire function whose set of zeros is precisely this sequence.

---

[2] *Translator's note*: The entire function

$$f(z) = \sum \left(\frac{\sigma e \rho}{n}\right)^{n/\rho} z^n$$

has order ρ and type σ.

We can assume that these points have been arranged in order of increasing moduli (if several different points a_n have the same modulus, then we take them in any order). Choose a sequence of natural numbers p_n such that the series

$$\sum_{n=1}^{\infty} \left| \frac{z}{a_n} \right|^{p_n+1} \tag{1.12}$$

converges uniformly in each bounded domain. Such a choice is possible since, for $|z| \leqslant R \ (R > 0)$ the inequality

$$\left| \frac{z}{a_n} \right| < q < 1$$

is satisfied for all sufficiently large values of n, and thus, for example, one can choose $p_n = n$. We form the infinite product

$$\Pi(z) = \prod_{n=1}^{\infty} G\left(\frac{z}{a_n};\ p_n\right), \tag{1.13}$$

in which

$$G(u;\ p) = (1-u)\, e^{u + \frac{u^2}{2} + \ldots + \frac{u^p}{p}}, \qquad G(u;\ 0) = 1-u.$$

The functions $G(u; p)$ are called *primary factors*.

Let us show that the product (1.13) converges uniformly on each closed bounded set that contains none of the points a_n, and therefore defines an entire function that vanishes at the points a_n and only at them. To this end we estimate the quantity $|\ln G(u; p)|$ for $|u| \leqslant q < 1$ and $|\arg(1-u)| \leqslant \pi$. From the expansion

$$\ln G(u;\ p) = -\frac{u^{p+1}}{p+1} - \frac{u^{p+2}}{p+2} - \ldots,$$

which is valid under the conditions indicated, we obtain

$$|\ln G(u;\ p)| < \frac{|u|^{p+1}}{1-|u|} \leqslant \frac{1}{1-q} |u|^{p+1}.$$

It follows from this inequality that for $|z| \leqslant R$ and $n > n(q, R)$ we will have

$$\left| \ln G\left(\frac{z}{a_n};\ p_n\right) \right| < \frac{1}{1-q} \left| \frac{z}{a_n} \right|^{p_n+1},$$

and this implies, because of the uniform convergence of the series (1.12) in the circle $|z| \leqslant R$, that the series

$$\sum_{n=1}^{\infty} \ln G\left(\frac{z}{a_n};\ p_n\right),$$

and therefore also the product (1.13), converges uniformly in each closed subset of this circle that contains none of the points a_n.

THEOREM 3. *Every entire function $f(z)$ can be represented in the form*

$$f(z)=z^m e^{g(z)}\prod_{n=1}^{\omega} G\left(\frac{z}{a_n};\, p_n\right) \qquad (\omega \leqslant \infty), \tag{1.14}$$

where $g(z)$ is an entire function, a_n are the nonzero roots of $f(z)$, and m is the order of the zero of $f(z)$ at the origin.

PROOF. The nonzero roots a_n of the entire function $f(z)$ $(n = 1, 2, \ldots, \omega \leqslant \infty)$ have no finite limit point. We form the product

$$\Pi(z)=\prod_{n=1}^{\omega} G\left(\frac{z}{a_n};\, p_n\right),$$

in which $p_n = 0$, if ω is finite. The function

$$\varphi(z)=\frac{f(z)}{z^m\Pi(z)}$$

is entire and has no zeros. Hence the function

$$g(z)=\ln\varphi(z)$$

is also entire and we obtain the desired representation

$$f(z)=z^m e^{g(z)}\Pi(z).$$

In the representation (1.14) the sequence of numbers p_n is not uniquely determined, and therefore the function $g(z)$ is not determined uniquely either.

The representation of the function $f(z)$ is considerably simpler if the numbers a_n satisfy the following supplementary condition: the series

$$\sum_{n=1}^{\infty}\frac{1}{|a_n|^{\lambda}} \tag{1.12'}$$

converges for some positive λ. In this case let p denote the smallest integer for which the series

$$\sum_{n=1}^{\infty}\frac{1}{|a_n|^{p+1}}$$

converges. Clearly, $0 \leqslant p < \lambda$. The series (1.12) will converge uniformly if we put all $p_n = p$.

The uniformly convergent infinite product

$$\Pi(z)=\prod_{n=1}^{\infty} G\left(\frac{z}{a_n};\, p\right) \tag{1.15}$$

is called a *canonical product*, and the number p is called the *genus of the canonical product.*

In the case just considered it is customary in the representation (1.14) to choose the canonical product for the infinite product. Then the function $g(z)$ is uniquely determined.

If $g(z)$ is a polynomial, $f(z)$ is said to be an *entire function of finite genus.* The larger of the numbers p and q, where q is the degree of the polynomial $g(z)$, is called the genus of the entire function $f(z)$.

If $g(z)$ is not a polynomial, or if the series (1.12′) diverges for all values of λ, then the genus is said to be infinite.

4. Estimates for Canonical Products

The representation of an entire function as an infinite product makes it possible to establish a very important dependence between the growth of the function and the density of distribution of its zeros. As a measure of the density of the sequence of points a_n, having no finite limit point, we introduce the convergence exponent. By the *convergence exponent* of the sequence

$$a_1,\ a_2,\ \ldots,\ a_n,\ \ldots \qquad (a_n \neq 0, \qquad \lim_{n\to\infty} a_n = \infty)$$

we mean the greatest lower bound of numbers λ for which the series

$$\sum_{n=1}^{\infty} \frac{1}{|a_n|^{\lambda}} \tag{1.16}$$

converges. Clearly the more rapidly the sequence of numbers $|a_n|$ increases, the smaller will be the convergence exponent. In particular, the convergence exponent can be zero.

If the series (1.16) diverges for all $\lambda > 0$, then the convergence exponent is said to be infinite. If λ is equal to the convergence exponent, the series (1.16) may or may not converge, depending on the sequence $\{a_n\}$.

For example, the sequences $a_n = n^{1/\rho}$ and $a_n = (n \ln^2 n)^{1/\rho}$ have the same convergence exponent ρ, but in the first case the series (1.16) diverges when $\lambda = \rho$, while in the second case it converges. The sequences $\{e^n\}$ and $\{\ln n\}$ have convergence exponents 0 and ∞, respectively.

Let us recall again the following obvious relation between the convergence exponent ρ_1 of the sequence $\{a_n\}$ and the genus p of the corresponding canonical product (1.15)

$$p \leqslant \rho_1 \leqslant p + 1.$$

If ρ_1 is an integer, then $p = \rho_1$ when the series (1.16) diverges for $\lambda = \rho_1$, while $\rho_1 = p + 1$ means that the series converges.

A more precise description of the density of the sequence $\{a_n\}$ than the convergence exponent is given by the growth of the function $n(r)$ equal to the number of points of the sequence in the circle $|z| < r$. By the *order* of this monotone function we mean the number

$$\rho_1 = \overline{\lim_{r\to\infty}} \frac{\ln n(r)}{\ln r}, \tag{1.17}$$

and by the *upper density* of the sequence $\{a_n\}$ we mean the number

$$\Delta = \overline{\lim_{r\to\infty}} \frac{n(r)}{r^{\rho_1}}. \tag{1.18}$$

If the limit exists, then Δ is simply called the density.

LEMMA 1. *The convergence exponent of the sequence*

$$a_1, a_2, \ldots, a_n, \ldots \ (\lim_{n\to\infty} |a_n| = \infty)$$

is equal to the order of the corresponding function $n(r)$.

PROOF. The series (1.16) can be rewritten as a Stieltjes integral in the following form:

$$\int_0^\infty \frac{dn(t)}{t^\lambda}.$$

Integrating by parts we have[3]

$$\int_0^r \frac{dn(t)}{t^\lambda} = \frac{n(r)}{r^\lambda} + \lambda \int_0^r \frac{n(t)}{t^{\lambda+1}} dt. \tag{1.19}$$

If the series (1.16) converges, then both the positive terms on the right side of (1.19) are bounded, and since the second term is monotone increasing, the integral

$$\int_0^\infty \frac{n(t)}{t^{\lambda+1}} dt \tag{1.20}$$

converges. From this convergence it follows that for arbitrary $\varepsilon > 0$ and $r > r_0(\varepsilon)$

$$\varepsilon > \lambda \int_r^\infty \frac{n(t)}{t^{\lambda+1}} dt \geqslant \lambda n(r) \int_r^\infty \frac{dt}{t^{\lambda+1}} = \frac{n(r)}{r^\lambda},$$

or

$$\lim_{r\to\infty} \frac{n(r)}{r^\lambda} = 0.$$

Thus the order of the function $n(r)$ is not greater than the convergence exponent of the sequence. Also, from the above reasoning we see that conversely the

[3] Here we use the fact that $n(t) = 0$ for $0 \leqslant t < |a_1|$.

convergence of the integral (1.20) implies the convergence of the series (1.16).

Let now ρ_1 be the order of the function $n(r)$. Then for $\varepsilon > 0$ we have asymptotically

$$n(t) < t^{\rho_1 + \frac{\varepsilon}{2}}.$$

Therefore for $\lambda = \rho_1 + \varepsilon$ the integral (1.20) converges, and therefore the series (1.16) also converges. Thus the convergence exponent is not greater than the order of the function $n(r)$. The lemma is proved.

To estimate the canonical product (1.15) we first estimate the primary factors.

LEMMA 2. *For $p > 0$ and all complex numbers u*

$$\ln |G(u;\ p)| < A_p \frac{|u|^{p+1}}{1+|u|}, \tag{1.21}$$

where

$$A_p = 3e(2 + \ln p).$$

For $p = 0$

$$\ln |G(u;\ 0)| \leqslant \ln(1 + |u|). \tag{1.22}$$

PROOF. The second assertion of the lemma is obvious. Let $p > 0$. If $|u| \leqslant p/(p+1)$, then

$$\ln |G(u;\ p)| = \operatorname{Re}\left\{\ln(1-u) + u + \frac{u^2}{2} + \dots + \frac{u^p}{p}\right\}$$
$$= -\operatorname{Re}\sum_{p+1}^{\infty}\frac{u^k}{k} \leqslant \sum_{p+1}^{\infty}\frac{|u|^k}{k} < \frac{|u|^{p+1}}{(p+1)(1-|u|)} \leqslant |u|^{p+1}.$$

For $|u| > p/(p+1)$ the inequality $\ln(1+|u|) \leqslant |u|$ leads to the following:

$$\ln |G(u;\ p)| \leqslant 2|u| + \frac{|u|^2}{2} + \dots + \frac{|u|^p}{p}$$
$$= |u|^p\left(\frac{1}{p} + \frac{1}{p-1}\left|\frac{1}{u}\right| + \dots + \frac{1}{2}\left|\frac{1}{u}\right|^{p-2} + 2\left|\frac{1}{u}\right|^{p-1}\right)$$
$$< |u|^p\left[2\left(1+\frac{1}{p}\right)^{p-1} + \sum_{k=2}^{p}\frac{1}{k}\left(1+\frac{1}{p}\right)^{p-k}\right]$$
$$< |u|^p\left(1+\frac{1}{p}\right)^p\left(2 + \sum_{k=2}^{p}\frac{1}{k}\right) < |u|^p\left(1+\frac{1}{p}\right)^p\left(2 + \int_1^p\frac{dx}{x}\right)$$
$$< |u|^p e(2 + \ln p) < \frac{|u|^{p+1}}{1+|u|}\, 3e(2 + \ln p).$$

LEMMA 3. *If the series*

$$\sum_{n=1}^{\infty}\frac{1}{|a_n|^{p+1}} \tag{1.23}$$

converges, then the infinite product

$$\Pi(z)=\prod_{n=1}^{\infty} G\left(\frac{z}{a_n};\, p\right) \tag{1.15}$$

satisfies in the entire complex plane the inequality

$$\ln|\Pi(z)|<k_p r^p\left(\int_0^r \frac{n(t)}{t^{p+1}}\,dt+r\int_r^{\infty}\frac{n(t)}{t^{p+2}}\,dt\right)\quad (r=|z|), \tag{1.24}$$

in which

$$k_p=3e(p+1)(2+\ln p)\quad \textit{for}\quad p>0\quad \textit{and}\ k_0=1.$$

PROOF. From inequality (1.21) with $p>0$ we have

$$\ln|\Pi(z)|<A_p\sum_{n=1}^{\infty}\frac{r^{p+1}}{|a_n|^p(|a_n|+r)}=A_p r^{p+1}\int_0^{\infty}\frac{dn(t)}{t^p(t+r)}.$$

As was proved above, it follows from the convergence of the series (1.23) that

$$\lim_{r\to\infty}\frac{n(r)}{r^{p+1}}=0.$$

Integrating by parts we have

$$\ln|\Pi(z)|<A_p r^{p+1}\int_0^{\infty}\frac{pr+(p+1)t}{t^{p+1}(t+r)^2}\,n(t)\,dt$$

$$<A_p(p+1)r^{p+1}\int_0^{\infty}\frac{n(t)\,dt}{t^{p+1}(t+r)}$$

or

$$\ln|\Pi(z)|<A_p(p+1)r^p\left(\int_0^r\frac{n(t)}{t^{p+1}}\,dt+r\int_r^{\infty}\frac{n(t)}{t^{p+2}}\,dt\right).$$

The case $p=0$ is verified in a similar manner, replacing the estimate (1.21) by (1.22).

THEOREM 4 (BOREL). *The order ρ of the canonical product*

$$\Pi(z)=\prod_{n=1}^{\infty} G\left(\frac{z}{a_n};\, p\right)$$

does not exceed the convergence exponent ρ_1 of the sequence $\{a_n\}$.

PROOF. The number p is the smallest integer for which the series (1.23) converges.

Thus for the convergence exponent ρ_1 we have the inequality

$$p \leqslant \rho_1 \leqslant p+1.$$

Suppose that $\rho_1 < p + 1$ and $\rho_1 < \lambda < p + 1$. Then by (1.17) there exists a constant C_λ such that for all values $t > 0$

$$n(t) < C_\lambda t^\lambda.$$

From Lemma 3 we have

$$\ln|\Pi(z)| < C_\lambda k_p r^p \left(\int_0^r t^{\lambda-p-1} dt + r\int_r^\infty t^{\lambda-p-2} dt\right) = B_\lambda C_\lambda r^\lambda,$$

where

$$B_\lambda = k_p\left[\frac{1}{\lambda-p} + \frac{1}{p+1-\lambda}\right],$$

that is, the order[4] of $\Pi(z)$ does not exceed λ, and therefore, it does not exceed ρ_1. If $\rho_1 = p + 1$, then, as was shown in the proof of Lemma 1,

$$\lim_{t\to\infty} \frac{n(t)}{t^{p+1}} = 0$$

and the integral

$$\int_0^\infty \frac{n(t)}{t^{p+2}}\, dt$$

converges. Using this we have from the lemma that asymptotically

$$\ln|\Pi(z)| < \varepsilon r^{p+1},$$

that is, $\Pi(z)$ is at most of order $p + 1$ and minimal type.

REMARK. If ρ_1 is not an integer and if the upper density of the sequence $\{a_n\}$ is finite, then $\Pi(z)$ is at most of order ρ_1 and normal type; if ρ_1 is not an integer and if the density Δ of the sequence $\{a_n\}$ is zero, then $\Pi(z)$ is at most of order ρ_1 and minimal type.

This assertion follows easily from the asymptotic inequality

$$n(t) < (\Delta + \varepsilon)\, t^{\rho_1},$$

which, for ρ_1 not an integer, leads to the asymptotic inequality

$$\ln|\Pi(z)| < (\Delta + \varepsilon)\, B_{\rho_1} r^{\rho_1}. \tag{1.25}$$

Note that we cannot obtain this inequality from Lemma 3 in case ρ_1 is an integer. Later on we shall see that for integral ρ_1 the assertion ceases to be true.

For the further investigation of the connection between the growth of the function $n(r)$ and the growth of $\ln M_\Pi(r)$ we require a series of general theorems that play a fundamental role in the theory of entire functions.

[4] Clearly in this inequality one can replace $|\Pi(z)|$ by $M_\Pi(r)$.

5. Jensen's Theorem

THEOREM 5. *Let $f(z)$ be holomorphic in a circle of radius R with center at the origin, and $f(0) \neq 0$. Then*

$$\int_0^R \frac{n_f(t)}{t}\,dt = \frac{1}{2\pi}\int_0^{2\pi} \ln|f(Re^{i\theta})|\,d\theta - \ln|f(0)|, \tag{1.26}$$

where $n_f(t)$ is the number of zeros of $f(z)$ in the circle $|z| < t$.

Note that if the function has no zeros in the circle $|z| < R$, then $n(t) = 0$ and equation (1.26) expresses a well-known property of harmonic functions. If there are zeros, then it follows from (1.26) that

$$\ln|f(0)| < \frac{1}{2\pi}\int_0^{2\pi} \ln|f(Re^{i\theta})|\,d\theta.$$

PROOF. If the function has no zeros on the circle $|z| = t$, then by the argument principle we have

$$n_f(t) = \frac{1}{2\pi}\int_0^{2\pi} \frac{d\,[\arg f(te^{i\theta})]}{d\theta}\,d\theta.$$

Using the Cauchy-Riemann equations

$$\frac{\partial\,[\arg f(te^{i\theta})]}{t\partial\theta} = \frac{\partial \ln|f(te^{i\theta})|}{\partial t},$$

we have

$$\frac{n_f(t)}{t} = \frac{1}{2\pi}\int_0^{2\pi} \frac{d}{dt}\ln|f(te^{i\theta})|\,d\theta$$

for all except a finite set of values of t in the interval $0 < t < R$. Integrating both sides of this equation from 0 to R we obtain (1.26). Here, of course, we are using the continuity of the function

$$\int_0^{2\pi} \ln|f(te^{i\theta})|\,d\theta.$$

It is easy to establish this continuity if we represent the function $\ln|f(z)|$ in the form $\ln|f_1(z)| + \sum_{|a_n| \leqslant R} \ln|z - a_k|$, where a_k are the zeros of the function $f(z)$, and $f_1(z)$ has no zeros in the circle $|z| \leqslant R$.

The first term is continuous, and

$$\int_0^{2\pi} \ln|Re^{i\theta} - a_k|\,d\theta = \begin{cases} 2\pi \ln R & \text{for } R > |a_k|, \\ 2\pi \ln|a_k| & \text{for } R < |a_k|, \end{cases}$$

so that this is also a continuous function. Thus the theorem is proved.

Jensen's formula can be generalized to include the case when $f(z)$ has a zero of order λ at the origin. Then, applying formula (1.26) to the function $f(z)z^{-\lambda}$, we obtain

$$\lambda \ln R + \int_0^R \frac{n_f(t)}{t}\,dt = \frac{1}{2\pi}\int_0^{2\pi} \ln|f(Re^{i\theta})|\,d\theta - \ln\frac{|f^{(\lambda)}(0)|}{\lambda!}.$$

Here $n_f(t)$ is defined to be the number of zeros of $f(z)$, different from the origin, in the circle $|z| < t$.

The following lemma gives an important estimate for the number of zeros of $f(z)$ in a circle.

LEMMA 4. *If $f(z)$ is holomorphic in the circle*

$$|z| \leqslant er$$

and if $|f(0)|$ 1, = *then*

$$n_f(r) \leqslant \ln M_f(er). \tag{1.27}$$

PROOF. From Jensen's formula and the monotonicity of $n_f(t)$ we have at once·

$$n_f(r) \leqslant \int_r^{er} \frac{n_f(t)}{t}\,dt \leqslant \frac{1}{2\pi}\int_0^{2\pi} \ln|f(ere^{i\theta})|\,d\theta \leqslant \ln M_f(er). \tag{1.28}$$

Let us investigate when the equality sign can hold in (1.27). For this it is necessary and sufficient that equality hold everywhere in (1.28).

The equation

$$n_f(r) = \int_r^{er} \frac{n_f(t)}{t}\,dt$$

says that the function $f(z)$ has no zeros in the annulus $r < |z| < er$.

The equation

$$\int_r^{er} \frac{n_f(t)}{t}\,dt = \frac{1}{2\pi}\int_0^{2\pi} \ln|f(ere^{i\theta})|\,d\theta$$

says, by (1.26) since $|f(0)| = 1$, that $f(z)$ has no zeros in the circle $|z| < r$.

Finally, the equation

$$\frac{1}{2\pi}\int_0^{2\pi} \ln|f(ere^{i\theta})|\,d\theta = \ln M_f(er)$$

says that on the whole circumference $|z| = er$

$$|f(z)| = M_f(er).$$

All these conditions are fulfilled for functions of the form[5]

$$\varphi(z) = e^{n+i\alpha} \prod_{k=1}^{n} \frac{er(z - a_k)}{e^2 r^2 - \bar{a}_k z}$$

$$(\,|a_k| = r, \ \alpha \text{ is real}\,).$$

Indeed, each factor in this product maps the circle $|z| \leqslant er$ onto the unit circle, and therefore has modulus one everywhere on the circumference $|z| = er$.

Thus on the circumference $|z| = er$ we have $|\varphi(z)| \equiv e^n$. In addition, all roots of $\varphi(z)$ are on the circle $|z| = r$, and $|\varphi(0)| = 1$.

From Lemma 4 we have at once

THEOREM 6. *The convergence exponent of the zeros of an arbitrary entire function does not exceed its order.*

PROOF. Without loss of generality we may assume that $f(0) = 1$. Otherwise in place of $f(z)$ one considers

$$f_1(z) = \lambda!\, z^{-\lambda} \frac{f(z)}{f^{(\lambda)}(0)},$$

which has the same order and convergence exponent as $f(z)$. From (1.27) it follows that

$$\rho_1 = \overline{\lim_{r\to\infty}} \frac{\ln n_f(r)}{\ln r} \leqslant \overline{\lim_{r\to\infty}} \frac{\ln \ln M_f(er)}{\ln er} = \rho.$$

The theorem is proved. Note that from (1.27) one can also obtain

$$\Delta_f = \overline{\lim_{r\to\infty}} \frac{n(r)}{r^{\rho_1}} \leqslant \overline{\lim_{r\to\infty}}\, e^{\rho_1} \frac{\ln M(er)}{(er)^{\rho_1}},$$

and if $\rho_1 = \rho$ we will have

$$\Delta_f \leqslant e^{\rho} \sigma_f.$$

Comparing this with Theorem 4 and the remarks following it we have the following theorem.

THEOREM 7. *For canonical products the convergence exponent of the zeros is equal to the order of the function. In addition, if the convergence exponent is not an integer, then the canonical product is of maximal, minimal or normal type according to whether the upper density of the set of zeros*

$$\Delta_f = \overline{\lim_{r\to\infty}} \frac{n_f(r)}{r^{\rho}}$$

is equal to infinity, to zero, or to a number different from zero and infinity.

---

[5] It is not difficult to show that this is the general form of the functions for which equality holds in (1.27).

Later we shall investigate more fully the connection between the type of an entire function and the quantity Δ_f.

6. The Relation between the Maximum Modulus of a Holomorphic Function and the Maximum of Its Real Part

We denote the maximum of the real part of a function $f(z)$, holomorphic in the circle $|z| \leqslant r$, by $A_f(r)$. From the maximum principle for harmonic functions it follows that $A_f(r)$ is a monotone increasing function of r. Also we clearly have

$$|A_f(r)| \leqslant M_f(r).$$

For $R > r$ there is an inequality between $A_f(R)$ and $M_f(r)$, which is a sort of converse to the inequality above.

THEOREM 8. *If $f(z)$ is any function holomorphic in the circle*[6] $|z| \leqslant R$, *then*

$$M_f(r) \leqslant [A_f(R) - \operatorname{Re} f(0)] \frac{2r}{R-r} + |f(0)| \quad (r < R). \tag{1.29}$$

This inequality is called the *Carathéodory inequality for the circle.*

PROOF. Our starting point is the formula of Schwarz,

$$f(z) = \frac{1}{2\pi} \int_{-\pi}^{\pi} u(Re^{i\psi}) \frac{Re^{i\psi} + z}{Re^{i\psi} - z} d\psi + iv(0) \tag{1.30}$$

$$(f(z) = u(z) + iv(z)),$$

which represents a function holomorphic in the the circle $|z| \leqslant R$ in terms of the boundary values of its real part.[7]

Let us add to the right side of (1.30) the quantity

$$u(0) - \frac{1}{2\pi} \int_{-\pi}^{\pi} u(Re^{i\psi})\, d\psi,$$

which is equal to zero. We obtain the equation

$$f(z) = \frac{1}{\pi} \int_{-\pi}^{\pi} u(Re^{i\psi}) \frac{z}{Re^{i\psi} - z} d\psi + f(0).$$

In particular, for $f(z) \equiv 1$ we obtain

$$\frac{1}{\pi} \int_{-\pi}^{\pi} \frac{z}{Re^{i\psi} - z} d\psi = 0.$$

[6] This hypothesis can be weakened slightly. It is sufficient to require that $f(z)$ be holomorphic in the circle $|z| < R$ and that its real part have a limiting value at each boundary point, and that these boundary values be bounded above.

[7] The derivation of formula (1.30) may be found, for example, in Titchmarsh [1].

From the last two equations we obtain the formula

$$-f(z)=\frac{1}{\pi}\int_{-\pi}^{\pi}[A_f(R)-u(Re^{i\psi})]\frac{z}{Re^{i\psi}-z}\,d\psi-f(0).$$

Since the quantity $A_f(R) - u(\mathrm{Re}^{i\psi})$ is non-negative we obtain the inequality

$$|f(z)|\leqslant\frac{2r}{R-r}[A_f(R)-u(0)]+|f(0)|.$$

Clearly it is possible to replace $|f(z)|$ by $M_f(r)$ and obtain (1.29). If the function $w = f(z)$ takes values lying in the left half-plane ($\mathrm{Re}\, w < 0$) when z is in the circle $|z| < R$, then $A_f(R) < 0$ and

$$|f(z)|\leqslant-\frac{2r}{R-r}u(0)+|f(0)| \quad (|z|\leqslant r<R).$$

By mapping the circle $|z| < R$ onto the half-plane one can obtain from this inequality the corresponding inequality for functions analytic in a half-plane.

THEOREM 8′. *A function $f(z)$ that is holomorphic in the half-plane* $\mathrm{Im}\, z > 0$ *and takes on values in the upper half-plane* $\mathrm{Im}\, f(z) > 0$ *satisfies, for* $\mathrm{Im}\, z > 0$ *and* $|z| > 1$, *the inequalities*

$$\frac{1}{5}|f(i)|\frac{\sin\theta}{r}<|f(z)|<5|f(i)|\frac{r}{\sin\theta} \tag{1.31}$$

$$(z=re^{i\theta},\quad 0<\theta<\pi).$$

This inequality will be called the *Carathéodory inequality for the half-plane.*

PROOF. Map the upper half-plane $\mathrm{Im}\, z > 0$ onto the unit circle

$$u=\frac{z-i}{z+i}\quad\text{or}\quad z=-i\frac{u+1}{u-1}.$$

The function

$$F(u)=if\left(-i\frac{u+1}{u-1}\right)$$

is defined in the unit circle and satisfies $\mathrm{Re}\, F(u) \leqslant 0$. Consequently, $A_F(R) \leqslant 0$, and from Carathéodory's inequality for the circle $|u| \leqslant 1$ we have

$$|f(z)|\leqslant|f(i)|-\mathrm{Re}\,(if(i))\frac{2|z-i|}{|z+i|-|z-i|}\qquad(\mathrm{Im}\, z>0).$$

Note that for $|z| \geqslant 1$

$$|z-i|\leqslant 2r,\ |z+i|\leqslant 2r\ \text{ and }\ |z+i|^2-|z-i|^2=4r\sin\theta,$$[8]

[8] It is easy to obtain the last equality, if we consider the triangles with vertices at 0, z, i and at 0, z, $-i$.

we obtain

$$|f(z)| < |f(i)|\left(1 + \frac{4r}{\sin\theta}\right) < 5|f(i)|\frac{r}{\sin\theta}. \qquad (1.31')$$

To obtain the left side of the inequality note that the function $f(z)$ does not vanish in the open upper half-plane. Writing the inequality (1.31') for the function $f^{-1}(z)$ we obtain the left inequality in (1.31).

Theorem 8 enables us to estimate from below the modulus of the holomorphic function $f(z)$ having no zeros in the circle $|z| \leqslant R$. Indeed, assume that $f(0) = 1$, and write inequality (1.29) for the holomorphic function $\ln f(z)$. We obtain the inequality

$$|\ln f(z)| \leqslant \frac{2r}{R-r}\ln M_f(R) \qquad (0 < r < R;\ |z| \leqslant r),$$

from which follows

$$\ln\left|\frac{1}{f(z)}\right| \leqslant \frac{2r}{R-r}\ln M_f(R),$$

or

$$\ln|f(z)| \geqslant -\frac{2r}{R-r}\ln M_f(R).$$

Thus we obtain the following theorem.

THEOREM 9. *If the function $f(z)$ is holomorphic in the circle $|z| \leqslant R$ and has no zeros in this circle, and if $f(0) = 1$, then its modulus in the circle $|z| \leqslant r < R$ satisfies the inequality*

$$\ln|f(z)| \geqslant -\frac{2r}{R-r}\ln M_f(R) \qquad (|z| \leqslant r < R). \qquad (1.32)$$

If the function $f(z)$ has zeros, such an estimate is, of course, impossible. However one can give an analogous estimate in the domain obtained from the original circle by removing certain small circles containing the zeros. To obtain this precise estimate, which plays an important role in the theory of entire functions, we first estimate from below the modulus of a polynomial.

7. Lower Bounds for the Modulus of a Polynomial

The following theorem of H. Cartan [1] provides an estimate from below for the modulus of a polynomial.

THEOREM 10. *Given any number $H > 0$ and complex numbers $a_1, a_2, \ldots, a_n$, there is a system of circles in the complex plane, with the sum of the radii equal to $2H$, such that for each point z lying outside these circles one has the inequality*

$$|z - a_1|\cdot|z - a_2|\ldots|z - a_n| > \left(\frac{H}{e}\right)^n. \qquad (1.33)$$

PROOF. We divide the proof into several stages.

1. We choose the quantity H/n as the unit of measurement and show that there are closed circles in the complex plane, having radius equal to the number of points $\{a_k\}$ contained within the circle. Indeed, form the smallest convex polygon containing all the points $\{a_k\}$, and choose any vertex a_j of this polygon. Clearly there are circles of arbitrary radius that contain this point but do not contain any other of the points $\{a_k\}$. In particular, the radius can be chosen to equal the multiplicity of the point a_j.

2. From among all circles with radius equal to the number of points $\{a_k\}$ lying inside the circle choose one with the largest radius, $\lambda_1 H/n$, and call it C_1. Note that no circle in the plane with radius greater than or equal to $\lambda_1 H/n$ can contain more points of the set $\{a_k\}$ than the number of units of measurement in its radius. Indeed, suppose a circle of radius $\lambda H/n$, with $\lambda \geqslant \lambda_1$, contains $\lambda' > \lambda$ points of the set $\{a_k\}$. The concentric circle of radius $\lambda' H/n$ either contains λ' points or $\lambda'' > \lambda'$ points. In the second case we consider the concentric circle of radius $\lambda'' H/n$ and so forth. Since the set $\{a_k\}$ is finite, we eventually come to a circle of radius $\lambda H/n$, larger than $\lambda_1 H/n$, that contains λ points. This is impossible, since C_1 is the largest circle having this property. The points of the set $\{a_k\}$ that lie inside C_1 will be said to be of rank λ_1.

3. We remove the points of rank λ_1, and for the remaining $n - \lambda_1$ points we construct the largest circle C_2 that contains the same number of points as there are units in its radius. Let its radius be $\lambda_2 H/n$. We shall show that $\lambda_2 < \lambda_1$.

Indeed, if this were not the case then the circle C_2 would have a radius larger than $\lambda_1 H/n$, and would contain at least as many of the original n points as there are units in its radius. But this contradicts the result of §2. The points in C_2 will be said to be of rank λ_2. Now remove these points too, and for the remaining $n - \lambda_1 - \lambda_2$ points we find the largest circle C_3 containing the same number of points as there are units in its radius. Suppose that its radius is $\lambda_3 H/n$. Clearly $\lambda_3 \leqslant \lambda_2$. The points in C_3 will be said to be of rank λ_3, and so forth.

We thus obtain a sequence of circles

$$C_1, C_2, \ldots, C_p$$

with radii that contain $\lambda_1, \lambda_2, \ldots, \lambda_p$ units of measurement, respectively, where $\lambda_1 \geqslant \lambda_2 \geqslant \ldots \geqslant \lambda_p$ and

$$\frac{H}{n}(\lambda_1 + \lambda_2 + \ldots + \lambda_p) = H.$$

4. We now form circles

$$\Gamma_1, \Gamma_2, \ldots, \Gamma_p$$

concentric with the circles $C_1, C_2, \ldots, C_p$ but with radii twice as large. Let z be an arbitrary point lying outside all of the new circles. Describe about z the circle C_z of radius $\lambda H/n$, where λ is some natural number. This circle does not intersect any of the circles C_j that have a radius larger than or equal to λ. Thus this circle can only contain points whose rank is less than λ.

From the definition of rank it follows that after removing all points of rank greater than or equal to λ, no circle of radius greater than or equal to $\lambda H/n$ can contain as many of the remaining points as there are units of measurement in its radius.

It follows from §2 that the circle C_z can contain at most $\lambda - 1$ points.

5. Enumerating the points $\{a_k\}$ in the order of increasing distance from z we have

$$|z - a_k| > k\frac{H}{n}$$

and

$$|z-a_1||z-a_2|\ldots|z-a_n| > \left(\frac{H}{n}\right)^n n! > \left(\frac{H}{e}\right)^n.$$

8. Lower Bounds for the Modulus of a Holomorphic Function

THEOREM 11. *Let $f(z)$ be holomorphic in the circle $|z| \leqslant 2eR$ $(R > 0)$ with $f(0) = 1$, and let η be an arbitrary positive number not exceeding $3e/2$. Then inside the circle $|z| \leqslant R$, but outside of a family of excluded circles the sum of whose radii is not greater than $4\eta R$, we have*

$$\ln|f(z)| > -H(\eta)\ln M(2eR) \tag{1.34}$$

for

$$H(\eta) = 2 + \ln\frac{3e}{2\eta}.$$

PROOF. Construct the function

$$\varphi(z) = \frac{(-2R)^n}{a_1a_2\ldots a_n}\prod_{k=1}^{n}\frac{2R(z-a_k)}{(2R)^2 - \bar{a}_k z},$$

where $a_1, a_2, \ldots, a_n$ are the roots of $f(z)$ in the circle $|z| < 2R$. We have

$$\varphi(0) = 1 \quad\text{and}\quad |\varphi(2Re^{i\theta})| = \frac{(2R)^n}{|a_1a_2\ldots a_n|}.$$

The function

$$\psi(z) - \frac{f(z)}{\varphi(z)}$$

has no roots in the circle $|z| \leqslant 2R$, and therefore by the corollary to the Carathéodory theorem for the circle (Theorem 9) we have for $|z| \leqslant R$

$$\begin{aligned}\ln|\psi(z)| &\geqslant -2\ln M_f(2eR) + 2\ln\frac{(2R)^n}{|a_1a_2\ldots a_n|}\\ &\geqslant -2\ln M_f(2eR).\end{aligned} \tag{1.35}$$

Now we estimate $|\varphi(z)|$ from below. For $|z| \leqslant R$

$$\prod_{k=1}^{n}|(2R)^2 - \bar{a}_k z| < (6R^2)^n.$$

Applying the Cartan estimate (see Theorem 10) to the polynomial in the numerator of $\varphi(z)$ we see that, outside of a family of circles with the sum of the radii equal to $4\eta R$, we have

$$\left|\prod_{k=1}^{n} 2R(z-a_k)\right| > \left(\frac{2\eta R}{e}\right)^n (2R)^n$$

and consequently

$$|\varphi(z)| > \frac{(2R)^{2n}}{|a_1 a_2 \ldots a_n|}\left(\frac{2\eta R}{e}\right)^n \frac{1}{(6R^2)^n} \geqslant \left(\frac{2\eta}{3e}\right)^n.$$

By Lemma 4

$$n = n(2R) \leqslant \ln M_f(2eR),$$

and for $|\varphi(z)|$ this leads to the estimate

$$\ln|\varphi(z)| > \ln\left(\frac{2\eta}{3e}\right) \ln M_f(2eR).$$

Combining this with the estimate (1.35) we see that in the circle $|z| \leqslant R$, but outside of a family of circles, the sum of whose radii does not exceed $4\eta R$, we have

$$\ln|f(z)| > -\left(2 + \ln\frac{3e}{2\eta}\right) \ln M(2eR).$$

9. The Growth of the Product of Two Entire Functions

It follows at once from the definition of the order of a function that the order of a product does not exceed the larger of the orders of the factors.

With the aid of the lower bound for the modulus of an entire function that was established in the previous section, we can obtain a much more precise result. We have the following theorem.

THEOREM 12. (a) *The order of the product of two entire functions of different orders is equal to the larger of the orders of the factors, and the type is equal to the type of the function that has the larger order.*

(b) *The product of two entire functions of the same order, one of which has normal type σ and the other has minimal type, is an entire function of the same order and of type σ.*

(c) *The product of two entire functions of the same order ρ, one having maximal type and the other having at most normal type, is an entire function of order ρ and of maximal type.*

This theorem can be formulated more briefly by introducing the notion of category. We shall say that the functions $f(z)$ and $\varphi(z)$ belong to the same *category* if they have the same order, and if they are both of maximal type, or both of normal type, or both of minimal type. Otherwise we shall say that the function having the larger order, or if the orders are equal the one having the larger type, is in the larger category.

Theorem 12 may now be formulated as follows:

If the function $f(z)$ and $\varphi(z)$ are of different categories, then the product $f(z)\,\varphi(z)$ has the same order and type as the factor belonging to the larger category.

PROOF. We limit ourselves to the proof of (b), since the other two parts are proven similarly. Let $f(z)$ be an entire function of order ρ and type $\sigma > 0$, and let $\varphi(z)$ have order ρ and minimal type. From the asymptotic inequalities

$$M_f(r) < e^{\left(\sigma+\frac{\varepsilon}{2}\right) r^\rho} \quad \text{and} \quad M_\varphi(r) < e^{\frac{\varepsilon}{2} r^\rho}$$

we have the asymptotic inequality

$$M_{f\varphi}(r) < e^{(\sigma+\varepsilon) r^\rho}. \tag{1.36}$$

To obtain the necessary estimates from below for the function $M_{f\varphi}(r)$ on some sequence of values of r tending to infinity, we apply to $\varphi(z)$ the lower bound found in § 8 (Theorem 11). Here we may assume without loss of generality that $\varphi(0) = 1$, since multiplication of $\varphi(z)$ by cz^{-n} changes neither the order nor the type of the function.

First we find a number R_1 (arbitrarily large) such that

$$M_f(R_1) > e^{\left(\sigma-\frac{\varepsilon}{2}\right) R_1^\rho}$$

and for all $R \geqslant R_1$

$$M_\varphi(R) < e^{\delta R^\rho}.$$

Such a choice is possible for arbitrarily given positive ε and δ since the first inequality is satisfied on some sequence tending to infinity, and the second holds asymptotically.

We shall assume that $0 < \delta < 1$. Choose $\eta = \delta/8$ and inside the circle $|z| \leqslant R\ (R = R_1(1-\delta)^{-1})$ form the excluded circles described in Theorem 11. The sum of the diameters of all the excluded circles is less than δR. Therefore in the interval (R_1, R) there is a number r_1 such that the circumference $|z| = r_1$ does not meet any of the excluded circles.

From Theorem 11 on this circumference we have

$$\ln|\varphi(z)| > -\left(2+\ln\frac{12e}{\delta}\right)\ln M_\varphi(2eR). \tag{1.37}$$

Since $R_1 < r_1 < R_1(1-\delta)^{-1}$, we have

$$M_f(r_1) > M_f(R_1) > e^{\left(\sigma-\frac{\varepsilon}{2}\right) R_1^\rho} > e^{\left(\sigma-\frac{\varepsilon}{2}\right)(1-\delta)^\rho r_1^\rho}.$$

Comparing this with (1.37) we have

$$\ln M_{f\varphi}(r_1) > \left[\left(\sigma-\frac{\varepsilon}{2}\right)(1-\delta)^\rho r_1^\rho - \left(2+\ln\frac{12e}{\delta}\right)\ln M_\varphi(2eR)\right],$$

and since

$$\ln M_\varphi(2eR) < \delta(2e)^\rho R^\rho < \delta(1-\delta)^{-\rho}(2e)^\rho r_1^\rho,$$

we have

$$\ln M_{f\varphi}(r_1) > \left[\left(\sigma - \frac{\varepsilon}{2}\right)(1-\delta)^\rho - \left(2 + \ln\frac{12e}{\delta}\right)(1-\delta)^{-\rho}(2e)^\rho\delta\right] r_1^\rho.$$

For given $\varepsilon > 0$ one can choose δ so small that the expression in the square brackets is not less than $\sigma - \varepsilon$. Consequently,

$$M_{f\varphi}(r_1) > e^{(\sigma-\varepsilon)r_1^\rho}$$

for a sequence of values r_1 tending to infinity, and the theorem is proved.

COROLLARY. *If the quotient of two entire functions $f(z)$ and $\varphi(z)$ is an entire function $\psi(z)$, then its category does not exceed the larger of the categories of the functions $f(z)$ and $\varphi(z)$; here the categories of $f(z)$ and $\varphi(z)$ may be the same. If they have different categories, then the category of the function $\psi(z)$ equals the larger of the categories of $f(z)$ and $\varphi(z)$.*

PROOF. We are given

$$\varphi(z)\psi(z) = f(z).$$

If the category of $\psi(z)$ is larger than the category of $\varphi(z)$, then the category of $f(z)$ is equal to the category of $\psi(z)$. Thus the category of $\psi(z)$ cannot exceed the categories of $f(z)$ and $\varphi(z)$, and the first part of the corollary is proved. Clearly also, if the category of $f(z)$ is larger than the category of $\varphi(z)$, then the categories of $f(z)$ and $\psi(z)$ are equal. On the other hand, the category of $f(z)$ can be smaller than the category of $\varphi(z)$ only in case the categories of $\varphi(z)$ and $\psi(z)$ are equal.

10. Hadamard's Theorem

The theorems of the preceding sections enable us to refine considerably the theorem on the representation of an entire function as an infinite product. This refinement, which is due to Hadamard, concerns the representation of entire functions of finite order, and is one of the classical theorems of the theory of entire functions.

THEOREM 13. *The entire function $f(z)$ of finite order ρ can be represented in the form*

$$f(z) = z^m e^{P(z)} \prod_{n=1}^{\omega} G\left(\frac{z}{a_n};\ p\right) \quad (\omega \leqslant \infty), \tag{1.38}$$

where a_n are the nonzero roots of $f(z)$, $p \leqslant \rho$, $P(z)$ is a polynomial whose degree q does not exceed ρ, and m is the multiplicity of the zero at the origin.

This theorem can also be formulated as follows:

The genus of an entire function does not exceed its order (for the *definition of genus*, see the next to the last paragraph in § 3).

PROOF. The convergence exponent ρ_1 of the zeros of the entire function does not exceed its order ρ, and the integer p in the associated canonical product

$$\Pi(z) = \prod_{n=1}^{\omega} G\left(\frac{z}{a_n};\ p\right) \qquad (\omega \leqslant \infty)$$

does not exceed ρ_1. Consequently, $p \leqslant \rho$.

It remains to show that the function $g(z)$ in (1.14) is a polynomial of degree q not exceeding ρ. To this end we recall that the order of the canonical product is equal to the convergence exponent $\rho_1 \leqslant \rho$. By the corollary to Theorem 12 the entire function.

$$\psi(z) = f(z)\,\Pi^{-1}(z)\,z^{-m}$$

has order at most ρ, and therefore satisfies the asymptotic inequality

$$\ln|\psi(z)| < r^{\rho+\varepsilon}.$$

In addition, the function $\psi(z)$ has no zeros. Thus

$$g(z) = \ln \psi(z)$$

is an entire function, and we have asymptotically

$$A_g(r) < r^{\rho+\varepsilon}.$$

By the Carathéodory inequality it follows that asymptotically

$$M_g(r) < r^{\rho+2\varepsilon}.$$

By Theorem 1 the function $g(z)$ is a polynomial of degree at most ρ. The theorem is proved.

For functions of nonintegral order the degree of the polynomial in the exponential factor is less than ρ. Therefore the order of the function coincides with the order of the canonical product, and therefore with the convergence exponent. Thus in this case

$$p < \rho < p + 1,$$

and the genus of the function is $p = [\rho]$.

For integral ρ the genus of the function is either equal to ρ or to $\rho - 1$. Indeed, if the genus is less than ρ, then $q < \rho$ and the order of the function coincides with the order of the canonical product, which is equal to the convergence exponent ρ_1.

On the other hand, $p \leqslant \rho_1 \leqslant p + 1$, so that $p \leqslant \rho \leqslant p + 1$, and since by assumption $p < \rho$, we have $p = \rho - 1$.

EXAMPLES. The function

$$f(z) = \frac{\sin \pi\sqrt{z}}{\pi\sqrt{z}}$$

is entire, of order one-half, and has the zeros $1^2, 2^2, \ldots$. By Hadamard's theorem it is of genus zero and therefore has a representation

$$f(z) = c \prod_{n=1}^{\infty} \left(1 - \frac{z}{n^2}\right).$$

From the equation $f(0) = 1$ we have $c = 1$. Replacing z by z^2 we obtain

$$\sin \pi z = \pi z \prod_{n=1}^{\infty} \left(1 - \frac{z^2}{n^2}\right),$$

from which we also have

$$\sin \pi z = \pi z \prod_{-\infty}^{\infty}{}' \left(1 - \frac{z}{n}\right) e^{\frac{z}{n}}.$$

Thus $\sin \pi z$ is not only an entire function of order one, but also of genus one. Similarly, for the entire function of order one

$$f(z) = \frac{1}{n}\left(e^{\omega_1 z} + e^{\omega_2 z} + \ldots + e^{\omega_n z}\right),$$

where ω_j $(j = 1, 2, \ldots, n)$ are all the roots of the equation $\omega^n + 1 = 0$, we obtain the expansion

$$f(z) = \prod_{k=1}^{\infty} \left(1 - \frac{z^n}{\alpha_k^n}\right),$$

in which α_k are the zeros of $f(z)$. It is easy to show that this function also has genus one.

The function

$$\varphi(z) = \prod_{n=1}^{\infty} \left(1 - \frac{z}{n \ln^2 n}\right)$$

is entire of genus zero. On the other hand the convergence exponent is one, and therefore it is of order one.

In general, if the convergence exponent ρ_1 of the sequence $\{a_n\}$ is an integer and if the series

$$\sum_{k=1}^{\infty} \frac{1}{|a_k|^{\rho_1}}$$

diverges, then the genus p of the canonical product is equal to its order $\rho = \rho_1$. If the series converges, then the genus of the canonical product is equal to $\rho - 1$.

We have already observed that functions of nonintegral order have an infinite set of zeros, and the convergence exponent of the set of zeros is equal to the order of the function. In this case, by assertion (a) of Theorem 12, the function $f(z)$ has the same order and type as the canonical product. From this remark and Theorem 7 we have the following more precise theorem.

THEOREM 14. *If the order ρ of the entire function $f(z)$ is not an integer, and if Δ_f is the upper density of the set of its zeros, then for $\Delta_f = 0$ the function is of minimal type, for $\Delta_f \neq 0, \infty$ the function is of normal type, while for $\Delta_f = \infty$ it is of maximal type.*

Since the order and type of the function $f(z) - a$ coincide with the order and type of $f(z)$ for all complex a, it follows from Theorem 14 that a function of nonintegral order takes on all complex values "equally often."

11. Entire Functions of Integral Order

The conclusion of Theorem 14 is no longer valid when the order of the entire function is equal to an integer.

Indeed, in this case the order can be determined by the exponential factor, and the convergence exponent may be less than the order of the function. In particular, the function need not have any zeros at all.

But this is not the only peculiarity of functions of integral order. It turns out that in this case the type of the canonical product is not determined just by the upper density Δ of the zeros, but depends also on the distribution of the arguments of the zeros.

Two examples that illustrate this are the functions

$$\sin \frac{\pi}{2} z = \frac{\pi}{2} z \prod_{n=1}^{\infty} \left(1 - \frac{z^2}{(2n)^2}\right)$$

and

$$\varphi(z) = \frac{1}{\Gamma(z)} = z e^{\gamma z} \prod_{n=1}^{\infty} \left(1 + \frac{z}{n}\right) e^{-\frac{z}{n}}.$$

For both functions we have

$$n(r) \sim r,$$

that is, the density of the zeros is the same. However, $\sin(\pi z/2)$ is of order one and normal type, while $\varphi(z)$ is of order one and maximal type. This is easily seen for the latter function by using Stirling's formula[9]

$$-\ln \Gamma(z) = \left(z - \frac{1}{2}\right) \ln z - z + \frac{1}{2} \ln 2\pi + O\left(\frac{1}{r}\right),$$

from which it follows that

$$\ln M_\varphi(r) \sim r \ln r.$$

---

[9] This formula is usually established for all complex values of z lying in some angle $|\arg z| \leqslant \pi - \delta$ (see, for example, Titchmarsh [1]). However, from the relation $\Gamma(z)\Gamma(1 - z) = \pi/\sin \pi z$ one easily sees that it is valid outside of circles of arbitrarily small constant radius, about the points $0, -1, -2, \ldots$. This is sufficient for our conclusion.

In the investigation of functions of integral order ρ the quantity

$$\delta_f(r) = \alpha_\rho + \frac{1}{\rho} \sum_{|a_n| < r} a_n^{-\rho}$$

plays an essential role, where ρ is the order of $f(z)$, and α_ρ is the coefficient of z^ρ in the exponential factor in the representation (1.38).

Let

$$\delta_f = \overline{\lim_{r\to\infty}} |\delta_f(r)| \text{ and } \gamma_f = \max(\Delta_f, \delta_f).$$

The question of the type of an entire function of integral order is settled by the following theorem of Lindelöf [**1**].

THEOREM 15. (a) *If the genus p of the canonical product of the entire function $f(z)$ of integral order represented by* (1.38) *is equal to its order $\rho \geqslant 1$, then for $\gamma_f = 0$ the function $f(z)$ is of minimal type, for $0 < \gamma_f < \infty$ of normal type, and for $\gamma_f = \infty$ of maximal type.*

(b) *If the genus of the entire function is less than its order, then its type is equal to the coefficient of z^ρ in the exponential factor standing in front of the canonical product.*

PROOF. Consider first the case $p = \rho$. We introduce the function

$$f_R(z) = \prod_{|a_n| \leqslant R} G\left(\frac{z}{a_n}; \rho - 1\right) \prod_{|a_n| > R} G\left(\frac{z}{a_n}; \rho\right), \tag{1.39}$$

with the aid of which $f(z)$ can be represented in the form

$$f(z) = z^m e^{P_{\rho-1}(z)} e^{\delta_f(R) z^\rho} f_R(z), \tag{1.40}$$

where $P_{\rho-1}(z)$ is a polynomial of degree at most $\rho - 1$. This important representation will be used frequently in the study of entire functions of integral order.

To estimate the growth of the function $f_R(z)$ we use inequalities (1.21) and (1.22). Assume that $\rho > 1$. Then, if we put

$$M_R(r) = \max_{0 \leqslant \theta < 2\pi} |f_R(re^{i\theta})|,$$

we have

$$\ln M_R(R) < A_\rho \left[R^\rho \int_0^R \frac{dn(t)}{t^{\rho-1}(t+R)} + R^{\rho+1} \int_R^\infty \frac{dn(t)}{t^\rho (t+R)} \right].$$

Just as in the proof of Lemma 3, if we integrate by parts we will obtain the inequality

$$\ln M_R(R) < k_\rho \left[R^{\rho-1} \int_0^R \frac{n(t)}{t^\rho} dt + R^{\rho+1} \int_R^\infty \frac{n(t)}{t^{\rho+2}} dt \right], \tag{1.41}$$

in which k_ρ is a constant. In case $\rho = 1$, inequality (1.41) can be obtained in a similar manner.

If Δ_f is the upper density of the set of zeros of $f(z)$, then asymptotically for $\varepsilon > 0$

$$n(t) < (\Delta_f + \varepsilon) t^{\rho},$$

and from (1.41) we have the asymptotic inequality

$$\ln M_R(R) < 2k_\rho (\Delta_f + \varepsilon) R^\rho. \tag{1.42}$$

Also, from (1.40) we have

$$\ln M_f(R) < |\delta_f(R)| R^\rho + \ln M_R(R) + O(R^{\rho-1}), \tag{1.43}$$

and, taking account of (1.42), we obtain the asymptotic inequality[10]

$$\frac{\ln M_f(R)}{R^\rho} < |\delta_f(R)| + 2k_\rho (\Delta_f + \varepsilon) + O\left(\frac{1}{R}\right)$$

or

$$\sigma_f \leqslant (1 + 2k_\rho) \gamma_f. \tag{1.44}$$

To prove the reverse inequality we note that by Lemma 4

$$\overline{\lim_{r\to\infty}} \frac{n_f(r)}{r^\rho} \leqslant \overline{\lim_{r\to\infty}} e^\rho \frac{\ln M_f(er)}{(er)^\rho}$$

and, consequently,

$$\Delta_f \leqslant e^\rho \sigma_f. \tag{1.45}$$

It remains to give an analogous estimate for δ_f. For this purpose we must estimate the function $f_R(z)$ from below. This can be done using Theorem 11. Choose $\eta = \frac{3}{2} e^{-3}$. Then inside the circle $|z| \leqslant r$ but outside a set of excluded circles, the sum of whose radii is less than $6re^{-3}$, we have the inequality

$$\ln |f_R(z)| > -H\left(\frac{3}{2} e^{-3}\right) \ln M_R(R) = -6 \ln M_R(R),$$

where $R = 2er$.

Since $6e^{-3} < \frac{1}{2}$, there is a number r_1 between $r/2$ and r such that the circumference $|z| = r_1$ does not meet any of the excluded circles. From the representation (1.40) we see that on this whole circumference we have the inequality

$$\mathrm{Re}\,[\delta_f(R) z^\rho] < 6 \ln M_R(R) + \ln M_f(r_1) + O(R^{\rho-1}).$$

Choose the argument of z so that z^ρ and $\delta_f(R)$ have arguments that cancel one another. Then we will have

$$|\delta_f(R)| < 6 \frac{\ln M_R(R)}{r_1^\rho} + \frac{\ln M_f(r_1)}{r_1^\rho} + O(R^{-1}). \tag{1.46}$$

---

[10] The direct application of (1.24) to the estimation of a canonical product of integral order ρ leads to the estimate

$$\ln M_f(R) < 2k_\rho (\Delta_f + \varepsilon) R^\rho \ln R.$$

Since $r_1 > (4e)^{-1}R$, it follows from (1.42) that

$$\overline{\lim_{R\to\infty}} \frac{\ln M_R(R)}{r_1^\rho} \leqslant 2k_\rho \Delta_f (4e)^\rho. \tag{1.47}$$

It follows from (1.45), (1.46) and (1.47) that

$$\delta_f \leqslant C\sigma_f, \tag{1.48}$$

where

$$C = 12k_\rho (4e^2)^\rho + 1.$$

Combining (1.45) and (1.48) we obtain

$$\sigma_f \geqslant C^{-1}\gamma_f. \tag{1.49}$$

The assertion of the theorem in case (a) now follows from inequalities (1.44) and (1.49).

Now suppose that $p < \rho$. In this case the series

$$\sum \frac{1}{|a_n|^\rho}$$

converges. Therefore the integral

$$\int_0^\infty \frac{n(t)}{t^{\rho+1}}\,dt \tag{1.50}$$

also converges and, in addition,

$$\lim_{t\to\infty} \frac{n(t)}{t^\rho} = 0. \tag{1.51}$$

Applying Lemma 3 to the canonical product $\Pi(z)$ of the function $f(z)$ we obtain the estimate

$$\ln|\Pi(z)| < kr^{\rho-1}\left(\int_0^r \frac{n(t)}{t^\rho}\,dt + r\int_r^\infty \frac{n(t)}{t^{\rho+1}}\,dt\right).$$

From this estimate, equation (1.51) and the convergence of the integral (1.50), it is easy to obtain the asymptotic inequality

$$\ln|\Pi(z)| < \varepsilon r^\rho,$$

which means that the function $\Pi(z)$ grows no faster than a function of order ρ and minimal type. By Theorem 12 (b), it follows from this that the type of the function $f(z)$ coincides with the type of the exponential factor, that is, it is equal to the coefficient of z^ρ in this factor. Thus if the genus of an entire function is less than its order, then (when the order is an integer) it must be of minimal type.

In conclusion, we note the meaning of the conditions

$$\overline{\lim_{r\to\infty}} |\delta_f(r)| < \infty, \qquad \lim_{r\to\infty} \delta_f(r) = 0,$$

that play an essential role for the case $p = \rho$. In the presence of these equations the series

$$\sum \frac{1}{|a_n|^{\rho}}$$

diverges. Hence $\delta_f(r)$ remains bounded only if there is some special symmetry in the distribution of the roots a_n. In the absence of such symmetry $\limsup |\delta_f(r)| = \infty$. For example, if all the roots a_n lie on one ray, then $\delta_f(r) \to \infty$. In this case $f(z)$ is of maximal type, by what has been proven. This was the case with the example we considered before of the function $\Gamma^{-1}(z)$. Thus, for functions of nonintegral order, the type is determined by the modulus of the zeros, while, for functions of integral order, the type also depends on the distribution of the arguments of the zeros.

12. Proximate Orders

The growth scale that we have used in the preceding sections can be further refined. Thus, for example, Lindelöf compared $\ln M_f(r)$ with functions of the form

$$r^{\rho} \ln^{\alpha_1} r \ln_2^{\alpha_2} r \ldots \ln_p^{\alpha_p} r,$$

where $\ln_j r = \ln \ln_{j-1} r$ and $\alpha_1, \alpha_2, \ldots, \alpha_p$ are real numbers.

Clearly each successive logarithm grows more slowly than any power of the preceding one. The generalized order of the function $f(z)$ is defined to be a system of numbers $\rho, \alpha_1, \alpha_2, \ldots, \alpha_p$ such that for arbitrary $\varepsilon > 0$

$$r^{\rho} \ln^{\alpha_1} r \ln_2^{\alpha_2} r \ldots \ln_p^{\alpha_p - \varepsilon} r < \ln M_f(r) < r^{\rho} \ln^{\alpha_1} r \ln_2^{\alpha_2} r \ldots \ln_p^{\alpha_p + \varepsilon} r$$
$$(\varepsilon > 0),$$

in which the right inequality is asymptotic, while the left inequality is satisfied for some sequence of values of r tending to infinity. The order $\rho', \alpha_1', \alpha_2', \ldots, \alpha_p'$ of the function $f_1(z)$ is said to be greater than the order $\rho, \alpha_1, \alpha_2, \ldots, \alpha_p$ of the function $f(z)$, if $\rho' > \rho$, or if $\rho' = \rho$ and $\alpha_1' > \alpha_1$, or more generally if $\rho' = \rho$, $\alpha_1' = \alpha_1, \ldots, \alpha_{j-1}' = \alpha_{j-1}$ but $\alpha_j' > \alpha_j$. All the theorems of the preceding sections can be generalized to this more precise growth scale.

In particular, it can be shown that if ρ is not an integer, then the asymptotic inequalities

$$\ln M_f(r) < r^{\rho} \ln^{\alpha_1} r \ldots \ln_p^{\alpha_p + \varepsilon} r$$

and

$$n_f(r) < r^{\rho} \ln^{\alpha_1} r \ldots \ln_p^{\alpha_p + \varepsilon} r$$

are equivalent, and therefore in this case the generalized orders of the functions $M_f(r)$ and $n_f(r)$ are equal.

We shall not prove these theorems, as we shall soon prove more general theorems. The growth scale can be further refined by introducing as comparison functions monotone functions that grow more slowly than any of the iterated logarithms $\ln_j r$, etc. However, instead of such further refinements, it is more natural to define in a general manner a certain class of "slowly increasing" functions $L(r)$ and then to compare $\ln M_f(r)$ with functions of the form $r^\rho L(r)$.

This was the path taken by Valiron [1], who introduced the concept of proximate order of growth.

A function $\rho(r)$ that satisfies the conditions[11]

$$\lim_{r\to\infty} \rho(r) = \rho \geqslant 0 \quad \text{and} \quad \lim_{r\to\infty} r\rho'(r)\ln r = 0 \tag{1.52}$$

is called a *proximate order.*

If for the entire function $f(z)$ the quantity

$$\sigma_f = \overline{\lim_{r\to\infty}} \frac{\ln M_f(r)}{r^{\rho(r)}}$$

is different from zero and infinity, then $\rho(r)$ is called a *proximate order of the given entire function* $f(z)$, *and* σ_f *is called the type of the function* $f(z)$ *with respect to the proximate order* $\rho(r)$. Clearly, the proximate order and the corresponding type of the given function are not uniquely determined. For example, if we add $\ln c/\ln r$ to the proximate order, then we obtain a new proximate order for the given function, and now the type has been divided by c.

With respect to an arbitrary proximate order, a function may be of minimal, normal, or maximal type. With respect to its own proximate order it is always of normal type.

A positive function will be called *slowly increasing*, and will be denoted by $L(r)$, if

$$\lim_{r\to\infty} \frac{L(kr)}{L(r)} = 1$$

uniformly on each interval $0 < a \leqslant k \leqslant b < \infty$.

The following lemma plays a fundamental role in the study of proximate orders.

LEMMA 5. *If* $\rho(r)$ *is a proximate order, then the function*

$$r^{\rho(r)-\rho}$$

is slowly increasing.[12]

PROOF. Let

$$L(r) = r^{\rho(r)-\rho},$$

then

$$\ln \frac{L(kr)}{L(r)} = [\rho(kr) - \rho] \ln k + [\rho(kr) - \rho(r)] \ln r.$$

---

[11] Here and throughout this book the assertion $\lim \varphi(r) = A$ means "the limit exists and is equal to A."

[12] The converse is true in the presence of certain supplementary restrictions on the function $L(r)$ (for example, if $L(r)$ is logarithmically convex).

To fix ideas let us assume that $0 < \sigma \leqslant k < 1$. Then by the second of conditions (1.52), for arbitrary $\eta > 0$ and all sufficiently large values of r we have by Lagrange's theorem:

$$|\rho(r) - \rho(kr)| < \eta \frac{r - kr}{kr \ln kr} \leqslant \eta \left(\frac{1}{\sigma} - 1\right) \frac{1}{\ln(\sigma r)}.$$

Also, for r sufficiently large we have by the first of conditions (1.52)

$$|\rho(kr) - \rho| < \eta$$

and thus uniformly for $0 < \sigma \leqslant k \leqslant 1$

$$\lim_{r\to\infty} \ln \frac{L(kr)}{L(r)} = 0 \quad \text{or} \quad \lim_{r\to\infty} \frac{L(kr)}{L(r)} = 1.$$

The case $k > 1$ is treated similarly.

We now establish some simple properties of proximate orders that we shall need frequently.

(1) *For* $\rho > 0$ *the function* $r^{\rho(r)}$ *is monotone increasing for all sufficiently large values of* r.

Indeed,

$$(r^{\rho(r)})' = \rho(r) r^{\rho(r)-1} + r^{\rho(r)} \rho'(r) \ln r,$$

and since the quantity $r\rho'(r) \ln r$ tends to zero we have asymptotically

$$(r^{\rho(r)})' > (\rho - \varepsilon) r^{\rho(r)-1} > 0 \qquad (0 < \varepsilon < \rho).$$

Note that the assertion remains true for $\rho = 0$, provided that $\rho(r) > 0$ and the second condition of (1.52) is replaced by the condition $\rho'(r) r \ln r/\rho(r) \to 0$ as $r \to \infty$ (for $\rho > 0$ these conditions are equivalent).

Indeed, in this case $\rho(r) \ln r \to \infty$ and

$$(r^{\rho(r)})' = \rho(r) r^{\rho(r)-1} \left(1 + \frac{\rho'(r)}{\rho(r)} r \ln r\right) > \rho(r) r^{\rho(r)-1} (1 - \varepsilon) > 0.$$

Since in the study of asymptotic properties the behavior of the proximate order on a finite interval plays no role, we shall in the future assume that $r^{\rho(r)}$ is a monotone function for $r \geqslant 0$, and we shall define it to be zero at the origin. This may be done without loss of generality.

(2) *For* $r \to \infty$ *and* $0 < a \leqslant k \leqslant b < \infty$ *the asymptotic inequality*

$$(1 - \varepsilon) k^{\rho} r^{\rho(r)} < (kr)^{\rho(kr)} < (1 + \varepsilon) k^{\rho} r^{\rho(r)}$$

holds uniformly in k.

This inequality follows at once from Lemma 5 and the definition of slowly increasing function.

Note that for $k > 1$ *we have the asymptotic inequality*

$$(kr)^{\rho(kr)} < k^{\rho+\varepsilon} r^{\rho(r)}.$$

(3) *For* $\lambda < \rho + 1$

$$\int_a^r t^{\rho(t)-\lambda}\,dt = \frac{1}{\rho+1-\lambda}\,r^{\rho(r)+1-\lambda} + o\left(r^{\rho(r)+1-\lambda}\right). \tag{1.53}$$

Indeed, integrating by parts we have

$$\int_a^r t^{\rho-\lambda} t^{\rho(t)-\rho}\,dt = \frac{1}{\rho+1-\lambda}\,t^{\rho(t)+1-\lambda}\Big|_a^r$$

$$-\frac{1}{\rho+1-\lambda}\int_a^r \left[t^{\rho(t)-\lambda}(\rho(r)-\rho) + t^{\rho(t)+1-\lambda}\rho'(t)\ln t\right] dt.$$

From the definition of proximate order we have asymptotically

$$|\rho(r)-\rho| < \frac{\varepsilon}{2}, \quad |t\rho'(t)\ln t| < \frac{\varepsilon}{2}.$$

Using this we have

$$(1+o(1))\int_a^r t^{\rho(t)-\lambda}\,dt = \frac{1}{\rho+1-\lambda}\,r^{\rho(r)+1-\lambda} + O(1)$$

and, finally,

$$\int_a^r t^{\rho(t)-\lambda}\,dt = \frac{1}{\rho+1-\lambda}\,r^{\rho(r)+1-\lambda} + o\left(r^{\rho(r)+1-\lambda}\right).$$

For $\lambda > \rho + 1$ one obtains in a similar manner the formula

$$\int_r^\infty t^{\rho(t)-\lambda}\,dt = \frac{1}{\lambda-\rho-1}\,r^{\rho(r)-\lambda+1} + o\left(r^{\rho(r)-\lambda-1}\right). \tag{1.53'}$$

(4) *If* $\varphi(r)$ *is a bounded function on each finite interval, then for* $\lambda < \rho + 1$

$$\overline{\lim_{r\to\infty}}\left\{r^{-\rho(r)+\lambda-1}\int_a^r \frac{\varphi(t)}{t^\lambda}\,dt\right\} \leqslant \frac{\bar{\Delta}}{\rho-\lambda+1} \qquad (a>0),$$

where

$$\bar{\Delta} = \overline{\lim_{r\to\infty}}\,\frac{\varphi(r)}{r^{\rho(r)}}.$$

Indeed, for $\varepsilon > 0$ and $r > r_\varepsilon$

$$\varphi(t) < (\bar{\Delta}+\varepsilon)\,t^{\rho(t)}$$

and consequently

$$\int_a^r \frac{\varphi(t)}{t^\lambda}\,dt \leqslant O(1) + (\bar{\Delta}+\varepsilon)\int_{r_\varepsilon}^r t^{\rho(t)-\lambda}\,dt.$$

Applying Rule (3), we obtain

$$\int_a^r \frac{\varphi(t)}{t^\lambda}\,dt \leqslant (\bar{\Delta}+\varepsilon)\frac{r^{\rho(r)+1-\lambda}}{\rho+1-\lambda}+o\left(r^{\rho(r)+1-\lambda}\right).$$

From this we have at once

$$\varlimsup_{r\to\infty}\left\{r^{-\rho(r)-1+\lambda}\int_a^r \frac{\varphi(t)}{t^\lambda}\,dt\right\} \leqslant \frac{\bar{\Delta}}{\rho+\lambda-1}.$$

Similarly one sees that

$$\varliminf_{r\to\infty}\left\{r^{-\rho(r)-1+\lambda}\int_a^r \frac{\varphi(t)}{t^\lambda}\,dt\right\} \geqslant \frac{\underline{\Delta}}{\rho+1-\lambda},$$

where

$$\underline{\Delta} = \varliminf_{r\to\infty}\frac{\varphi(r)}{r^{\rho(r)}}.$$

From Rule (4) and the last remark we have at once the following assertion:

(5) *If $\varphi(t)$ is a bounded function on each finite interval and if the limit*

$$\Delta = \lim_{r\to\infty}\frac{\varphi(r)}{r^{\rho(r)}}$$

exists, then for $\lambda < \rho + 1$ the limit

$$\lim_{r\to\infty}\left\{r^{-\rho(r)-1+\lambda}\int_a^r \frac{\varphi(t)}{t^\lambda}\,dt\right\} = \frac{\Delta}{\rho+1-\lambda}$$

exists.

Valiron established the basic fact that every entire function has a proximate order. Thus after we have introduced the concept of proximate order, every function of finite order can be regarded as a function of normal type.

We prove a somewhat more general theorem.

THEOREM 16. *If $f(r)$ is any continuous function that is positive for $r > 0$ and satisfies the condition*

$$\rho = \varlimsup_{r\to\infty}\frac{\ln f(r)}{\ln r} < \infty,$$

then the proximate order $\rho(r)$ can be chosen so that for all positive values of r

$$f(r) \leqslant r^{\rho(r)}$$

and for some sequence of values r_n $(n = 1, 2, \dots)$ tending to infinity

$$f(r_n) = r_n^{\rho(r_n)}.$$

PROOF. 1. In place of the given function $f(r)$ it will be more convenient for us to consider the function

$$\varphi(r) = f(r)\,r^{-\rho},$$

whose growth is less than any power of r. Clearly

$$\overline{\lim_{r\to\infty}} \frac{\ln \varphi(r)}{\ln r} = 0.$$

Passing to the logarithmic scale, that is, putting $x = \ln r$ and $y = \ln \varphi(r)$, we have

$$y = \varphi_1(x),$$

where $\varphi_1(x) = \ln \varphi(e^x)$. At the same time the equation

$$\overline{\lim_{x\to\infty}} \frac{\varphi_1(x)}{x} = 0$$

is valid, from which it follows that for arbitrary $\varepsilon > 0$ and for all values of x that exceed a certain x_ε, the entire curve $y = \varphi_1(x)$ lies below the line

$$y = \varepsilon x.$$

On the other hand there are points on this curve with arbitrarily large abscissas lying above the line $y = -\varepsilon x$.

2. First suppose that

$$\overline{\lim_{x\to+\infty}} \varphi_1(x) = +\infty,$$

and form the smallest convex domain containing all the points of the curve $y = \varphi_1(x)$ and the positive ray of the x-axis. The part of the boundary of this domain that lies above the abscissa-axis is a continuous curve (Fig. 1) $y = \psi(x)$.

Fig. 1

It is easy to establish the following properties of this curve:

(a) the curve is convex from above;[13]

(b) $\lim_{x\to+\infty} \frac{\psi(x)}{x} = 0$;

(c) $\varphi_1(x) \leqslant \psi(x)$;

(d) each extreme point (that is, not lying inside any line segment of the curve) of the curve $y = \psi(x)$ also lies on the curve $y = \varphi_1(x)$, that is, at the extreme points $\varphi_1(x) = \psi(x)$;

(e) the curve $y = \psi(x)$ has a sequence of extreme points that tend to infinity.

[13] Such functions are called concave.

By making inessential changes in the curve $y = \psi(x)$ in the neighborhood of each angular point we may assume that $\psi(x)$ is everywhere differentiable. It follows at once from (a) and (b) that

$$\lim_{x\to+\infty} \psi'(x) = 0. \tag{1.54}$$

Returning to our previous variables, inequality (c) can be written in the form

$$\ln \varphi(r) \leqslant \psi(\ln r),$$

or

$$f(r) \leqslant r^{\rho} e^{\psi(\ln r)}$$

and finally

$$f(r) \leqslant r^{\rho + \frac{\psi(\ln r)}{\ln r}}.$$

Putting $\rho(r) = \rho + \psi(\ln r)/\ln r$ we have the inequality

$$f(r) \leqslant r^{\rho(r)} \quad (r > 0). \tag{1.55}$$

Also, by properties (d) and (e), there exists a sequence of values r_n tending to infinity for which equality holds in (1.55). It remains to verify that the function $\rho(r)$ that we have constructed is a proximate order, that is, that it satisfies the conditions (1.52) of Valiron. Clearly, from (b) it follows that $\lim_{r\to\infty}\rho(r) = \rho$. Also,

$$r\rho'(r)\ln r = \psi'(\ln r) - \frac{\psi(\ln r)}{\ln r}.$$

Hence, from (b) and (1.54), we see that $\lim_{r\to\infty} r\rho'(r)\ln r = 0$. We constructed the function $\rho(r)$ under the condition that $\limsup_{x\to+\infty}\varphi_1(x) = \infty$.

3. We now show how the general case can be reduced to this case. For this purpose we construct a concave function $y = \psi_1(x)$ so that

$$\lim_{x\to\infty} \frac{\psi_1(x)}{x} = 0, \qquad \lim_{x\to\infty} \psi_1'(x) = 0$$

and so that $\limsup_{x\to\infty}[\varphi_1(x) + \psi_1(x)] = \infty$. We construct the curve $\psi_1(x)$ in the following manner: pass a segment d_1 of the line

$$y = -\varepsilon_1 x$$

from the origin to the point x_1 (Fig. 2) at which

$$\varphi_1(x_1) > -\varepsilon_1 x_1 + 1.$$

Choose a positive number $\varepsilon_2 < \varepsilon_1$ and from the point $(x_1, -\varepsilon_1 x_1)$ pass a segment d_2 of the line

$$y + \varepsilon_1 x_1 = -\varepsilon_2(x - x_1)$$

to a point $x_2 > x_1$, at which

$$\varphi_1(x_2) > -\varepsilon_1 x_1 - \varepsilon_2(x_2 - x_1) + 2.$$

From the point $(x_2, -\varepsilon_1 x_1 - \varepsilon_2(x_2 - x_1))$ pass a segment d_3 having slope $-\varepsilon_3$ $(0 < \varepsilon_3 < \varepsilon_2)$ etc.

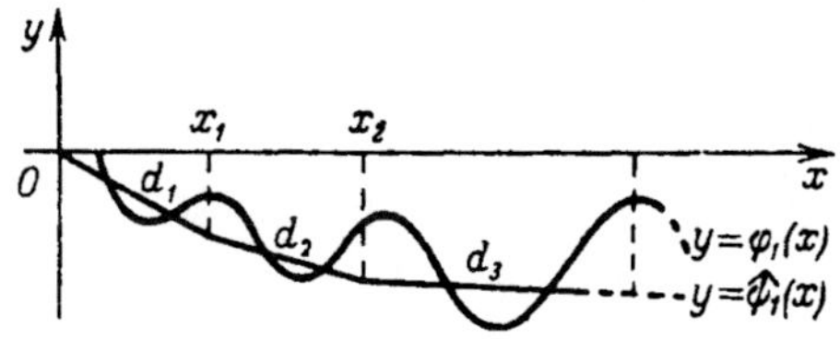

Fig. 2

The positive numbers $\varepsilon_1, \varepsilon_2, \varepsilon_3, \ldots, \varepsilon_n, \ldots$ will be chosen so that $\varepsilon_1 > \varepsilon_2 > \varepsilon_3 > \ldots > \varepsilon_n > \ldots$ and $\varepsilon_n \to 0$, and the points $x_1, x_2, \ldots, x_n, \ldots$ are chosen so that $x_n \to \infty$. Clearly the polygonal function $y = \hat{\varphi}_1(x)$ that we have constructed satisfies the condition

$$\lim_{x\to+\infty} \frac{\hat{\psi}_1(x)}{x} = 0.$$

By changing the function $\hat{\varphi}_1(x)$ in an inessential manner in a neighborhood of each angular point, we can make it everywhere differentiable. Let $\psi_1(x)$ denote the function with the opposite sign. This function $y = \psi_1(x)$ has the required properties.

Now construct a convex majorant $\psi_2(x)$ for the function $\varphi_1(x) + \psi_1(x)$, just as in § 2.

Then putting

$$\psi(x) = \psi_2(x) - \psi_1(x),$$

we obtain

$$\varphi_1(x) \leqslant \psi(x)$$

and, on some sequence $\{x_n'\}$ of extreme points tending to infinity, we have

$$\varphi_1(x_n') = \psi(x_n') \quad (n = 1, 2, \ldots).$$

Also

$$\lim_{x\to+\infty} \psi'(x) = 0 \text{ and } \lim_{x\to+\infty} \frac{\psi(x)}{x} = 0. \tag{1.56}$$

Putting, just as before,

$$\rho(r) = \rho + \frac{\psi(\ln r)}{\ln r},$$

we obtain from (1.56)

$$\lim_{r\to\infty} \rho(r) = \rho,$$

$$\lim_{r\to+\infty} r\rho'(r)\ln r = \lim_{r\to+\infty} \left[\psi'(\ln r) - \frac{\psi(\ln r)}{\ln r}\right] = 0,$$

that is, $\rho(r)$ is a proximate order. Also

$$f(r) \leqslant r^{\rho(r)}$$

and for some sequence r_n tending to infinity

$$f(r_n) = r_n^{\rho(r_n)}.$$

The theorem is proved.

Note that the function $\exp(\psi(\ln r))$ is a slowly increasing function $L(r)$.

In Chapter II we shall need the following remark.

REMARK. The slowly increasing function that we have constructed can be represented, as is easily seen, in the form

$$L(r) = e^{\psi_2(\ln r) - \psi_1(\ln r)},$$

where $\psi_1(x)$ and $\psi_2(x)$ are unbounded increasing concave functions satisfying condition (c).

We shall show that *in the expression for $L(r)$ the functions $\psi_1(x)$ and $\psi_2(x)$ can be replaced by two other concave functions $\vartheta_1(x)$ and $\vartheta_2(x)$ that also satisfy the conditions*

$$\text{(a)}\qquad \lim_{x\to+\infty} \vartheta_i(x) = \infty,$$

$$\text{(b)}\qquad \lim_{x\to+\infty} \frac{\vartheta_i(x)}{x} = 0, \qquad (i = 1,\ 2)$$

and that satisfy, in addition, the condition

$$\text{(b')}\qquad \lim_{x\to+\infty} \frac{\vartheta_i''(x)}{\vartheta_i'(x)} = 0.$$

To prove this we make the following construction. Let $y = \psi(x)$ be a concave function satisfying the conditions $\lim_{x\to+\infty} \psi(x) = \infty$, $\lim_{x\to+\infty} \psi(x)/x = 0$, and let (l_0) be a line of support[14] of this curve. On the line (l_0) choose a point (x_0, y_0) and consider the curve (l_1)

$$y = c_0^{(1)} + c_1^{(1)}(x - x_0) - c_2^{(1)} e^{-\varepsilon_1(x-x_0)},$$

that is tangent to the line (l_0) at the point (x_0, y_0). In the equation for this curve, ε_1 is a positive number, and the coefficients $c_0^{(1)}$, $c_2^{(1)}$ are determined by the condition that (l_0) and (l_1) be tangent at the point (x_0, y_0)

$$c_2^{(1)} = \frac{1}{\varepsilon_1}(y_0' - c_1^{(1)}), \qquad c_0^{(1)} = y_0 + \frac{1}{\varepsilon_1}(y_0' - c_1^{(1)}),$$

where y_0' is the slope of the line (l_0). The parameter $c_1^{(1)}$ will be chosen to be positive, but less than y_0'. Then the coefficient $c_2^{(1)}$ will also be positive.

The curve (l_1) is clearly concave and approaches asymptotically the line $y = c_0^{(1)} + c_1^{(1)}(x - x_0)$.

---

[14] A line of support of a given curve is a line having points in common with the curve, but such that the entire curve lies in one of the closed half-planes determined by the line. (For further details on lines of support, see § 19.)

Also, from the equation

$$\frac{y''(x)}{y'(x)} = -\frac{c_2^{(1)}\varepsilon_1^2}{c_2^{(1)}\varepsilon_1 + c_1^{(1)}e^{\varepsilon_1(x-x_0)}}$$

it follows that on the whole curve (l_1)

$$\left|\frac{y''(x)}{y'(x)}\right| < \varepsilon_1. \tag{1.57}$$

If the abscissa x_0 is sufficiently large, then that part of the curve l_1 that lies to the right of x_0 is above the curve $y = \psi(x)$. Choosing the curve (l_1) in this manner and then decreasing $c_1^{(1)}$ while keeping the point (x_0, y_0) and the quantity ε_1 fixed, we can cause this curve to touch the curve $y = \psi(x)$ from above (Fig. 3). Since the curve (l_1) is concave and contains no line segments, the point of contact (x_0', y_0') must be an extreme point of the curve $y = \psi(x)$.

Now choose numbers $\varepsilon_2 < \varepsilon_1$ and $c_1^{(2)} < c_1^{(1)}$ and choose a point (x_1, y_1) on the curve (l_1) far enough out so that the part of the curve (l_2)

$$y = c_0^{(2)} + c_1^{(2)}(x - x_1) - c_2^{(2)}e^{-\varepsilon_2(x-x_1)},$$

(this curve is tangent to the curve (l_1) at the point (x_1, y_1)), lying to the right of this point, lies above the curve $y = \psi(x)$. Without changing the point (x_1, y_1) or the parameter ε_2, we decrease the parameter $c_1^{(2)}$ until the curve (l_2) has a point of contact with the curve $y = \psi(x)$. Next we choose numbers ε_3 and $c_1^{(3)} < c_1^{(2)}$

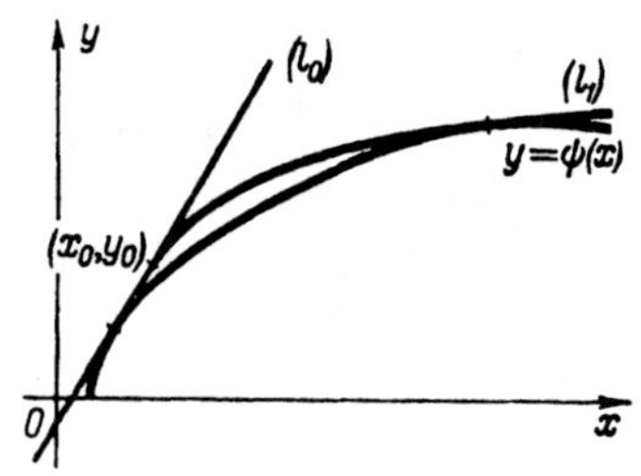

Fig. 3

and a point (x_2, y_2) on l_2, and form a curve (l_3), etc. This process is continued indefinitely, and the numbers $\varepsilon_1 > \varepsilon_2 > \varepsilon_3 > \ldots$, $c_1^{(1)} > c_1^{(2)} > c_1^{(3)} > \ldots$ are chosen so that $\varepsilon_n \to 0$ and $c_1^{(n)} \to 0$.

Now we form a smooth concave curve $y = \vartheta(x)$ from the segments of the curves $(l_0), (l_1), \ldots, (l_n), \ldots$ taken between the points of contact. Clearly

$$\lim_{x\to+\infty} \vartheta(x) = \infty, \qquad \lim_{x\to+\infty} \frac{\vartheta(x)}{x} = 0$$

and $\vartheta(x) \geqslant \psi(x)$, with equality holding for a sequence of values $x_1', x_2', \ldots, x_n', \ldots, x_n' \to \infty$, corresponding to extreme points of the curve $y = \psi(x)$. In addition, $\vartheta(x)$ is everywhere twice differentiable, except at the points of contact.

At these points of contact the first derivative $y' = \vartheta'(x)$ has angular points. By changing the function $\vartheta(x)$ in an inessential manner these angles can be smoothed out so that the second derivative $\vartheta''(x)$ in each interval lies between the upper and lower limits of the function $\vartheta''(x)$ at the corresponding points of contact. On the basis of inequality (1.57) we can assert that

$$\lim_{x \to +\infty} \frac{\vartheta''(x)}{\vartheta'(x)} = 0.$$

Now choose a majorant $\vartheta_1(x)$ for the concave function $\psi_1(x)$. Let $\tilde{\psi}_2(x)$ be the smallest convex majorant for the function $\varphi_1(x) = \varphi(x) + \vartheta_1(x)$, and let $\vartheta_2(x)$ be a majorant for $\tilde{\psi}_2(x)$. Then we have everywhere $\vartheta_2(x) \geqslant \tilde{\psi}_2(x) \geqslant \varphi(x) + \vartheta_1(x)$, while on some sequence of points $x_1', x_2', \ldots, x_n', \ldots$ tending to infinity and corresponding to extreme points of the curve $y = \tilde{\psi}(x)$ we have the equality $\vartheta_2(x_n') = \tilde{\psi}_2(x_n')$. On the other hand, since at the extreme points of the curve $y = \tilde{\psi}_2(x)$ the equality $\tilde{\psi}_2(x) = \varphi(x) + \vartheta_1(x)$ holds, we have

$$\vartheta_2(x_n') = \varphi(x_n') + \vartheta_1(x_n').$$

Thus we have constructed functions $\vartheta_1(x)$ and $\vartheta_2(x)$ satisfying conditions (a), (b), (b′) and such that, in addition,

$$\varphi(x) \leqslant \vartheta_2(x) - \vartheta_1(x), \quad \varphi(x_n') = \vartheta_2(x_n') - \vartheta_1(x_n') \quad (x_n' \to \infty).$$

In the future $L^*(r)$ will denote a function having a representation

$$\ln L^*(r) = \vartheta_2(\ln r) - \vartheta_1(\ln r),$$

in which the functions $\vartheta_1(x)$ and $\vartheta_2(x)$ are concave functions satisfying conditions (a), (b), (b′).

By $\rho^*(r)$ we mean the function defined by

$$r^{\rho^*(r)} = r^{\rho} L^*(r),$$

that is,

$$\rho^*(r) = \rho + \frac{\vartheta_2(\ln r) - \vartheta_1(\ln r)}{\ln r}.$$

The functions $\rho^*(r)$ clearly form a smaller class than the proximate orders $\rho(r)$; however, as we have seen, this class is sufficient to construct a growth scale for the functions of finite order. More precisely, in the statement of Theorem 16 the proximate order can be replaced by $\rho^*(r)$. The function $\rho^*(r)$ will be called a *strong proximate order*.

13. Extension of the Classical Theorems to Proximate Orders

The classical theorems that were presented in the preceding sections on the connection between the growth of the function

$$\ln M_f(r)$$

and the distribution of the zeros of the function $f(z)$ can be extended to the more precise description of the growth that is given by the proximate orders. The theorem relating the type of the function to the rate of decrease of the coefficients in the Taylor expansion can also be generalized. To formulate this theorem precisely, we introduce the function $\varphi(t)$, defined to be the unique (for $t > t_0$) solution of the equation $t = r^{\rho(r)}$.

THEOREM 2′. *The type σ_f of the entire function*

$$f(z) = \sum_{n=0}^{\infty} c_n z^n$$

with the proximate order $\rho(r)$ ($\rho > 0$)[14a] *is given by the equation*

$$\overline{\lim_{n\to\infty}}\ \varphi(n)\sqrt[n]{|c_n|} = (\sigma_f \rho e)^{\frac{1}{\rho}}. \qquad (1.06')$$

PROOF. We first show that

$$\lim_{t\to\infty} \frac{\varphi(kt)}{\varphi(t)} = k^{\frac{1}{\rho}}. \qquad (1.58)$$

Indeed, differentiating the equation $\ln t = \rho(r) \ln r$, with respect to $\ln e$, we obtain

$$\frac{d \ln t}{d \ln r} = \rho'(r)\, r \ln r + \rho(r)$$

and consequently

$$\lim_{r\to\infty} \frac{d \ln t}{d \ln r} = \rho.$$

This equation can also be written in the form

$$\lim_{t\to\infty} \frac{d \ln \varphi(t)}{d \ln t} = \frac{1}{\rho}.$$

Thus we have the asymptotic inequality

$$\left(\frac{1}{\rho} - \varepsilon\right) d \ln t < d \ln \varphi(t) < \left(\frac{1}{\rho} + \varepsilon\right) d \ln t.$$

Integrating from t to kt yields the asymptotic inequality

$$\left(\frac{1}{\rho} - \varepsilon\right) \ln k < \ln \frac{\varphi(kt)}{\varphi(t)} < \left(\frac{1}{\rho} + \varepsilon\right) \ln k,$$

from which we have at once

$$\lim_{t\to\infty} \frac{\varphi(kt)}{\varphi(t)} = k^{\frac{1}{\rho}}.$$

[14a] When $\rho = 0$ or $\rho = \infty$, the connection between the rate of growth of the function and the rate of decrease of the coefficients in its power series is rather complicated. There have been many papers on this problem; we mention Fridman [1; 2] and Šeremeta [1; 2]. The investigations of Videnskiĭ [1] and Evgrafov [1] belong to the same circle of ideas. We also mention investigations of entire functions of order zero by Bratiščev and Korobeĭnik [1] and by Rubel and Taylor [1].

The remainder of the proof of Theorem 2′ closely follows the proof of Theorem 2.

From inequality (1.07) we have asymptotically, for $\sigma > \sigma_f$,

$$\ln|c_n| < \sigma r^{\rho(r)} - n \ln r.$$

Choosing r to be the root of the equation

$$n = \sigma\rho r^{\rho(r)} \qquad (\sigma > \sigma_f),$$

we have

$$\ln|c_n| < \frac{n}{\rho} - n \ln \varphi\left(\frac{n}{\rho\sigma}\right),$$

or

$$\ln\left(\varphi(n)\sqrt[n]{|c_n|}\right) < \frac{1}{\rho} + \ln \frac{\varphi(n)}{\varphi\left(\frac{n}{\rho\sigma}\right)},$$

and passing to the limit we have

$$\overline{\lim_{n\to\infty}} \left(\varphi(n)\sqrt[n]{|c_n|}\right) \leqslant (\sigma\rho e)^{\frac{1}{\rho}},$$

and, since $\sigma > \sigma_f$ was arbitrary,

$$\overline{\lim_{n\to\infty}} \varphi(n)\sqrt[n]{|c_n|} \leqslant (\sigma_f\rho e)^{\frac{1}{\rho}}. \tag{1.09′}$$

We must show that equality holds in (1.09′). We define σ by the equation

$$\overline{\lim_{n\to\infty}} \varphi(n)\sqrt[n]{|c_n|} = (\sigma\rho e)^{\frac{1}{\rho}}$$

and show that the assumption $\sigma < \sigma_f$ leads to a contradiction. Choose any number σ_1 between σ and σ_f ($\sigma < \sigma_1 < \sigma_f$). We have for all sufficiently large n

$$|c_n| < \left\{\frac{(\sigma_1\rho e)^{\frac{1}{\rho}}}{\varphi(n)}\right\}^n.$$

Using (1.58) this asymptotic inequality can be put into the form

$$|c_n| < \left\{\frac{e^{\frac{1}{\rho}}}{\varphi\left(\frac{n}{\sigma_1\rho}\right)}\right\}^n.$$

Thus, for all $n > n_0$,

$$|c_n r^n| < \left\{\frac{e^{\frac{1}{\rho}}}{\varphi\left(\frac{n}{\sigma_1\rho}\right)}\right\}^n r^n$$

and the maximal term $\mu_f(r)$ of the Maclaurin series of $f(r)$ satisfies the inequality

$$\mu_f(r) < \max_{n>n_0} \left\{ \frac{e^{\frac{1}{\rho}}}{\varphi\left(\frac{n}{\sigma_1 \rho}\right)} \right\}^n r^n$$

The maximum of the right-hand side is attained for $n = [\sigma, \rho r^{\rho(r)}]$. Hence we have

$$\mu_f(r) < e^{\sigma_1 r^{\rho(r)}}.$$

Hence from (1.10) we see that for all $\rho_1 > \rho$

$$M_f(r) < (2 + 2^{\rho_1} e \sigma_f \rho_1 r^{\rho_1}) e^{\sigma_1 r^{\rho(r)}},$$

and this leads to the contradiction $\sigma_f \leqslant \sigma_1$. Thus the inequality sign in (1.09′) cannot occur, and the theorem is proved.

We pass now to the theorem on the relation between the type of an entire function with a given proximate order and the distribution of its zeros.

Let $f(z)$ be an entire function with proximate order $\rho(r)$, that is,

$$\sigma_f = \overline{\lim_{r\to\infty}} \frac{\ln M_f(r)}{r_{\rho(r)}} \neq 0, \ \infty. \tag{1.59}$$

Then for $\varepsilon > 0$ we have asymptotically

$$\ln M_f(r) < \left(\sigma_f + \frac{\varepsilon}{2}\right) r^{\rho(r)},$$

and by Lemma 4

$$n_f(r) < \left(\sigma_f + \frac{\varepsilon}{2}\right) (er)^{\rho(er)}.$$

Using the asymptotic inequality (see (2) following Lemma 5, § 12)

$$(er)^{\rho(er)} = e^\rho r^\rho L(er) < (1+\eta) e^\rho r^\rho L(r) = (1+\eta) e^\rho r^{\rho(r)},$$

which is valid for arbitrary $\eta > 0$, we have from (1.59)

$$\overline{\lim_{r\to\infty}} \frac{n_f(r)}{r^{\rho(r)}} \leqslant e^\rho \sigma_f.$$

Retaining the notation

$$\Delta_f = \overline{\lim_{r\to\infty}} \frac{n_f(r)}{r^{\rho(r)}},$$

we write this inequality in the form

$$\Delta_f \leqslant e^\rho \sigma_f.$$

If ρ is not an integer we can obtain a converse estimate. In this case, from the representation (1.38) of the entire function and from Lemma 3, we have

$$\ln M_f(r) \leqslant O(r^p) + k_p r^p \left(\int_0^r \frac{n_f(t)\,dt}{t^{p+1}} + r \int_r^\infty \frac{n_f(t)\,dt}{t^{p+2}} \right) \qquad (p < \rho).$$

From the asymptotic inequality

$$n_f(r) < (\Delta_f + \varepsilon)\, r^{\rho(r)}$$

and property (3) of proximate orders we have

$$\ln M_f(r) \leqslant k_p (\Delta_f + \varepsilon) \left[\frac{1}{\rho - p} + \frac{1}{p + 1 - \rho}\right] r^{\rho(r)} + o(r^{\rho(r)}),$$

from which follows

$$\sigma_f \leqslant c_\rho \Delta_f,$$

where c_ρ is a constant not depending on the function $f(z)$.

Thus we have obtained a theorem that generalizes Theorem 14.

THEOREM 17. *Let $f(z)$ be an entire function of nonintegral order ρ and let $\rho(r)$ be a proximate order with* lim $\rho(r) = \rho$. *If with this exponent $\Delta_f = 0$, then $f(z)$ is of minimal type with respect to $\rho(r)$, if $0 < \Delta_f < \infty$, then of normal type, and if $\Delta_f = \infty$, then f is of maximal type.*

The case of integral order is exceptional here too. We have already seen in § 10 that in this case the value of the upper density Δ_f is not enough by itself to determine whether the function is of minimal, normal or maximal type.

We define the quantity δ_f for entire functions of order $\rho(r)$. Whether or not the series

$$\sum_{n=1}^{\infty} |a_n|^{-\rho}$$

converges, we represent the entire function $f(z)$ of integral order ρ in the form

$$f(z) = z^m e^{P(z)} \prod_{n=1}^{\omega} G\left(\frac{z}{a_n};\ \rho\right) \qquad (\omega \leqslant \infty), \tag{1.38$'$}$$

where $P(z) = \alpha_\rho z^\rho + \alpha_{\rho-1} z^{\rho-1} + \dots + \alpha_1 z + \alpha_0$. We put

$$\delta_f(r) = \alpha_\rho + \frac{1}{\rho} \sum_{|a_n| < r} \frac{1}{a_n^\rho}$$

and

$$\delta_f = \varlimsup_{r \to \infty} \frac{|\delta_f(r)|}{L(r)},$$

where $L(r) = r^{\rho(r) - \rho}$, and we introduce the quantity

$$\gamma_f = \max(\Delta_f,\ \delta_f).$$

The following generalization of Lindelöf's theorem answers the question of the type of a function $f(z)$ with respect to a given proximate order.

THEOREM 18. *Let $f(z)$ be an entire function of integral order ρ and let $\rho(r)$ be a proximate order with* $\lim_{r\to\infty} \rho(r) = \rho \geqslant 1$.

If $\gamma_f = 0$, then $f(z)$ is of minimal type with respect to $\rho(r)$, if $0 < \gamma_f < \infty$, then $f(z)$ is of normal type, and if $\gamma_f = \infty$, then $f(z)$ is of maximal type.

We give the proof briefly, using the estimates that we found in the proof of Theorem 15(a).

The function $f(z)$ can be represented in the form (1.40) whether the genus of the canonical product coincides with the order of the function or is less than it. From the bound (1.41), (1.53) and (1.53′), it follows easily that asymptotically

$$\ln M_R(R) < 2k_\rho (\Delta_f + \varepsilon) R^{\rho(R)}. \tag{1.42′}$$

Dividing the asymptotic inequality (1.43) by $R^{\rho(r)}$ and taking the upper limit as $R \to \infty$ we have

$$\sigma_f \leqslant \left(1 + 2k_\rho\right) \gamma_f. \tag{1.44′}$$

To obtain the reverse inequality we note that from the estimate $n(r) \leqslant \ln M(er)$ we have the inequality

$$\Delta_f \leqslant e^\rho \sigma_f \tag{1.45′}$$

for the proximate order. Also, from inequality (1.46) we have, using (1.42′) and (1.45′),

$$\delta_f \leqslant c_1 \sigma_f,$$

where c_1 is a constant. From this last inequality and (1.45′) we have:

$$\sigma_f \geqslant c^{-1} \gamma_f. \tag{1.49′}$$

The assertion of the theorem follows from (1.44′) and (1.49′).

The concept of the category of a function with respect to a given proximate order can be introduced, just as in § 9 for the usual order.

Two functions $f(z)$ and $\varphi(z)$ are said to be of the same category if σ_f and σ_φ are both infinite, or both finite and positive, or both equal to zero. Theorem 12 can be extended to these generalized categories, namely:

The category of the product of two functions $f(z)$ and $\varphi(z)$ does not exceed the larger of the categories of the factors, and if the categories of $f(z)$ and $\varphi(z)$ are different, then it coincides with the larger of these categories.

The proof can be given just as in § 9. However, it is simpler to reason as follows, using Theorems 17 and 18. From the equation

$$n_{f\varphi}(r) = n_f(r) + n_\varphi(r)$$

we have

$$\Delta_{f\varphi} \leqslant \Delta_f + \Delta_\varphi,$$

while if either of the terms on the right side is zero or infinity, then equality holds.

For functions of nonintegral order the quantities σ_f and Δ_f are either both zero, or both finite and positive, or both infinite. The assertion of the theorem follows at once for this case. For integral $\rho = \lim_{n\to\infty} \rho(r)$ it is necessary to consider the relation

$$\delta_{f\varphi}(r) = \delta_f(r) + \delta_\varphi(r).$$

The assertion of the theorem for integral ρ can be obtained easily from these relations if we note that by Theorem 18 the category of a function is completely determined by the quantity $\gamma_f = \max(\delta_f, \Delta_f)$. The corollaries to Theorem 12 can also be carried over to the case of generalized categories.

The crude characterization of the growth of a function that is given by the asymptotic behavior of $M_f(r)$ is insufficient for a deeper study of the relation between the growth of the function and the distribution of its zeros. In §15 we shall introduce and study a more precise growth characteristic, in which the growth of the function is studied in its dependence on the direction in which z tends to infinity.

It will then turn out that the growth of the function depends in an essential manner on the distribution of the arguments of the zeros, even for the case of nonintegral order.

To study this growth characteristic we need some general theorems of the theory of functions which we shall present in the following sections.

14. The Phragmén-Lindelöf Principle

The maximum principle states that a function $f(z)$ that is holomorphic in a domain and continuous on the closure of the domain attains the maximum value of its modulus on the boundary. This important principle was extended by Phragmén and Lindelöf [**1**] to the case when the function is not continuous at certain exceptional boundary points, under the assumption, however, that as these points are approached the modulus of the function does not grow too fast. The following theorem is fundamental.

THEOREM 19. *Let $f(z)$ be holomorphic in a domain G, and suppose there exists a holomorphic function $\omega(z)$ in this domain such that $f(z)[\omega(z)]^\delta$ has, for all $\delta > 0$, a limit at each boundary point of G (including the point at infinity, if it is a boundary point), and that on the entire boundary*

$$|f(z)[\omega(z)]^\delta| \leqslant M, \tag{1.60}$$

where M does not depend on δ.

Then

$$|f(z)| \leqslant M$$

throughout the domain G.

PROOF. From the usual maximum principle it follows that if (1.60) holds on the boundary of G, then it also holds throughout the interior of the domain.[15]

[15] If the function $\omega(z)$ has roots in the domain G, then $f(z)[\omega(z)]^\delta$ is not holomorphic in G, but the modulus of this multiple-valued function is single-valued in G. The maximum modulus principle can be extended to this case without essential modification of the proof.

Let the point z_0 belong to the domain G and let $\omega(z_0) \neq 0$. Then for arbitrary $\delta > 0$

$$|f(z_0)| \leqslant M\,|\,\omega(z_0)|^{-\delta}$$

and consequently

$$|f(z_0)| \leqslant M. \tag{1.61}$$

The roots of the function $\omega(z)$ are isolated points and, by the maximum principle inequality (1.61), must continue to hold at these points, that is, throughout the domain G.

Note that equality cannot be attained at any interior point, unless $f(z)$ is a constant. From this general principle we obtain a number of important theorems that are frequently applied in various questions.

THEOREM 20. *Let $f(z)$ be holomorphic in the domain G, whose boundary contains the point at infinity, and suppose that at all finite boundary points*

$$|f(z)| \leqslant M. \tag{1.62}$$

If $|f(z)|$ is bounded in G, then inequality (1.62) *holds throughout G.*

PROOF. Remove from the domain G a disc C of radius r with center at some point z_0. Put

$$\omega(z) = \frac{r}{z - z_0}.$$

The function $f(z)[\omega(z)]^\delta$ tends to zero at infinity, and since $|\omega(z)| < 1$ for $|z - z_0| > r$, we will have on the boundary of G

$$|f(z)[\omega(z)]^\delta| \leqslant M.$$

In addition, on the circumference $|z - z_0| = r$, we have

$$|f(z)[\omega(z)]^\delta| \leqslant M_1,$$

where $M_1 = \max_{|z-z_0|=r} |f(z)|$.

By the general Phragmén-Lindelöf principle we have, throughout the domain formed by removing the disc C,

$$|f(z)| \leqslant \max(M, M_1) \quad (z \in G - C).$$

Now let r tend to zero. Then $\lim_{r\to 0} M_1 = |f(z_0)|$ and

$$|f(z)| \leqslant \max(M, |f(z_0)|) \quad (z \in G).$$

If $|f(z_0)| > M$, then we would have throughout G

$$|f(z)| \leqslant |f(z_0)|.$$

But at the interior point z_0 the function $|f(z_0)|$ cannot attain a maximum. Therefore $|f(z_0)| \leqslant M$ and consequently

$$|f(z)| \leqslant M \quad \text{for} \quad z \in G.$$

The theorem on the growth of a function inside an angle will be very important for us. This theorem was obtained by Phragmén and Lindelöf from the general principle.

First we introduce the following concepts. Let

$$M_f(r, \alpha, \beta) = \max_{\alpha \leqslant \theta \leqslant \beta} |f(re^{i\theta})|$$

and define the *order* ρ and the *type* σ_f of a function $f(z)$ analytic in the angle $\alpha \leqslant \arg z \leqslant \beta$ to be the quantities

$$\rho = \overline{\lim_{r \to \infty}} \frac{\ln^+ \ln^+ M_f(r, \alpha, \beta)}{\ln r} \quad \text{and} \quad \sigma_f = \overline{\lim_{r \to \infty}} \frac{\ln^+ M_f(r, \alpha, \beta)}{r^\rho}.$$

THEOREM 21. *Let the function $f(z)$ be holomorphic inside an angle of opening π/α and continuous on the boundary. Assume that on the sides of the angle*

$$|f(z)| \leqslant M$$

and that the order ρ of the function is less than α. Then

$$|f(z)| \leqslant M$$

throughout the angle.

PROOF. Without loss of generality we can consider the angle $|\arg z| \leqslant \pi/2\alpha$. Let

$$\omega(z) = e^{-z^\gamma} \qquad \left(|\arg z| \leqslant \frac{\pi}{2\alpha}\right),$$

where $\rho < \gamma < \alpha$. Then asymptotically inside the angle $|\arg z| \leqslant \pi/2$

$$|f(z)[\omega(z)]^\delta| = |f(z)|\, e^{-\delta r^\gamma \cos \gamma\theta} \qquad (z = re^{i\theta}, \ \delta > 0)$$

and consequently for $\rho < \rho_1 < \gamma$

$$|f(z)[\omega(z)]^\delta| < e^{r^{\rho_1} - \delta r^\gamma \cos \gamma\theta}.$$

Thus we have shown that the function

$$f(z)[\omega(z)]^\delta$$

tends uniformly to zero inside the angle $|\arg z| \leqslant \pi/2\alpha$ as $r \to \infty$. Further, since the modulus of this function is bounded by M on the sides of the angle, we have by the general Phragmén-Lindelöf principle

$$|f(z)| \leqslant M$$

throughout the angle.

The theorem is sharp in the sense that there are functions of order ρ inside an angle of opening π/ρ that are bounded on the sides of the angle. An example of such a function is

$$e^{z^\rho},$$

taken inside the angle $|\arg z| \leqslant \pi/2\rho$. However, for the case of an angle of opening π/ρ, one can give an estimate of the modulus of the function inside the angle that depends on the type.

THEOREM 22. *If $f(z)$ is a function of order ρ and type $\sigma \geqslant 0$, holomorphic inside the angle $|\arg z| \leqslant \pi/2\rho$, and if on the sides of this angle*

$$|f(re^{\pm i\frac{\pi}{2\rho}})| \leqslant M,$$

then throughout the angle we have the inequality

$$|f(re^{i\theta})| \leqslant Me^{\sigma r^\rho \cos \rho\theta} \qquad \left(|\theta| \leqslant \frac{\pi}{2\rho}\right).$$

The angle $|\arg z| \leqslant \pi/2\rho$ can, of course, be replaced by any other angle of opening π/ρ.

PROOF. The function

$$\varphi_\varepsilon(z) = e^{-(\sigma+\varepsilon)z^\rho} f(z)$$

is bounded on the sides of the angle $|\arg z| \leqslant \pi/2\rho$ and, in addition, is bounded on the positive real axis.

We apply Theorem 21 to this function inside each of the angles $0 \leqslant \arg z \leqslant \pi/2\rho$ and $-\pi/2\rho \leqslant \arg z \leqslant 0$, and thus we see that the function $\varphi_\varepsilon(z)$ is bounded throughout the full angle. Hence, by Theorem 20, inside this angle we have

$$|\varphi_\varepsilon(z)| \leqslant M$$

or

$$|f(z)| \leqslant Me^{(\sigma+\varepsilon)r^\rho \cos \rho\theta} \qquad \left(|\theta| \leqslant \frac{\pi}{2\rho}\right)$$

and finally, since ε was arbitrary,

$$|f(z)| \leqslant Me^{\sigma r^\rho \cos \rho\theta}.$$

In particular, if the function $f(z)$ is of order 1 and type σ in the upper half-plane ($\operatorname{Im} z \geqslant 0$) and is bounded on the real axis

$$|f(x)| \leqslant M \qquad (-\infty < x < \infty),$$

then throughout the half-plane $\operatorname{Im} z \geqslant 0$

$$|f(z)| \leqslant Me^{\sigma \operatorname{Im} z}. \tag{1.63}$$

For $\sigma = 0$ Theorem 22 can be regarded as a strengthening of Theorem 21. Indeed, in this case, it differs from Theorem 21 in that the requirement that the order ρ of the function $f(z)$ inside the angle be less than α is replaced by the weaker requirement that the function be at most of minimal type with respect to the order $\rho = \alpha$ inside the angle.

COROLLARY. *If the entire function $f(z)$ is at most of first order and minimal type and if its modulus is bounded on some line, then it is a constant.*

Indeed, in this case, $|f(z)|$ is bounded in each of the half-planes determined by the line, that is, $|f(z)|$ is bounded in the entire plane, and hence, by Liouville's theorem, it is a constant.

We note that by completely analogous reasoning we can establish the following Phragmén-Lindelöf theorem for the strip.

Let $f(z)$ be analytic in the strip $|y| \leqslant \gamma$, with

$$|f(x \pm i\gamma)| \leqslant M \text{ and } |f(x+iy)| \leqslant Ne^{e^{k|x|}} \quad (M > 0 \text{ and } N > 0)$$

and $0 \leqslant k < \pi/2\gamma$. Then $|f(z)| \leqslant M$ throughout the strip.

Indeed, choosing

$$\omega(z) = e^{-e^{\sigma z}} \quad \left(k < \sigma < \frac{\pi}{2\gamma}\right),$$

we have

$$|\omega(z)|^{\delta} = e^{-\delta e^{\sigma x} \cos \sigma y} \quad (\delta > 0).$$

Putting $M_1 = \max|f(iy)|$ for $-\gamma \leqslant y \leqslant \gamma$ and $M_2 = \max(M, M_1)$ we have, by the maximum modulus principle,

$$|f(z)[\omega(z)]^{\delta}| \leqslant M_2$$

in the half-strip $x \geqslant 0$, $|y| \leqslant \gamma$. By the general Phragmén-Lindelöf principle we then have $|f(z)| \leqslant M_2$ for $x \geqslant 0$, $|y| \leqslant \gamma$.

In the same way one proves this inequality for the left half strip $x \leqslant 0$, $|y| \leqslant \gamma$. Now from Theorem 20 we have

$$|f(z)| \leqslant M$$

for $|y| \leqslant \gamma$.

15. The Indicator Function

To characterize the dependence of the growth of a function $f(z)$ of finite order ρ in an angle $\theta_1 \leqslant \arg z \leqslant \theta_2$ on the direction in which z tends to infinity, Phragmén and Lindelöf [1] introduced the function

$$h_f(\theta) = \overline{\lim_{r \to \infty}} \frac{\ln|f(re^{i\theta})|}{r^{\rho}} \quad (\theta_1 \leqslant \theta \leqslant \theta_2), \tag{1.64}$$

called the *indicator function of* $f(z)$. Note that for functions of normal type inside an angle the indicator is bounded above, and for entire functions of normal type[16] it is bounded below because of Theorem 11.

It follows at once from this definition that the indicator of the product of two functions is less than or equal to the sum of the indicators of the factors

$$h_{\varphi f}(\theta) \leqslant h_{\varphi}(\theta) + h_f(\theta) \tag{1.65}$$

[16] *Translator's note*: We assume $f \not\equiv 0$.

and the indicator of the sum of two functions is less than or equal to, for each value of θ, the larger of the two indicators

$$h_{f+\varphi}(\theta) \leqslant \max\left(h_f(\theta),\ h_\varphi(\theta)\right). \tag{1.66}$$

Let us verify the second of these assertions:

$$h_{f+\varphi}(\theta) = \overline{\lim_{r\to\infty}} \frac{\ln|f(re^{i\theta}) + \varphi(re^{i\theta})|}{r^\rho}$$
$$\leqslant \overline{\lim_{r\to\infty}} \frac{\ln\{2\max[|f(re^{i\theta})|;\ |\varphi(re^{i\theta})|]\}}{r^\rho} \leqslant \max[h_f(\theta);\ h_\varphi(\theta)].$$

It is easy to see that if, for some value of θ, the indicators of the two summands are not equal, then equality holds in (1.66).

For the function

$$f(z) = e^{(a-ib)z^\rho},$$

holomorphic in any angle $\alpha \leqslant \theta \leqslant \beta$, $\beta - \alpha \leqslant 2\pi$, we have

$$|f(re^{i\theta})| = e^{(a\cos\rho\theta + b\sin\rho\theta)r^\rho}$$

and its indicator is equal to

$$a\cos\rho\theta + b\sin\rho\theta.$$

Such an indicator will be called "trigonometric," and the expression $a\cos\rho\theta + b\sin\rho\theta$ will be denoted by $H(\theta)$. It is easy to see that the function $H(\theta)$ that takes the values h_1 and h_2 at the points θ_1 and θ_2 is given by the formula

$$H(\theta) = \frac{h_1\sin\rho(\theta_2-\theta) + h_2\sin\rho(\theta-\theta_1)}{\sin\rho(\theta_2-\theta_1)}. \tag{1.67}$$

LEMMA 6. *Let $h(\theta)$ be the indicator for a function of order ρ, holomorphic in the angle $\theta_1 \leqslant \theta \leqslant \theta_2$, $\theta_2 - \theta_1 < \pi/\rho$, and*

$$h_f(\theta_1) \leqslant h_1, \qquad h_f(\theta_2) \leqslant h_2;$$

let the function $H(\theta)$ be defined by (1.67). *Then for all θ in the interval* (θ_1, θ_2)

$$h(\theta) \leqslant H(\theta). \tag{1.68}$$

This property is called *trigonometric convexity.*

PROOF. Construct the function

$$H_\delta(\theta) = a_\delta\cos\rho\theta + b_\delta\sin\rho\theta,$$

that takes the values $h_1 + \delta$ and $h_2 + \delta$ at the points θ_1 and θ_2, respectively ($\delta > 0$). The function

$$\varphi(z) = f(z)\,e^{-(a_\delta - ib_\delta)z^\rho}$$

has the indicator[17]

$$h_\varphi(\theta) = h_f(\theta) - H_\delta(\theta).$$

Thus, along the rays $\arg z = \theta_1, \theta_2$, the function $\varphi(z)$ tends to zero.

By the Phragmén-Lindelöf theorem $\varphi(z)$ is bounded throughout the angle $\theta_1 \leqslant \arg z \leqslant \theta_2$ and consequently its indicator is less than or equal to zero. Thus for all $\delta > 0$

$$h_f(\theta) \leqslant H_\delta(\theta), \qquad \theta_1 \leqslant \theta \leqslant \theta_2,$$

from which (1.68) follows.

It follows from Lemma 6 that if the indicator function is finite for $\theta = \theta_1, \theta_2$ $(\theta_2 - \theta_1 < \pi/\rho)$, then it is bounded above for all values of θ in the interval $\theta_1 \leqslant \theta \leqslant \theta_2$. Further, if for some value $\theta = \theta_0$ in this interval

$$h(\theta_0) = -\infty,$$

then $h(\theta) = -\infty$ in the whole interval (θ_1, θ_2). This last is easy to obtain if we take θ_0 as one end point of the interval and change the reasoning of Lemma 6 slightly. Sometimes the following equivalent formulation of Lemma 6 is convenient:

If $\theta_1 < \theta_2 < \theta_3$, $\theta_3 - \theta_1 < \pi/\rho$ and $h(\theta_1) \leqslant H(\theta_1)$, $h(\theta_2) \geqslant H(\theta_2)$, then

$$h(\theta_3) \geqslant H(\theta_3).$$

It follows from this formulation that if $h(\theta_2) = \infty$, then $h(\theta) = \infty$ inside some angle containing the ray $\theta = \theta_2$.

16. The Fundamental Relations and Analytic Properties of the Indicator Function

Using equation (1.67) we can reformulate Lemma 6 on the trigonometric convexity of the indicator in the following symmetric form.

THEOREM 23. *The indicator $h_f(\theta)$ of a function $f(z)$ holomorphic and of order ρ inside the angle $\theta_1 \leqslant \arg z \leqslant \theta_3$, $\theta_3 - \theta_1 < \pi/\rho$, satisfies the relation*

$$h(\theta_1)\sin\rho(\theta_2 - \theta_3) + h(\theta_2)\sin\rho(\theta_3 - \theta_1) + h(\theta_3)\sin\rho(\theta_1 - \theta_2) \leqslant 0 \qquad (1.69)$$

for $\theta_1 < \theta_2 < \theta_3$.

Inequality (1.69) will be called the *fundamental relation of the indicator.* This inequality is unchanged by cyclic permutation of the quantities $\theta_1, \theta_2, \theta_3$ and changes sign if any two of these quantities are interchanged.

From the fundamental relation we obtain a number of analytical properties of trigonometrically convex functions in a similar manner to the way in which they are obtained for ordinary convex functions.

---

[17] Here we are using the following general fact: if $\lim_{r\to\infty} \varphi(r)$ exists, then

$$\overline{\lim_{r\to\infty}}\,[f(r) \pm \varphi(r)] = \overline{\lim_{r\to\infty}}\, f(r) \pm \lim_{r\to\infty} \varphi(r).$$

For this we first transform the fundamental relation as follows:

$$h(\theta)\sin\rho(\theta_3-\theta_1)+h(\theta_3)\sin\rho(\theta_1-\theta)$$
$$+h(\theta_1)[\sin\rho(\theta_1-\theta_3)-\sin\rho(\theta_1-\theta)]$$
$$\leqslant 4h(\theta_1)\sin\rho\frac{\theta_3-\theta_1}{2}\sin\rho\frac{\theta-\theta_1}{2}\sin\rho\frac{\theta_3-\theta}{2}\quad(\theta_1<\theta<\theta_3). \tag{1.70}$$

Dividing both parts of the inequality by $\sin\rho(\theta-\theta_1)\sin\rho(\theta_3-\theta_1)$ we obtain

$$\frac{h(\theta)-h(\theta_1)}{\sin\rho(\theta-\theta_1)}\leqslant\frac{h(\theta_3)-h(\theta_1)}{\sin\rho(\theta_3-\theta_1)}$$
$$+h(\theta_1)\sin\rho\frac{\theta_3-\theta}{2}\sec\rho\frac{\theta_3-\theta_1}{2}\sec\rho\frac{\theta-\theta_1}{2}. \tag{1.71}$$

Transposing the quantities θ_1 and θ, and taking account of the sign of the denominator, it can be shown that inequality (1.71) holds also for $\theta<\theta_1<\theta_3$.

For $\theta_3<\theta_1<\theta$ and for $\theta_3<\theta<\theta_1$ the sign in inequality (1.71) is reversed.

We now establish the following property of the trigonometrically convex function $h(\theta)$ that is bounded on some interval:

(a) *The function $h(\theta)$ is continuous.*

It is clearly sufficient to prove this for intervals of length less than π/ρ.

Considering (1.71) for $\theta_1<\theta<\theta_3$ and for $\theta_3<\theta_1<\theta$, we see that for $\theta>\theta_1$ the ratio

$$\frac{h(\theta)-h(\theta_1)}{\sin\rho(\theta-\theta_1)}$$

is bounded. Considering (1.71) in the other two cases we see that this ratio is also bounded for $\theta<\theta_1$. From the boundedness of this ratio the continuity of $h(\theta)$ clearly follows.

(b) *$h(\theta)$ has a derivative from the left and from the right at every point.*

PROOF. Let

$$k=\max\left|\frac{\rho}{2}h(\theta)\sec^2\rho\frac{q}{2}\right|$$

for $|\theta_3-\theta_1|\leqslant q<\pi/\rho$ and $\theta_1<\theta<\theta_3$. From inequality (1.71) one can obtain

$$\frac{h(\theta)-h(\theta_1)}{\sin\rho(\theta-\theta_1)}\leqslant\frac{h(\theta_3)-h(\theta_1)}{\sin\rho(\theta_3-\theta_1)}+k[(\theta_3-\theta_1)-(\theta-\theta_1)], \tag{1.72}$$

and consequently the bounded quantity

$$r(\theta,\theta_1)=\frac{h(\theta)-h(\theta_1)}{\sin\rho(\theta-\theta_1)}+k(\theta-\theta_1)$$

decreases monotonically as $\theta\to\theta_1+0$. From this follows the existence of the limit

$$h'_+(\theta_1)=\lim_{\theta\to\theta_1+0}[\rho r(\theta,\theta_1)].$$

Similarly one shows the existence of the derivative from the left.

(c) *The right hand derivative is greater than or equal to the left hand derivative at each point:*

$$h'_+(\theta_1) \geqslant h'_-(\theta_1).$$

This inequality is immediate if one writes (1.71) with $\theta < \theta_1 < \theta_3$ and passes to the limit as $\theta \to \theta_1 - 0$ and $\theta_3 \to \theta_1 + 0$.

(d) *The derivative $h'_+(\theta)$ is continuous from the right and the derivative $h'_-(\theta)$ is continuous from the left.*

PROOF. Choose $\theta_1 < \theta < \theta_3$. Just as we obtained (1.71) we can obtain

$$\frac{h(\theta) - h(\theta_3)}{\sin \rho(\theta - \theta_3)}$$
$$\geqslant \frac{h(\theta_1) - h(\theta_3)}{\sin \rho(\theta_1 - \theta_3)} - h(\theta_3) \sin \rho \frac{\theta - \theta_1}{2} \sec \rho \frac{\theta_1 - \theta_3}{2} \sec \rho \frac{\theta - \theta_3}{2}$$

or

$$\frac{h(\theta) - h(\theta_3)}{\sin \rho(\theta - \theta_3)} \geqslant \frac{h(\theta_1) - h(\theta_3)}{\sin \rho(\theta_1 - \theta_3)} - k(\theta - \theta_1). \tag{1.73}$$

Passing to the limit as $\theta \to \theta_3 - 0$ we obtain

$$h'_-(\theta_3) \geqslant \rho r(\theta_1, \theta_3) - 2\rho k(\theta_3 - \theta_1).$$

Similarly to (1.72) we obtain

$$h'_+(\theta_1) \leqslant \rho r(\theta_1, \theta_3).$$

It follows from the last two inequalities that

$$h'_-(\theta_3) - h'_+(\theta_1) \geqslant -2k\rho(\theta_3 - \theta_1); \tag{1.74}$$

and comparing this with property (c) we see that the functions

$$h'_+(\theta) + 2k\rho\theta \text{ and } h'_-(\theta) + 2k\rho\theta$$

are nondecreasing on the interval $[\theta_1, \theta_1 + q]$. Thus the limits $h'_-(\theta - 0)$ and $h'_+(\theta + 0)$ both exist.

Passing to the limit as $s \to +0$ in the equations

$$\frac{h(\theta + s) - h(\theta)}{s} = \frac{1}{s} \int_\theta^{\theta+s} h'_+(\theta)\, d\theta$$

and

$$\frac{h(\theta - s) - h(\theta)}{-s} = \frac{1}{s} \int_{\theta-s}^{\theta} h'_-(\theta)\, d\theta,$$

we obtain

$$h'_+(\theta) = h'_+(\theta + 0) \text{ and } h'_-(\theta) = h'_-(\theta - 0).$$

(e) *The function $h(\theta)$ has a derivative at all points except possibly on a countable set.*

For the proof we write (1.74) in the form

$$h'_+(\theta_1) - h'_-(\theta_1) - 2k\rho(\theta_3 - \theta_1) \leqslant h'_-(\theta_3) - h'_-(\theta_1).$$

If θ_1 is a point of continuity for $h'_-(\theta)$, then

$$h'_+(\theta_1) \leqslant h'_-(\theta_1).$$

Comparing with property (c) we see that $h(\theta)$ has a derivative at all points of continuity of the monotone function $h'_-(\theta) + 2k\rho\theta$, that is, at all points except possibly on a countable set.

(f) *If θ_0 is a local maximum or minimum point for the function $h(\theta)$, then for $|\theta - \theta_0| \leqslant \pi/\rho$*

$$h(\theta) \geqslant h(\theta_0)\cos\rho(\theta - \theta_0).$$

Indeed, at a maximum point $h'(\theta_0) = 0$, while at a minimum $h'_+(\theta_0) \geqslant 0 \geqslant (h'_-\theta_0)$. Putting $\theta_1 = \theta_0$ in (1.71) and passing to the limit as $\theta \to \theta_0 + 0$, we obtain

$$\frac{h(\theta_3) - h(\theta_0)}{\sin\rho(\theta_3 - \theta_0)} + h(\theta_0)\sin\rho\frac{\theta_3 - \theta_0}{2}\sec\rho\frac{\theta_3 - \theta_0}{2} \geqslant \frac{1}{\rho}h'_+(\theta_0) \geqslant 0$$

or

$$h(\theta_3) \geqslant h(\theta_0)\cos\rho(\theta_3 - \theta_0)$$

for $0 < \theta_3 - \theta_0 < \pi/\rho$.

Similarly one obtains an estimate to the left of the point θ_0.

Note that if θ_0 is a local maximum for $h(\theta)$ and if $h(\theta_0) = 0$, then, by what has been shown, $h(\theta) \equiv 0$ in some neighborhood of the point θ_0.

(g) *For arbitrary α and β $(\alpha < \beta)$*

$$h'_-(\beta) - h'_+(\alpha) + \rho^2\int_\alpha^\beta h(\theta)\,d\theta \geqslant 0, \tag{1.75}$$

and equality holds only for

$$h(\theta) = a\cos\rho\theta + b\sin\rho\theta \quad (\alpha \leqslant \theta \leqslant \beta),$$

where a and b are constants.

PROOF. From inequality (1.71) it follows that as $\theta \to \theta_1$:

$$\frac{1}{\rho}h'_+(\theta_1) \leqslant \frac{h(\theta_3) - h(\theta_1)}{\sin\rho(\theta_3 - \theta_1)} + h(\theta_1)\operatorname{tg}\rho\frac{\theta_3 - \theta_1}{2}.$$

Interchanging θ_1 and θ_3 we have

$$\frac{1}{\rho}h'_-(\theta_3) \geqslant \frac{h(\theta_3) - h(\theta_1)}{\sin\rho(\theta_3 - \theta_1)} - h(\theta_3)\operatorname{tg}\rho\frac{\theta_3 - \theta_1}{2}$$

and, after subtracting these inequalities,

$$\frac{1}{\rho}[h'_-(\theta_3) - h'_+(\theta_1)] + [h(\theta_1) + h(\theta_3)]\operatorname{tg}\rho\frac{\theta_3 - \theta_1}{2} \geqslant 0. \tag{1.76}$$

Subdivide the interval (α, β) by means of points $\theta_1, \theta_2, \ldots, \theta_{n-1}$ at which the derivative $h'(\theta)$ exists; write inequality (1.76) for each of these subintervals and sum the results. We obtain

$$h'_-(\beta) - h'_+(\alpha) + \rho \sum_{j=0}^{n-1} [h(\theta_j) + h(\theta_{j+1})] \operatorname{tg} \rho \frac{\theta_{j+1} - \theta_j}{2} \geqslant 0$$

$$(\alpha = \theta_0, \ \beta = \theta_n).$$

To obtain inequality (1.75) it is necessary to pass to the limit as $\max |\theta_{j+1} - \theta_j| \to 0$.

Suppose that for some α and β $(\beta > \alpha)$

$$s(\alpha, \beta) = h'_-(\beta) - h'_+(\alpha) + \rho^2 \int_\alpha^\beta h(\theta)\, d\theta = 0.$$

Since for $\alpha < \theta < \beta$

$$s(\alpha, \beta) = s(\alpha, \theta) + s(\theta, \beta) + h'_+(\theta) - h'_-(\theta)$$

and

$$s(\alpha, \theta) \geqslant 0, \quad s(\theta, \beta) \geqslant 0, \quad h'_+(\theta) - h'_-(\theta) \geqslant 0,$$

we have

$$h'_-(\theta) + \rho^2 \int_\alpha^\theta h(\theta)\, d\theta = h'_+(\alpha),$$

or

$$h''(\theta) + \rho^2 h(\theta) = 0.$$

Thus equality holds in (1.75) only for

$$h(\theta) = a \cos \rho\theta + b \sin \rho\theta.$$

Property (g) is *characteristic for trigonometrically convex functions.* In other words, *the necessary and sufficient condition that the function $h(\theta)$ be trigonometrically convex for some ρ on the interval (α, β) is that the function*

$$s(\theta) = h'(\theta) + \rho^2 \int^\theta h(\varphi)\, d\varphi \tag{1.77}$$

be nondecreasing on that interval.

It is only necessary to prove the sufficiency.

First assume that $s(\theta)$ is a differentiable function. Then

$$h''(\theta) + \rho^2 h(\theta) = s'(\theta). \tag{1.77$'$}$$

The Green's function[18] for the differential operator $h'' + \rho^2 h$ on the interval $\theta_1 < \theta < \theta_3$ $(0 < \theta_1 < \theta_3 < \pi/\rho)$, with the boundary conditions $G(\theta_1, \theta) = G(\theta_3, \theta) = 0$, has the form

$$G(\psi, \theta) = \begin{cases} \dfrac{\sin \rho(\theta_3 - \theta) \sin \rho(\psi - \theta_1)}{\rho \sin \rho(\theta_1 - \theta_3)} & \text{for } \psi < \theta, \\[2ex] \dfrac{\sin \rho(\theta - \theta_1) \sin \rho(\theta_3 - \psi)}{\rho \sin \rho(\theta_1 - \theta_3)} & \text{for } \psi > \theta. \end{cases}$$

[18] For the Green's function see, for example, G. Sansone **[1]**.

The solution of equation (1.77) can be written in the form

$$h(\theta) = \frac{h(\theta_1)\sin\rho(\theta_3 - \theta) + h(\theta_3)\sin\rho(\theta - \theta_1)}{\sin\rho(\theta_3 - \theta_1)} + \int_{\theta_1}^{\theta_3} G(\psi, \theta)\, ds(\psi). \qquad (1.78)$$

If $s(\theta)$ is not a differentiable function then we approximate it by nondecreasing differentiable functions and pass to the limit in (1.78), and thus in this case too we obtain a representation of $h(\theta)$ in the form (1.78).

Since $G(\psi, \theta) < 0$ for $\theta_1 < \psi$, $\theta < \theta_3$ and since $s(\psi)$ is a nondecreasing function, we have from (1.78)

$$h(\theta) \leqslant \frac{h(\theta_1)\sin\rho(\theta_3 - \theta) + h(\theta_3)\sin\rho(\theta - \theta_1)}{\sin\rho(\theta_3 - \theta_1)}; \qquad (1.79)$$

$h(\theta)$ is a trigonometrically convex function in every interval in which the function $h'(\theta) + \rho^2 \int^{\theta} h(\theta)\, d\theta$ is nondecreasing.

Equality in (1.79) for even a single value of $\theta \neq \theta_1$, θ_3 is only possible if the integral in (1.78) is equal to zero, that is, if $s(\theta)$ is constant on the interval $\theta_1 < \psi < \theta_3$, or in other words, if $h(\theta) = A\cos\rho\theta + B\sin\rho\theta$.

Formula (1.78) remains valid for $\theta_3 - \theta_1 > \pi/\rho$. It is only necessary that $\theta_3 - \theta_1$ should not be an integral multiple of π/ρ, since then the Green's function does not exist.

We now show that every nondecreasing function $S(\theta)$ determines a periodic, trigonometrically convex function $h(\theta)$. For this we must construct the Green's function for the differential operator $h'' + \rho^2 h$ with the periodic boundary conditions

$$G(0, \theta) = G(2\pi, \theta) \text{ and } G'_{\psi}(0, \theta) = G'_{\psi}(2\pi, \theta).$$

For ρ not an integer the function

$$G(\psi, \theta) = \frac{\pi}{2\rho\sin\pi\rho}\cos\rho(\theta - \psi - \pi) \qquad (\theta - 2\pi < \psi \leqslant \theta)$$

satisfies the conditions

$$G(\theta - 2\pi, \theta) = G(\theta, \theta) \text{ and } G'_{\psi}(\theta - 2\pi, \theta) - G'_{\psi}(\theta, \theta) = 1.$$

Continuing this function periodically with the period 2π we obtain the desired Green's function. If the function $s(\psi)$ is differentiable, then the function

$$h(\theta) = \int_0^{2\pi} G(\psi, \theta)\, ds(\psi)$$

is the unique periodic solution of equation (1.77′). If $s(\psi)$ is nondifferentiable, then we approximate it by differentiable functions and pass to the limit, and thus we see that in this case too this representation is valid. If θ is a point of continuity of $s(\psi)$, then the integration from 0 to 2π can be replaced by the integration from $\theta - 2\pi$ to θ if we put $s(\theta + 2\pi) = s(2\pi) - s(0) + s(\theta)$. Then we shall have

$$h(\theta) = \frac{1}{2\rho \sin \pi\rho} \int_{\theta-2\pi}^{\theta} \cos \rho (\theta - \psi - \pi)\, ds(\psi). \tag{1.80}$$

Formulas (1.77) and (1.80) are inverse to one another and establish a connection between all periodic, trigonometrically convex functions for a given, nonintegral ρ, and all nondecreasing functions on the interval $(0, 2\pi)$.

The case of integral ρ is somewhat more complicated. First we note that in this case if the trigonometrically convex function $h(\theta)$ with period 2π is given, then the function $s(\theta)$ defined by (1.77) must satisfy the condition

$$\int_0^{2\pi} e^{i\rho\psi}\, ds(\psi) = 0. \tag{1.81}$$

Indeed

$$\int_0^{2\pi} e^{i\rho\psi}\, ds(\psi) = \rho^2 \int_0^{2\pi} e^{i\rho\psi} h(\psi)\, d\psi + \int_0^{2\pi} e^{i\rho\psi} d\{h'(\psi)\},$$

and, transforming the second integral by integrating by parts twice, we obtain (1.81). We now show that to every monotone function $s(\psi)$, satisfying (1.81), there corresponds a trigonometrically convex function of period 2π satisfying equation (1.77). However, this solution of equation (1.77) cannot be constructed as in the case of nonintegral ρ. Indeed, for ρ an integer the homogeneous equation $h'' + \rho^2 h = 0$ has the periodic solutions $\cos \rho\theta$, $\sin \rho\theta$, and therefore there is no Green's function satisfying periodic boundary conditions. However, in this case one can construct a generalized Green's function that gives the periodic solution of the inhomogeneous equation (1.77′) when $s'(\theta)$ is orthogonal to the solutions of the homogeneous equation $\cos \rho\theta$, $\sin \rho\theta$.

To construct this generalized Green's function we note that the function

$$G_1(\psi, \theta) = \frac{1}{2\pi\rho} (\theta - \psi) \sin \rho (\theta - \psi) \qquad (\theta - 2\pi \leqslant \psi \leqslant \theta),$$

satisfies the equation

$$G''_{1\theta\theta} + \rho^2 G_1 = 2 \cos \rho (\theta - \psi).$$

Also,

$$G_1(\theta - 2\pi, \theta) - G_1(\theta, \theta) = 0, \quad G'_{1\psi}(\theta - 2\pi, \theta) - G'_{1\psi}(\theta, \theta) = 1.$$

If we continue the function $G(\psi, \theta)$ periodically with period 2π, we obtain the generalized Green's function. Thus if $s(\psi)$ is a differentiable, nondecreasing function,[18a] satisfying (1.81), then every periodic solution of (1.77′) can be represented in the form

$$h(\theta)=\frac{1}{2\pi\rho}\int_{\theta-2\pi}^{\theta}(\theta-\psi)\sin\rho(\theta-\psi)\,ds(\psi)+A\cos\rho\theta+B\sin\rho\theta. \tag{1.82}$$

It remains to consider the case of a nondifferentiable monotone function $s(\theta)$, satisfying condition (1.81). In this case we must approximate $s(\theta)$ by differentiable functions satisfying (1.81). For example, for the approximating function we can take

$$s_k(\theta)=\frac{1}{k}\int_{\theta}^{\theta+k}s(\varphi)\,d\varphi.$$

(Here we suppose that $s(\theta+2\pi)=s(\theta)+s(2\pi)-s(0)$.) Then one must pass to the limit in formula (1.82).

Thus for integral ρ formulas (1.77) and (1.82) are inverse to one another and establish a correspondence between all periodic and trigonometrically convex (for the given ρ) functions $h(\theta)$ and all nondecreasing functions $s(\theta)$ on the interval $(0, 2\pi)$ that satisfy condition (1.81). We formulate the result as a separate theorem.

THEOREM 24. *The general form of periodic functions $h(\theta)$ with period 2π is given, for nonintegral ρ, by the formula*

$$h(\theta)=\frac{1}{2\rho\sin\pi\rho}\int_{\theta-2\pi}^{\theta}\cos\rho(\theta-\psi-\pi)\,ds(\psi), \tag{1.80}$$

where $s(\psi)$ is an arbitrary nondecreasing function on $(\theta-2\pi\leqslant\psi\leqslant\theta)$, and for integral ρ by the formula

$$h(\theta)=\frac{1}{2\pi\rho}\int_{\theta-2\pi}^{\theta}(\theta-\psi)\sin\rho(\theta-\psi)\,ds(\psi)+A\cos\rho\theta+B\sin\rho\theta, \tag{1.82}$$

where A and B are constants and $s(\psi)$ is an arbitrary nondecreasing function satisfying the condition

$$\int_0^{2\pi}e^{i\rho\psi}\,ds(\psi)=0. \tag{1.81}$$

[18a] Here and later a Stieltjes integral from $\theta-2\pi$ to θ is to be taken as an integral from $\theta-2\pi-0$ to $\theta-0$.

The function $s(\theta)$ is expressed in terms of $h(\theta)$ by the formula[19]

$$s(\theta) = h'(\theta) + \rho^2 \int^{\theta} h(\varphi)\, d\varphi.$$

We shall use Theorem 24 in Chapter II.

In conclusion we note an almost self-evident property of the indicator, which we shall frequently need:

$$\text{(h)} \qquad h\left(\theta + \frac{\pi}{\rho}\right) + h(\theta) \geqslant 0.$$

This property follows at once from the fundamental inequality (1.69) if we put $\theta_1 = \theta$, $\theta_2 = \theta + \pi/2\rho$ and $\theta_3 = \theta + \pi/\rho$.

17. Auxiliary Functions

In this section we investigate some special entire functions that we shall need in the study of the indicator for proximate orders. These functions will play in the theory of proximate orders the same role as was played by the functions

$$e^{Az^\rho}$$

in the study of entire functions of order ρ and normal type.[20] In addition, these functions will play a fundamental role in Chapter II in the discovery of asymptotic formulas for functions with very regular distribution of zeros.

We shall have to consider logarithms of canonical products, and to choose a single-valued branch we shall define the function

$$\ln G(u;\ p)$$

in the u-plane cut along the positive real axis from 1 to ∞, with $\arg G(u; p) = -\pi$ along the upper side of the cut. If in the z-plane we introduce cuts along rays emanating from the zeros of the canonical product $\Pi(z)$ (such that the backward extension of each ray would pass through the origin), then in the resulting star domain the function

$$\ln \Pi(z) = \sum_{1}^{\infty} \ln G\left(\frac{z}{a_k};\ p\right)$$

is defined and single-valued.

---

[19] One can verify directly that the functions represented by formulas (1.80) and (1.82) satisfy equation (1.77). This gives an independent proof of Theorem 24, since for ρ nonintegral the periodic solution of (1.77) is unique, while for ρ integral it is defined up to the addition of terms of the form $A \cos \rho\theta + B \sin \rho\theta$.

[20] These functions were first constructed by Valiron [3].

We first give an asymptotic expression for the function $\ln \Pi(z)$ in the special case when all the zeros of $\Pi(z)$ lie on one ray, and the set of zeros has a density Δ and ρ is not an integer.

Note that it is not difficult to construct a sequence $\{a_k\}$ of positive numbers having a given density Δ with respect to a given proximate order $\rho(r)$. For this it is sufficient to choose a_k to be the unique positive root of the equation $k = \Delta r^{\rho(r)}$. To obtain the asymptotic formula we need two lemmas.

LEMMA 7. *If the set of points $\{a_k\}$ is such that for some $C > 0$ and proximate order $\rho(r)$*

$$n(r) < Cr^{\rho(r)} \tag{1.83}$$

and

$$f_\sigma(z) = \prod_{|a_k|<\sigma r} G\left(\frac{z}{a_k};\ p\right) \qquad (p < \rho;\ 0 < \sigma \leqslant 1/2), \tag{1.84}$$

then asymptotically

$$|\ln f_\sigma(z)| < C_p \sigma^{\rho - p} r^{\rho(r)}, \tag{1.85}$$

where the constant C_p does not depend on σ.[20a]

PROOF. For $p > 0$ it follows from inequality (1.21) that

$$\ln|f_\sigma(z)| \leqslant A_p r^{p+1} \int_0^{\sigma r} \frac{dn(t)}{t^p(t+r)}$$

or

$$\ln|f_\sigma(z)| \leqslant A_p \frac{r^{p+1} n(\sigma r)}{(\sigma r)^p (1+\sigma) r} + A_p (p+1) r^{p+1} \int_0^{\sigma r} \frac{n(t)\, dt}{t^{p+1}(t+r)}$$

$$< A_p C \frac{(\sigma r)^{\rho(\sigma r)}}{\sigma^p (1+\sigma)} + A_p (p+1) c r^p \int_0^{\sigma r} t^{\rho(t) - p - 1}\, dt.$$

The last integral can be calculated by formula (1.53) and thus asymptotically

$$\ln|f_\sigma(z)| < C'_p \sigma^{\rho - p} r^{\rho(r)},$$

where C'_p does not depend on σ. On the other hand, for $\sigma \leqslant 1/2$ and $|u| \geqslant 1/\sigma$ we have $\ln|1 - u| \geqslant 0$, and consequently

$$\ln|G(u,p)| \geqslant -|u| - \frac{|u|^2}{2} - \cdots - \frac{|u|^p}{p} > -|u|^p(1 + \ln \rho),$$

whence

$$\ln|G(u,p)| > -A_p|u|^{p+1}/(1 + |u|).$$

[20a] To obtain a single-valued branch of $\arg f(z)$ we cut the plane along the ray $\arg(z/a_k) = 0$ from a_k to infinity, and take $\arg f_\sigma(0) = 0$.

If we use this inequality instead of (1.21), we obtain, by reasoning similar to that used above,

$$\ln|f_\sigma(z)| > c_p'\sigma^{\rho-p}r^{\rho(r)}$$

and finally

$$|\ln|f_\sigma(z)|\,| < c_p'\sigma^{\rho-p}r^{\rho(r)} \tag{1.86}$$

For $p = 0$ and $\rho > 0$ inequality (1.86) is obtained similarly from (1.22).
Now let us estimate $\arg f_\sigma(z)$. We have

$$|\arg G(u;\ p)| < 2\pi + \sum_{k=1}^{p} \frac{|u|^k}{k} < A_p' |u|^p \quad \text{for } |u| \geqslant 1$$

from which we obtain

$$|\arg f_\sigma(z)| \leqslant A_p' r^p \int_0^{\sigma r} \frac{dn(t)}{t^p}.$$

Estimating this integral as we did the preceding one, we have

$$|\arg f_\sigma(z)| < C_p'' \sigma^{\rho-p} r^{\rho(r)}. \tag{1.88}$$

The assertion of the lemma follows from (1.86) and (1.88).

LEMMA 8. *Let the sequence* $\{a_k\}$ *satisfy* (1.83) *and let*

$$_\tau f(z) = \prod_{|a_k| > \tau r} G\left(\frac{z}{a_k};\ p\right) \qquad (\text{for} \quad \tau > 2, \quad \rho < p+1).$$

Then asymptotically

$$|\ln {_\tau f(z)}| < C^{(p)} \tau^{\rho-p-1} r^{\rho(r)},$$

where the constant $C^{(p)}$ *does not depend on* τ.

PROOF. From the expansion

$$\ln G(u;\ p) = -\frac{u^{p+1}}{p+1} - \frac{u^{p+2}}{p+2} - \cdots$$

it follows that for $|u| < \frac{1}{2}$,

$$|\ln G(u;\ p)| < \frac{2}{p+1} |u|^{p+1}. \tag{1.89}$$

It follows from this inequality that

$$|\ln {_\tau f(z)}| < \frac{2r^{p+1}}{p+1} \int_{\tau r}^{\infty} \frac{dn(t)}{t^{p+1}} = -\frac{2}{p+1} r^{p+1} \frac{n(\tau r)}{(\tau r)^{p+1}} + 2r^{p+1} \int_{\tau r}^{\infty} \frac{n(t)}{t^{p+2}}\, dt.$$

In other words

$$|\ln {_\tau f(z)}| < 2Cr^{p+1} \int_{\tau r}^{\infty} t^{\rho(t)-p-2}\, dt,$$

and by (1.53′) [see property (3), § 12]

$$\int_{\tau r}^{\infty} t^{\rho(t)-p-2}\,dt = \frac{1}{p+1-\rho}(\tau r)^{\rho(\tau r)-p-1} + o\left((\tau r)^{\rho(\tau r)-p-1}\right).$$

It follows that asymptotically

$$|\ln {}_{\tau} f(z)| < \frac{3C}{p+1-\rho}\tau^{\rho-p-1} r^{\rho(r)}.$$

THEOREM 25. *If the sequence $\{a_k\}$ of points on the positive axis has a density*

$$\Delta = \lim_{r\to\infty} \frac{n(r)}{r^{\rho(r)}}. \qquad (1.90)$$

for some proximate order $\rho(r)$, if ρ is not an integer, and if

$$V(z) = \prod_{k=1}^{\infty} G\left(\frac{z}{a_k};\ p\right) \qquad (p < \rho < p+1)$$

then for $z = re^{i\theta}$, $0 < \theta < 2\pi$,

$$\lim_{r\to\infty} \frac{\ln V(re^{i\theta})}{r^{\rho(r)}} = \frac{\pi\Delta}{\sin \pi\rho} e^{i\rho(\theta-\pi)},$$

and the limit is uniform in any interval $0 < \eta < \theta < 2\pi - \eta$.

PROOF. It follows from Lemmas 7 and 8 that for arbitrary $\varepsilon > 0$ and for sufficiently small $\sigma > 0$ and sufficiently large $\tau > 0$ we have asymptotically

$$\left|\ln V(z) - \sum_{\sigma r < a_k < \tau r} \ln G\left(\frac{z}{a_k};\ p\right)\right| < \frac{\varepsilon}{4} r^{\rho(r)}. \qquad (1.91)$$

The sum on the left can be written as a Stieltjes integral

$$\sum_{\sigma r < a_k < \tau r} \ln G\left(\frac{z}{a_k};\ p\right) = \int_{\sigma r}^{\tau r} \ln G\left(\frac{z}{t};\ p\right) dn(t).$$

After integrating by parts we have

$$\sum_{\sigma r < a_k < \tau r} \ln G\left(\frac{z}{a_k};\ p\right) = n(\tau r)\ln G\left(\frac{e^{i\theta}}{\tau};\ p\right)$$

$$- n(\sigma r)\ln G\left(\frac{e^{i\theta}}{\sigma};\ p\right) - z^{p+1}\int_{\sigma r}^{\tau r} \frac{n(t)\,dt}{t^{p+1}(t-z)} \quad (\theta = \arg z). \qquad (1.92)$$

We use the asymptotic estimate

$$n(\tau r) < (\Delta + \varepsilon)\tau^{\rho} r^{\rho(r)},$$

which follows from (1.90) and property (2) of proximate orders. From this estimate and inequality (1.89) it follows that for sufficiently large τ we have the asymptotic inequality

$$n(\tau r)\left|\ln G\left(\frac{e^{i\theta}}{\tau};\ p\right)\right| < \frac{\varepsilon}{8} r^{\rho(r)}.$$

Similarly from the asymptotic estimate

$$n(\sigma r) < (\Delta + \varepsilon)\sigma^{\rho} r^{\rho(r)}$$

and from inequalities (1.21) and (1.87) we have the asymptotic inequality

$$n(\sigma r)\left|\ln G\left(\frac{e^{i\theta}}{\sigma};\ p\right)\right| < \frac{\varepsilon}{8} r^{\rho(r)},$$

which is valid for all sufficiently small values of σ. Returning to (1.92), we can now write

$$\left|\sum_{\sigma r < a_k < \tau r} \ln G\left(\frac{z}{a_k};\ p\right) + z^{p+1}\int_{\sigma r}^{\tau r} \frac{n(t)\,dt}{t^{p+1}(t-z)}\right| < \frac{\varepsilon}{4} r^{\rho(r)}. \tag{1.93}$$

It follows from (1.90) that for arbitrary $\delta > 0$ and for all sufficiently large values of t

$$|n(t) - \Delta t^{\rho} L(t)| < \delta t^{\rho} L(t)$$

and therefore for $0 < \theta < 2\pi$

$$\left|z^{p+1}\int_{\sigma r}^{\tau r} \frac{n(t)}{t^{p+1}(t-z)}\,dt - \Delta z^{p+1}\int_{\sigma r}^{\tau r} \frac{t^{\rho-p-1}L(t)}{t-z}\,dt\right| < \delta r^{p+1}\int_{\sigma r}^{\tau r} \frac{t^{\rho-p-1}L(t)\,dt}{|t-re^{i\theta}|}.$$

For $\sigma r \leqslant t \leqslant \tau r$

$$\lim_{r\to\infty} \frac{L(t)}{L(r)} = 1 \tag{1.94}$$

and therefore asymptotically

$$r^{p+1}\int_{\sigma r}^{\tau r} \frac{t^{\rho-p-1}L(t)\,dt}{|t-re^{i\theta}|} < 2r^{p+1}L(r)\int_{\sigma r}^{\tau r} \frac{t^{\rho-p-1}\,dt}{|t-re^{i\theta}|} = 2r^{\rho}L(r)\int_{\sigma}^{\tau} \frac{u^{\rho-p-1}\,du}{|u-e^{i\theta}|}.$$

Thus, for $\eta > 0$ and $\eta \leqslant \theta \leqslant 2\pi - \eta$ and sufficiently small $\delta > 0$, we obtain the asymptotic inequality

$$\left|z^{p+1}\int_{\sigma r}^{\tau r} \frac{n(t)\,dt}{t^{p+1}(t-re^{i\theta})} - \Delta z^{p+1}\int_{\sigma r}^{\tau r} \frac{t^{\rho-p-1}L(t)\,dt}{t-re^{i\theta}}\right| < \frac{\varepsilon}{8} r^{\rho(r)}. \tag{1.95}$$

Using (1.94) once more we obtain

$$\left|\Delta z^{p+1}\int_{\sigma r}^{\tau r} \frac{t^{\rho-p-1}L(t)}{t-re^{i\theta}}\,dt - \Delta e^{i(p+1)\theta}\left\{\int_{\sigma}^{\tau} \frac{u^{\rho-p-1}\,du}{u-e^{i\theta}}\right\} r^{\rho}L(r)\right| < \frac{\varepsilon}{8} r^{\rho(r)}. \tag{1.96}$$

Finally, for sufficiently small $\sigma > 0$ and sufficiently large τ,

$$\left| \int_\sigma^\tau \frac{u^{\rho-p-1}\,du}{u-e^{i\theta}} - \int_0^\infty \frac{u^{\rho-p-1}}{u-e^{i\theta}}\,du \right| < \frac{\varepsilon}{8\Delta}. \tag{1.97}$$

From inequalities (1.91), (1.93), (1.95), (1.96) and (1.97) it follows that asymptotically and uniformly in θ for $\eta \leqslant \theta \leqslant 2\pi - \eta$

$$\left| \ln V(re^{i\theta}) - \Delta r^{\rho(r)} e^{i(p+1)\theta} \int_0^\infty \frac{u^{\rho-p-1}}{u-e^{i\theta}}\,du \right| < \varepsilon r^{\rho(r)}.$$

To complete the proof it only remains to compute this integral.[21]

The reasoning we have used is not applicable in case ρ is an integer. In this case Lemmas 7 and 8 cannot be used to derive the asymptotic inequality (1.91). Also, (1.97) is not valid in this case since then the integral from zero to infinity does not exist. To study the case of integral ρ we prove the following lemma.

LEMMA 9. *Let the set $\{a_k\}$ of points on the positive axis have density Δ with respect to the proximate order $\rho(r)$ and let ρ be an integer ($\rho \geqslant 1$). Let*

$$V_r(z) = \prod_{a_k \leqslant r} G\left(\frac{z}{a_k};\, \rho - 1\right) \prod_{a_k > r} G\left(\frac{z}{a_k};\, \rho\right).$$

Then for $z = re^{i\theta}$, $0 < \theta < 2\pi$,

$$\lim_{r\to\infty} \frac{\ln V_r(re^{i\theta})}{r^{\rho(r)}} = -\Delta\left[\frac{1}{\rho} - i(\theta - \pi)\right] e^{i\rho\theta}.$$

The limit is uniform in each interval $0 < \eta \leqslant \theta \leqslant 2\pi - \eta$.

PROOF. We start from the equation

$$\ln V_r(z) = \sum_{a_k \leqslant r} \ln G\left(\frac{z}{a_k};\, \rho - 1\right) + \sum_{a_k > r} \ln G\left(\frac{z}{a_k};\, \rho\right).$$

It follows from Lemmas 7 and 8 that, for sufficiently small $\sigma > 0$, sufficiently large τ and all sufficiently large r,

$$\left| \ln V_r(z) - \int_{\sigma r}^{r} \ln G\left(\frac{z}{t};\, \rho - 1\right) dn(t) \right.$$
$$\left. - \int_r^{\tau r} \ln G\left(\frac{z}{t};\, \rho\right) dn(t) \right| < \frac{\varepsilon}{4} r^{\rho(r)}$$

[21] It can be calculated in exactly the same way as the classical Euler integral

$$\int_0^\infty \frac{u^{p-1}\,du}{1+u}.$$

or, after integrating by parts over $0 < \theta < 2\pi$,

$$\left| \ln V_r(z) + \frac{1}{\rho} e^{i\rho\theta} n(r) + z^{\rho} \int_{\sigma r}^{r} \frac{n(t)}{t^{\rho}(t-z)} dt \right.$$

$$\left. + z^{\rho+1} \int_{r}^{\tau r} \frac{n(t)}{t^{\rho+1}(t-z)} dt \right| < \frac{\varepsilon}{2} r^{\rho(r)}.$$

Just as in the proof of Theorem 25, we obtain from the properties of the function

$$t^{\rho(t)} = t^{\rho} L(t),$$

for $0 < \eta \leqslant \theta \leqslant 2\pi - \eta$, the inequality

$$\left| z^{\rho} \int_{\sigma r}^{r} \frac{n(t)\, dt}{t^{\rho}(t-z)} - \Delta r^{\rho(r)} e^{i\rho\theta} \int_{\sigma}^{1} \frac{dt}{t - e^{i\theta}} \right| < \frac{\varepsilon}{8} r^{\rho(r)}$$

and

$$\left| z^{\rho+1} \int_{r}^{\tau r} \frac{n(t)\, dt}{t^{\rho+1}(t-z)} - \Delta r^{\rho(r)} e^{i(\rho+1)\theta} \int_{1}^{\tau} \frac{dt}{t(t-e^{i\theta})} \right| < \frac{\varepsilon}{8} r^{\rho(r}.$$

Also,

$$\left| \int_{\sigma}^{1} \frac{dt}{t - e^{i\theta}} - \ln(1 - e^{-i\theta}) \right| < \frac{\varepsilon}{8} \text{[22]}$$

and

$$\left| e^{i\theta} \int_{1}^{\tau} \frac{dt}{t(t - e^{i\theta})} + \ln(1 - e^{i\theta}) \right| < \frac{\varepsilon}{8}.$$

It follows from these inequalities that for $0 < \eta \leqslant \theta \leqslant 2\pi - \eta$ we have asymptotically

$$\left| \ln V_r(re^{i\theta}) + \Delta \left[\frac{1}{\rho} - i(\theta - \pi) \right] e^{i\rho\theta} r^{\rho(r)} \right| < \varepsilon r^{\rho(r)}.$$

The function $V_r(z)$ is not entire. It has discontinuities for $|z| = a_k$. However, we shall use Lemma 9 to construct an entire function of integral order with an asymptotic representation inside an angle analogous to the representation given by Theorem 25.

[22] Indeed

$$\int_{\sigma}^{1} \frac{dt}{t - e^{i\theta}} = \ln \frac{1 - e^{i\theta}}{\sigma - e^{i\theta}} = \ln(1 - e^{-i\theta}) - \ln(1 - \sigma e^{-i\theta}),$$

and the inequality is valid for sufficiently small σ.

THEOREM 26. *Let the set $\{a_k\}$ of points on the positive axis have density Δ with respect to some proximate order $\rho(r)$, where ρ is an integer ($\rho \geqslant 1$). Let*

$$u^{(1)}(z)=\prod_{k=1}^{\infty} G\left(\frac{z}{a_k};\ \rho\right), \qquad u^{(2)}(z)=\prod_{k=1}^{\infty} G\left(\frac{ze^{-i\frac{\pi}{\rho}}}{a_k};\ \rho\right)$$

and

$$V(z)=u^{(1)}(z)\,u^{(2)}(z).$$

Then, uniformly on each closed interval that is interior to one of the intervals $0<\theta<\pi/\rho$ or $\pi/\rho<\theta<2\pi$, we have

$$\lim_{r\to\infty}\frac{\ln V(re^{i\theta})}{r^{\rho(r)}}=\frac{i\pi}{\rho}\Delta e^{i\rho\theta}.$$

PROOF. Assume that $\rho=p$ and represent the functions $u^{(1)}(z)$ and $u^{(2)}(z)$ in the form

$$u^{(1)}(z)=\exp\Big(\frac{1}{\rho}z^{\rho}\sum_{a_k\leqslant r}a_k^{-\rho}\Big)\cdot u_r^{(1)}(z)$$

and

$$u^{(2)}(z)=\exp\Big(-\frac{1}{\rho}z^{\rho}\sum_{a_k\leqslant r}a_k^{-\rho}\Big)\cdot u_r^{(2)}(z),$$

where

$$u_r^{(1)}(z)=\prod_{a_k\leqslant r}G\left(\frac{z}{a_k};\ \rho-1\right)\prod_{a_k>r}G\left(\frac{z}{a_k};\ \rho\right)$$

and

$$u_r^{(2)}(z)=\prod_{a_k\leqslant r}G\left(\frac{z}{a_k}e^{-\frac{i\pi}{\rho}};\ \rho-1\right)\prod_{a_k>r}G\left(\frac{z}{a_k}e^{-\frac{i\pi}{\rho}};\ \rho\right).$$

We then obtain

$$V(z)=u_r^{(1)}(z)\,u_r^{(2)}(z),$$

and to complete the proof it is sufficient to quote Lemma 9.

For $\rho=p+1$ it is necessary to represent the functions $u^{(1)}(z)$ and $u^{(2)}(z)$ in the form

$$u^{(1)}(z)=\exp\Big(-\frac{1}{\rho}z^{\rho}\sum_{a_k>r}a_k^{-\rho}\Big)u_r^{(1)}(z)$$

and

$$u^{(2)}(z)=\exp\Big(\frac{1}{\rho}z^{\rho}\sum_{a_k>r}a_k^{-\rho}\Big)u_r^{(2)}(z)$$

and then to use the same lemma.

In conclusion, we construct a function that will play the fundamental role in the study of the generalized indicator function.

LEMMA 10. *Let $\rho(r)$ be a given proximate order ($\rho > 0$) and let the opening of the angle*

$$\theta_1 \leqslant \arg z \leqslant \theta_2$$

be less than the smaller of the numbers π/ρ and 2π. Then for arbitrary real numbers a and b a function $W(z)$ can be constructed that is holomorphic and has no zeros in this angle, and such that for $\theta_1 \leqslant \theta \leqslant \theta_2$ we have uniformly

$$\lim_{r\to\infty} \frac{\ln|W(re^{i\theta})|}{r^{\rho(r)}} = a\cos\rho\theta + b\sin\rho\theta.$$

PROOF. First consider the case of nonintegral order ρ. For arbitrary ψ_1 and ψ_2 satisfying the inequality $0 < \psi_2 - \psi_1 < \pi/\rho$, the system of equations

$$\left.\begin{aligned} A_1\cos\rho(\psi_1+\pi)+A_2\cos\rho(\psi_2+\pi) &= a,\\ A_1\sin\rho(\psi_1+\pi)+A_2\sin\rho(\psi_2+\pi) &= b, \end{aligned}\right\} \tag{1.98}$$

has the solution A_1, A_2, so that the determinant of the system

$$\sin\rho(\psi_2-\psi_1) \neq 0.$$

Choose ψ_1 and ψ_2 outside the interval $\theta_1 \leqslant \theta \leqslant \theta_2$. Let

$$\Delta_i = \left|A_i\frac{\sin\pi\rho}{\pi}\right| \text{ and } \varepsilon_i = \operatorname{sign}(A_i\sin\pi\rho) \qquad (i=1,\ 2).$$

Construct sets $\{a_{k,1}\}$ and $\{a_{k,2}\}$ with respective densities Δ_1 and Δ_2, form the functions

$$V_1(z) = \prod_{k=1}^{\infty} G\left(\frac{z}{a_{k,1}};\ p\right), \qquad V_2(z) = \prod_{k=1}^{\infty} G\left(\frac{z}{a_{k,2}};\ p\right)$$

and, finally, form the function

$$W(z) = V_1^{\varepsilon_1}(ze^{-i\psi_1})\,V_2^{\varepsilon_2}(ze^{-i\psi_2})$$

that is holomorphic in the angle $\theta_1 \leqslant \theta \leqslant \theta_2$.

From Theorem 25 and equations (1.98) it follows that uniformly for $\theta_1 \leqslant \theta \leqslant \theta_2$

$$\lim_{r\to\infty} \frac{\ln|W(re^{i\theta})|}{r^{\rho(r)}} = a\cos\rho\theta + b\sin\rho\theta.$$

If ρ is an integer, then in place of (1.98) we consider the system

$$\left.\begin{aligned} A_1\cos\rho\psi_1+A_2\cos\rho\psi_2 &= b,\\ A_1\sin\rho\psi_1+A_2\sin\rho\psi_2 &= a, \end{aligned}\right\} \tag{1.98'}$$

where $0 < \psi_2 - \psi_1 < \pi/\rho$ and the four points ψ_1, $\psi_1 + \pi/\rho$, ψ_2, $\psi_2 + \pi/\rho$ lie outside the interval $\theta_1 \leqslant \theta \leqslant \theta_2$ (for $\rho \geqslant 1$ and $\theta_2 - \theta_1 < \pi/\rho$ such a choice is possible). Then put

$$\Delta_i = \left| \frac{A_i \rho}{\pi} \right| \text{ and } \varepsilon_i = \operatorname{sign} A_i \qquad (i = 1,\ 2). \tag{1.99}$$

Construct sets $\{a_{k,1}\}$ and $\{a_{k,2}\}$ of points on the positive axis with densities Δ_1 and Δ_2, respectively, and construct the corresponding functions $V_1(z)$ and $V_2(z)$ as in Theorem 26.

From these functions we form the further function

$$W(z) = V_1^{\varepsilon_1}(ze^{-i\psi_1})\, V_2^{\varepsilon_2}(ze^{-i\psi_2}),$$

holomorphic inside the angle $\theta_1 \leqslant \arg z \leqslant \theta_2$.

It follows from Theorem 26 and equations (1.98′) and (1.99) that uniformly for $\theta_1 \leqslant \theta \leqslant \theta_2$

$$\lim_{r \to \infty} \frac{\ln |W(re^{i\theta})|}{r^{\rho(r)}} = a \cos \rho\theta + b \sin \rho\theta. \tag{1.100}$$

18. The Generalized Indicator

The concept of the indicator of a function can be generalized by replacing the usual order by a proximate order. *By a generalized indicator of the function* $f(z)$, holomorphic in the angle $\theta_1 \leqslant \arg z \leqslant \theta_2$, we mean the function

$$h_f(\theta) = \overline{\lim_{r \to \infty}} \frac{\ln |f(re^{i\theta})|}{r^{\rho(r)}} \qquad (\theta_1 \leqslant \theta \leqslant \theta_2),$$

where $\rho(r)$ is some proximate order.

Using the auxiliary functions of the preceding section we can obtain the following result.

THEOREM 27. *If $\rho > 0$, then the generalized indicator satisfies the fundamental relation*

$$h(\theta_1) \sin \rho (\theta_2 - \theta_3) + h(\theta_2) \sin \rho (\theta_3 - \theta_1) + h(\theta_3) \sin \rho (\theta_1 - \theta_2) \leqslant 0 \tag{1.69′}$$

for $\theta_1 < \theta_2 < \theta_3$ and $\theta_3 - \theta_1 < \pi/\rho$.

PROOF. Let $\theta_1 < \theta < \theta_3$, $\theta_3 - \theta_1 < \pi/\rho$, and let

$$H_\varepsilon(\theta) = \frac{[h(\theta_1) + \varepsilon] \sin \rho (\theta - \theta_1) + [h(\theta_3) + \varepsilon] \sin \rho (\theta_3 - \theta)}{\sin \rho (\theta_3 - \theta_1)} \quad (\varepsilon > 0).$$

By Lemma 10 there exists a function $W(z)$ having the proximate order $\rho(r)$, holomorphic and without zeros in the angle $\theta_1 \leqslant \arg z \leqslant \theta_3$, and such that the function

$$\frac{\ln |W(re^{i\theta})|}{r^{\rho(r)}}$$

tends uniformly to the function $H_\epsilon(\theta)$ on the interval $\theta_1 \leqslant \theta \leqslant \theta_3$. Thus the function $f(z)W^{-1}(z)$ tends to zero along the rays $\arg z = \theta_j$ $(j = 1, 3)$ and by the Phragmén-Lindelöf theorem it is bounded inside the angle $\theta \leqslant \arg z \leqslant \theta_3$. It follows from this that, for $\theta_1 \leqslant \theta \leqslant \theta_3$ and for arbitrary $\epsilon > 0$, $h(\theta) \leqslant H_\epsilon(\theta)$. It remains to pass to the limit as $\epsilon \to 0$.

It follows from the fundamental relation that the generalized indicator $h(\theta)$ satisfies properties (a)—(h) of § 16 which were derived from the fundamental relation. Now we shall derive some general properties of the growth of functions along rays, where the function has the proximate order $\rho(r)$.

The results will, of course, be true also in the special case $\rho(r) = \rho$, that is, for functions of normal type with respect to the usual order ρ.[22a]

THEOREM 28. *If $f(z)$ is holomorphic and has the proximate order $\rho(r)$ $(\rho > 0)$ inside the angle $\alpha \leqslant \arg z \leqslant \beta$, then to each $\varepsilon > 0$ there corresponds a number r_ε such that the inequality*

$$\ln |f(re^{i\theta})| < [h(\theta) + \varepsilon]\, r^{\rho(r)} \tag{1.101}$$

is valid for $r > r_\varepsilon$ for all values of θ in the interval $[\alpha, \beta]$.

PROOF. Divide the interval $[\alpha, \beta]$ into subintervals with points of subdivision $\theta_1, \theta_2, \ldots, \theta_{n-1}$ $(\alpha = \theta_0, \beta = \theta_n)$ and for each subinterval (θ_j, θ_{j+1}) construct the trigonometric function $H_j(\theta)$ of the form (1.67), taking at the points θ_j and θ_{j+1} the respective values

$$h(\theta_j) + \frac{\varepsilon}{3} \text{ and } h(\theta_{j+1}) + \frac{\varepsilon}{3}.$$

The intervals (θ_j, θ_{j+1}) can be chosen so small that the oscillation of the functions $h(\theta)$ and $H_j(\theta)$ on each of these intervals is less than $\varepsilon/3$. Thus in each interval

$$H_j(\theta) < h(\theta) + \frac{2\varepsilon}{3}.$$

In each of these intervals we construct the function $W_j(z)$ with indicator $H_j(\theta)$, whose existence was asserted in Lemma 10. The modulus of the function

$$f(z)\, W_j^{-1}(z)$$

is bounded on the sides of the angle $\theta_j \leqslant \arg z \leqslant \theta_{j+1}$. By the Phragmén-Lindelöf theorem it is also bounded inside this angle. Hence, for sufficiently

---

[22a] If we divide (1, 69′) by ρ and formally take the limit as $\rho \to 0$, we obtain the inequality

$$h(\theta_1)(\theta_2 - \theta_3) + h(\theta_2)(\theta_3 - \theta_1) + h(\theta_3)(\theta_1 - \theta_2) < 0$$

that defines ordinary convexity. Grišin [1] showed that the generalized indicator of a function that is holomorphic and of order zero inside an angle $\alpha < \theta < \beta$ is a convex function of θ. Conversely, every convex function $h(\theta)$, $\alpha < \theta < \beta$, is the indicator of a function that is holomorphic and of order zero inside the angle $\alpha < \theta < \beta$. Since the indicator of an entire function is a 2π-periodic function of θ, Grišin's result implies Gol′dberg's result [3] that the generalized indicator of an entire function of order 0 is constant.

large $R_j(\varepsilon)$ and $r > R_j(\varepsilon)$,

$$\ln|f(re^{i\theta})| < \left[H_j(\theta) + \frac{\varepsilon}{3}\right] r^{\rho(r)} \qquad (\theta_j \leqslant \theta \leqslant \theta_{j+1}),$$

and if $R(\varepsilon) = \max R_j(\varepsilon)$, then for $r > R(\varepsilon)$

$$\ln|f(re^{i\theta})| < [h(\theta) + \varepsilon] r^{\rho(r)}.$$

The following theorem on the connection between the type and the indicator follows at once from this theorem.

THEOREM 29. *The maximum value of the indicator $h_f(\theta)$ of the function $f(z)$ on the interval $\alpha \leqslant \theta \leqslant \beta$ is equal to the type σ_f of this function inside the angle $\alpha \leqslant \arg z \leqslant \beta$.*

Indeed, it follows from (1.101) that $\max h_f(\theta) \geqslant \sigma_f$, and on the other hand it is obvious from the definition of indicator that

$$\max_{\alpha \leqslant \theta \leqslant \beta} h_f(\theta) \leqslant \sigma_f.$$

THEOREM 30 (WIMAN). *If the order ρ of an entire function is less than one, then there exists a sequence $r_1 < r_2 < r_3 < \ldots$ ($r_n \to \infty$) such that for arbitrary $\varepsilon > 0$ and $n > n_\varepsilon$*

$$m_f(r_n) > [M_f(r_n)]^{\cos \pi\rho - \varepsilon},$$

where

$$m_f(r) = \min_{0 < \theta < 2\pi} |f(re^{i\theta})|.$$

PROOF. Let

$$f(z) = Cz^m \prod_{k=1}^{\infty} \left(1 - \frac{z}{a_k}\right)$$

be an entire function of order $\rho < 1$. Clearly

$$Cr^m \prod_{k=1}^{\infty} \left(1 - \frac{r}{|a_k|}\right) \leqslant |f(z)| \leqslant Cr^m \prod_{k=1}^{\infty} \left(1 + \frac{r}{|a_k|}\right) \quad (|z| = r).$$

Let

$$f_1(z) = Cz^m \prod_{k=1}^{\infty} \left(1 - \frac{z}{|a_k|}\right)$$

and let $h_1(\theta)$ be the indicator of the function $f_1(z)$. Since $\max h_1(\theta) = h_1(\pi)$, it follows from property (f) of the indicator (§ 16) that if $0 < \rho < 1/2$ we have $h_1(0) \geqslant h_1(\pi) \cos \pi\rho$. Hence $h_1(0) > 0$ if $\rho < 1/2$, and consequently we have, on a sequence $r_n \uparrow \infty$, the inequality $\ln|f_1(r_n)| > 0$ and also

$$\lim_{r_n \to \infty} \frac{\ln|f_1(r_n)|}{r_n^{\rho(r_n)}} = h_1(0).$$

Hence in this case we obtain

$$\varlimsup_{r\to\infty} \frac{\ln m(r)}{\ln M(r)} \geqslant \lim_{r_n\to\infty} \frac{\ln|f_1(r_n)|}{\ln|f_1(-r_n)|} \geqslant \lim_{r_n\to\infty} \frac{\ln|f_1(r_n)|}{h_1(x)r_n^{\rho(r_n)}} = \frac{h_1(0)}{h_1(\pi)} = \cos x\rho.$$

Now let $1/2 \leqslant \rho < 1$ and let $|f(z_0)| = m(r)$. Then

$$\ln[m(r)M(r)] \geqslant \ln|f(z_0)f(-z_0)| \geqslant \ln\left|\prod_1^\infty\left(1 - \frac{r^2}{|a_k|^2}\right)\right|.$$

The substitution $r^2 = R$ returns us to the case of a function of order $\rho_1 = \rho/2 < 1/2$, and by the preceding discussion we find that

$$\ln\left|\prod_1^\infty\left(1 - \frac{r^2}{|a_k|^2}\right)\right| > 0$$

on some sequence $r_n \uparrow \infty$. Therefore

$$\varlimsup_{r\to\infty} \frac{\ln m(r)M(r)}{\ln M(r)} \geqslant \varlimsup_{r\to\infty} \frac{\ln|f_1(r)f_1(-r)|}{h_1(\pi)r^{\rho(r)}}.$$

Hence, by the preceding discussion,

$$\varlimsup_{r\to\infty} \frac{\ln m(r)}{\ln M(r)} + 1 \geqslant \cos\frac{\pi\rho}{2} \varlimsup_{r\to\infty} \frac{\ln\prod_1^\infty\left(1 + \dfrac{r^2}{|a_k|^2}\right)}{h_1(\pi)r^{\rho(r)}}$$

$$= \cos\frac{\pi\rho}{2}\lim\frac{2\ln|f_1(ir)|}{h_1(\pi)r^{\rho(r)}}$$

Again applying property (f) of the indicator, we obtain

$$\varlimsup_{r\to\infty} \frac{\ln m(r)}{\ln M(r)} \geqslant 2\cos^2\frac{\pi\rho}{2} - 1 = \cos \pi\rho.$$

This completes the proof of the theorem.

The special case $\rho < \frac{1}{2}$ is especially interesting. In this case the theorem tells us that there is a sequence of circles $|z| = r_n$ on which $m(r_n) \to \infty$, and the order of growth of $m(r_n)$ is the same as the order of growth of $M(r_n)$, while the type is $\geqslant \sigma_f \cos \pi\rho$.[22b]

V. Bernstein proved the following precise theorem, giving an estimate of the density of the set of those r on which for fixed θ and $\varepsilon > 0$ the inequality $\ln |f(re^{i\theta})| > [h(\theta) - \varepsilon]\, r^{\rho(r)}$ is valid.

THEOREM 31 (V. BERNSTEIN [2]). *Let the function $f(z)$ be holomorphic and of order $\rho(r)$ in the angle $\alpha \leqslant \arg z \leqslant \beta$. To arbitrary positive numbers $\varepsilon > 0$ and $0 < \omega < 1$, and each fixed ray $\arg z = \theta$, there correspond a number $\delta > 0$ and a sequence of intervals $r_n \leqslant r \leqslant r_n(1 + \delta)$ $(r_n \to \infty)$, on each of which the inequality*

$$\ln |f(re^{i\theta})| > [h(\theta) - \varepsilon]\, r^{\rho(r)}$$

is satisfied except perhaps on a set of measure not exceeding $\omega\delta r_n$.

PROOF. Without loss of generality we may assume that $\theta = 0$ and $h(0) = 0$. To each $\gamma > 0$ there is a sequence $r_n \to \infty$ such that

$$\ln |f(r_n)| > -\gamma r_n^{\rho(r_n)}.$$

Also, if we choose $\delta > 0$ sufficiently small, then for $|\theta| < \arcsin(2e\delta)$ and $r > r_{\gamma,\delta}$, depending only on δ and γ,

$$\ln |f(re^{i\theta})| < [h(\theta) + \gamma]\, r^{\rho(r)} < 2\gamma r^{\rho(r)}.\text{[22c]}$$

Put

$$\varphi_n(z) = \frac{f(r_n + z)}{f(r_n)}.$$

Then $\varphi_n(0) = 1$ and for $|z| \leqslant 2e\delta r_n$

$$\ln |\varphi_n(z)| \leqslant 3\gamma (r_n + |z|)^{\rho(r_n + |z|)}.$$

It follows from Theorem 11 that the inequality

$$\ln |\varphi_n(z)| > -3\gamma H\left(\frac{\omega}{4}\right)(r_n + 2e\delta r_n)^{\rho(r_n + 2e\delta r_n)}$$

is satisfied in $|z| \leqslant \delta r_n$ but outside exceptional circles the sum of whose radii is not greater than $\omega\delta r_n$. Returning to the function $f(z)$ we see that the asymptotic inequality

$$\ln|f(r)| > -\gamma\left[1 + 3H\left(\frac{\omega}{4}\right)(1 + 2e\delta)^{2\rho}\right] r_n^{\rho(r_n)}$$

is satisfied on the whole interval $(1 - \delta)r_n < r < (1 + \delta)r_n$, except perhaps for intervals the sum of whose lengths is less than $2\omega\delta r_n$. Using properties of the

[22b] There are many papers about Wiman's theorem. For a generalization of Wiman's theorem to meromorphic functions see Gol'dberg and Ostrovskiĭ [1].

[22c] The continuity of the indicator when $\rho = 0$ follows from a theorem of Grišin (see footnote at the end of §6 in Appendix VIII).

function $r^{\rho(r)}$ one can verify easily that for $(1-\delta)r_n < r < (1+\delta)r_n$ we have asymptotically

$$r_n^{\rho(r_n)} > (1+\delta)^{-2\rho} r^{\rho(r)}.$$

To complete the proof it is sufficient to choose γ and δ so small that

$$\gamma\left[1 + 3H\left(\frac{\omega}{4}\right)(1+2\varepsilon\delta)^{2\rho}\right] < \varepsilon(1+\delta)^{2\rho}.$$

19. Plane Convex Sets

For $\rho = 1$ the indicator $h(\theta)$ has a simple geometric interpretation. To present it we shall need the basic properties of plane convex sets.

A *convex set* is a nonempty closed set which contains the entire line segment joining each pair of points in the set. In particular, a line segment is a convex set. A single point is also considered to be a convex set. Clearly a nonempty intersection of convex sets is again convex. The intersection of all the convex sets containing a given set is thus the smallest convex set containing the given set. It is sometimes called the *convex hull* of the given set. It is easy to see that the convex hull is also equal to the intersection of all the half-planes containing the given set.

We introduce the operation of arithmetic sum of convex sets. By the sum $G_1 + G_2$ of two convex sets is meant the set of all points of the form $z = z_1 + z_2$, where $z_1 \in G_1$ and $z_2 \in G_2$. Let us show that $G_1 + G_2$ is a convex set. Let $z_1' + z_2'$ and $z_1'' + z_2''$ be two points of this sum. Since

$$z = \lambda(z_1' + z_2') + \mu(z_1'' + z_2'') = \lambda z_1' + \mu z_1'' + \lambda z_2' + \mu z_2''$$
$$(\lambda + \mu = 1,\ \lambda \geqslant 0,\ \mu \geqslant 0),$$

and since $\lambda z_1' + \mu z_1'' \in G_1$ and $\lambda z_2' + \mu z_2'' \in G_2$ by the convexity of G_1 and G_2, we have $z \in G_1 + G_2$. It is clear that $G_1 + G_2$ is closed.

We shall always assume in this section, without explicitly mentioning it each time, that our convex sets are bounded.

An important concept is that of the supporting function of a convex set. By the *supporting function* of a convex set G we mean the function

$$k(\theta) = \sup_{x+iy \in G} (x\cos\theta + y\sin\theta) = \sup_{z \in G} \operatorname{Re}(ze^{-i\theta}). \qquad (1.102)$$

Since G is closed and bounded, this supremum is attained at some point of the convex set G. The line

$$x\cos\theta + y\sin\theta - k(\theta) = 0 \qquad (1.103)$$

will be denoted by l_θ. Clearly the line l_θ has a point in common with G. Further, all points of G lie on one side of each such line, since from the definition of support function $k(\theta)$ we have, for $x + iy \in G$, $x\cos\theta + y\sin\theta - k(\theta) \leqslant 0$.

The lines (l_θ) are called the *supporting lines* of the convex figure. A line of support is clearly a generalization of a tangent. The points that a support line has in common with the convex set are called the *supporting points* of the given line of support. A line of support has either a single point of support, or a whole line segment of supporting points. Those boundary points of G that are not interior to any straight line segment in the boundary are called *extreme points* of the convex set.

It is easy to see the geometric meaning of the supporting function. It is equal to the distance from the supporting line to the origin. Clearly the supporting function of a single point $z_0 = \rho \exp (i\theta_0)$ has the form $k(\theta) = \rho \cos (\theta - \theta_0)$. The supporting function of the segment $-H \leqslant y \leqslant H$ on the imaginary axis is $k(\theta) = H |\sin \theta|$.

Denoting by $k(\theta)$ the supporting function of the sum $G_1 + G_2 + \ldots + G_n$ of convex sets, we easily verify that

$$k(\theta) = k_1(\theta) + k_2(\theta) + \ldots + k_n(\theta), \tag{1.104}$$

where $k_j(\theta)$ $(j = 1, 2, \ldots, n)$ are the supporting functions of the convex sets $G_1, G_2, \ldots, G_n$. For the intersection of convex sets we have[23]

$$k(\theta) \leqslant \inf \{k_1(\theta),\ k_2(\theta),\ \ldots,\ k_n(\theta)\}. \tag{1.105}$$

For the convex hull of the union of the convex sets $G_1, G_2, \ldots, G_n$ the supporting function is given by

$$k(\theta) = \max \{k_1(\theta),\ k_2(\theta),\ \ldots,\ k_n(\theta)\}. \tag{1.106}$$

Let us prove this last assertion. Any half-plane that contains all the sets G_j $(j = 1, 2, \ldots, n)$ is given by an inequality

$$x \cos \theta + y \sin \theta - \tilde{k}(\theta) \leqslant 0,$$

where $\tilde{k}(\theta) \geqslant \max \{k_1(\theta), k_2(\theta), \ldots, k_n(\theta)\}$. Since the convex hull is the intersection of all such half-planes, we obtain (1.106).

The following theorem gives a characterization of supporting functions of convex domains.

THEOREM 32.. *A necessary and sufficient condition that the function $k(\theta)$ be the supporting function of a bounded convex set is that the following conditions be fulfilled:*

---

[23] *Translator's note*: this should be an inequality, $k(\theta) \leqslant \inf \{k_i(\theta)\}$. For example, consider the intersection of the two convex sets $-1 \leqslant y \leqslant 1$ and $-1 \leqslant x \leqslant 1$. The supporting functions are $|\sin \theta|$ and $|\cos \theta|$, whereas the supporting function for the intersection is 0.

(a) $k(\theta + 2\pi) = k(\theta) \qquad (-\infty < \theta < \infty)$,

(b) $\begin{vmatrix} k(\theta_1) \cos\theta_1 & \sin\theta_1 \\ k(\theta_2) \cos\theta_2 & \sin\theta_2 \\ k(\theta_3) \cos\theta_3 & \sin\theta_3 \end{vmatrix} \geqslant 0$

for all $\theta_1 \leqslant \theta_2 \leqslant \theta_3$, $\theta_3 - \theta_2 < \pi/2$, $\theta_2 - \theta_1 < \pi/2$.

The expanded form of condition (b) *can be written as follows:*

$$k(\theta_1)\sin(\theta_2 - \theta_3) + k(\theta_2)\sin(\theta_3 - \theta_1) + k(\theta_3)\sin(\theta_1 - \theta_2) \leqslant 0.$$

PROOF. Let G be a convex set, let $k(\theta)$ be its supporting function, and let $z = x + iy$ be a point of the boundary of G belonging to the supporting line l_{θ_2}. Then we have three relations:

$$x\cos\theta_1 + y\sin\theta_1 - k(\theta_1) \leqslant 0,$$
$$x\cos\theta_2 + y\sin\theta_2 - k(\theta_2) = 0,$$
$$x\cos\theta_3 + y\sin\theta_3 - k(\theta_3) \leqslant 0.$$

Multiplying the first of these by

$$\begin{vmatrix} \cos\theta_2 & \sin\theta_2 \\ \cos\theta_3 & \sin\theta_3 \end{vmatrix} = \sin(\theta_3 - \theta_2) > 0, \tag{1.107}$$

the second by

$$\begin{vmatrix} \cos\theta_3 & \sin\theta_3 \\ \cos\theta_1 & \sin\theta_1 \end{vmatrix} = \sin(\theta_1 - \theta_3), \tag{1.108}$$

the third by

$$\begin{vmatrix} \cos\theta_1 & \sin\theta_1 \\ \cos\theta_2 & \sin\theta_2 \end{vmatrix} = \sin(\theta_2 - \theta_1) > 0 \tag{1.109}$$

and adding, we obtain (b). Condition (a) is obvious.

Let us prove the sufficiency. We choose θ_2 arbitrarily and construct the line l_{θ_2}:

$$x\cos\theta_2 + y\sin\theta_2 - k(\theta_2) = 0. \tag{1.110}$$

Without loss of generality we can assume that $\theta_2 = \pi/2$.

The open half-planes

$$x\cos\theta_1 + y\sin\theta_1 - k(\theta_1) > 0 \qquad \left(-\frac{\pi}{2} < \theta_1 < \frac{\pi}{2}\right) \tag{1.111}$$

meet $l_{\pi/2}$ in the intervals $(b_1, +\infty)$, whose union is the interval $(b, +\infty)$.

Similarly, the union of the intersections of the half-planes

$$x \cos \theta_3 + y \sin \theta_3 - k(\theta_3) > 0 \qquad \left(\frac{\pi}{2} < \theta_3 < \frac{3\pi}{2}\right) \tag{1.112}$$

with the line $l_{\pi/2}$ is the interval $(-\infty, a)$.

We now show that no point of the line $l_{\pi/2}$ can belong to both of the intervals $(-\infty, a)$ and $(b, +\infty)$.

Indeed, if there were such a point then its coordinates would satisfy (1.110), (1.111) and (1.112), and multiplying them, respectively, by (1.108), (1.107) and (1.109) and adding we would obtain a relation contradicting (b).

Consequently, $a \leqslant b$, and all points of the interval $a \leqslant x \leqslant b$ on the line $l_{\pi/2}$ belong to all the half-planes

$$x \cos \theta + y \sin \theta - k(\theta) \leqslant 0.$$

Thus the intersection of these half-planes is not empty, and therefore it is some convex set G. This set lies on one side of each of the lines $x \cos \theta + y \sin \theta - k(\theta) = 0$, and has points of contact with each of these lines. Thus these lines are supporting lines for the convex set G, and $k(\theta)$ is the supporting function of this set.

It follows from Theorems 23 and 32 that the indicator function of an arbitrary entire function of the first order (with respect to some proximate order $\rho(r) \to 1$) is the supporting function of some bounded convex set. This convex set is called the *indicator diagram* of the given function.

An arc will be called *convex* if it lies entirely in the boundary of its convex hull, and *locally convex* if each of its points has a convex neighborhood on the arc. From the preceding considerations it is clear that condition (b) is necessary and sufficient for the convexity of an arc. A locally convex arc can intersect itself. Clearly the indicator function of a function that is analytic and of the first order inside an angle of opening less than or equal to π is the support function of some convex arc. In the general case, if $h_f(\theta)$ is the indicator function of some function holomorphic and of order ρ in some angle, then by (1.69) the function $h_f(\theta/\rho)$ is the supporting function of some locally convex arc.

If $f(z)$ is an entire function of integral order ρ, then the corresponding locally convex curve is closed, and in traversing this curve the supporting line makes ρ complete rotations.

Let us study convex sets in more detail. We first show that through each boundary point there is a supporting line. Since the assertion is obvious for sets consisting of a single point or of a line segment, we limit ourselves to sets having interior points.

If the extension of some ray through the boundary point z_0 of the convex set G passes through the interior point z_1, then clearly there is some angle at z_0, containing this ray, and containing no points of G (Fig. 4). A boundary ray of this set of rays and its extension form a line l that can contain no interior

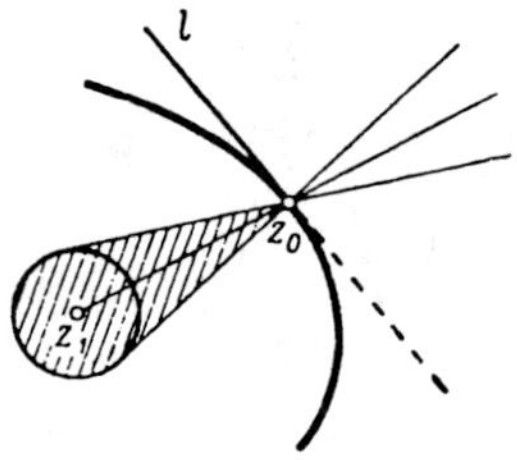

Fig. 4

points of G. The entire set G lies on one side of l, since if there were interior points of G on one side of the line l and if there were also points of G on the other side, then there would be interior points of G on l. Thus l is a supporting line to G through z_0.

The boundary Γ of a convex set has a finite length. This follows from the fact that the perimeter of any polygon (inscribed in the boundary of the set and having no self-intersections) does not exceed the perimeter of any triangle formed by three supporting lines.

If the origin is inside the domain and if arc length on the boundary is measured counterclockwise from some fixed point to the point of support, then the angle θ is a nondecreasing function of arc length s.

Indeed, let z_1 and z_2 be two points on the boundary of G, with $\arg z_2 > \arg z_1$. The line segment joining z_1 and z_2 belongs to G and therefore lies on the same side of the supporting lines at the points z_1 and z_2 as does the origin. It follows that the supporting line at z_2 has been rotated through a non-negative angle with respect to the supporting line through z_1, that is, $\theta_2 \geqslant \theta_1$. Clearly at angular points of the boundary the function $\theta = \theta(s)$ has jumps, while on line segments in the boundary $\theta(s)$ is constant. The inverse function $s(\theta)$ has jumps when the ray $\arg z = \theta$ is perpendicular to some line segment in the boundary and is continuous for all other values of θ.

The geometric interpretation of trigonometrically convex functions enables us to obtain the analytic properties of these functions that were established earlier. We shall obtain geometric interpretations of the quantities

$$k'_+(\theta),\ k'_-(\theta) \text{ and } s(\theta_1, \theta_2) = k'_-(\theta_2) - k'_+(\theta_1) + \int_{\theta_1}^{\theta_2} k(\theta)\,d\theta$$

of which we spoke in § 16.

Let us show that $k(\theta)$ is a continuous function. Without loss of generality we can assume that the origin lies inside the convex set G. We denote by $\delta(\Delta\theta)$ the length of the segment of the ray $\arg z = \theta + \Delta\theta$ between the supporting lines l_θ and $l_{\theta+\Delta\theta}$ (Fig. 5). We then have

$$k(\theta + \Delta\theta) = k(\theta)\sec(\Delta\theta) + \delta(\Delta\theta). \tag{1.113}$$

To establish continuity it is sufficient to observe that

$$|\delta(\Delta\theta)| < d\,\mathrm{tg}\,(\Delta\theta),$$

where d is the diameter of the domain. We now investigate the differentiability of $k(\theta)$ and we find the geometric significance of $k'(\theta)$.

Let $p(\theta)$ denote the segment of the line l_θ from the point of support to the foot of the perpendicular from the origin to the line l_θ. It is easy to see that the radius-vector to a point of the curve Γ is a continuous function of s and consequently, at the extreme points, it is a continuous function of θ. It follows that $p(\theta)$ is continuous at the extreme points.

If the direction θ is perpendicular to some segment on the boundary Γ, then we have

$$p(\theta+0)-p(\theta-0)=d(\theta),$$

where $d(\theta)$ is the length of this segment.

From simple geometric considerations (see Fig. 5) we have the inequality

$$\mathrm{tg}\,(\Delta\theta)\,[p(\theta+\Delta\theta)-\Delta s] \leqslant \delta(\Delta\theta) \leqslant p\,(\theta+\Delta\theta)\,\mathrm{tg}\,(\Delta\theta).$$

From (1.113) and this inequality we see that at each extreme point

$$k'(\theta)=p(\theta).$$

At points of discontinuity one establishes in the same way the existence of left and right derivatives of $k(\theta)$, with

$$k'_+(\theta)=p(\theta+0),$$
$$k'_-(\theta)=p(\theta-0)\ .$$

We now find an expression for the length s of the arc between the points of support of the lines l_α and l_β in terms of the supporting function $k(\theta)$. We inscribe in this arc a polygonal line Σ with such small segments that

$$|k(\theta)-R(\theta)|<\varepsilon,$$

where $R(\theta)$ is the support function of the inscribed polygon Σ.

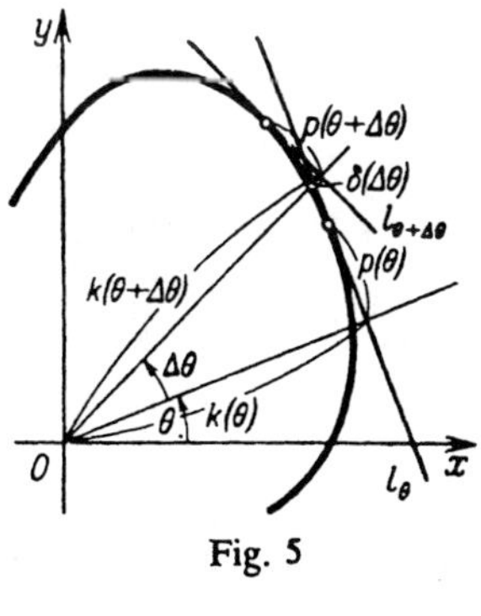

Fig. 5

Letting d_k denote the lengths of the segments, we have

$$\begin{aligned} s_\Sigma &= \sum_{k=1}^{n} d_k = \sum_{k=0}^{n} [R'_+(\theta_k) - R'_-(\theta_k)] \\ &= R'_+(\theta_n) - R'_-(\theta_0) - \sum_{k=1}^{n-1} [R'_-(\theta_{k+1}) - R'_+(\theta_k)]. \end{aligned}$$

Here θ_k are the directions perpendicular to the segments of the polygonal line. Note that in the intervals (θ_k, θ_{k+1}) the function $R(\theta)$ has the form $A_k \cos(\theta - \theta_k)$; we have

$$s_\Sigma = R'_+(\theta_n) - R'_-(\theta_0) - \sum_{k=1}^{n-1} \int_{\theta_k}^{\theta_{k+1}} R''(\theta)\, d\theta$$

or

$$s_\Sigma = R'_+(\theta_n) - R'_-(\theta_0) + \int_{\theta_0}^{\theta_n} R(\theta)\, d\theta$$

and, passing to the limit as $\max d_k \to 0$, we have

$$s(\beta \pm 0) - s(\alpha \pm 0) = k'_\pm(\beta) - k'_\pm(\alpha) + \int_\alpha^\beta k(\theta)\, d\theta. \tag{1.114}$$

Note that for the length of the entire boundary of the convex set we obtain the formula[24]

$$s = \int_0^{2\pi} k(\theta)\, d\theta. \tag{1.115}$$

From the monotonicity of $s(\theta)$ it follows that the radius of curvature $R = ds/d\theta$ exists at all points θ, except perhaps for a set of measure zero, and that

$$R = k''(\theta) + k(\theta).$$

Comparing formula (1.114) with (1.77) of § 16 we see that if $\rho = 1$ the function $s(\theta)$ defined by (1.77) is the length on the boundary of the indicator diagram of the function from some fixed point to the point of support of the line l_θ. If $\rho \neq 1$ and if $\sigma(\theta)$ denotes the length of the locally convex arc with support function $k(\theta/\rho)$, then from (1.114) we have the following equation:

$$\sigma(\rho\beta \pm 0) - \sigma(\rho\alpha \pm 0) = \frac{1}{\rho}\left\{k'_\pm(\beta) - k'_\pm(\alpha) + \rho^2 \int_\alpha^\beta k(\theta)\, d\theta\right\}.$$

---

[24] If the curve Γ has constant width, that is, if $k(\theta) + k(\theta + \pi) = d\,(0 \leqslant \theta < 2\pi)$, then from (1.115) we have $s = \pi d$. This is a well-known theorem of Barbier.

Comparing this with (1.77) we see that for $\rho > 0$

$$s(\theta) = \rho\sigma(\rho\theta).$$

We now solve the converse problem of expressing the support function $k(\theta)$ in terms of the arc length $s(\theta)$. Put the origin at the point of support of some supporting line l_0 and inscribe a polygonal line between this point of support and the point of support of the line l_θ. Clearly the support function of this polygonal line is

$$R(\theta) = \sum_{k=1}^{n} d_k \sin(\theta - \theta_k),$$

where θ_k are the directions perpendicular to the segments in the polygonal line, d_k are the lengths of the segments and $\theta_n \leqslant \theta < \theta_{n+1}$.

Passing to the limit as $\max d_k \to 0$, we have

$$k(\theta) = \int_0^\theta \sin(\theta - \psi)\, ds(\psi).$$

Translating the origin to an arbitrary point in the plane and using the formula for the support function of the sum of two convex sets we have

$$k(\theta) = A\cos(\theta - \theta_0) + \int_0^\theta \sin(\theta - \psi)\, ds(\psi). \tag{1.116}$$

Formula (1.116) enables us to associate a convex set to each monotone function $s(\psi)$. The closure condition $k(\theta + 2\pi) = k(\theta)$ can be written in the form

$$\int_0^{2\pi} \sin(\theta - \psi)\, ds(\psi) = 0,$$

or[25]

$$\int_0^{2\pi} e^{i\psi}\, ds(\psi) = 0.$$

Similarly one can construct a locally convex curve L with support function $k(\theta/\rho)$ for a given monotone function

$$s(\theta) = k'(\theta) + \rho^2 \int_0^\theta k(\varphi)\, d\varphi.$$

[25] In Theorem 24 a different representation of the function $k(\theta)$ than (1.116) was given for a closed convex curve

$$k(\theta) = A\cos(\theta - \theta_0) + \int_{\theta-2\pi}^{\theta} (\theta - \psi)\sin(\theta - \psi)\, ds(\psi).$$

In this case formula (1.116) is rewritten in the form

$$k(\theta) = A\cos\rho(\theta - \theta_0) + \int_0^\theta \sin\rho(\theta - \psi)\,ds(\psi),$$

and the condition of closure of the curve for integral ρ takes the form

$$\int_0^{2\pi} e^{i\rho\psi}\,ds(\psi) = 0. \qquad (1.117)$$

Using this geometric interpretation of condition (1.117) we can solve the following problem.

Let a nondecreasing function $s(\psi)$ be given such that for some integer ρ

$$\int_0^{2\pi} e^{i\rho\psi}\,ds(\psi) = 0.$$

Find a nondecreasing, piecewise constant function $s_p(\psi)$, approximating $s(\psi)$ to within $\varepsilon > 0$, that is,

$$|s(\psi) - s_p(\psi)| < \varepsilon \qquad (0 \leqslant \psi < 2\pi).$$

We require further that the jumps of the function $s_p(\psi)$ be rational numbers, not exceeding a given positive number δ and such that the function $s_p(\psi)$ satisfies condition (1.117).

To solve this problem we construct the closed, locally convex curve L corresponding to the given nondecreasing function $s(\psi)$, and inscribe a closed polygonal path L_p in it in such a way that the segments in L_p have lengths not exceeding δ, the length of L_p differs from the length of L by less than ε, and the length of each piece of the path whose endpoints are at points of L_p is less than ε. By shifting the vertices of L_p slightly we can require in addition that the lengths of the segments be rational. The function $s_p(\psi)$ corresponding to L_p gives the solution to the problem. In Chapter II we shall use this solution.

To conclude this section we note some almost obvious properties of the indicator diagram of entire functions of finite order.

1. *If the entire function $f(z)$ is of order one and normal type (that is, $\rho(r) = 1$), then the indicator diagram of the function*

$$\varphi(z) = e^{az} f(z) \qquad (a \text{ is a complex number})$$

is obtained from the indicator diagram I_f of the function $f(z)$ by a translation through the vector $\bar{a}$.

Indeed, clearly

$$h_\varphi(\theta) = |a|\cos(\theta + \alpha) + h_f(\theta) \qquad (\alpha = \arg a).$$

The function $|a|\cos(\theta + \alpha)$ is the support function for the convex set consisting of the single point $\bar{a}$.

Thus $h_\varphi(\theta)$ is the support function of the arithmetic sum of the convex sets I_f and $\bar{a}$, or

$$I_\varphi = I_f + \bar{a}.$$

2. *The indicator diagram of the product of two functions is contained in the arithmetic sum of their indicator diagrams.*[26]

This follows at once from the fact that the indicator of the product of two functions does not exceed the sum of their indicators.

3. *The indicator diagram of the sum of several functions is contained within the smallest convex set G containing the indicator diagrams of all the summands.*

This follows from the inequality

$$h_{f_1+f_2+\cdots+f_n}(\theta) \leqslant \max\,[h_{f_1}(\theta),\ h_{f_2}(\theta),\ \ldots,\ h_{f_n}(\theta)].$$

We now sharpen this theorem somewhat. Let E denote the set of those extreme points of G that belong to just one of the indicator diagrams. Clearly, each isolated point of this set is the common end point of two straight line segments in the boundary of G. The smallest convex set containing E will be denoted by $I(E)$.

4. *The indicator diagram of the sum of several functions contains the set $I(E)$.*

This result follows easily from the following fact noted in § 15: if for some value of θ the indicator of one of the summands is larger than the indicators of all the others, then for this value of θ the indicator of the sum function is equal to this largest indicator. Now suppose that $z_0 \in E$ is not the end point of some line segment in the boundary of $I(E)$, and let θ_0 be a direction for which z_0 is a point of support for the line l_{θ_0}. By the definition of the set E the indicator of one of the summands is larger than all the others for $\theta = \theta_0$, and hence z_0 is in the indicator diagram of the sum. If z_0 is the common end point of two straight line segments in the boundary of $I(E)$, then from among all the directions θ corresponding to the angle point z_0 we can choose one direction θ_0 for which the indicator of one of the summands exceeds the indicators of the others, and hence z_0 is contained in the indicator diagram of the sum. Finally, points of E that are end points of just a single straight line segment in the boundary of $I(E)$ are limit points of other points of E. Hence the closed convex hull of the points of E of the first two types contains these points, and thus coincides with $I(E)$.

Thus the indicator diagram of a sum of functions contains $I(E)$.

In particular, if G and $I(E)$ coincide, then the indicator diagram of the sum is equal to G, namely, the smallest convex set containing the indicator diagrams of all the summands. Such will be the case, for example, if the indicator diagrams of the summands are obtained from one another by means of nonzero

[26] Further on (Chapter III) we shall prove a much more precise theorem in a special case.

translations. In particular, if we multiply the function $f(z)$ of order one and normal type by a finite sum of the form

$$\varphi(z) = \sum_{k=1}^{n} a_k e^{\lambda_k z},$$

where a_k and λ_k $(k = 1, 2, \ldots, n)$ are complex numbers, $a_k \neq 0$, and all λ_k are distinct, then the indicator diagram of the product

$$\varphi(z) f(z)$$

is the arithmetic sum of the indicator diagram of $f(z)$ and the smallest convex polygon containing the points $\bar{\lambda}_1, \bar{\lambda}_2, \bar{\lambda}_3, \ldots, \bar{\lambda}_n$.

20. Entire Functions of Exponential Type

Entire functions of at most the first order and normal type are called *entire functions of exponential type*,[27] and the *exponential type of the function* is the quantity

$$\sigma = \overline{\lim_{r\to\infty}} \frac{\ln M_f(r)}{r}.$$

Thus functions of order less than one, or of order one and minimal type, are said to be of exponential type zero.

Functions of exponential type occur frequently in the applications, especially in harmonic analysis and in boundary value problems in differential equations.

To each entire function of exponential type

$$f(z) = \sum_{n=0}^{\infty} \frac{a_n}{n!} z^n \tag{1.118}$$

there corresponds a function

$$\varphi(z) = \sum_{n=0}^{\infty} \frac{a_n}{z^{n+1}}, \tag{1.119}$$

called the *Borel transform* of the function $f(z)$.

If σ is the exponential type of the function $f(z)$, then by formula (1.06)

$$\sigma = \overline{\lim_{n\to\infty}} \sqrt[n]{|a_n|}$$

and thus the series (1.119) represents a function holomorphic in the domain $|z| > \sigma$.

The smallest convex domain $\bar{I}_f$ containing all the singularities of $\varphi(z)$ is called the *conjugate diagram of the function* $f(z)$. It clearly lies inside the circle $|z| \leqslant \sigma$. Pólya [1] established the following remarkable theorem.

[27] *Translator's note*: In the Russian literature such functions are said to be of "finite degree."

THEOREM 33. *The conjugate diagram of an entire function of exponential type is the reflection in the real axis of its indicator diagram.*

PROOF. The proof depends on two integral formulas. The first represents $f(z)$ in terms of its Borel transform

$$f(z) = \frac{1}{2\pi i} \int_C e^{\zeta z} \varphi(\zeta)\, d\zeta, \tag{1.120}$$

where C is any contour containing the conjugate diagram. Indeed, the integral on the right is unchanged if we replace the contour C by any contour containing the disc $|z| \leqslant \sigma$. But for such a contour $\varphi(z)$ can be represented by the series (1.119) and integrating termwise we obtain (1.120). The second formula gives an integral representation for $\varphi(z)$,

$$\varphi(z) = \int_0^\infty e^{-z\zeta} f(\zeta)\, d\zeta, \tag{1.121}$$

where the integration is along the ray $\zeta = te^{-i\theta}$, $t > 0$.

It is easy to verify that this integral converges absolutely and uniformly in the domain

$$\operatorname{Re} ze^{-i\theta} > h_f(-\theta) + \varepsilon. \tag{1.122}$$

The convergence of the integral follows at once from the asymptotic inequalities

$$\left| e^{-zte^{-i\theta}} \right| < e^{-[h_f(-\theta)+\varepsilon]\,t}$$

and

$$|f(te^{-i\theta})| < e^{\left[h_f(-\theta)+\frac{\varepsilon}{2}\right]\,t}.$$

To prove (1.121) it is sufficient to show that the equation is valid in a part of the domain (1.122), for example, for

$$\operatorname{Re} ze^{-i\theta} > 3\sigma.$$

We show that for these values of z the series (1.118) can simply be substituted into (1.121) and integrated termwise. Indeed, in that half-plane

$$\left| e^{-zte^{-i\theta}} \right| < e^{-3\sigma t}. \tag{1.123}$$

On the other hand, from the inequality

$$\frac{|a_n|}{n!} < \frac{M_f(r)}{r^n}$$

it follows that the remainder

$$R_n(t) = \sum_{k=n+1}^{\infty} \frac{a_k}{k!} t^k$$

of the series (1.118) satisfies the inequality

$$|R_n(t)| < \frac{M_f(r)}{1-\frac{t}{r}}\left(\frac{t}{r}\right)^{n+1},$$

and if we put $r = 2t$, then

$$|R_n(t)| < \left(\frac{1}{2}\right)^n e^{2(\sigma+\varepsilon)t}. \qquad (1.124)$$

From inequalities (1.123) and (1.124) it follows that the series

$$\sum_{k=0}^{\infty} e^{-z\zeta}\frac{a_k\zeta^k}{k!}$$

converges uniformly on the whole ray $t > 0$, $\zeta = te^{-i\theta}$. Integrating termwise we obtain equation (1.121).

We now use formulas (1.120) and (1.121) to prove the theorem. Let $k_f(\theta)$ be the support function of the conjugate diagram $\bar{I}_f$. In formula (1.120) choose the contour of integration so that it is no further than sóme $\varepsilon > 0$ from the conjugate diagram. Then we obtain the estimate

$$|f(re^{i\theta})| < M_\varepsilon e^{r\max_{\zeta\in C}[\operatorname{Re}(\zeta e^{i\theta})]} \leqslant M_\varepsilon e^{[k_f(-\theta)+\varepsilon]r}$$

$$\left(M_\varepsilon = \frac{c}{2\pi}\max_{\zeta\in C}|\varphi(\zeta)|\right)$$

where c is the length of C. Hence

$$h_f(\theta) \leqslant k_f(-\theta).$$

On the other hand, from the convergence of the integral (1.121) for

$$\operatorname{Re}(ze^{-i\theta}) > h_f(-\theta) + \varepsilon$$

it follows that $\varphi(z)$ has no singularities in this domain, and hence

$$k_f(\theta) \leqslant h_f(-\theta).$$

Thus

$$k_f(-\theta) = h_f(\theta),$$

and the theorem is proved. We present some examples.

If

$$f(z) = \sum_{k=1}^{n} a_k e^{\lambda_k z},$$

then

$$\varphi(z) = \sum_{k=1}^{n}\frac{a_k}{z-\lambda_k}$$

and the conjugate diagram is the smallest convex set (polygon) containing all the points λ_k. If

$$f(z)=\sum_{k=1}^{n} e^{\lambda_k z}(a_{k0}+a_{k1}z+\ldots+a_{kp_k}z^{p_k}),$$

then

$$\varphi(z)=\sum_{k=1}^{n}\left[\frac{a_{k0}}{z-\lambda_k}+\frac{a_{k1}}{(z-\lambda_k)^2}+\ldots+\frac{a_{kp_k}\, p_k!}{(z-\lambda_k)^{p_k+1}}\right]$$

and the conjugate diagram is the same as in the first example.

We construct an entire function of exponential type having for indicator diagram a convex set I given in advance.[28] Let $\bar{I}$ denote the set symmetric to I with respect to the real axis. Choose a countable set of points λ_k, dense in the boundary of $\bar{I}$, and form the function

$$\varphi(z)=\sum_{k=1}^{\infty}\frac{a_k}{z-\lambda_k},$$

with

$$\sum_{k=1}^{\infty}|a_k|<\infty.$$

This function is holomorphic outside the set $\bar{I}$. We now show that all the points λ_k are singular points of this function. Indeed, let z approach the point λ_p along the perpendicular to one of the lines of support through the point λ_p; then, clearly,

$$|z-\lambda_p|<|z-\lambda_k| \quad \text{for} \quad p\neq k.$$

Choose the number N_p so that

$$\sum_{k>N_p}|a_k|<\frac{1}{2}|a_p|.$$

We have

$$|\varphi(z)|\geqslant\frac{|a_p|}{|z-\lambda_p|}-\sum_{k=1,\,k\neq p}^{N_p}\frac{|a_k|}{|z-\lambda_k|}-\sum_{k=N_p+1,\,k\neq p}^{\infty}\frac{|a_k|}{|z-\lambda_k|},$$

or

$$|\varphi(z)|>\frac{|a_p|}{2|z-\lambda_p|}-\sum_{k=1,\,k\neq p}^{N_p}\frac{|a_k|}{|z-\lambda_k|}.$$

[28] In Chapter II we solve the corresponding problem for functions having an arbitrary proximate order.

Clearly as $z \to \lambda_p$ in the manner described, $|\varphi(z)| \to \infty$. Thus $\bar{I}$ is the smallest convex set containing all the singularities of the Borel transform $\varphi(z)$ of the entire function

$$f(z) = \sum_{k=1}^{\infty} a_k e^{\lambda_k z}.$$

Note also that Pólya's theorem enables us to sharpen assertion (4) about the indicator diagram of the sum of entire functions, provided that the functions are of exponential type. Namely, we can assert that the indicator diagram of a sum of entire functions of exponential type contains the smallest convex set containing all those extreme points of each of the indicator diagrams that do not lie in any of the other diagrams. Indeed, each extreme point of one of the conjugate diagrams is a singularity of the corresponding Borel transform. If such a point does not lie in any of the other diagrams then it is a singularity for the sum of the transforms, and hence it is in the conjugate diagram of the sum of the given entire functions of exponential type.

CHAPTER II

ENTIRE FUNCTIONS WHOSE SETS OF ZEROS HAVE AN ANGULAR DENSITY

1. Basic Results

In this chapter we shall consider a special class of entire functions whose zero distributions have a certain type of regularity.

This class is met quite frequently in various applications and plays an important role in the following chapters of this book. We shall give a precise asymptotic representation of the functions of this class. The regularity of the distribution of the zeros will be characterized by a certain type of density of the set of zeros.[1]

The proofs of the majority of the theorems of this chapter are somewhat involved. In order that the reader may not lose the perspective of this chapter as a whole, we shall present all the basic results of this chapter in the first section without proofs. During the first reading of this book the reader may confine himself to reading the first section only; he should, however, return later to the other sections that contain the proofs.

We shall say that the point set $\mathfrak{N}$ of the complex plane has an *angular density of index* $\rho(r)$ if for all but a denumerable set of values ϑ and θ ($0 < \vartheta < \theta \leqslant 2\pi$) there exists the limit[2]

(A) $$\Delta(\vartheta, \theta) = \lim_{r \to \infty} \frac{n(r, \vartheta, \theta)}{r^{\rho(r)}}, \tag{2.00}$$

where $n(r, \vartheta, \theta)$ is the number of points of the set $\mathfrak{N}$ within the sector $|z| \leqslant r$, $\vartheta < \arg z < \theta$. The quantity $\Delta(\vartheta, \theta)$ will be called the angular density of the set $\mathfrak{N}$ in the angle $\vartheta < \arg z < \theta$, or, more simply, in the angle (ϑ, θ). For a fixed ϑ, the equation

$$\Delta(\theta) - \Delta(\vartheta) = \Delta(\vartheta, \theta) \tag{2.01}$$

determines, to within an additive constant, a nondecreasing function $\Delta(\theta)$. This function is defined for all values of θ except, possibly, for a denumerable set of values. The exceptional values of ϑ and θ for which there does not exist

[1] The principal results of this chapter have been published earlier by the author [**1**; **2**]. Analogous results were obtained independently by A. Pfluger [**1**; **2**; **4**].

[2] When we say that a limit exists, we shall always mean a finite limit.

an angular density $\Delta(\vartheta + 0, \theta - 0)$ must be points of discontinuity of the function $\Delta\theta$. Indeed, if ϑ and $\theta (\theta > \vartheta)$ are two arbitrary numbers, and if the positive number ε is such that $\vartheta \pm \varepsilon$ and $\theta \pm \varepsilon$ are points of continuity of the function $\Delta\theta$, then

$$\Delta(\theta - \varepsilon) - \Delta(\vartheta + \varepsilon) \leqslant \varliminf_{r \to \infty} \frac{n(r, \vartheta, \theta)}{r^{\rho(r)}} \leqslant \varlimsup_{r \to \infty} \frac{n(r, \vartheta, \theta)}{r^{\rho(r)}}$$
$$\leqslant \Delta(\theta + \varepsilon) - \Delta(\vartheta - \varepsilon).$$

If ϑ and θ are points of continuity of $\Delta\theta$, then the lower limit is equal to the upper limit. A set will be said to be *regularly distributed relative to* $\rho(r)$ if it has an angular density $\rho(r)$ with ρ not integral.

We shall show in this chapter that the regularity of the distribution of the zeros of an entire function implies a certain regularity of the growth of the function. In order that we may be able to describe precisely the nature of this regularity of growth, we introduce now still another concept.

The linear density of a set C *of circles* (C_j) in the complex plane is defined as the limit

$$\rho^*(C) = \lim_{r \to \infty} \left\{\frac{1}{r} \sum^r r_j\right\},$$

where r_j is the radius of the circle C_j, and the symbol Σ^r indicates summation over all circles whose centers lie in the circle $|z| < r$.

If the limit $\rho^*(C)$ does not exist then we shall consider the upper and lower limits of the same quantity, and we will denote these upper and lower limits by $\bar{\rho}^*(C)$ and $\underline{\rho}^*(C)$, respectively. We shall call them the *upper and lower linear densities of the set of circles C*. A set whose linear density is zero will be called a C^0*-set*.

The asymptotic equalities which we shall establish are related to the order of growth. By the asymptotic equation

$$f(r) \approx \varphi(r)$$

we shall mean the fulfillment of the following condition:

$$\lim_{r \to \infty} r^{-\rho(r)} [f(r) - \varphi(r)] = 0.$$

THEOREM 1. *Let $\rho(r)$ be a proximate order of growth, let $\rho = \lim \rho(r)$ be a nonintegral number, and let $\{a_n\}$ be a point set of the complex plane whose angular density $\Delta(\psi)$ has the index $\rho(r)$. Then for every z not belonging to some C^0-set, the entire function*

$$\left.\begin{array}{l} f(z) = z^m e^{P(z)} \prod_1^\infty G\left(\frac{z}{a_k}; p\right) \\ (P(z) = c_0 + c_1 z + \dots + c_p z^p, \quad p = [\rho]) \end{array}\right\} \qquad (2.02)$$

will satisfy the asymptotic equation

$$\ln|f(re^{i\theta})| \approx H(\theta)\, r^{\rho(r)}, \tag{2.03}$$

in which

$$H(\theta) = \frac{\pi}{\sin \pi\rho} \int_{\theta-2\pi}^{\theta} \cos\rho(\theta - \psi - \pi)\, d\Delta(\psi). \tag{2.04}$$

Here the function

$$h_{f,r}(\theta) = r^{-\rho(r)} \ln|f(re^{i\theta})|$$

will tend to the limit $H(\theta)$ *uniformly relative to* θ *when* $0 \leqslant \theta \leqslant 2\pi$.

Hence, for a function of a nonintegral order, the existence of an angular density for the set of zeros implies a special type of regularity of growth which is expressed by the uniform convergence to a limit of the function $h_{f,r}(\theta)$; the function $H(\theta)$ does not depend on the factor in front of the canonical product, and is therefore completely determined by the regularly distributed set of zeros.

For functions of integral order, the existence of an angular density of the set of zeros is, however, not a sufficient condition for such a regularity of the growth of the function. In this case there is also required a special type of symmetry in the distribution of the zeros of the function which is characterized by the existence of the limit

$$\text{(B)} \qquad \delta_f = \lim_{r\to\infty} \frac{\delta_f(r)}{L(r)} \qquad \left(\delta_f(r) = c_\rho + \frac{1}{\rho} \sum_{|a_n|<r} a_n^{-\rho}\right),$$

where $L(r) = r^{\rho(r)-\rho}$, and c_ρ is the coefficient of the term of highest degree in the exponential factor in front of the canonical product in the representation (1.38). We shall refer to this condition as condition (B). If $p = \rho = \rho(r)$, then this condition indicates a special conditional convergence of the series

$$\sum_1^\infty a_k^{-\rho},$$

namely, the existence of the principal value

$$\lim_{r\to\infty} \left(\sum_{|a_k|<r} a_k^{-\rho} \right)$$

of the sum of this series. For functions of integral order we shall prove the following theorem.

THEOREM 2. *Let* $\rho(r)$ *be an order of growth, let* $\rho = \lim_{r\to\infty} \rho(r)$ *be a positive integer, and assume that the set* $\{a_k\}$ *of the zeros of the entire function*

$$f(z) = z^m e^{P(z)} \prod_1^\infty G\left(\frac{z}{a_k}; \rho\right)$$

has an angular density with index $\rho(r)$, *and satisfies the condition* (B).

Then for all z not belonging to a certain C^0-set, the following asymptotic equation will hold

$$\ln|f(re^{i\theta})| \approx H(\theta) r^{\rho(r)}, \tag{2.05}$$

where

$$H(\theta) = -\int_{\theta-2\pi}^{\theta} (\psi-\theta)\sin\rho(\psi-\theta)\,d\Delta(\psi) + \tau_f \cos\rho(\theta-\theta_f) \tag{2.06}$$

and $\delta_f = \tau_f e^{i\theta_f}$.

Here the function

$$h_{f,r}(\theta) = r^{-\rho(r)} \ln|f(re^{i\theta})|$$

will tend to the limit $H(\theta)$ uniformly relative to θ when $0 \leqslant \theta \leqslant 2\pi$.[3]

In view of this theorem, it is natural to define a *regularly distributed set with index $\rho(r)$, when ρ is not an integer*, as a planar point set which has an angular density for this $\rho(r)$ and which is such that for some constant $C_{\mathfrak{N}}$ the following limit exists:

(B′)
$$\delta = \lim_{r\to\infty} \frac{1}{L(r)}\left\{C_{\mathfrak{N}} + \frac{1}{\rho}\sum_{|a_n|\leqslant r} a_n^{-\rho}\right\}$$

$$\left(a_n \in \mathfrak{N},\quad L(r) = r^{\rho(r)-\rho}\right).$$

If $L(r) \to \infty$, or has a limit different from zero when $r \to \infty$, then in the condition (B′) the quantity $C_{\mathfrak{N}}$ may be replaced by an arbitrary number. In the opposite case, the number $C_{\mathfrak{N}}$ is uniquely determined.

From the entire functions whose zero sets coincide with $\mathfrak{N}$ we select the canonical functions $f(z;\mathfrak{N})$. If ρ is not an integer we shall say that a function is a canonical function if it has the form

$$f(z;\mathfrak{N}) = \prod_1^\infty G\left(\frac{z}{a_n};\ p\right) \qquad (p = [\rho]),$$

while if ρ is an integer, a function is said to be a canonical function if it is of the form

$$f(z;\mathfrak{N}) = e^{C_{\mathfrak{N}} z^\rho} \prod_1^\infty G\left(\frac{z}{a_n};\ \rho\right).$$

Here, if the quantity $C_{\mathfrak{N}}$ in condition (B′) is arbitrary, we take $C_{\mathfrak{N}} = 0$. The quantity $H(\theta)$, defined in the statements of Theorems 1 and 2, is completely determined by the set $\mathfrak{N}$ if ρ is not an integer, or if ρ is an integer but the quantity $C_{\mathfrak{N}}$ is uniquely determined. In general, we shall denote the function $H(\theta)$ for canonical functions by $H_{\mathfrak{N}}(\theta)$, and we shall call it the *indicator* of the set $\mathfrak{N}$.

A function $f(z)$ whose roots are regularly distributed can be evaluated within

---

[3] B. Ja. Levin [1, pp. 1121–1127], A. Pfluger [4].

the exceptional circles by the principle of maximum modulus, and one obtains for the indicator $h_f(\theta)$ of an entire function the equation

$$h_f(\theta) = H(\theta).$$

The formulas (2.04) and (2.06) thus yield expressions for the indicators of entire functions with regularly distributed zeros in terms of the angular density and of the quantity δ_f. In Chapter 1, § 16, we derived similar forms for indicators; but it is important to note that in formulas (2.04) and (2.06) in place of the nondecreasing function $s_f(\psi)$ we have $\Delta_f(\psi)$, the density of the set of zeros of the entire function $f(z)$ in the angle $(0, \psi)$.

From the inversion formulas of § 16, Chapter 1, it follows that

$$\Delta_f(\psi \pm 0) = \frac{1}{2\pi\rho} s_f(\psi \pm 0) + C \quad (C \text{ is a constant}) \tag{2.07}$$

or

$$\Delta_f(\theta \pm 0) - \Delta_f(\vartheta \pm 0) = \frac{1}{2\pi\rho} [s_f(\theta \pm 0) - s_f(\vartheta \pm 0)].$$

Starting from the geometric interpretation of the function $s_f(\theta)$ (see § 19, Chapter I), we find that if $\rho = 1$, the angular density of the set of zeros of the function $f(z)$ in the angle (ϑ, θ) is equal to $\frac{1}{2}\pi$ times the length of the arc of the indicator diagram between the points of support of the supporting lines l_ϑ and l_θ, which are perpendicular to the sides of the angle. In the general case, the right hand side of (2.07) is proportional to the length of some locally convex arc (Fig. 6).

The formulas (2.07), (2.04) and (2.06) permit one to construct an entire function of any given proximate order $\rho(r)$ in such a manner that its indicator will coincide with an arbitrary trigonometrically convex function $h(\theta)$ of period 2π. The essence of the method is the construction of a nondecreasing function $\Delta(\theta)$ for the given function $h(\theta)$ with the aid of equations (2.07) and (1.77). By next constructing a set $\{a_k\}$ with angular density $\Delta(\theta)$ for the index $\rho(r)$,

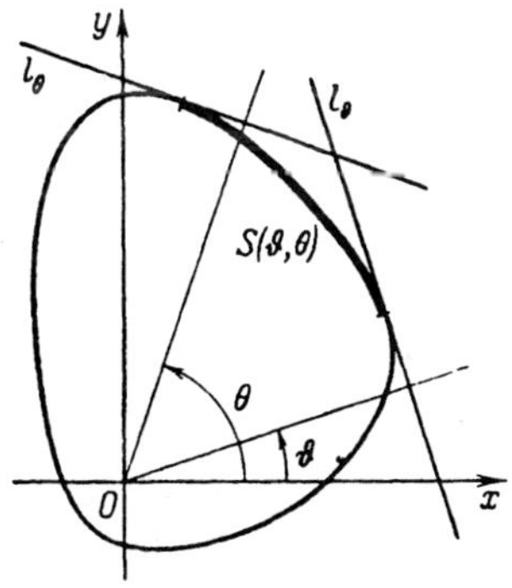

Fig. 6

and then forming the canonical product with these points, we obtain the required entire function. The construction of the entire function with a given indicator is carried out in § 4 of this chapter.

The results may be formulated as in the following theorem.

THEOREM 3.[4] *For every given strong proximate order* $\rho^*(r)$ ($\lim_{r\to\infty}\rho^*(r) = \rho > 0$), *and for every trigonometrically convex (with this* ρ*) function* $h(\theta)$ *of period* 2π, *there exists an entire function of proximate order* $\rho^*(r)$ *with indicator* $\rho^*(\theta)$. *An analogous statement is true for an arbitrary proximate order if* ρ *is a nonintegral number.*

We note that the entire function with a given indicator which we shall construct in § 4 possesses a special regularity of growth.

When the zeros of an entire function are regularly distributed, one can obtain an asymptotic expression not only for the function $\ln|f(re^{i\theta})|$ but also for the function $\arg f(z)$, i.e., one can obtain a representation for the entire function $f(z)$ itself. § 5 of the present chapter is devoted to this topic. In order that the expression $\arg f(z)$ may have a definite meaning for the canonical product

$$f(z)=\prod_{k=1}^{\infty} G\left(\frac{z}{a_k};\ p\right) \qquad (p=[\rho]),$$

we define the function $\arg G(u;p)$ in the u plane, cut along the ray $(1, +\infty)$, to be zero on the upper side of the cut and then we extend in a continuous manner over the entire plane. The function $\arg G(u;p)$ that is defined uniquely in this way over the entire cut plane will have a jump equal to 2π at each point of the ray $(1, +\infty)$.

We shall say that a ray $\arg z = \theta$ is an "ordinary" ray if it satisfies the condition

$$\lim_{h\to 0}\left\{\overline{\lim_{r\to\infty}}\,\frac{n(r,\theta-h,\theta+h)}{r^{\rho(r)}}\right\}=0.$$

All other rays will be called "exceptional" rays. The monotone nature of the function $\Delta(\theta)$ implies that if the zeros of the functions are regularly distributed then the set of exceptional rays is at most denumerable. The basic result of § 5 can be formulated in the following way.

THEOREM 4. *If* $\mathfrak{N}$ *is a regularly distributed set with index* $\rho(r)$ *and if* $f(z, \mathfrak{N})$ *is its canonical function, then for any* z *belonging to an ordinary ray, and not belonging to some* C^0*-set, the following asymptotic relation will hold*:

$$\ln f(re^{i\theta}, \mathfrak{N}) \approx J(\theta)\, r^{\rho(r)}, \tag{2.08}$$

[4] This theorem is a generalization of a theorem of V. Bernstein [1; 3] on functions of normal type for $\rho(r) = \rho$.

where

$$J(\theta)=\frac{\pi}{\sin \pi\rho}\int_{\theta-2\pi}^{\theta} e^{i\rho(\theta-\psi-\pi)}\,d\Delta(\psi) \tag{2.09}$$

for a nonintegral ρ, *and*

$$J(\theta)=-i\int_{\theta-2\pi}^{\theta} e^{i\rho(\theta-\psi)}(\theta-\psi)\,d\Delta(\psi)+Ae^{i\rho(\theta-\theta_0)} \tag{2.10}$$

for a ρ *that is an integer.*

Here the convergence of $r^{-\rho(r)}\ln f(re^{i\theta})$ *to the limit* $J(\theta)$ *will be uniform in* θ *on any closed set* $\mathfrak{M}$ *of ordinary rays.*

We note that the multiplication of the function $f(z, \mathfrak{N})$ by an exponential factor $e^{P(z)}$ in which the exponent $P(z)$ is less than ρ will not invalidate the asymptotic equation (2.08).

The exclusion of the C^0-sets in Theorems 1, 2, 4 and 5 is based on the Cartan estimate (see Chapter I, § 7). In this estimate the set of exceptional circles is not effectively constructed. It is for this reason that Theorems 1, 2, 4 and 5 assert only that the asymptotic equations are valid outside of certain exceptional regions (C^0-sets) but do not indicate how these regions can be constructed. However, with certain auxiliary limitations on the roots of the functions, one can give more precise information about the position of the circles of the exceptional set. We assume that the points can not come arbitrarily close to each other. More precisely, we assume that one of the following conditions (C) or (C′) holds.

(C) *There exists a number* $d > 0$ *such that the circles of radii*

$$r_n = d\,|a_n|^{1-\frac{\rho(|a_n|)}{2}}$$

with centers at the points a_n *do not intersect.*

(C′) *The points* a_n *lie inside angles with a common vertex at the origin but with no other points in common, which are such that if one arranges the points of the set* $\{a_n\}$ *within any one of these angles in the order of increasing moduli, then for all points which lie inside the same angle it is true that*

$$|a_{k+1}|-|a_k| > d\,|a_k|^{1-\rho(|a_k|)}$$

for some $d > 0$.

In the conditions (C) and (C′) the proximate order $\rho(r)$ has the same meaning as in condition (A).

A regular point set $\{a_n\}$ satisfying one of the conditions (C) or (C′) will be called a regular, or more briefly an *R-set*, while the circles $|z - a_n| \leqslant r_n$ in the first case, and $|z - a_n| \leqslant d\,|a_n|^{1-\rho(|a_n|)}$ in the second case, will be called the *exceptional circles of the R-set* (C_R*-circles*).

In § 6 we will prove the following theorem.

THEOREM 5. *If the set* $\mathfrak{N}$ *is an R-set of order* $\rho(r)$, *then outside the exceptional* C_R-*circles the canonical function* $f(z, \mathfrak{N})$ *satisfies the asymptotic equality*

$$\ln |f(re^{i\theta}, \mathfrak{N})| \approx H_{\mathfrak{N}}(\theta)\, r^{\rho(r)}, \tag{2.05'}$$

and if $\arg z = \theta$ *is an ordinary ray, then*

$$\ln f(re^{i\theta}, \mathfrak{N}) \approx J_{\mathfrak{N}}(\theta)\, r^{\rho(r)}, \tag{2.08'}$$

where $H_{\mathfrak{N}}(\theta)$ *and* $J_{\mathfrak{N}}(\theta)$ *are defined by the equations* (2.04), (2.06), (2.09) *and* (2.10). *The convergence of the functions* $r^{-\rho(r)} \ln |f(re^{i\theta}, \mathfrak{N})|$ *and* $r^{-\rho(r)} \ln f(re^{i\theta}, \mathfrak{N})$ *to their limits is uniform in the same sense as that of Theorems* 1, 2, *and* 4.

In this theorem, the exceptional set is not a C^0-set; however, if the condition (C′) is satisfied, then the upper linear density of this set can be made arbitrarily small by the proper choice of d. When condition (C) holds, one can make another "density" arbitrarily small, namely, the upper limit of the ratio of the sum of the areas of the exceptional circles lying in some circle $|z| \leqslant R$ to the area of this circle arbitrarily small when $R \to \infty$.

We shall need the following definition.

The relative measure of a set of positive numbers is defined to be the limit

$$m^*(E) = \lim_{r \to \infty} \frac{\operatorname{mes}(E^r)}{r},$$

where E^r denotes the intersection of the set E with the interval $(0, r)$. If this limit does not exist, then we shall consider the upper and lower relative measures

$$\overline{m}^*(E) = \overline{\lim_{r \to \infty}} \frac{\operatorname{mes}(E^r)}{r}, \qquad \underline{m}^*(E) = \underline{\lim_{r \to \infty}} \frac{\operatorname{mes}(E^r)}{r}.$$

A set of zero relative measure will be called an E^0-set. It is obvious that the set of those values of r for which the circle $|z| = r$ intersects a C^0-set is an E^0-set. Hence, in the formulation of Theorems 1 and 2 one may replace the requirement that z should not take on the values of some C^0-set by the requirement that $|z|$ must not take on values from some E^0-set.

As has been mentioned before, it follows from Theorems 1 and 2 that if the zeros of the function are regularly distributed one can omit from the values of r an E^0-set such that the functions of θ

$$h_{f,r}(\theta) = \frac{\ln |f(re^{i\theta})|}{r^{\rho(r)}}$$

will form an equicontinuous family. This property can be extended to arbitrary entire functions of finite order, for which it takes on the following form.

THEOREM 6. *To every entire function* $f(z)$ *of proximate order* $\rho(r)$ *and to every positive number* η *there corresponds a set* E_η *of positive numbers with an upper relative measure less than* η *such that if* $r \,\overline{\in}\, E_\eta$, *the family of functions* $h_{f,r}(\theta)$ *is equicontinuous.*

Theorem 6 can be extended to functions that are holomorphic within some angle $\alpha < \arg z < \beta$. This generalization can be formulated in the following way.

THEOREM 7. *Let us assume that the function $F(z)$ is holomorphic and of proximate order $\rho(r)$ within the angle $\alpha < \arg z < \beta$, and that for some positive $l < 1$, for some positive N, and for all sufficiently large values of r, every circle*

$$|z - re^{i\frac{\alpha+\beta}{2}}| < lr \sin\frac{\beta-\alpha}{2}$$

contains a point $z_1 = r_1 e^{i\theta_1}$ at which

$$\ln|F(z_1)| > -Nr_1^{\rho(r_1)}. \tag{2.12}$$

Then for every positive η there exists a set E_η of positive numbers with an upper relative measure less than η such that if $r \bar{\in} E_\eta$, the family of functions of θ

$$h_{F,r}(\theta) = r^{-\rho(r)} \ln|F(re^{i\theta})|$$

is equicontinuous within the angle $\alpha + \varepsilon \leqslant \theta \leqslant \beta - \varepsilon$.[4]

This theorem, which we shall prove in § 7 of the present chapter, is quite useful in the investigation of various questions in the theory of the growth of functions of finite order.[4a]

2. Entire Functions of Nonintegral Order with Regular Distribution of Zeros (Proof of Theorem 1)

For functions of nonintegral order the exponential factor in front of the canonical product does not affect the indicator of the function. The indicator is completely determined in this case by the canonical product or, what is the same thing, by the set of zeros of the function. Therefore, we shall consider in this section the canonical product only, which will not restrict the generality of our considerations.

Let us first consider some particular cases of regular distribution of the zeros. We have the simplest case when all zeros of the canonical product $\Pi(z)$ lie on one ray $\arg z = \psi$ with some density

$$\Delta = \lim_{r\to\infty} \frac{n(r)}{r^{\rho(r)}}.$$

This case has been considered separately in Theorem 25 of Chapter I. Indeed, if one rotates the plane through an angle ψ, i.e., if one substitutes for θ

---

[4] The auxiliary condition given by the inequality (2.12) for the case of an entire function is satisfied in view of a known estimate of an entire function from below (see Chapter I, pages 21–22).

[4a] A far-reaching development of Theorems 6 and 7 has been obtained by Grišin [2–4] and Krasičkov [1; 3].

the quantity $\theta - \psi$, then for $\eta > 0$ one obtains the limit

$$\lim_{r\to\infty} \frac{\ln \Pi(re^{i\theta})}{r^{\rho(r)}} = \frac{\pi\Delta}{\sin \pi\rho} e^{i\rho(\theta-\psi-\pi)}$$

where the convergence is uniform relative to θ in $(\psi + \eta \leqslant \theta \leqslant \psi + 2\pi - \eta)$.

Equating the real parts, we find that

$$\lim_{r\to\infty} \frac{\ln |\Pi(re^{i\theta})|}{r^{\rho(r)}} = \frac{\pi\Delta}{\sin \pi\rho} \cos \rho(\theta - \psi - \pi).$$

The case next in complexity occurs when the zeros of the canonical product $\Pi(z)$ are located on a finite number of rays $\arg z = \psi_j$ $(j = 0, 1, \ldots, m)$, and the density of the set of zeros exists on each ray, that is, when we have the limits

$$\Delta_j = \lim_{r\to\infty} \frac{n_j(r)}{r^{\rho(r)}} \qquad (j = 0, 1, \ldots, m),$$

where $n_j(r)$ is the number of zeros of the product $\Pi(z)$ in the interval $(0, r)$ on the ray $\arg z = \psi_j$. In this case the function $\Pi(z)$ is the product of functions $\Pi_j(z)$ whose roots lie on a single ray $\arg z = \psi_j$. This case can be reduced to the preceding one, and we have the limits

$$\lim_{r\to\infty} \frac{\ln |\Pi(re^{i\theta})|}{r^{\rho(r)}} = \frac{\pi}{\sin \pi\rho} \sum_{j=0}^{m} \Delta_j \cos \rho(\theta - \psi_j - \pi)$$

where the convergence is uniform relative to θ if $|\theta - \psi_j| \geqslant \eta > 0$ $(j = 0, 1, \ldots, m)$.

If one writes the sum of the right hand side of this equation by means of a Stieltjes integral, one obtains formula (2.04) for this particular case of regular distribution of zeros. The transition to the general case is based on the following lemma.

LEMMA 1. *Let us assume that the set $\{a_n\}$ of the zeros of the canonical product*

$$\Pi(z) = \prod_1^\infty G\left(\frac{z}{a_n}; p\right)$$

has a density with index $\rho(r)$, i.e., *there exists the limit*

$$\Delta = \lim \frac{n(r)}{r^{\rho(r)}},$$

and suppose that $\rho = \lim_{r\to\infty} \rho(r)$ is not an integer. Let us denote by $\Pi^\delta(z)$ another canonical product

$$\Pi^\delta(z) = \prod_1^\infty G\left(\frac{z}{a'_n}; p\right),$$

in which $|a'_n| = |a_n|$ and $|\arg a'_n - \arg a_n| < \delta$. Then for every $\varepsilon > 0$ and $\eta > 0$

there exists a $\delta > 0$ *such that*

$$|\ln|\Pi(z)| - \ln|\Pi^{\delta}(z)|| < \varepsilon r^{\rho(r)}$$

for all z that do not belong to some exceptional set of circles C with upper linear density less than η. *(Here* δ *depends only on the numbers* ε *and* η, *while the exceptional set of circles C depends on the sets* $\{a_n\}$ *and* $\{a_n'\}$.)

This lemma makes it possible to replace, with an arbitrary degree of asymptotic precision, a given canonical product by another one the roots of which lie on a finite number of rays. The proof of this lemma is somewhat cumbersome and we shall give it later in order not to interrupt the train of thought. Let us now continue the interrupted proof by making use of the lemma. Thus let us assume that $f(z)$ is an entire function of integral order ρ, and that the set of its zeros has an angular density $\Delta_f\theta$ with index $\rho(r)$.

Let the quantity $H(\theta)$ be determined by formula (2.04). We construct the sum

$$S_m(\theta) = \frac{\pi}{\sin \pi\rho} \sum_{j=0}^{m-1} \cos \rho(\theta - \psi_j - \pi)[\Delta(\psi_{j+1}) - \Delta(\psi_j)]$$

$$(\psi_m = \psi_0 + 2\pi).$$

It is obvious that for every given positive number ε there exists a positive number δ such that if $\max_j |\psi_{j+1} - \psi_j| < \delta$ $(j = 0, 1, 2, \ldots, m-1)$, then

$$|H(\theta) - S_m(\theta)| < \frac{\varepsilon}{3} \qquad (0 \leqslant \theta < 2\pi). \tag{2.13}$$

We assume, as has been mentioned above, that the function $f(z)$ is a canonical product

$$f(z) = \prod_1^{\infty} G\left(\frac{z}{a_n};\ p\right) \qquad (\rho - 1 < p < \rho).$$

Now we construct another function $f^{\delta}(z)$ the zeros of which lie on the rays $\arg z = \psi_j$ $(j = 0, 1, \ldots, m-1)$, and which asymptotically differs little from $f(z)$. For this purpose we select the points a_n' so that $|a_n'| = |a_n|$ and $\arg a_n' = \psi_j$, where $\psi_j \leqslant \arg a_n < \psi_{j+1}$. By Lemma 1, the function

$$f^{\delta}(z) = \prod_1^{\infty} G\left(\frac{z}{a_n'};\ p\right)$$

will satisfy, for any given $\varepsilon > 0$, $\eta > 0$, and for a sufficiently small positive δ, the inequality

$$|\ln|f(z)| - \ln|f^{\delta}(z)|| < \frac{\varepsilon}{3} r^{\rho(r)} \qquad (|z| = r) \tag{2.14}$$

for all values of z that do not belong to an exceptional set C of circles whose upper linear density satisfies the inequality $\bar{\rho}^*(C) < \eta/2$.

The zeros of the function $f^{\delta}(z)$ are regularly distributed, and they lie on a finite number of rays $\arg z = \psi_j$ $(j = 0, 1, \ldots, m-1)$. In such a case, as was indicated above, the quantity

$$r^{-\rho(r)} \ln |f^{\delta}(re^{i\theta})|$$

will converge uniformly to $S_n(\theta)$ for any $\sigma > 0$, and $|\theta - \psi_j| \geqslant \sigma$. In other words, there exists a number $r_{\varepsilon,\sigma}$ such that if $r > r_{\varepsilon,\sigma}$ and $|\theta - \psi_j| \geqslant \sigma$ $(j = 0, 1, \ldots, m-1)$, then

$$|r^{-\rho(r)} \ln |f^{\delta}(re^{i\theta})| - S_n(\theta)| < \frac{\varepsilon}{3}. \tag{2.15}$$

From (2.13), (2.14) and (2.15) it follows that if $|\theta - \psi_j| \geqslant \sigma > 0$ $(j = 0, 1, \ldots, m-1)$, $re^{i\theta} \bar{\in} C$ and $r > r_{\varepsilon,\sigma}$, then

$$|r^{-\rho(r)} \ln |f(re^{i\theta})| - H(\theta)| < \varepsilon. \tag{2.16}$$

The requirement $r > r_{\varepsilon,\sigma}$ may be dropped if one includes the circle $|z| \leqslant r_{\varepsilon,\sigma}$ in the set C.

We have proved that the inequality (2.16) holds outside certain angles, i.e., if $|\theta - \psi_j| \geqslant \sigma > 0$. In order to free ourselves from this auxiliary requirement, we shall make a second subdivision of the interval $(0, 2\pi + \delta)$ by means of points ψ_j' $(j = 0, 1, \ldots, m-1, m)$ such that $\psi_m' = \psi_0' + 2\pi$, and $|\psi_{j+1}' - \psi_j'| < \delta$, and by selecting $\sigma > 0$ so that the sets $|\theta - \psi_j| \geqslant \sigma$ and $|\theta - \psi_j'| \geqslant \sigma$ $(j = 0, 1, \ldots, m-1, m)$ will cover the interval $[0, 2\pi]$ completely. Making use of this second subdivision, we find that the inequality (2.16) is valid if $re^{i\theta}$ does not belong to a certain exceptional set C' with upper linear density $\bar{\rho}^*(C') < \eta/2$, and if $|\theta - \psi_j'| \geqslant \sigma$ $(j = 0, 1, \ldots, m)$. Because the sets $|\theta - \psi_j| \geqslant \sigma$ and $|\theta - \psi_j'| \geqslant \sigma$ cover the entire interval $[0, 2\pi]$, we see that the inequality (2.16) is satisfied when $re^{i\theta} \bar{\in} \hat{C} = C + C'$, where $\bar{\rho}^*(\hat{C}) < \eta$.

One can thus construct an exceptional set $\hat{C}$ of circles, with an arbitrarily small upper linear density, outside of which the inequality (2.16) will be satisfied for any preassigned positive ε. Making use of this fact, we shall construct a set of circles, of zero linear density, outside of which set the required asymptotic equation will hold. For this purpose, we select two arbitrary null sequences of positive numbers $\{\varepsilon_p\}$ and $\{\eta_p\}$ and construct a sequence of sets C_p $(p = 1, 2, \ldots)$ of circles such that $\bar{\rho}^*(C_p) < \eta_p$ and that, for $re^{i\theta} \bar{\in} C_p$,

$$|r^{-\rho(r)} \ln |f(re^{i\theta})| - H(\theta)| < \varepsilon_p. \tag{2.17}$$

We shall denote by $\Sigma^r r_j^p$ the sum of the radii of those circles of the set C_p whose centers lie in the circle $|z| < r$. Since $\bar{\rho}^*(C_p) < \eta_p$, it follows that for r sufficiently large we have the inequality

$$\sum^r r_i^p < 2\eta_p r. \tag{2.18}$$

Let us select a sequence of positive numbers R_p $(p = 1, 2, \ldots)$ such that $R_{p+1} > (p+1)R_p$ and that the inequality (2.18) is satisfied for $r > R_p$.

From each set C_p we pick out those circles whose centers lie in the ring

$$R_p \leqslant |z| < R_{p+1},$$

and we denote the set of these circles by C. This set is a C^0-set. Indeed, if we reorder the circles of C and denote by r_j the radius of the jth circle, we find that for

$$R_p \leqslant r < R_{p+1}$$

we have the inequality

$$\frac{1}{r}\sum^{r} r_j < \frac{2\eta_1}{2\cdot 3\ldots p} + \frac{2\eta_2}{3\cdot 4\ldots p} + \ldots + 2\eta_p < \frac{2\eta_1 e}{p} + 2\eta_p.$$

Therefore,

$$\rho^*(C) = 0.$$

Furthermore, if $re^{i\theta} \bar{\in} C$, and if $r > R_p$, the inequality (2.17) is satisfied. In other words, if $z \bar{\in} C$, then the quantity

$$h_{f,r}(\theta) = r^{-\rho(r)} \ln |f(re^{i\theta})|$$

converges uniformly to $H(\theta)$. This completes the proof of the theorem.[6]

We shall now prove Lemma 1.

1. Let the positive numbers σ, τ and β be given such that $\sigma < 1$, $\tau > 1$, and β does not exceed either $1 - \sigma$ or $\tau - 1$. We shall represent the infinite product

$$f(z) = \prod_1^{\infty} G\left(\frac{z}{a_n};\ p\right)$$

as a product of the following functions:

$$f_\sigma(z) = \prod_{|a_n| \leqslant \sigma r} G\left(\frac{z}{a_n};\ p\right), \quad {}_\tau f(z) = \prod_{|a_n| > \tau r} G\left(\frac{z}{a_n};\ p\right),$$

$$f_{\sigma,\beta}(z) = \prod_{\sigma r < |a_n| < (1-\beta) r} G\left(\frac{z}{a_n};\ p\right),$$

$${}_{\tau,\beta}f(z) = \prod_{(1+\beta) r < |a_n| < \tau r} G\left(\frac{z}{a_n};\ p\right),$$

$$\Phi_\beta(z) = \prod_{(1-\beta) r \leqslant |a_n| < (1+\beta) r} G\left(\frac{z}{a_n};\ p\right).$$

---

[6] It is useful to note still another formula for the function $H(\theta)$,

$$H(\theta) = \frac{\pi}{\sin \pi\rho} \int_0^{2\pi} \cos \rho\,(|\theta - \psi| - \pi)\, d\Delta(\psi),$$

which can be obtained directly from (2.04) if one notices that if $\theta - 2\pi < \psi < \theta$, then $\cos \rho(\theta - \psi - \pi) = \cos \rho(|\theta - \psi| - \pi)$ and $\cos \rho\{|\theta - (\psi + 2\pi)| - \Pi\} = \cos \rho(|\theta - \psi| - \pi)$.

In the sequel we shall consider, in addition to these functions, also those functions which we obtain if we replace in each product the points a_n by points a'_n such that $|a_n| = |a'_n|$ and $|\arg a_n - \arg a'_n| < \delta$ $(\delta > 0)$. For these new functions we shall retain the earlier notation except for the superscript δ (for example, $f^\delta(z), f^\delta_\sigma(z)$, and so on). We have the inequality

$$\begin{aligned}|\ln|f(z)| - \ln|f^\delta(z)|| &\leqslant |\ln|f_\sigma(z)| - \ln|f^\delta_\sigma(z)|| + |\ln|_\tau f(z)| - \ln|_\tau f^\delta(z)|| \\ &+ |\ln|f_{\sigma,\beta}(z)| - \ln|f^\delta_{\sigma,\beta}(z)|| + |\ln|_{\tau,\beta}f(z)| - \ln|_{\tau,\beta}f^\delta(z)|| \\ &+ |\ln|\Phi_\beta(z)| - \ln|\Phi^\delta_\beta(z)||. \qquad (2.19)\end{aligned}$$

Let us next estimate the value of each of the terms on the right hand side of this inequality.

2. From Lemmas 7 and 8 of Chapter I, it follows that for sufficiently small $\sigma > 0$ and for sufficiently large $\tau > 1$ we have the asymptotic relations

$$|\ln|f_\sigma(z)|| < \frac{\varepsilon}{10} r^{\rho(r)}, \quad |\ln|f^\delta_\sigma(z)|| < \frac{\varepsilon}{10} r^{\rho(r)},$$

$$|\ln|_\tau f(z)|| < \frac{\varepsilon}{10} r^{\rho(r)}, \quad |\ln|_\tau f^\delta(z)|| < \frac{\varepsilon}{10} r^{\rho(r)}$$

and hence,

$$|\ln|f_\sigma(z)| - \ln|f^\delta_\sigma(z)|| < \frac{\varepsilon}{5} r^{\rho(r)},$$

$$|\ln|_\tau f(z)| - \ln|_\tau f^\delta(z)|| < \frac{\varepsilon}{5} r^{\rho(r)}. \qquad (2.20)$$

3. To estimate the third and fourth term we shall make use of the uniform continuity of the function

$$\ln|G(se^{-i\theta}; p)| \qquad (s \geqslant 0)$$

in s and θ within the regions

$$(1-\beta)^{-1} \leqslant s \leqslant \sigma^{-1}, \qquad \tau^{-1} \leqslant s \leqslant (1+\beta)^{-1}.$$

From the existence of the density Δ of the set of zeros $\{a_n\}$ $(a_n \neq 0)$ it follows that for some $c > \Delta$

$$n(r) < cr^{\rho(r)}. \qquad (2.21)$$

If we select $\delta > 0$ so small that for $(1 - \beta)^{-1} \leqslant s \leqslant \sigma^{-1}$, or $\tau^{-1} \leqslant s \leqslant (1 + \beta)^{-1}$, and for $|\theta_1 - \theta_2| < \delta$ the inequality

$$|\ln|G(se^{i\theta_1}; p)| - \ln|G(se^{i\theta_2}; p)|| < \frac{\varepsilon}{10c\tau^\rho}$$

is satisfied, then we obtain

$$\begin{aligned}&\left|\ln|f_{\sigma,\beta}(z)|-\ln|f^{\delta}_{\sigma,\beta}(z)|\right|\\&\leqslant \sum_{\sigma r<|a_n|<(1-\beta)r}\left|\ln\left|G\left(\frac{z}{a_n};p\right)\right|-\ln\left|G\left(\frac{z}{a'_n};p\right)\right|\right|<\frac{\varepsilon}{10c\tau^{\rho}}n(r),\\&\left|\ln|_{\tau,\beta}f(z)|-\ln|_{\tau,\beta}f^{\delta}(z)|\right|\\&<\sum_{(1+\beta)r<|a_n|<\tau r}\left|\ln\left|G\left(\frac{z}{a_n};p\right)\right|-\ln\left|G\left(\frac{z}{a'_n};p\right)\right|\right|<\frac{\varepsilon}{10c\tau^{\rho}}n(\tau r).\end{aligned}$$

From (2.21) it can easily be seen that for sufficiently large r

$$n(\tau r)<2c\tau^{\rho}r^{\rho(r)}$$

and the asymptotic inequalities hold:

$$\left.\begin{aligned}&\left|\ln|f_{\sigma,\beta}(z)|-\ln|f^{\delta}_{\sigma,\beta}(z)|\right|<\frac{\varepsilon}{5}r^{\rho(r)},\\&\left|\ln|_{\tau,\beta}f(z)|-\ln|_{\tau,\beta}f^{\delta}(z)|\right|<\frac{\varepsilon}{5}r^{\rho(r)}.\end{aligned}\right\}\tag{2.22}$$

It remains to show that for a small enough positive β and for z not belonging to a certain exceptional set, the last term in (2.19) is asymptotically less than $\varepsilon r^{\rho(r)}/5$. This is the most difficult step of the proof.

4. In place of the function $\Phi_{\beta}(z)$, we first consider the function

$$\psi_{R,\beta}(z)=\prod_{(1-\beta)R\leqslant|a_n|<(1+\beta)R}\left(1-\frac{z}{a_n}\right)\quad\left(0<\beta<\frac{1}{2},R>0\right)\tag{2.23}$$

and estimate $\left|\ln|\psi_{R,\beta}(z)|\right|$ in the ring $\left||z|-R\right|<\beta R$. Here we do not make immediate use of the existence of the density of the point set $\{a_n\}$, but only assume that the set has a limit point at infinity. Thus we obtain a general estimate (see (2.27)), and then we utilize the density and obtain the required inequality. First of all let us find an upper bound for the function $|\psi_{R,\beta}(z)|$. From the inequality

$$\left|1-\frac{z}{a_n}\right|\leqslant\left(1+\frac{r}{|a_n|}\right)\leqslant\frac{2}{1-\beta}\qquad(r=|z|),$$

which is true when $\left||a_n|-R\right|\leqslant\beta R$ and $|r-R|\leqslant\beta R$, it follows that if $\left||z|-R\right|<\beta R$, then

$$\ln|\psi_{R,\beta}(z)|<N(R,\beta)\ln\frac{2}{1-\beta}.\tag{2.24}$$

Here $N(R,\beta)$ denotes the number of points of the set $\{a_n\}$ that lie in the ring $(1-\beta)R\leqslant|z|<(1+\beta)R$.

In order to obtain a lower bound of $\psi_{R,\beta}(z)$, we start with the Cartan estimate (Chapter I, § 7) according to which for every $H > 0$ the polynomial

$$P(z) = \prod_{(1-\beta)R \leqslant |a_n| < (1+\beta)R} (z - a_n)$$

satisfies the inequality

$$\ln |P(z)| > N(R, \beta) \ln \left(\frac{H}{e}\right)$$

everywhere outside circles the sum of whose radii is equal to $2H$.

Noting that in the product (2.23) $|a_n| < (1 + \beta)R$ and setting $H = \beta^2 R$, we deduce from this inequality that outside the exceptional circles with the sum of radii equal to $2\beta^2 R$ the following inequality holds:

$$\ln |\psi_{R,\beta}(z)| > N(R, \beta) \ln \frac{\beta^2}{e(1+\beta)}.$$

Let us now construct the exceptional set C in the entire plane. For this purpose we divide the positive half-line into semi-intervals I_j,

$$R_j(1-\beta) \leqslant r < R_j(1+\beta) \qquad (j = 0, 1, 2, \ldots),$$

where

$$R_j = \left(\frac{1+\beta}{1-\beta}\right)^j,$$

while the part of the plane outside the circle $|z| \leqslant 1 - \beta$ is divided into the rings K_j,

$$R_j(1-\beta) \leqslant |z| \leqslant R_j(1+\beta) \qquad (j = 0, 1, 2, \ldots).$$

We shall denote by n_j the number of the points a_n lying in the ring K_j.

With each ring K_j one can associate the exceptional circles, with sum of radii equal to $2\beta^2 R_j$, outside of which the following inequality holds:

$$\ln |\psi_{R_j,\beta}(z)| > n_j \ln \frac{\beta^2}{e(1+\beta)} > n_j [2 \ln \beta - \ln 2 - 1]. \qquad (2.25)$$

Since each exceptional circle contains a point a_n from the ring K_j, the centers of these circles lie either in K_j or in one of the adjacent rings K_{j-1} and K_{j+1}. Let us denote by C the set of all exceptional circles that correspond to the various rings K_j ($j = 1, 2, \ldots$) and by r_k ($k = 1, 2, \ldots$) the radii of these circles. We now shall give an estimate of the upper linear density of this set.

Let R be an arbitrary number greater than $1 - \beta$, and let m be such $R_m < R < R_{m+1}$. Then

$$\sum^R r_k \leqslant \sum_{j=0}^{m+2} 2\beta^2 R_j < \beta(1+\beta) R_{m+2} \leqslant \beta(1+\beta)^3(1-\beta)^{-2} R.$$

On the left side of this inequality stands the sum of the radii of those circles of C whose centers lie in the circle $|z| \leqslant R$. Therefore,

$$\bar{\rho}^*(C) \leqslant \beta(1+\beta)^3(1-\beta)^{-2}.$$

The inequality (2.25) yields a lower bound of the function $|\psi_{R_j,\beta}(z)|$ inside the circle K_j but outside the set C which we have constructed. Let us now proceed to estimate the function $|\psi_{R,\beta}(z)|$ outside of this set for an arbitrary R. For this purpose we express the function $\psi_{R,\beta}(z)$, for $R_m \leqslant R \leqslant R_{m+1}$, in the form

$$\psi_{R,\beta}(z) = \psi_{R_m,\beta}(z)\,\psi_{R_{m+1},\beta}(z)\left(\psi^{(1)}_{R,\beta}(z)\,\psi^{(2)}_{R,\beta}(z)\right)^{-1},$$

where

$$\psi^{(1)}_{R,\beta}(z) = \prod_{(1+\beta)R \leqslant |a_n| < (1+\beta)R_{m+1}} \left(1 - \frac{z}{a_n}\right),$$

$$\psi^{(2)}_{R,\beta}(z) = \prod_{(1-\beta)R_m \leqslant |a_n| < (1-\beta)R} \left(1 - \frac{z}{a_n}\right).$$

The functions $\psi_{R_m,\beta}(z)$ and $\psi_{R_{m+1},\beta}(z)$ are estimated by means of the inequality (2.25). The functions $\psi^{(1)}_{R,\beta}(z)$ and $\psi^{(2)}_{R,\beta}(z)$ have to be estimated from above, i.e., we have to find upper bounds for them. In order to do this we note that in each one of these functions

$$|a_n| > R_m(1-\beta) \geqslant R(1-\beta)^2(1+\beta)^{-1},$$

and hence for $|z| < (1+\beta)R$

$$\left|1 - \frac{z}{a_n}\right| \leqslant 1 + \frac{|z|}{|a_n|} < 1 + \frac{(1+\beta)^2}{(1-\beta)^2} < 10.$$

Therefore, if $|z| < (1+\beta)R$

$$\ln|\psi^{(1)}_{R,\beta}(z)| < n_{m+1}\ln 10, \qquad \ln|\psi^{(2)}_{R,\beta}(z)| < n_m \ln 10.$$

Making use of the bounds of all four of the functions, we obtain, when $||z| - R| < \beta R$ and z lies outside the circles of the set C, the inequality

$$\ln|\psi_{R,\beta}(z)| > (n_m + n_{m+1})[2\ln\beta - 1 - \ln 10].$$

Combining this result with (2.24) we see that

$$|\ln|\psi_{R,\beta}(z)|| < (n_m + n_{m+1})\left[2\ln\frac{1}{\beta} + 4\right]. \tag{2.26}$$

Let us now find bounds for the function

$$\Phi_{R,\beta}(z) = \prod_{(1-\beta)R \leqslant |a_n| < (1+\beta)R} G\left(\frac{z}{a_n};\ p\right).$$

It satisfies the equation

$$\ln|\Phi_{R,\beta}(z)| = \ln|\psi_{R,\beta}(z)| + \operatorname{Re}\left\{\sum_{(1-\beta)R \leqslant |a_n| < (1+\beta)R} \left(\frac{z}{a_n} + \frac{z^2}{2a_n^2} + \dots + \frac{z^p}{pa_n^p}\right)\right\}.$$

A bound for the first summand is given by the inequality (2.26), while for an estimate of the second term one needs only to note that if $(1-\beta)R \leqslant a_n < (1+\beta)R$ and $\big||z|-R\big| < \beta R$, then

$$\left|\frac{z}{a_n}\right| < \frac{1+\beta}{1-\beta}$$

and hence

$$\sum_{(1-\beta)R \leqslant |a_n| < (1+\beta)R} \left|\left(\frac{z}{a_n}+\frac{z^2}{2a_n^2}+\cdots+\frac{z^p}{pa_n^p}\right)\right|$$

$$< p\left(\frac{1+\beta}{1-\beta}\right)^p N(R,\beta) \leqslant 3^p p(n_m+n_{m+1}).$$

Combining these bounds we find that if $z \overline{\in} C$ and $\big||z|-R\big| < \beta R$, then

$$\big|\ln|\Phi_{R,\beta}(z)|\big| < (n_m+n_{m+1})\left[2\ln\frac{1}{\beta}+B_p\right], \tag{2.27}$$

where

$$B_p = 4+3^p p.$$

5. We now make use of the fact that in our case the set has a density, i.e., the limit

$$\Delta = \lim_{r\to\infty}\frac{n(r)}{r^{\rho(r)}}$$

exists. If $R_m/(1-\beta) \leqslant R \leqslant R_{m+1}/(1-\beta)$ and $|r-R| \leqslant \beta R$, we have

$$\overline{\lim_{R\to\infty}}\frac{n_m}{r^{\rho(r)}} \leqslant \lim_{m\to\infty}\frac{n_m}{R_m^{\rho(R_m)}} = [(1+\beta)^\rho-(1-\beta)^\rho]\Delta$$

and

$$\overline{\lim_{R\to\infty}}\frac{n_{m+1}}{r^{\rho(r)}} \leqslant \lim_{m\to\infty}\frac{n_{m+1}}{R_m^{\rho(R_m)}} = \left(\frac{1+\beta}{1-\beta}\right)^\rho[(1+\beta)^\rho-(1-\beta)^\rho]\Delta.$$

From these inequalities and (2.27) we obtain the asymptotic inequality

$$\big|\ln|\Phi_{R,\beta}(z)|\big| < 2\left(2\ln\frac{1}{\beta}+B_p\right)\left(\frac{1+\beta}{1-\beta}\right)^\rho[(1+\beta)^\rho-(1-\beta)^\rho]\Delta r^{\rho(r)},$$

in which $\big||z|-R\big| < \beta R$ and $z \overline{\in} C$.

It is now obvious that for any given pair of positive numbers ξ and ζ one can find a positive number β and a set C of exceptional circles such that $\bar{\rho}^*(C) < \xi$ and that outside these circles

$$\big|\ln|\Phi_{R,\beta}(z)|\big| < \zeta r^{\rho(r)} \qquad (|z|=r).$$

6. Let us select $\xi = \eta/2$ and $\zeta = \varepsilon/10$, and construct the sets of circles C_1 and C_1^δ corresponding to the sequences $\{a_n\}$ and $\{a_n'\}$ of Lemma 1. Then for

$R = r = |z|$, for sufficiently small β, and for any $z \overline{\in} C_1 + C_1^\delta$, we have the asymptotic relations

$$\left|\ln|\Phi_{r,\beta}(z)|\right| < \frac{\varepsilon}{10} r^{\rho(r)}, \qquad \left|\ln|\Phi^{\delta}_{r,\beta}(z)|\right| < \frac{\varepsilon}{10} r^{\rho(r)}$$

and hence

$$\left|\ln|\Phi_{r,\beta}(z)| - \ln|\Phi^{\delta}_{r,\beta}(z)|\right| < \frac{\varepsilon}{5} r^{\rho(r)}.$$

Furthermore,

$$\overline{\rho}^*(C) < \eta.$$

Thus we have obtained the required bound for the last summand in the inequality (2.19), and we have completed the proof of Lemma 1. This establishes Theorem 1.

In the proof of Lemma 1, subsections 4 and 5, we established two propositions which we shall use in the sequel. For this reason we restate them here as separate lemmas.

LEMMA 2. *Suppose that $\{a_n\}$ is a sequence of complex numbers with a single limit point at infinity, and let*

$$\Phi_{R,\beta}(z) = \prod_{(1-\beta)R \leqslant |a_n| < (1+\beta)R} G\left(\frac{z}{a_n};\ p\right), \tag{2.28}$$

where p is a natural number, $0 < \beta < \frac{1}{2}$, and $R > 1 - \beta$. Let

$$R_j = \left(\frac{1+\beta}{1-\beta}\right)^j$$

and n_j be the number of points of $\{a_n\}$ which lie in the ring

$$(1-\beta)R_j \leqslant |z| < (1+\beta)R_j.$$

Under the above conditions one can select a set C of circles, with an upper linear density

$$\rho^*(C) \leqslant \beta(1+\beta)^3(1-\beta)^{-2},$$

such that if $z \overline{\in} C$, $\big||z| - R\big| < \beta R$ and $R_m \leqslant R \leqslant R_{m+1}$, then

$$\left|\ln|\Phi_{R,\beta}(z)|\right| < (n_m + n_{m+1})\left(2\ln\frac{1}{\beta} + B_p\right),$$

where $B_p = 4 + 3^\nu p$.

LEMMA 3. *If the sequence $\{a_n\}$ of points of the complex plane has density Δ with index $\rho(r)$, then for every pair of given positive numbers ξ and ζ one can select a set C of circles, with upper linear density less than ξ, and a number β such that if $R \to \infty$, $\big||z| - R\big| < \beta R$, and $z \overline{\in} C$, then the function $\Phi_{R,\beta}(z)$ given by* (2.28) *will satisfy the asymptotic inequality*

$$\left|\ln|\Phi_{R,\beta}(z)|\right| < \zeta r^{\rho(r)} \qquad (|z| = r).$$

We call attention to the fact that in Lemma 2 we did not impose any restrictions on the sequence $\{a_n\}$, while in Lemma 3 we required only the existence of the density. It was unessential whether $\rho = \lim_{r\to\infty} \rho(r)$ was an integer or not.

3. Entire Functions of Integral Order with Regular Distribution of Zeros (Proof of Theorem 2)

For functions of integral order the indicator is not completely determined by the set $\{a_n\}$ of the zeros of the function, but may depend also on the exponential factor which stands in front of the canonical product. Furthermore, in this case, for a given proximate order $\rho(r)$ the quantity Δ_f, as was shown in § 13 of Chapter I, does not determine the type of the canonical product. For us it is important that in the case of an integral order the considerations of the preceding section do not carry over even for the simplest case when all the zeros of the function $f(z)$ lie on a single ray. The fact was already mentioned in the proof of Theorem 25, Chapter I, § 17.

In order to obtain an asymptotic equation in this case, we express the function $f(z)$ (as was done in § 11, Chapter I) in the form

$$f(z) = z^q e^{P_{\rho-1}(z)+\delta_f(r)z^\rho} f_r(z), \tag{2.29}$$

where

$$f_r(z) = \prod_{|a_n|\leqslant r} G\left(\frac{z}{a_n};\ \rho-1\right) \prod_{|a_n|>r} G\left(\frac{z}{a_n};\ \rho\right).$$

Here, $P_{\rho-1}(z) = c_0 + c_1 z + \ldots + c_{\rho-1} z^{-1}$ and

$$\delta_f(r) = c_\rho + \frac{1}{\rho} \sum_{|a_n|<r} a_n^{-\rho}.$$

If the zeros a_n and the quantity c_ρ satisfy the condition (B), i.e., if we have the existence of the limit

$$\lim_{r\to\infty} \frac{1}{L(r)} \delta_f(r) = \tau_f e^{-i\rho\theta_f},$$

then the quantity

$$\gamma(r,\ \theta) = r^{-\rho(r)} [q \ln z + P_{\rho-1}(z) + \delta_f(r) z^\rho]$$

will converge uniformly in θ $(0 \leqslant \theta \leqslant 2\pi)$ to the limit

$$\lim_{r\to\infty} \gamma(r,\ \theta) = \tau_f e^{i\rho(\theta-\theta_f)}. \tag{2.30}$$

This circumstance explains the meaning of the condition (B). The entire problem is now reduced to the study of the second factor $f_r(z)$, for which we shall need only the existence of the angular density of the set of zeros. As before, let us consider first the simple case when all the sets $\{a_n\}$ that possess angular densities are located on a finite number of rays $\arg z = \psi_j\,(j = 0, 1, \ldots, m-1)$.

Now applying Lemma 9, § 17 of Chapter I, instead of Theorem 25, we find that if $|\theta - \psi_j| \geqslant \sigma > 0$,

$$\lim_{r\to\infty} \frac{\ln|f_r(re^{i\theta})|}{r^{\rho(r)}} = -\sum_{j=0}^{m-1} \Delta_j \left[(\theta - \psi_j - \pi)\sin\rho(\theta-\psi_j) + \frac{1}{\rho}\cos\rho(\theta-\psi_j)\right],$$

where the convergence is uniform and Δ_j stands for the density of the set of zeros on the ray $\arg z = \psi_j$. One might expect that also in the general case of the existence of an angular density of the set of zeros outside some exceptional set one would have the equation

$$\lim_{r\to\infty} \frac{\ln|f_r(re^{i\theta})|}{r^{\rho(r)}} = \int_{\theta-2\pi}^{\theta} \left[(\theta-\psi-\pi)\sin\rho(\theta-\psi) - \frac{1}{\rho}\cos\rho(\theta-\psi)\right] d\Delta(\psi).$$

In order to prove that this is actually the case we need a lemma which is analogous to Lemma 1.

LEMMA 4. *If ρ is an integer, if the set $\{a_n\}$ of zeros of the function*

$$f_r(z) = \prod_{|a_n|\leqslant r} G\left(\frac{z}{a_n}; \rho-1\right) \prod_{|a_n|>r} G\left(\frac{z}{a_n}; \rho\right)$$

has a density with index $\rho(r) \to \rho$, and if

$$f_r^\delta(z) = \prod_{|a'_n|\leqslant r} G\left(\frac{z}{a'_n}; \rho-1\right) \prod_{|a'_n|>r} G\left(\frac{z}{a'_n}; \rho\right),$$

where $|a'_n| = |a_n|$ and $|\arg a'_n - \arg a_n| < \delta$, then for every $\eta > 0$ and $\varepsilon > 0$ there exists a $\delta > 0$ such that we have the asymptotic relation

$$\big|\ln|f_r^\delta(z)| - \ln|f_r(z)|\big| < \varepsilon r^{\rho(r)}$$

for all z not belonging to a certain exceptional set C of circles with upper linear density less than η. (Here δ depends only on ε and η, while the exceptional set C depends only on the sets $\{a_n\}$ and $\{a'_n\}$.)[6a]

The proof of this lemma is analogous to the proof of Lemma 1, but it can be considerably shortened because we use the established Lemma 3.

We give the proof here.

[6a] Lemmas 1 and 3 of this chapter have been developed further by Krasičkov [1; 4]. See also Gol'dberg [5].

Setting

$$f_{r,\sigma}(z)=\prod_{|a_n|\leqslant\sigma r} G\left(\frac{z}{a_n};\ \rho-1\right) \qquad (0<\sigma<1),$$

$$_\tau f_r(z)=\prod_{|a_n|>\tau r} G\left(\frac{z}{a_n};\ \rho\right) \qquad (1<\tau<\infty),$$

$$f_{r,\sigma,\beta}(z)=\prod_{\sigma r<|a_n|\leqslant(1-\beta)r} G\left(\frac{z}{a_n};\ \rho-1\right) \qquad \left(0<\beta<\frac{1}{2}\right),$$

$$_{\tau,\beta}f_r(z)=\prod_{(1+\beta)r\leqslant|a_n|<\tau r} G\left(\frac{z}{a_n};\ \rho\right) \qquad (\tau\geqslant 1+\beta),$$

$$\Phi_{r,\beta}(z)=\prod_{(1-\beta)r<|a_n|<(1+\beta)r} G\left(\frac{z}{a_n};\ \rho-1\right),$$

$$\psi_{r,\beta}(z)=\exp\frac{z^\rho}{\rho}\sum_{r\leqslant|a_n|<(1+\beta)r} a_n^{-\rho},$$

we obtain

$$\begin{aligned}
&\left|\ln|f_r(z)|-\ln|f_r^\delta(z)|\right| \\
&\leqslant\left|\ln|f_{r,\sigma}(z)|-\ln|f_{r,\sigma}^\delta(z)|\right|+\left|\ln|_\tau f_r(z)|-\ln|_\tau f_r^\delta(z)|\right| \\
&\qquad+\left|\ln|f_{r,\sigma,\beta}(z)|-\ln|f_{r,\sigma,\beta}^\delta(z)|\right| \\
&\qquad+\left|\ln|_{\tau,\beta}f(z)|-\ln|_{\tau,\beta}f^\delta(z)|\right| \\
&\qquad+\left|\ln|\Phi_{r,\beta}(z)|-\ln|\Phi_{r,\beta}^\delta(z)|\right| \\
&\qquad\qquad+\left|\ln|\psi_{r,\beta}(z)|-\ln|\psi_{r,\beta}^\delta(z)|\right|.
\end{aligned} \tag{2.31}$$

From Lemmas 7 and 8 of Chapter I it follows that for any given $\varepsilon>0$, for a sufficiently small $\sigma>0$, and for a sufficiently large $\tau\geqslant 1$, we have the asymptotic relations

$$\left|\ln|f_{r,\sigma}(z)|\right|<\frac{\varepsilon}{12}r^{\rho(r)}, \qquad \left|\ln|f_{r,\sigma}^\delta(z)|\right|<\frac{\varepsilon}{12}r^{\rho(r)},$$

$$\left|\ln|_\tau f_r(z)|\right|<\frac{\varepsilon}{12}r^{\rho(r)}, \qquad \left|\ln|_\tau f_r^\delta(z)|\right|<\frac{\varepsilon}{12}r^{\rho(r)}.$$

Therefore

$$\left.\begin{aligned}
\left|\ln|f_{r,\sigma}(z)|-\ln|f_{r,\sigma}^\delta(z)|\right|&<\frac{\varepsilon}{6}r^{\rho(r)}, \\
\left|\ln|_\tau f_r(z)|-\ln|_\tau f_r^\delta(z)|\right|&<\frac{\varepsilon}{6}r^{\rho(r)}.
\end{aligned}\right\} \tag{2.32}$$

From the existence of the density Δ of the set of roots $\{a_n\}$ of the function $f(z)$ it follows that for some $c > \Delta$

$$n(r) < cr^{\rho(r)} \qquad (r > 0).$$

Making use of the continuity of the functions

$$\ln|G(se^{-i\theta};\ \rho-1)| \quad \text{if} \quad (1-\beta)^{-1} \leqslant s \leqslant \sigma^{-1}$$

and

$$\ln|G(se^{-i\theta};\ \rho)| \quad \text{if} \quad \tau^{-1} \leqslant s \leqslant (1+\beta)^{-1},$$

we can (just as in subsection 3 of § 2 of this chapter) select a positive δ so small that

$$\left.\begin{aligned} &\left|\ln|f_{r,\sigma,\beta}(z)| - \ln|f^{\delta}_{r,\sigma,\beta}(z)|\right| < \frac{\varepsilon}{6} r^{\rho(r)}, \\ &\left|\ln|{}_{\tau}f_{r,\beta}(z)| - \ln|{}_{\tau}f^{\delta}_{r,\beta}(z)|\right| < \frac{\varepsilon}{6} r^{\rho(r)}. \end{aligned}\right\} \tag{2.33}$$

By Lemma 3 one can choose a number β $(0 < \beta < \frac{1}{2})$ and a set C of circles, with upper linear density less than the given number $\eta > 0$, such that if $z \overline{\in} C$ then

$$\left|\ln|\Phi_{r,\beta}(z)|\right| < \frac{\varepsilon}{12} r^{\rho(r)}, \qquad \left|\ln|\Phi^{\delta}_{r,\beta}(z)|\right| < \frac{\varepsilon}{12} r^{\rho(r)},$$

i.e.,

$$\left|\ln|\Phi_{r,\beta}(z)| - \ln|\Phi^{\delta}_{r,\beta}(z)|\right| < \frac{\varepsilon}{6} r^{\rho(r)}. \tag{2.34}$$

Furthermore,

$$\left|\ln|\psi_{r,\beta}(z)|\right| < \frac{1}{\rho}\{n[r(1+\beta)] - n(r)\}$$

and hence we have the asymptotic relations

$$\begin{aligned} &\left|\ln|\psi_{r,\beta}(z)|\right| < \frac{\Delta+\varepsilon}{\rho}[(1+\beta)^{\rho}-1]\, r^{\rho(r)}, \\ &\left|\ln|\psi^{\delta}_{r,\beta}(z)|\right| < \frac{\Delta+\varepsilon}{\rho}[(1+\beta)^{\rho}-1]\, r^{\rho(r)}, \end{aligned} \qquad (\varepsilon > 0).$$

For $\beta > 0$ sufficiently small,

$$\left|\ln|\psi_{r,\beta}(z)| - \ln|\psi^{\delta}_{r,\beta}(z)|\right| < \frac{\varepsilon}{6} r^{\rho(r)}. \tag{2.35}$$

From the inequalities (2.31), (2.32), (2.33), (2.34) and (2.35) it follows that for a sufficiently small positive δ and for z outside of the set C we have the inequality

$$\left|\ln|f_r(z)| - \ln|f^{\delta}_r(z)|\right| < \varepsilon r^{\rho(r)}.$$

This proves the lemma. Let us now consider the asymptotic representation of the function $\ln|f_r(z)|$. Such a representation is given by the next lemma.

LEMMA 5. *If the point set* $\{a_n\}$ *has angular density* $\Delta(\psi)$ *with index* $\rho(r)$, *where* ρ *is an integer, and if*

$$f_r(z) = \prod_{|a_n|<r} G\left(\frac{z}{a_n};\ \rho-1\right) \prod_{|a_n|>r} G\left(\frac{z}{a_n};\ \rho\right),$$

then for any z exterior to a certain C^0*-set the following asymptotic equation is valid*

$$\ln|f_r(re^{i\theta})| \approx H_1(\theta)\, r^{\rho(r)},$$

in which

$$H_1(\theta) = -\int_{\theta-2\pi}^{\theta}\left[(\theta-\psi-\pi)\sin\rho(\theta-\psi) + \frac{1}{\rho}\cos\rho(\theta-\psi)\right] d\Delta(\psi).$$

PROOF. For the integral $H_1(\theta)$ we construct the sum

$$S_n(\theta) = -\sum_{j=0}^{n}\Big[(\theta-\psi_j-\pi)\sin\rho(\theta-\psi_j) + \frac{1}{\rho}\cos\rho(\theta-\psi_j)\Big][\Delta(\psi_{j+1})-\Delta(\psi_j)],$$

$$\psi_0<\psi_1<\dots<\psi_n<\psi_{n+1},$$
$$\psi_{j+1}-\psi_j<\delta \quad (j=0,\ 1,\ 2,\ \dots,\ n), \quad \psi_{n+1}=\psi_0+2\pi.$$

Here we choose the positive number δ so small that

$$|H_1(\theta)-S_n(\theta)| < \frac{\varepsilon}{3}. \tag{2.36}$$

Next we select the numbers a_k' so that $|a_k'| = |a_k|$, $\arg a_k' = \psi_j$ if $\psi_j \leqslant \arg a_k < \psi_{j+1}$, and we construct the function

$$f_r^\delta(z) = \prod_{|a_k'|<r} G\left(\frac{z}{a_k'};\ \rho-1\right) \prod_{|a_k'|>r} G\left(\frac{z}{a_k'};\ \rho\right).$$

By Lemma 4, for every given pair of numbers $\varepsilon>0$, $\eta>0$, and for a sufficiently small positive δ, the inequality

$$\big|\ln|f_r(z)| - \ln|f_r^\delta(z)|\big| < \frac{\varepsilon}{3} r^{\rho(r)} \tag{2.37}$$

will be satisfied by all values of z which do not belong to an exceptional set C of circles with upper linear density less than $\eta/2$. For the function $f_r^\delta(z)$ it was established that if $\sigma>0$ and $|\arg z-\psi_j| \geqslant \sigma$ $(j=0, 1, \dots, n)$, one can find a number $r_{\varepsilon,\sigma}$ such that for $r \geqslant r_{\varepsilon,\sigma}$

$$|r^{-\rho(r)}\ln|f_r^\delta(re^{i\theta})| - S_n(\theta)| < \frac{\varepsilon}{3}. \tag{2.38}$$

From (2.36), (2.37) and (2.38) it follows that if $z \,\bar{\in}\, C$, and if $|\arg z-\psi_j| \geqslant \sigma$ $(j=0, 1, 2, \dots, n)$, then

$$|r^{-\rho(r)}\ln|f_r(re^{i\theta})| - H_1(\theta)| < \varepsilon. \tag{2.39}$$

After making a second subdivision of the interval $(0, 2\pi + \delta)$ by means of the points ψ_j' $(j = 0, 1, 2, \ldots, n)$ in such a manner that $\psi_n' = \psi_0' + 2\pi$, $|\psi_{j+1}' - \psi_j'| < \delta$, and after choosing a positive σ sufficiently small, we get the result that the sets $|\theta - \psi_j| \geqslant \sigma$ and $|\theta - \psi_j'| > \sigma$ $(j = 0, 1, \ldots, n)$ cover the entire interval $[0, 2\pi)$. Making use of this second subdivision, we find that the inequality (2.39) is satisfied if $z = re^{i\theta}$ does not lie in a certain set C' of circles whose upper density is less than $\eta/2$ and if $|\theta - \psi_j'| \geqslant \sigma$ $(j = 0, 1, \ldots, n)$. In this manner the inequality (2.39) will hold everywhere except for a set of circles $\hat{C} = C + C'$, where

$$\bar{\rho}^*(\widehat{C}) < \eta.$$

In exactly the same way as in § 2 for the proof of Theorem 1 (pages 100–101), one next constructs the C^0-set outside of which the relation

$$\lim_{r\to\infty} \frac{\ln|f_r(re^{i\theta})|}{r^{\rho(r)}} = H_1(\theta) \tag{2.40}$$

holds uniformly relative to θ.

We note once more that in this derivation we have made use only of the existence of the angular density of the set of roots. Lemma 5 has thus been proved. Let us now proceed to the proof of Theorem 2.

By hypothesis, the point set $\{a_n\}$ satisfies the condition (B) (see page 91). If one makes use of this condition, the representation (2.29) and the equations (2.30) and (2.40) yield at once for $z \bar{\in} C^0$ the uniform convergence of the limit

$$\lim_{r\to\infty} \frac{\ln|f(re^{i\theta})|}{r^{\rho(r)}} = H(\theta), \tag{2.05}$$

where

$$H(\theta) = \tau_f \cos\rho(\theta - \theta_f)$$
$$- \int_{\theta-2\pi}^{\theta} \left[(\theta - \psi - \pi)\sin\rho(\theta - \psi) + \frac{1}{\rho}\cos\rho(\theta - \psi)\right] d\Delta(\psi).$$

Let us transform this expression to the form (2.06). For this purpose we note that for an integer

$$\int_{\theta-2\pi}^{\theta} \left[\frac{1}{\rho}\cos\rho(\theta - \psi) + \pi\sin\rho(\theta - \psi)\right] d\Delta(\psi)$$
$$= \int_{0}^{2\pi} \left[\frac{1}{\rho}\cos\rho(\theta - \psi) + \pi\sin\rho(\theta - \psi)\right] d\Delta(\psi)$$
$$= A\cos\rho\theta + B\sin\rho\theta,$$

where A and B are constants, and we find that

$$H(\theta) = A_1 \cos \rho\theta + B_1 \sin \rho\theta + \int_{\theta-2\pi}^{\theta} (\theta - \psi) \sin \rho (\theta - \psi)\, d\Delta(\psi).$$

As was shown at the end of § 16, Chapter I, this inequality implies

$$\Delta(\psi) = \frac{1}{2\pi\rho}\left[H'(\psi) + \rho^2 \int_{\psi_0}^{\psi} H(\varphi)\, d\varphi\right],$$

and in view of the periodicity of the function $H(\theta)$[7]

$$\int_0^{2\pi} e^{i\rho\psi}\, d\Delta(\psi) = 0. \tag{2.41}$$

Taking this into consideration, we find that $A = B = 0$ and that

$$H(\theta) = \tau_f \cos \rho (\theta - \theta_f) - \int_{\theta-2\pi}^{\theta} (\theta - \psi) \sin \rho (\theta - \psi)\, d\Delta(\psi).$$

This completes the proof of Theorem 2. In § 1 we stated that if the zeros of the function $f(z)$ are regularly distributed its indicator coincides with the function $H(\theta)$ defined by the formulas (2.04) and (2.06). Let us examine this in greater detail. By Theorems 1 and 2, in the exterior of the exceptional C^0-set we have uniform convergence of the quantity $h_{f,r}(\theta)$ to the limit $H(\theta)$. We shall show that in the interior of the exceptional circles we have the asymptotic relation

$$h_{f,r}(\theta) < H(\theta) + \varepsilon \qquad (\varepsilon > 0).$$

Indeed, the ratio of the diameter d_Ω of an arbitrary region Ω, consisting of exceptional circles of the C^0-set, to the distance of this region from the origin tends to zero when this distance is increased. From the maximum principle it follows that the value of the function $\ln|f(z)|$ at any point of such an exceptional region does not exceed its maximum on the boundary of this region. But on the boundary of the exceptional region the asymptotic formula for $h_{f,r}(\theta)$ is valid, and, taking into consideration the continuity of the function $H(\theta)$, we find that for $re^{i\theta} \in \Omega$ the following asymptotic inequality applies:

$$\ln|f(re^{i\theta})| < \left(H(\theta) + \frac{\varepsilon}{2}\right)(r + d_\Omega)^{\rho(r+d_\Omega)} < (H(\theta) + \varepsilon)\, r^{\rho(r)}.$$

On each ray $\arg z = \theta$ there are points arbitrarily far away from the origin which are not covered by the circles of the exceptional C^0-set, and hence for every θ

$$\overline{\lim_{r\to\infty}} \frac{\ln|f(re^{i\theta})|}{r^{\rho(r)}} = H(\theta).$$

[7] The equation (2.41) can be obtained directly from the condition (B).

It is interesting to note that the regions Ω consisting of exceptional circles and not containing zeros of the function $f(z)$ can be omitted from the exceptional set.

Indeed, applying the minimum principle to the function $\ln|f(z)|$ in the interior of such a region, we obtain the asymptotic equation $\ln|f(re^{i\theta}) \approx H(\theta)r^{\rho(r)}|$ inside such a region. We shall therefore assume in what follows that each exceptional region contains zeros of the function $f(z)$. This remark permits us to state an important corollary to Theorems 1 and 2.

COROLLARY. *If the zeros of the entire function $f(z)$ of proximate order $\rho(r)$ are regularly distributed for the index $\rho(r)$, and if the density of the set of zeros within some angle (α, β) is equal to zero, then the indicator function within this angle is a trigonometric function, i.e.,*

$$h_f(\theta) = A\cos\rho\theta + B\sin\rho\theta \tag{2.42}$$

(here $\alpha \leqslant \theta \leqslant \beta$; A and B are constants).

If, however, inside this angle there are no zeros of the function, then for $\alpha < \theta < \beta$ there exists the limit

$$h_f(\theta) = \lim_{r\to\infty} \frac{\ln|f(re^{i\theta})|}{r^{\rho(r)}},$$

where the variable tends to the limit uniformly when $\alpha + \eta \leqslant \theta \leqslant \beta - \eta$ $(\eta > 0)$.

The second assertion follows directly from the fact that under the given hypothesis there are within the angle $(\alpha + \eta, \beta - \eta)$ no points belonging to the exceptional circles. The truth of the first assertion can be deduced from formulas (2.04) and (2.06). For example, formula (2.06) takes the form

$$h(\theta) = \int_{\beta-2\pi}^{\alpha} \psi \sin\rho(\theta - \pi)\, d\Delta(\psi) + \tau_f \cos\rho(\theta - \theta_f).$$

Expanding the sine of the difference of two angles and performing the integration, we obtain formula (2.42).

It is useful to consider also the following example of an entire function of integral order with regularly distributed zeros. Let $\{a_n\}$ be a set of angular density $\Delta(\psi)$ within some angle $\alpha < \psi < \alpha + \pi/\rho$, let ρ be an integer, and let

$$\Phi(z) = \prod_1^{\infty} \left(1 - \frac{z^{2\rho}}{a_k^{2\rho}}\right).$$

It is obvious that the set of zeros of the function $\Phi(z)$ possesses an angular density and that $\delta_f = 0$. The formula for $H_\Phi(\theta)$ can easily be obtained if one applies Theorem 1 to the function $\Phi(z^{1/2\rho})$ of fractional order. After some simple transformations one then obtains

$$H(\theta) = \pi \int_{\alpha}^{\alpha+\frac{\pi}{\rho}} \sin\rho|\theta - \psi|\, d\Delta(\psi).$$

4. Construction of an Entire Function with a Given Indicator (Proof of Theorem 3)

Theorem 1 shows that it is possible to construct an entire function of nonintegral proximate order $\rho(r)$ $(\rho > 0)$ with an indicator that coincides with a given trigonometrically convex (for the given ρ) function $h(\theta)$ of period 2π. Theorem 2 yields an analogous result for an integer ρ and for an arbitrarily strong proximate order. We shall first carry out this construction for the simpler case of an entire function of nonintegral order.

1. First of all we construct a nondecreasing function $\Delta\psi$ for a given trigonometrically convex function $h(\theta)$:

$$\Delta(\psi) = \frac{1}{2\pi\rho}\left[h'(\psi + 0) - h'(\psi - 0) + \rho^2\int_0^{\psi} h(\varphi)\, d\varphi\right].$$

By Theorem 24, Chapter I, the function $h(\theta)$ can be expressed in the form

$$h(\theta) = \frac{\pi}{\sin \pi\rho}\int_{\theta-2\pi}^{\theta} \cos\rho(\theta - \psi - \pi)\, d\Delta(\psi). \tag{2.43}$$

In view of Theorem 1, the problem can be reduced to the construction of a point set $\{a_n\}$ with an angular density $\Delta(\psi)$ for the index $\rho(r)$. After this set is constructed it is sufficient to write down the canonical product

$$f(z) = \prod_1^{\infty} G\left(\frac{z}{a_k};\, p\right) \qquad (p = [\rho]).$$

By Theorem 1, the indicator of this function can be expressed as

$$h_f(\theta) = \frac{\pi}{\sin \pi\rho}\int_{\theta-2\pi}^{\theta} \cos\rho(\theta - \psi - \pi)\, d\Delta(\psi)$$

and hence $h_f(\theta) = h(\theta)$.

2. In order to construct the point set $\{a_k\}$ with angular density $\Delta(\psi)$, we approximate the function $\Delta(\psi)$ uniformly by means of a sequence of piece-wise constant functions $\Delta_m\psi$ with a finite number of discontinuities, selecting these functions so that the magnitudes of the jumps $\Delta_{m,j}$ $(j = 1, 2, \ldots, s_m)$ are rational numbers. The points of discontinuity we denote by $\psi_{m,j}$ $(j = 1, 2, \ldots, s_m)$.

It is not difficult to construct point sets $\mathfrak{N}_m$ with angular densities $\Delta_m(\psi)$. For this purpose we carry out the following operations.

We select the values $r = r_{m,p}$ $(p = 1, 2, \ldots)$ for which the quantity $r^{\rho(r)}$ takes on values which are multiples of the common denominator d_m of all the rational numbers $\Delta_{m,j}$ $(j = 1, 2, \ldots, s_m)$. After that we mark the points that lie at the intersections of the circles $|z| = r_{m,p}$ $(p = 1, 2, \ldots)$ with the rays

$\arg z = \psi_{m,j}$ $(j = 1, 2, \ldots, s_m)$, and we assign to each marked point the multiplicity $q_{m,j} = d_m \Delta_{m,j}$ $(j = 1, 2, \ldots, s_m)$.

It is obvious that the number $n_m(r, \psi)$ of marked points within the sector $|z| \leqslant r$, $0 < \arg z \leqslant \psi$ is given (with multiplicity taken into account) by

$$n_m(r, \psi) = \left[\frac{r^{\rho(r)}}{d_m}\right] d_m \sum_{0 < \psi_j \leqslant \psi} \Delta_{m,j}$$

$$= \Delta_m(\psi) r^{\rho(r)} + c(r, \psi), \tag{2.44}$$

where

$$|c(r, \psi)| < d_m \Delta_m(\psi). \tag{2.45}$$

From this it follows, of course, that the constructed set possesses the angular density $\Delta_m(\psi)$. Let us now divide the entire plane into rings K_m,

$$r_{m-1} < |z| \leqslant r_m,$$

selecting for this purpose a sequence of numbers $r_0 < r_1 < r_2 < \ldots$ such that $r_0 = 0$ and $r_m \to \infty$ as $m \to \infty$. In each ring K_m we select the points of $\mathfrak{N}_m$. The union of all these point sets we denote by $\mathfrak{N}$. We shall prove that if the chosen sequence r_m $(m = 0, 1, \ldots)$ increases sufficiently fast, then the set $\mathfrak{N}$ will have angular density $\Delta(\psi)$.

In fact, let us select the sequence of numbers r_m $(m = 1, 2, \ldots)$ such that

$$\lim_{m\to\infty} \left\{ r_m^{-\rho(r_m)} \sum_{k=1}^{m+1} d_k \right\} = 0. \tag{2.46}$$

Denoting by $n(r, \psi)$ the number of points of the set $\mathfrak{N}$ that lie in the corresponding sector, we shall have, for $r_m < r < r_{m+1}$ and for some integer $\tau < m$, the equation

$$n(r, \psi) = n(r_\tau, \psi) + \sum_{\tau+1}^{m} [n_k(r_k, \psi) - n_k(r_{k-1}, \psi)]$$

$$+ n_{m+1}(r, \psi) - n_{m+1}(r_m, \psi). \tag{2.47}$$

For any given $\varepsilon > 0$ one can choose the number τ such that for $k > \tau$

$$\Delta(\psi) - \varepsilon < \Delta_k(\psi) < \Delta(\psi) + \varepsilon \qquad (0 \leqslant \psi < 2\pi).$$

Making use of the equation (2.44) and of the inequality (2.45) we find that for $k > \tau$

$$n_k(r_k, \psi) - n_k(r_{k-1}, \psi) < [\Delta(\psi) + \varepsilon](r_k^{\rho(r_k)} - r_{k-1}^{\rho(r_{k-1})} + 2d_k).$$

From this and from (2.47) we obtain

$$n(r, \psi) < n(r_\tau, \psi) + [\Delta(\psi) + \varepsilon]\left\{(r^{\rho(r)} - r_\tau^{\rho(r_\tau)}) + 2\sum_{\tau+1}^{m+1} d_k\right\},$$

and, in view of (2.46),

$$\overline{\lim_{r\to\infty}} \frac{n(r, \psi)}{r^{\rho(r)}} \leqslant \Delta(\psi) + \varepsilon.$$

Analogously we obtain

$$\varliminf_{r\to\infty} \frac{n(r,\psi)}{r^{\rho(r)}} \geqslant \Delta(\psi) - \varepsilon,$$

and, since ε is arbitrary, we see that

$$\lim_{r\to\infty} \frac{n(r,\psi)}{r^{\rho(r)}} = \Delta(\psi).$$

Thus we have constructed a set with the given angular density and have proved Theorem 3 for the case of nonintegral order.

We call attention to the fact that the function $f(z)$ which we have constructed not only possesses the given indicator $h(\theta)$ but also satisfies the asymptotic equation

$$\ln|f(re^{i\theta})| \approx h(\theta)\, r^{\rho(r)}$$

in the complement of some C^0-set.

3. The case of integral order is considerably more complicated. In this case it is not sufficient to construct a point set $\{a_n\}$ with the given angular density

$$\Delta(\psi) = \frac{1}{2\pi\rho}\left[h'(\psi+0) - h'(+0) + \rho^2 \int_0^{\psi} h(\varphi)\, d\varphi\right]. \tag{2.48}$$

It is necessary to construct the set so that the condition (B) is satisfied. Furthermore, comparing the expression (1.82) for the given function $h(\theta)$ with the expression (2.06) of the indicator, we see that for some c_ρ the limit

$$\tau_f e^{-i\rho\theta_f} = \lim_{r\to\infty} \frac{1}{L(r)} \left\{c_\rho + \frac{1}{\rho} \sum_{|a_k|<r} a_k^{-\rho}\right\} \tag{2.49}$$

must be equal to the given quantity. Namely, the following equation will have to be satisfied:

$$\tau_f \cos(\theta - \theta_f) = A\cos\rho\theta + B\sin\rho\theta,$$

where A and B have the same meaning as in formula (1.82).

The canonical function of this set with the indicated value c_ρ will satisfy the asymptotic equation $\ln|f(re^{i\theta})| \approx h(\theta)\, r^{\rho(r)}$.

We shall construct such a set for the given strong proximate order $\rho^*(r)$.[7a]

4. First we will construct for $h(\theta)$, with integral ρ and an arbitrary proximate order $\rho(r)$, a set $\{a_k\}$ with angular density $\Delta(\psi)$ given by equation (2.48) and such that

$$\sum_{|a_k|<r} a_k^{-\rho} = 0.$$

[7a] Recently Logvinenko [1] constructed such a set for an arbitrary proximate order and thus generalized V. Bernstein's theorem without restrictions on the proximate order. A simpler construction has been given by Boĭčuk [1].

For this construction we make use of the fact (established in Theorem 24, Chapter I) that the function $\Delta(\psi)$ given by equation (2.48) satisfies the condition

$$\int_0^{2\pi} e^{-i\rho\psi}\, d\Delta(\psi) = 0$$

if ρ is an integer. Next we approximate the function $\Delta(\psi)$ with a sequence of piecewise constant nondecreasing functions $\Delta_m(\psi)$ with a finite number of rational jumps $\Delta_{m,j}$ ($j = 1, 2, \ldots, s$) in such a way that these functions satisfy the condition

$$\int_0^{2\pi} e^{-i\rho\psi}\, d\Delta_m(\psi) = 0. \tag{2.50}$$

The possibility of making such an approximation was demonstrated at the end of § 19, Chapter I. With the aid of the function $\Delta_m(\psi)$ we construct the set $\mathfrak{N}$ (see subsection 2). This set will have angular density $\Delta(\psi)$. Also the points a_k will be located on concentric circles $|z| = r_{m,p}$, they will have the form $r_{m,p}\, e^{i\psi_{mj}}$ ($j = 1, 2, \ldots, s_m$), and they will have the multiplicity $q_{mj} = d_m \Delta_{mj}$ ($j = 1, 2, \ldots, s_m$) where Δ_{mj} is the jump of the function $\Delta_m(\psi)$ at the point ψ_j. Taking the sum

$$\delta_{m,p} = \sum a_k^{-\rho}$$

over all points which lie on a single circle, with every point a_k being repeated according to its multiplicity, we obtain

$$\delta_{m,p} = d_m r_{m,p}^{-\rho} \sum_{j=1}^{s_m} e^{-i\rho\psi_{mj}} \Delta_{mj} = 0.$$

This equality follows from (2.50), from which it is obvious that

$$\sum_{|a_k| < r} a_k^{-\rho} = 0.$$

For the construction of a set satisfying for some c_ρ the condition (2.49), and for given τ_f and θ_f, one must add to the set $\mathfrak{N}$ some set $\mathfrak{N}'$ with zero angular density and satisfying for some c_ρ the condition (2.49). The union set $\mathfrak{N} + \mathfrak{N}'$ will fulfill all requirements. It is obvious that a function $f(z)$ that satisfies in the complement of some C^0-set the asymptotic equation

$$\ln |f(re^{i\theta})| \approx h(\theta)\, r^{\rho(r)}$$

is equal to the product of the functions

$$f(z, \mathfrak{N}) = \prod_{k=1}^{\omega_1} G\left(\frac{z}{a_k}; \rho\right) \qquad (a_k \in \mathfrak{N},\ \omega_1 \leqslant \infty) \tag{2.51}$$

and

$$\varphi(z) = e^{c_\rho z^\rho} \prod_{k=1}^{\omega_2} G\left(\frac{z}{b_k}; \rho\right) \qquad (b_k \in \mathfrak{N}',\ \omega_2 \leqslant \infty). \tag{2.52}$$

If $\rho(r) = \text{const}$, then one may choose for $\mathfrak{N}'$ the empty set, and for the function $\varphi(z)$ the exponential function $\exp c_\rho z^\rho$ $(C_\rho = \tau_f e^{-i\rho\theta_f})$. In the general case, for an arbitrary $\rho^*(r)$ the function $\varphi(z)$ must (because of Theorem 2) satisfy in the complement of some C^0-set the asymptotic equation

$$\ln|\varphi(re^{i\theta})| \approx \tau_f \cos \rho(\theta - \theta_f) r^{\rho(r)}.$$

Hence, the case when ρ is an integer has been reduced to the construction of a set $\mathfrak{N}'$ with the indicated properties, or, what is the same thing, of a function $\varphi(z)$ which is the analog (for an arbitrary proximate order $\rho^*(r)$) of the function $\exp c_\rho z^\rho$.

For the construction of the desired function, we note that from the definition of a strong proximate order $\rho^*(r)$ and of the corresponding function $L^*(r)$ it follows that

$$r^{\rho+1}[L^*(r)]' = r^\rho[\vartheta_2'(\ln r) - \vartheta_1'(\ln r)]\, e^{\vartheta_2(\ln r) - \vartheta_1(\ln r)}.$$

If we set

$$\gamma_i(r) = r^{\rho^*(r)}\vartheta_i'(\ln r) \qquad (i = 1, 2),$$

we obtain

$$r^{\rho+1}[L^*(r)]' = \chi_2(r) - \chi_1(r), \tag{2.53}$$

where

$$\lim_{r\to\infty} \frac{\chi_i(r)}{r^{\rho^*(r)}} = 0.$$

We shall also show that the functions $\chi_i(r)$ can be made monotone increasing by subjecting them to an unessential transformation. Indeed, from the definition of $\chi_i(r)$ it follows that

$$\chi_i'(r) = r^{\rho-1}L^*(r)\,\vartheta_i'(\ln r)\left[\rho + \frac{\vartheta_i''(\ln r)}{\vartheta_i'(\ln r)} + \vartheta_2'(\ln r) - \vartheta_1'(\ln r)\right].$$

The quantity within the brackets tends to ρ as $r \to \infty$, and hence the derivative $\chi_i'(r)$ will be positive for sufficiently large values of r. The function $\chi_i(r)$ will therefore be monotone increasing after some value of r. For the construction of the set $\mathfrak{N}'$ we mark first those points b_{k1} on the ray $\arg z = 0$ at which the function $\chi_1(r)$ takes on integral values, and on the ray $\arg z = \pi/\rho$ those points b_{k2} at which the function $\chi_2(r)$ takes on integral values. Let us denote by $n_j(r)$ $(j = 1, 2)$ the number of the points b_{kj} whose moduli do not exceed r. Then it is obvious that

$$n_j(r) = [\chi_j(r)] \qquad (j = 1, 2)$$

and the angular density of the constructed set is zero. On the other hand,

$$\sum_{|b_{k1}|<r} b_{k1}^{-\rho} + \sum_{|b_{k2}|<r} b_{k2}^{-\rho}$$
$$= \sum_{|b_{k1}|<r} |b_{k1}|^{-\rho} - \sum_{|b_{k2}|<r} |b_{k2}|^{-\rho} = \int_0^r \frac{dn(t)}{t^\rho},$$

where $n(t) = n_2(t) - n_1(t)$. It is obvious that $n(t) = t^{\rho+1}[L^*(t)]' + O(1)$ and $n(t) = 0$ for some $\varepsilon > 0$ and $0 < t < \varepsilon$. Integration by parts yields

$$\frac{1}{\rho L^*(r)} \int_0^r \frac{dn(t)}{t^\rho} = \frac{n(r)}{\rho r^{\rho(r)}} + \frac{1}{L^*(r)} \left\{ \int_\varepsilon^r [L^*(t)]' \, dt + \int_\varepsilon^r \frac{\varphi(t)}{t^{\rho+1}} dt \right\}, \tag{2.54}$$

where $\varphi(t) = n(t) - t^{\rho+1}[L^*(t)]' = O(1)$. Let us include in the set of the points b_{kj} the point b defined by the equation

$$b^{-\rho} = L^*(\varepsilon) - \int_\varepsilon^\infty \frac{\varphi(t)}{t^{\rho+1}} dt.$$

Having constructed $\delta(r)$ for the corresponding canonical product, we obtain $\delta(r) = 1 + o(1)$. By multiplying each of the numbers b_{kj} and b by $\tau_j^{-1/\rho} e^{i\theta_j}$, we obtain a set $\{b_k\}$ of zero density such that

$$\lim_{r\to\infty} \left\{ \frac{1}{\rho L^*(r)} \sum_{|b_k| \leq r} b_k^{-\rho} \right\} = \tau_j e^{-i\rho\theta_j}.$$

We have thus constructed a set $\mathfrak{N}'$ whose canonical function has completely regular growth and has the required trigonometric indicator. This completes the proof of the theorem.

We note that the function constructed with the given indicator has completely regular growth.

REMARK. If the given trigonometrically convex function is non-negative,[8] then in the assertion of Theorem 3 one may replace, in the case also of integral order, the strong proximate order by an arbitrary proximate order.

In order to show this, we construct for the function $L^{-1}(r)$ a majorant $L_1^*(r)$ (the possibility of this construction was shown in § 12, Chapter I), and we set $L^*(r) = [L_1^*(r)]^{-1}$. Then we find that

$$\overline{\lim_{r\to\infty}} \frac{L^*(r)}{L(r)} = 1.$$

After this we construct the entire function $F(z)$ of strong proximate order $\rho^*(r)$ ($r^{\rho^*(r)} = r^\rho L^*(r)$) with the given indicator $h(\theta)$ and of completely regular growth. Let E be the exceptional E^0-set for this function. Then we obtain

$$\overline{\lim_{\substack{r\to\infty \\ r \overline{\in} E}}} \frac{\ln |F(re^{i\theta})|}{r^\rho L(r)} = \lim_{\substack{r\to\infty \\ r \overline{\in} E}} \frac{\ln |F(re^{i\theta})|}{r^\rho L^*(r)} \overline{\lim_{r\to\infty}} \frac{L^*(r)}{L(r)} = h(\theta).$$

Next, we can assert on the basis of a theorem of Bernstein (see § 18, Chapter I) that $h_F(\theta) = h(\theta)$.

---

[8] It may be that the requirement of non-negativeness is superfluous.

5. Asymptotic Representation of Entire Functions with Regular Distribution of Zeros (Proof of Theorem 4)

In order to obtain the asymptotic representation (2.08) of the function $f(re^{i\theta})$ with a regular distribution of zeros, one must investigate the function $\arg f(re^{i\theta})$. We shall restrict ourselves to the case when ρ is not an integer, because the integral case can be investigated in an analogous manner.[9] Our investigation will again be based, as in § 2, on Theorem 25 of Chapter I, but in order to complete the argument we shall need in this case the following extension of Lemma 1.[10]

LEMMA 6. *If the set $\{a_k\}$ has density*

$$\Delta = \lim_{r\to\infty} \frac{n(r)}{r^{\rho(r)}},$$

and the functions $f(z)$ and $f^{\delta}(z)$ are defined as in Lemma 1, *then for every $\varepsilon > 0$ and for any $z = re^{i\theta}$ belonging to some closed set of ordinary rays*[11] *one can find a $\delta > 0$ such that*

$$|\arg f(z) - \arg f^{\delta}(z)| < \varepsilon r^{\rho(r)}. \tag{2.55}$$

Here δ depends only on ε and on the set of rays $\mathfrak{M}$.

PROOF. Representing the canonical product

$$f(z) = \prod_1^{\infty} G\left(\frac{z}{a_k}; p\right)$$

in the form of the product of the three functions

$$f_\sigma(z) = \prod_{|a_k|\leq \sigma r} G\left(\frac{z}{a_k}; p\right),$$

$${}_\tau f(z) = \prod_{|a_k|>\tau r} G\left(\frac{z}{a_k}; p\right),$$

$${}_\tau f_\sigma(z) = \prod_{\sigma r<|a_k|<\tau r} G\left(\frac{z}{a_k}; p\right)$$

($0 < \sigma < 1 < \tau < \infty$, $|z| = r$) we can write the inequality

$$\begin{aligned} |\arg f(z) - \arg f^{\delta}(z)| &\leq |\arg f_\sigma(z) - \arg f_\sigma^{\delta}(z)| \\ &+ |\arg_\tau f(z) - \arg_\tau f^{\delta}(z)| + |\arg_\tau f_\sigma(z) - \arg_\tau f_\sigma^{\delta}(z)|. \end{aligned} \tag{2.56}$$

---

[9] Below we shall indicate the necessary changes in the argument in the case of integral order.
[10] When ρ is an integer one needs an analogous extension of Lemma 4.
[11] For the definition see § 1.

From Lemmas 7 and 8 of § 7, Chapter I, it follows that for any given positive ε, for a sufficiently small $\sigma > 0$, and for a sufficiently large $\tau > 1$,

$$\left.\begin{aligned} &|\arg f_\sigma(z) - \arg f_\sigma^\delta(z)| < \frac{\varepsilon}{3} r^{\rho(r)}, \\ &|\arg_\tau f(z) - \arg_\tau f^\delta(z)| < \frac{\varepsilon}{3} r^{\rho(r)}. \end{aligned}\right\} \tag{2.57}$$

For an estimate of the third summand in (2.56), we note that in the expression

$$\arg G(u;\ p) = \arg(1-u) + \arg\left(u + \frac{u^2}{2} + \dots + \frac{u^p}{p}\right)$$

the second summand is a uniformly continuous function of u when $1/\tau \leqslant |u| \leqslant 1/\sigma$ while the first term is a uniformly continuous function in the cut ring $(1/\tau \leqslant |u| \leqslant 1/\sigma,\ 0 \leqslant \arg u < 2\pi)$. In crossing over the cut $(1, +\infty)$ along the positive axis, this function increases by 2π.

Let us express ${}_\tau f_\sigma(z)$ as the product of the two functions

$$ {}_\tau f_{\sigma,\theta}(z) = \prod_{\substack{\sigma r < |a_k| < \tau r \\ |\arg a_k - \theta| \geqslant \delta}} G\left(\frac{z}{a_k};\ p\right),$$

$$ {}_\tau \widehat{f}_{\sigma,\theta}(z) = \prod_{\substack{\sigma r < |a_k| < \tau r \\ |\arg a_k - \theta| < \delta}} G\left(\frac{z}{a_k};\ p\right).$$

The existence of the density of the set of zeros implies that for some $C > 0$

$$n(r) < C r^{\rho(r)} \qquad (r > 0)$$

or that

$$n(\tau r) < C_\tau r^{\rho(r)}.$$

If $\delta > 0$ is sufficiently small, then for $|\psi| \geqslant \delta$ and $|\psi - \psi_1| < \delta$, and for $\sigma^{-1} \geqslant s \geqslant \tau^{-1}$ it is true that

$$|\arg G(se^{i\psi};\ p) - \arg G(se^{i\psi_1};\ p)| < \frac{\varepsilon}{6C_\tau},$$

and we shall have

$$|\arg {}_\tau f_{\sigma,\theta}(z) - \arg {}_\tau f_{\sigma,\theta}^\delta(z)| < \frac{\varepsilon}{6C_\tau} n(\tau r) < \frac{\varepsilon}{6} r^{\rho(r)}. \tag{2.58}$$

For an estimate of the difference $|\arg {}_\tau \widehat{f}_{\sigma,\theta}(re^{i\theta}) - \arg {}_\tau \widehat{f}_{\sigma,\theta}^\delta(re^{i\theta})|$ we note that for arbitrary sufficiently small $\delta > 0$ and $|\theta - \psi| < \delta$

$$|\arg G(se^{i\psi};\ p) - \arg G(se^{i\theta};\ p)| < 3\pi.$$

Therefore,

$$|\arg_\tau \widehat{f}_{\sigma,\theta}(re^{i\theta}) - \arg {}_\tau \widehat{f}_{\sigma,\theta}^\delta(re^{i\theta})| < 3\pi\,[n(\tau r,\ \theta + \delta) - n(\tau r,\ \theta - \delta)].$$

If $\arg z = \theta$ is an ordinary ray, then for a given ε_1 one can find a δ such that

$$n(\tau r, \theta + \delta) - n(\tau r, \theta - \delta) < \varepsilon_1 r^{\rho(r)}. \tag{2.59}$$

By the finite-covering theorem we can, for a given ε_1, choose a δ_1 such that the inequality (2.59) will be satisfied if $\delta < \delta_1$ for all $\theta \in \mathfrak{M}$. Having taken ε_1 and δ sufficiently small, we obtain

$$|\arg_\tau \widehat{f}_{\sigma,\theta}(re^{i\theta}) - \arg_\tau \widehat{f}^{\delta}_{\sigma,\theta}(re^{i\theta})| < \frac{\varepsilon}{6} r^{\rho(r)} \qquad (\theta \in \mathfrak{M})$$

and, combining this with (2.58), we obtain an estimate for the third summand in (2.56)

$$|\arg_\tau f_\sigma(re^{i\theta}) - \arg_\tau f^\delta_\sigma(re^{i\theta})| < \frac{\varepsilon}{3} r^{\rho(r)}$$

for $\theta \in \mathfrak{M}$ and δ sufficiently small.

This inequality in conjunction with (2.57) yields

$$|\arg f(re^{i\theta}) - \arg f^\delta(re^{i\theta})| < \varepsilon r^{\rho(r)}$$

for $\theta \in \mathfrak{M}$ and δ sufficiently small. Lemma 6 has thus been established.

Let us now proceed with the proof of the theorem. From Theorem 25 of Chapter I it follows that if the zeros of the function $f(z)$ are located on a finite number of rays $\arg z = \psi_j$ with densities Δ_j $(j = 0, 1, \ldots, n)$, then for $|\theta - \psi_j| \geqslant \eta > 0$ the following asymptotic relation will hold uniformly:

$$\arg f(re^{i\theta}) \approx \frac{\pi}{\sin \pi\rho} \sum_{j=0}^{n} \Delta_j \sin \rho(\theta - \psi_j - \pi). \tag{2.60}$$

In this case of regular distribution of the zeros we choose for a given positive ε a positive δ so small that, for $|\psi_{j+1} - \psi_j| < \delta$ $(j = 0, 1, \ldots, n)$ and $\psi_{n+1} = \psi_0 + 2\pi$, the sum on the right hand side of this equation will differ from the corresponding Stieltjes integral by less than $\varepsilon/2$. Furthermore, if δ is so small that for $\theta \in \mathfrak{M}$

$$|\arg f(re^{i\theta}) - \arg f^\delta(re^{i\theta})| < \frac{\varepsilon}{2} r^{\rho(r)},$$

then for $\theta \in \mathfrak{M}$ and θ not belonging to the interval $|\psi_j - \theta| < \eta$ we shall have for some $r_{\varepsilon\eta\mathfrak{M}}$ and $r > r_{\varepsilon\eta\mathfrak{M}}$ the inequality

$$\left|\arg f(re^{i\theta}) - \frac{\pi r^{\rho(r)}}{\sin \pi\rho} \int_{\theta-2\pi}^{\theta} \sin \rho(\theta - \psi - \pi)\, d\Delta(\psi)\right| < \varepsilon r^{\rho(r)}.$$

Selecting another subdivision of the interval $(0, 2\pi)$ by means of the points ψ_j' $(j = 0, 1, \ldots, n)$ such that $|\psi_{j+1}' - \psi_j'| < \delta$, and such that the intervals $|\theta - \psi_j'| < \eta$ have no points in common with the intervals $|\theta - \psi_j| < \eta$, we find that the displayed inequality holds for all $\theta \in \mathfrak{M}$ and all r exceeding some number $r_{\varepsilon\mathfrak{M}}$.

From this and from Theorem 1 there follows directly the truth of Theorem 4 when ρ is not an integer.

The case when ρ is an integer differs from the above only in that in place of the function $f(z)$ one must consider the function $f_r(z)$, and in place of Theorem 25, Chapter I, one has to use Lemma 9 of Chapter I. Also, one has to take into account the equation

$$\lim_{r\to\infty} \frac{e^{i\rho\theta}}{L(r)} \delta_r(r) = \tau_r e^{i\rho(\theta-\theta_r)}.$$

6. Entire Functions with Regular Sets of Zeros (Proof of Theorem 5)

The asymptotic formulas established in §§ 2, 3, and 5 are valid outside some exceptional C^0-set of circles, which is not effectively constructed. In this section we shall prove that the zeros constitute an R-set, and we shall show that the asymptotic formulas (2.03) and (2.05) are valid outside a well-defined C_R-set. The precise formulation is given in Theorem 5 (see § 1).

For the proof of Theorem 5 we shall need two lemmas.

LEMMA 7. *Let $\{a_n\}$ be an infinite set of points having a limit point at infinity, and let*

$$F_\delta(z) = \prod\left(1 - \frac{z}{a_n}\right),$$

where the product includes those factors in which the numbers a_n satisfy the inequality $|a_n - z| \leqslant \delta r$, $|z| = r$, and $0 < \delta < \frac{1}{2}$.

Then for $z \neq a_n$ we have the following inequality,

$$|\ln|F_\delta(z)|| < c_\delta n_z(\delta r) + \int_0^{\delta r} \frac{n_z(t)}{t}\, dt, \tag{2.61}$$

in which $n_z(t)$ is the number of the points a_n in the circle of radius with center at the point z, and where c_δ is a positive constant depending only on δ.

PROOF. From the form of the function $F_\delta(z)$ it follows at once that

$$|\ln|F_\delta(z)|| < - \sum_{|z-a_n|\leqslant\delta r} \ln\left|1 - \frac{z}{a_n}\right|.$$

It is easy to see that

$$\sum_{|a_n-z|\leqslant\delta r} \ln|a_n - z| = \int_0^{\delta r} \ln t\, dn_z(t) = n_z(\delta r)\ln \delta r - \int_0^{\delta r} \frac{n_z(t)}{t}\, dt$$

and

$$\ln\Big|\prod_{|a_n-z|\leqslant\delta r} a_n^{-1}\Big| > -n_z(\delta r)\ln(1+\delta) r.$$

From these inequalities it follows that

$$|\ln|F_\delta(z)|| < \ln\left(1+\frac{1}{\delta}\right)n_z(\delta r) + \int_0^{\delta r}\frac{n_z(t)}{t}\,dt.$$

LEMMA 8. *If the point set $\{a_n\}$ is an R-set and if $F_\delta(z)$ is the same as in Lemma 7, then for every positive ε, for a sufficiently small positive number δ, and for any z lying outside the exceptional C_R-circles and outside some circle $|z| \geqslant r_\varepsilon$, we have the inequality*

$$|\ln|F_\delta(z)|| < \varepsilon r^{\rho(r)}. \tag{2.62}$$

PROOF. Suppose the condition (C) (see § 1) is satisfied, and suppose that the point z does not belong to any of the circles mentioned in that condition. With z as center let us draw a circle K_t of arbitrary radius $t > 0$. Since the center of this circle does not lie in any of the exceptional circles, the radii of the exceptional circles with centers inside this circle are less than t. Since no two of the exceptional circles have a common point, the number $n_z(t)$ of their centers within the circle K_t is less than the ratio of the area of a circle of radius $2t$ to the area of the smallest one of the exceptional circles with centers in K_t. If we denote this smallest radius by $l_{t,z}$, then for $t < r$ we have

$$l_{t,z} > d(r-t)(r+t)^{-\frac{\rho(t+r)}{2}}$$

and consequently, for $t \leqslant \delta r$ and for sufficiently large r,

$$n_z(t) < \frac{4\pi t^2 (r+t)^{\rho(r+t)}}{\pi d^2 (r-t)^2} < \frac{8}{d^2}\frac{(1+\delta)^\rho t^2}{(1-\delta)^2} r^{\rho(r)-2}.$$

An analogous estimate can be obtained under condition (C′) by comparing the diameters of the circles. Substituting the derived estimate for $n_z(t)$ into (2.61) we obtain (2.62). This proves the lemma.[12] We shall now proceed with the proof of Theorem 5.

The function $f(z)$ satisfies the asymptotic equation (2.03) or (2.05). Here the inequality

$$\ln|f(re^{i\theta})| < [H(\theta)+\varepsilon]r^{\rho(r)} \qquad (0 \leqslant \theta < 2\pi)$$

will be satisfied for all r greater than a number r'_ε, and the inequality

$$\ln|f(re^{i\theta})| > [H(\theta)-\varepsilon]r^{\rho(r)}$$

will hold in the complement of some C^0-set.

[12] The conditions (C) and (C′) are required only to guarantee the validity of inequality (2.62), which is obtained from the estimate of $n_z(t)$ in view of Lemma 7. The conditions (C) and (C′) can therefore be replaced in Theorem 5 by any other conditions that will yield a corresponding estimate for $n_z(t)$. In particular, when $\rho(r) = 1$, one may, for example, require in place of condition (C) that for some $d > 0$ and for arbitrary n and k the inequality $|a_{n+k} - a_n| > dk$ be satisfied. Then $n_z(t) < (1/d)t$ and if $|z - a_n| > d_1$ $(d_1 > d,\ n = 1, 2, \ldots)$ the inequality (2.62) will hold.

We need to show that the second inequality is valid everywhere outside the C_R-circles when $r > r'_\varepsilon$. For this purpose we pick a z outside the C_R-circles such that $|z| > 2r'_\varepsilon$, and we define the function

$$F_\delta(z, u) = \prod_{|a_n - z| \leqslant \delta r} \left(1 - \frac{z+u}{a_n}\right).$$

It is obvious that $F_\delta(z, 0) = F_\delta(z)$, and that the inequality $\ln |F_\delta(z, u)| < 0$ will hold if $\delta < \frac{1}{4}$ and $|u| < \delta r$. This implies that within this circle but outside the exceptional C^0-set the function

$$\varphi_\delta(z, u) = \frac{f(z+u, \mathfrak{N})}{F_\delta(z, u)}$$

satisfies the inequality

$$\ln |\varphi_\delta(z, u)| > [H_{\mathfrak{N}}(\psi) - \varepsilon] \, |z+u|^{\rho(|z+u|)},$$

in which $\psi = \arg(z + u)$. From the continuity of $H_{\mathfrak{N}}(\psi)$ it follows that for a sufficiently small positive δ we have

$$\ln |\varphi_\delta(z, u)| > [H_{\mathfrak{N}}(\theta) - 2\varepsilon] \, r^{\rho(r)} \qquad (z = re^{i\theta}) \tag{2.63}$$

in the same region.

We shall now show that for sufficiently large values of r this inequality is valid when $u = 0$ even if z belongs to the C^0-set. Indeed, for a given $\delta > 0$, and a sufficiently large r, the sum of the diameters of the circles of the C^0-set which lie in the circle $|u| < \delta r$ is less than δr, and hence there exists a circle $|u| = \tau$ $(0 < \tau < \delta r)$ that does not intersect the C^0-set. The inequality (2.63) is satisfied on the circumference $|u| = \tau$, but since the function $\varphi_\delta(z, u)$ has no zeros in the circle $|u| \leqslant \tau$, it must also be satisfied at the center. Thus

$$\ln \frac{|f(z, \mathfrak{N})|}{|F_\delta(z)|} > [H_{\mathfrak{N}}(\theta) - \varepsilon] \, r^{\rho(r)}$$

and in view of the inequality (2.62)

$$\ln |f(z, \mathfrak{N})| > [H_{\mathfrak{N}}(\theta) - 3\varepsilon] \, r^{\rho(r)}.$$

This inequality, which we have proved for all sufficiently large r and z not lying in the C_R-circles, taken in conjunction with the asymptotic inequality

$$\ln |f(z, \mathfrak{N})| < [H_{\mathfrak{N}}(\theta) + \varepsilon] \, r^{\rho(r)} \qquad (z = re^{i\theta})$$

(see Chapter I, Theorem 28) establishes the first assertion of Theorem 5.

The second assertion concerning the argument of $f(z, \mathfrak{N})$ does not require any additional argument, since the asymptotic equation for $\arg f(re^{i\theta}, \mathfrak{N})$ is not connected with any exceptional sets of circles but is valid on any closed set of ordinary rays for all sufficiently large r.

EXAMPLE. Let $\Omega_{m,n} = m + in$ $(m, n = 0, \pm 1, \pm 2, \ldots)$

$$\sigma(z) = z\prod{}'\left(1 - \frac{z}{\Omega_{m,n}}\right)e^{\frac{z}{\Omega_{m,n}} + \frac{z^2}{2\Omega^2_{m,n}}}$$

(the prime indicating that the values $m = n = 0$ are excluded). The set of zeros of this function has angular density

$$\Delta(\psi) = \frac{\psi}{2}$$

with the index $\rho = 2$.

In view of the continuity of the function $\Delta(\psi)$ there are no exceptional rays. The condition (B) is satisfied because $\delta_\sigma(r) = 0$. Furthermore, the set of the zeros is an R-set because the condition (C) is satisfied when $d < \frac{1}{2}$. Thus, outside the circles with centers at the integral points $\Omega_{m,n}$, and with arbitrarily small radii, we have, in view of the formulas (2.08) and (2.10), the asymptotic equation

$$\ln|\sigma(re^{i\theta})| \approx \frac{\pi}{2}r^2$$

which is satisfied uniformly relative to θ in the interval $(0 \leqslant \theta \leqslant \pi)$.

7. Theorems on Equicontinuity (Proof of Theorems 6 and 7)

First of all we point out the close connection that exists between Theorem 6 and Lemma 1 proved above. In Lemma 1 we showed that if the set of zeros of an entire function $f(z)$ of proximate order $\rho(r)$ has a density, then under a sufficiently small change of the arguments of the zeros a_k, the function $\ln|f(re^{i\theta})|$ will undergo only a small change asymptotically. Since the original factors of the canonical product depend on the quantities

$$\frac{z}{a_k} = \frac{r}{|a_k|}e^{i(\theta - \psi_k)}$$

$$(\theta = \arg z,\ \psi_k = \arg a_k,\ k = 1, 2, 3, \ldots),$$

a small change in θ is equivalent to a small change of the arguments of the zeros a_k $(k = 1, 2, \ldots)$. Thus, if the set of zeros of the function $f(z)$ has a density, then to every positive ε there correspond a $\delta > 0$ and an $r_\varepsilon > 0$ such that if $|\theta_2 - \theta_1| < \delta$, $r > r_\varepsilon$, and $re^{i\theta_1}$, $re^{i\theta_2}$ do not belong to some C^0-set (or if r does not belong to an E^0-set that is independent of ε), then

$$|h_{f,r}(\theta_1) - h_{f,r}(\theta_2)| < \varepsilon.$$

For $r \leqslant r_\varepsilon$ and outside the E^0-set, the function $h_{f,r}(\theta)$ is uniformly continuous in r and θ, and hence for some $\delta_2 > 0$ and $|\theta_1 - \theta_2| < \delta_2$ the same inequality will hold. It is sufficient to take $\delta = \min(\delta_1, \delta_2)$. Thus if the set $\{a_k\}$ of the zeros of an entire function of proximate order $\rho(r)$ has a density, the assertion of Theorem 6 follows from Lemma 1, as has just been shown.[13]

In the proof of Lemma 1 we made use of the density of the set of zeros of the function only in the estimate of the quantity $\left|\ln|\Phi_{R,\beta}(z)|\right|$. For the proof of Theorem 6 without auxiliary hypotheses on the set of zeros, it is therefore natural to make use of the general estimate for $\left|\ln|\Phi_{R,\beta}(z)|\right|$ given in Lemma 2. This estimate involves the quantities n_m and n_{m+1}, namely, the number of zeros of $f(z)$ in certain rings K_m: $R_m(1-\beta) < |z| \leqslant R_m(1+\beta)$, and K_{m+1}: $R_{m+1}(1-\beta) < |z| < R_{m+1}(1+\beta)$. If the set of zeros of the function $f(z)$ does not have a density, then the roots can be strongly concentrated in certain rings; however, if one notes (see Lemma 4, Chapter I) that for some constant c

$$n(r) < cr^{\rho(r)}, \tag{2.64}$$

then it is easy to understand that the relative measure of such rings can not be great. Let us give an exact evaluation of the relative measure of such special rings.

We shall denote by I_q the intervals

$$R_q(1-\beta) < R \leqslant R_q(1+\beta)$$
$$\left(R_q = \left(\frac{1+\beta}{1-\beta}\right)^q,\ 0 < \beta < \frac{1}{2}\right)$$

and by K_q the rings

$$R_q(1-\beta) < |z| \leqslant R_q(1+\beta).$$

Denoting, as before, by n_q the number of points of the sequence $\{a_n\}$ in the ring K_q, we shall say that an interval I_q, or a ring K_q, is a *singular interval*, or a *singular ring*, if

$$n_q > \beta^{-\frac{1}{2}} \Delta_q R_q^{\rho(R_q)-1}, \tag{2.65}$$

where Δ_q denotes the length of the interval I_q. Let $n(t)$ denote the number of points of the sequence $\{a_n\}$ in the circle $|z| \leqslant t$.

We shall prove that the upper relative measure of the set of singular intervals will tend to zero when $\beta \to 0$. For this purpose we consider the sum

$$S_R = R^{-\rho(R)} \sum_{R_{q_j} \leqslant R} n_{q_j},$$

where the summation extends over those values of the index q which correspond to the singular intervals.

---

[13] In this case we have obtained the assertion of Theorem 6 in a stronger form, since in place of the exceptional set E of arbitrarily small upper relative measure, we have used an E^0-set.

From (2.65) we find that

$$S_R > \beta^{-\frac{1}{2}} \sum_{R_{q_j} < R} \Delta_{q_j} \frac{R_{q_j}^{\rho(R_{q_j})}}{R^{\rho(R)}} \frac{1}{R_{q_j}} \geqslant \beta^{-\frac{1}{2}} R^{-1} \sum_{\frac{R}{2} < R_{q_j} \leqslant R} \Delta_{q_j} \frac{R_{q_j}^{\rho(R_{q_j})}}{R^{\rho(R)}}.$$

From the monotone character of the function $r^{\rho(r)}$ it follows that if $R_{q_j} > R/2$, then

$$R_{q_j}^{\rho(R_{q_j})} \geqslant \left(\frac{R}{2}\right)^{\rho\left(\frac{R}{2}\right)}$$

and hence for $R_0 > 0$ and $R > R_0$

$$\frac{R_{q_j}^{\rho(R_{q_j})}}{R^{\rho(R)}} \geqslant \left(\frac{1}{2}\right)^{\rho} \frac{L\left(\frac{R}{2}\right)}{L(R)} > 2^{-\rho-1}.$$

In this manner we obtain the inequality

$$\sum_{\frac{R}{2} < R_{q_j} \leqslant R} \Delta_{q_j} < 2^{\rho+2} \beta^{\frac{1}{2}} S_R \frac{R}{2}. \tag{2.66}$$

Returning now to (2.64), we show that $S_R < c$, and that the inequality (2.66) implies that

$$\sum_{\frac{R}{2} < R_{q_j} \leqslant R} \Delta_{q_j} < 2^{\rho+2} c \beta^{\frac{1}{2}} \frac{R}{2}.$$

Substituting in this equation for R first $R/2$, then $R/4$, and so on, and adding the resulting inequalities, we find that for some $R_0 > 0$

$$\sum_{R_0 < R_{q_j} \leqslant R} \Delta_{q_j} < 2^{\rho+2} c \beta^{\frac{1}{2}} R$$

and finally

$$\sum_{R_{q_j} \leqslant R} \Delta_{q_j} < 2^{\rho+2} c \beta^{\frac{1}{2}} R + R_0.$$

We have thus shown that the upper relative measure of the set of singular intervals tends to zero when $\beta \to 0$.

Let us now proceed with the estimate of the function $|\ln|\Phi_{R,\beta}(z)||$ outside certain exceptional rings. We pick out from the sequence of the intervals I_j all singular intervals and all intervals adjacent to singular intervals and denote this set of intervals by E'. Then we shall have for $0 < \beta < \frac{1}{2}$ the following inequality:

$$\overline{m}^*(E') < 2^{\rho+5} c \beta^{\frac{1}{2}}. \tag{2.67}$$

Let us return to Lemma 2. If the number R does not belong to E' and if $R_m < R < R_{m+1}$, then the intervals I_m and I_{m+1} are not singular intervals, i.e.,

$$n_m < \beta^{-\frac{1}{2}} \Delta_m R_m^{\rho(R_m)-1}, \qquad n_{m+1} < \beta^{-\frac{1}{2}} \Delta_{m+1} R_{m+1}^{\rho(R_{m+1})-1}.$$

Hence, if z is such that $\big||z| - R\big| < \beta R$ and does not belong to the set C of exceptional circles, then it follows from Lemma 2 that

$$|\ln|\Phi_{R,\beta}(z)|| < 2\beta^{\frac{1}{2}}(R_m^{\rho(R_m)} + R_{m+1}^{\rho(R_{m+1})})\left(2\ln\frac{1}{\beta} + B_p\right),$$

where $B_p = 4 + 3^p p$, which implies that

$$|\ln|\Phi_{R,\beta}(z)|| < 2^{2p+2}\beta^{\frac{1}{2}}\left(2\ln\frac{1}{\beta} + B_p\right) r^{\rho(r)}. \tag{2.68}$$

Here the upper relative measure of the set E'' of those values of r for which the circumference $|z| = r$ does not intersect the exceptional circles of the set C will not exceed the quantity $\beta(1 + \beta)^3(1 - \beta)^{-2}$.

Let the numbers $\eta > 0$ and $\varepsilon > 0$ be given. We choose β so small that

$$\left.\begin{aligned} &2^{\rho+5}c\beta^{\frac{1}{2}} < \frac{\eta}{2}, \quad \beta(1+\beta)^3(1-\beta)^{-2} < \frac{\eta}{2}, \\ &2^{2\rho+2}\beta^{\frac{1}{2}}\left(2\ln\frac{1}{\beta} + B_p\right) < \frac{\varepsilon}{10}. \end{aligned}\right\} \tag{2.69}$$

If we now set $E = E' + E''$, then

$$\overline{m}^*(E) < \eta,$$

and if $|z| \bar{\in} E$,

$$|\ln|\Phi_{R,\beta}(z)|| < \frac{\varepsilon}{10} r^{\rho(r)}. \tag{2.70}$$

Let us now express the function $f(z)$ of the proximate order $\rho(r)$ in the form of a product of five functions as was done in subsection 1 of the proof of Lemma 1. We shall then have

$$\begin{aligned} |\ln|f(re^{i\theta_1})| - \ln|f(re^{i\theta_2})|| &\leqslant |\ln|f_\sigma(re^{i\theta_1})| - \ln|f_\sigma(re^{i\theta_2})|| \\ &+ |\ln|{}_\tau f(re^{i\theta_1})| - \ln|{}_\tau f(re^{i\theta_2})|| \\ &\qquad + |\ln|f_{\sigma\beta}(re^{i\theta_1})| - \ln|f_{\sigma\beta}(re^{i\theta_2})|| \\ &+ |\ln|{}_{\tau\beta} f(re^{i\theta_1})| - \ln|{}_{\tau\beta} f(re^{i\theta_2})|| \\ &\qquad + |\ln|\Phi_\beta(re^{i\theta_1})| - \ln|\Phi_\beta(re^{i\theta_2})|| \end{aligned}$$

and we must estimate the value of each summand of the right hand side.

From (2.70) it follows that if $r \bar{\in} E$,

$$|\ln|\Phi_\beta(re^{i\theta_1})| - \ln|\Phi_\beta(re^{i\theta_2})|| < \frac{\varepsilon}{5} r^{\rho(r)}. \tag{2.71}$$

Furthermore, from the inequality (2.20) we have the asymptotic relations

$$\left.\begin{aligned} &|\ln|f_\sigma(re^{i\theta_1})| - \ln|f_\sigma(re^{i\theta_2})|| < \frac{\varepsilon}{5} r^{\rho(r)}, \\ &|\ln|_\tau f(re^{i\theta_1})| - \ln|_\tau f(re^{i\theta_2})|| < \frac{\varepsilon}{5} r^{\rho(r)}. \end{aligned}\right\} \tag{2.72}$$

Finally, from the uniform continuity of the functions of θ

$$\ln|G(se^{i\theta};\ p)|$$

in the regions

$$(1-\beta)^{-1} \leqslant s \leqslant \sigma^{-1}, \qquad \tau^{-1} \leqslant s \leqslant (1+\beta)^{-1}$$

and from the inequalities (2.64) it follows that, for a sufficiently small $\delta > 0$ and for $|\theta_1 - \theta_2| < \delta$,

$$\left.\begin{aligned} &|\ln|f_{\sigma\beta}(re^{i\theta_1})| - \ln|f_{\sigma\beta}(re^{i\theta_2})|| < \frac{\varepsilon}{5} r^{\rho(r)}, \\ &|\ln|_{\tau\beta} f(re^{i\theta_1})| - \ln|_{\tau\beta} f(re^{i\theta_2})|| < \frac{\varepsilon}{5} r^{\rho(r)}. \end{aligned}\right\} \tag{2.73}$$

From the inequalities (2.71), (2.72) and (2.73) it follows that, for $r \,\overline{\in}\, E$ and $|\theta_1 - \theta_2| < \delta$,

$$|\ln|f(re^{i\theta_1})| - \ln|f(re^{i\theta_2})|| < \varepsilon r^{\rho(r)}. \tag{2.74}$$

We have thus proved that for every pair of positive numbers η and ε there corresponds a set E with an upper relative measure less than η and a number $\delta > 0$ such that if $r \,\overline{\in}\, E$ and $|\theta_1 - \theta_2| < \delta$, the inequality (2.74) will be satisfied. We must still prove that if one omits a certain fixed set E of values of r with an arbitrarily small upper relative measure, then for every $\varepsilon > 0$ there exists a $\delta > 0$ such that if $|\theta_1 - \theta_2| < \delta$, then (2.74) will hold.

For the construction of such a set E we choose for the given η two sequences of positive numbers: $\varepsilon_1, \varepsilon_2, \ldots$, converging to zero, and $\eta_1, \eta_2, \ldots$ such that

$$\sum_1^\infty \eta_k < \frac{\eta}{2}.$$

To each pair of numbers ε_k, η_k there corresponds a set E_k with upper relative measure less than η_k, and a positive number $\delta_k^{(1)}$ such that if $r \,\overline{\in}\, E_k$ and $|\theta_1 - \theta_2| < \delta_k^{(1)}$, then

$$|\ln|f(re^{i\theta_1})| - \ln|f(re^{i\theta_2})|| < \varepsilon_k r^{\rho(r)}. \tag{2.75}$$

We improve the sets E_k somewhat by dropping from each E_k the part belonging to a certain interval $(0, r_k)$, and we do this in such a way that the remaining set E_k satisfies the condition

$$\operatorname{mes}[E_k'^{\,r}] < 2\eta_k r \qquad (r > 0).$$

Then it is obvious that the upper relative measure of the set

$$E = E_1 + \sum_2^\infty E'_k$$

will not exceed η. The set E'_k and the number $\delta_k^{(1)}$ are selected in such a way that if $r > r_k$, $r \bar{\in} E$, and $|\theta_2 - \theta_1| < \delta_k^{(1)}$ the inequality (2.75) will hold. On the other hand, if $|z| < r_k$, and $|z| \bar{\in} E_1$, the function $\ln|f(z)|$ is uniformly continuous, and hence the inequality (2.75) will be satisfied for some $\delta_k^{(2)}$, $|\theta_1 - \theta_2| < \delta_k^{(2)}$, $r \leqslant r_k$, and $r \bar{\in} E$. If we choose $\delta_k = \min\{\delta_k^{(1)}, \delta_k^{(2)}\}$, we find that (2.75) is satisfied if $r \bar{\in} E$ and $|\theta_1 - \theta_2| < \delta_k$. This implies the equicontinuity of the family of functions $h_{f,r}(\theta)$ in θ if $r \bar{\in} E$. The theorem has thus been established.

Let us now proceed with the proof of the more general Theorem 7 on the equicontinuity of the family $h_{F,r}(\theta)$ of functions which are holomorphic inside some angle $\alpha \leqslant \arg z \leqslant \beta$. We shall first prove three lemmas.

LEMMA 9. *Let $\Phi(z)$ be a holomorphic function in the circle $|z| \leqslant R$, and let*

$$M_\Phi(R) = \max_{|z| \leqslant R} |\Phi(z)|.$$

Let us also suppose that at some point z_0 ($|z_0| \leqslant lR$, $l < 1$)

$$|\Phi(z_0)| = 1.$$

Then the following inequality will be true for the number $n_\Phi(tR)$ of zeros of the function $\Phi(z)$ in the circle $|z|R^{-1} < t < 1$ for $l < t < 1$:

$$n_\Phi(tR) \leqslant \nu(l, t) \ln M_\Phi(R) \quad \left(\nu(l, t) = \ln\left(\frac{1+lt}{l+t}\right)^{-1}\right).$$

LEMMA 10. *If under the conditions of Lemma 9 the function $\Phi(z)$ has no zeros in the circle $|z| \leqslant R$, then the minimum $m_\Phi(tR)$ of the function $|\Phi(z)|$ in the circle $|z| \leqslant tR$ satisfies the inequality*

$$\ln m_\Phi(tR) \geqslant -\nu(t) \ln M_\Phi(R) \quad \left(\nu(t) = \frac{t_1}{1-t_1},\ t_1 = \ln\frac{l+t}{1+tl}\right).$$

These lemmas differ from Lemma 4, § 5, Chapter I, and from the corollary of Theorem 8 in § 6, Chapter I, in that the requirement $|\Phi(0)| = 1$ has been replaced by $|\Phi(z_0)| = 1$, and $|z_0| \leqslant lR$. Both these lemmas reduce to the propositions of Chapter I if the circle $|z| \leqslant R$ is transformed into the unit circle and the point z_0 is replaced by the origin. We shall prove only Lemma 9. Lemma 10 can be proved in an analogous way.

PROOF. The function

$$z = \frac{R^2\zeta + Rz_0}{R + \zeta\bar{z}_0}$$

transforms the unit circle $|\zeta| \leqslant 1$ of the ζ-plane into the circle $|z| \leqslant R$. Here the point $\zeta = 0$ will go into the point z_0. Substituting the function

$$\varphi(\zeta) = \Phi\left(\frac{R^2\zeta + Rz_0}{R + \bar{z}_0\zeta}\right)$$

into Jensen's formula (see § 5, Chapter I) we obtain

$$\int_0^1 \frac{n_\varphi(\tau)}{\tau}\, d\tau = \frac{1}{2\pi}\int_0^{2\pi} \ln|\varphi(e^{i\vartheta})|\, d\vartheta \leqslant \ln M_\Phi(R)$$

and because of the monotone character of the function $n_\varphi(\tau)$ for $0 < t_1 < 1$, we have

$$n_\varphi(t_1)\ln\frac{1}{t_1} < \ln M_\Phi(R).$$

When $t_1 = (t + l)/(1 + tl)$ the image of the circle $|\zeta| \leqslant t_1$ contains the circle $|z| < tR$. Therefore, $n_\Phi(Rt) \leqslant n_\varphi(t_1)$ and hence

$$n_\Phi(tR) < \nu(l, t)\ln M_\Phi(R).$$

Lemma 9 makes it possible for us to find an estimate of the number of zeros within a sector of a function satisfying the conditions of Theorem 7. Namely, the following lemma is true.

LEMMA 11. *Suppose that $F(z)$ is a holomorphic function of proximate order $\rho(r)$ inside the angle $\alpha < \arg z < \beta$ and suppose that for some positive number N, and for some positive number $l < 1$, and for all sufficiently large values of r, each circle of the type*

$$\left|z - re^{i\frac{\alpha+\beta}{2}}\right| < l\sin\frac{\beta-\alpha}{2}r$$

contains a point $z_1 = r_1e^{i\theta_1}$ in which

$$\ln|F(z_1)| > -Nr^{\rho(r)}.$$

Then for $\xi > 0$ the number $n(r, \alpha + \xi, \beta - \xi)$ of the zeros of this function in the sector $\alpha + \xi < \arg z < \beta - \xi$, $|z| \leqslant r$, will satisfy the asymptotic inequality

$$n(r, \alpha+\xi, \beta-\xi) < \varkappa(\xi, \alpha, \beta, l)\,(h + 2N)\,r^{\rho(r)},$$

in which

$$h = \max_{\alpha \leqslant \theta \leqslant \beta} |h_F(\theta)|$$

and $\varkappa(\xi, \alpha, \beta, l)$ is a constant which depends only on the indicated quantities.

PROOF. We make the following preliminary remarks. The assertion of the lemma applies to an arbitrary angle inside the angle (α, β). Therefore, without loss of generality we may assume that $\beta - \alpha$ is not a multiple of π/ρ. By means of the transformation $z = e^{i\varphi}u^\lambda$ one can transform the angle (α, β)

into the right half-plane, whereby the function $F(z)$ will go over into a function of a nonintegral order in the right half-plane. Here one can select the positive number $l_1 < 1$ so that the region which will be the image of the circle

$$|z - re^{i\frac{\alpha+\beta}{2}}| < l \sin\frac{\beta-\alpha}{2} r$$

will be contained for some $R > 0$ in the circle

$$|z - R| < l_1 R.$$

Hence we may assume, without loss of generality, that $\alpha = -\pi/2$, $\beta = \pi/2$, and ρ is not an integer, and that for all sufficiently large values of r the circle

$$|z - r| < l_1 r$$

contains the point $z_{1,r} = r_1 e^{i\theta}$, at which

$$\ln |F(z_{1,r})| > -Nr_1^{\rho(r_1)} > -N(2r)^{\rho(2r)}. \tag{2.76}$$

On the other hand, by Theorem 28, Chapter I, the function $F(z)$ must satisfy the inequality

$$\ln |F(z)| < (h+\varepsilon)(2r)^{\rho(2r)} \qquad (\varepsilon > 0) \tag{2.77}$$

everywhere in the circle $|z - r| \leqslant lr$.

From the inequalities (2.76) and (2.77) it follows that for all sufficiently large values of r the function

$$\psi_r(z) = \frac{F(r+z)}{F(z_{1,r})}$$

will satisfy, in the circle $|z| \leqslant r$, the inequality

$$\ln |\psi_r(z)| < (h+N+\varepsilon)(2r)^{\rho(2r)} < (h+2N)(2r)^{\rho(2r)}$$
$$(0 < \varepsilon < N).$$

Also, for $z_0 = z_{1,r} - r$ we have the equation

$$\psi_r(z_0) = 1.$$

By Lemma 9, the number $n_\psi(kr)$ of zeros of the function $\psi_r(z)$ in the circle $|z| < kr$ $(l < k < 1)$ satisfies the inequality

$$n_\psi(kr) < \nu_1(l, k)(h+2N) r^{\rho(r)}, \tag{2.78}$$

in which $\nu_1(l, k)$ is a constant depending on l and k. With the aid of (2.78) we must now give an estimate of the number of zeros of the function $F(z)$ inside the sector $|z| \leqslant R$, and $|\arg z| \leqslant \sigma < \pi/2$. For this purpose we choose a number k $(\sin\sigma < k < 1)$ and draw a circle of radius kr with center at r. This circle will intersect the sides of the angle $\arg z = \pm\sigma$ at the points

$$|z| = r\left(\cos\sigma \pm \sqrt{k^2 - \sin^2\sigma}\right).$$

Let us set $R = r\cos\sigma$ and $s = \sqrt{k^2 - \sin^2\sigma}/\cos\sigma$, and let us denote by K_R the region

$$|\arg z| \leqslant \sigma, \quad R(1-s) \leqslant |z| < R(1+s).$$

The region K_R will lie entirely within the circle $|z - r| \leqslant kr$, and in view of (2.78) we obtain

$$n(K_R) < \nu_1(l, k)(h+2N)\, r^{\rho(r)},$$

where $n(K_R)$ is the number of zeros of the function $F(z)$ in the region K_R, while $\nu_1(l, k)$ is a constant. Making use of the equation $R = r\cos\sigma$ we can rewrite the last inequality in the form

$$n(K_R) < \nu_2(l, k, \sigma)(h+2N)\, R^{\rho(R)}, \tag{2.79}$$

where $\nu_2(l, k, \sigma)$ is a constant depending on the indicated quantities. For given $R > 0$, we take a sequence of numbers

$$R_0 = R(1+s)^{-1},\ R_1 = R_0 q,\ \ldots,\ R_p = R_0 q^p,\ \ldots \left(q = \frac{1-s}{1+s}\right). \tag{2.80}$$

From the definition of proximate order one can easily see[14] that for some r_0 and $r_2 > r_1 > r_0$

$$r_1^{\rho(r_1)} < \left(\frac{r_1}{r_2}\right)^{\frac{\rho}{2}} r_2^{\rho(r_2)}.$$

The sequence (2.80) will be assumed to be finite and to satisfy the condition $R_p > r_0$.

Writing down the inequality (2.79) for each one of the regions K_R and adding these inequalities we obtain

$$n(R, -\sigma, \sigma) < O(1) + (1 + q^{\frac{\rho}{2}} + \ldots + q^{\frac{p\rho}{2}})\,\nu_2(l, k, \sigma)(h+2N)\, R_0^{\rho(R_0)},$$

or asymptotically

$$n(R, -\sigma, \sigma) < \varkappa(l, \sigma)(h+2N)\, R^{\rho(R)}, \tag{2.81}$$

where $\varkappa(l, \sigma)$ is a constant depending on the indicated quantities. This completes the proof of Lemma 11. We now proceed with the proof of Theorem 7.

Let a_n be the zeros of the function $F(z)$ within the angle $|\arg z| \leqslant \sigma_1$ ($\sigma < \sigma_1 < \pi/2$). We assume that these roots have been arranged in the order of increasing magnitude. The canonical product

$$f(z) = \prod_1^\infty G\left(\frac{z}{a_n};\ p\right) \qquad (p = [\rho])$$

is an entire function of no higher than normal type for the proximate order $\rho(r)$. By Theorem 6, to every positive number η there corresponds a set E_η

[14] Indeed, $[\ln r^{\rho(r)}]' = \dfrac{\rho(r)}{r} + \rho'(r)\ln r = \dfrac{\rho}{r} + o\left(\dfrac{1}{r}\right) > \dfrac{\rho}{2r}$. Integrating both sides of this inequality from r_1 to r_2, we obtain the required result.

$[\overline{m}^*(E_\eta) < \eta]$ such that if $r \bar{\in} E_\eta$ the function $h_{f,r}(\theta)$ will constitute an equicontinuous family.

The fraction

$$\chi(z) = \frac{F(z)}{f(z)}$$

is a non-null holomorphic function of type no higher than normal and of proximate order $\rho(r)$ inside the angle $|\arg z| \leqslant \sigma_2$, $\sigma < \sigma_2 < \sigma_1$ (Chapter I, § 8). Therefore for some $N_\chi^{(1)} > 0$ we have the asymptotic inequality

$$\ln|\chi(z)| < N_\chi^{(1)} r^{\rho(r)}. \tag{2.82}$$

On the other hand, from the condition (2.12) of the theorem, and from the upper bound of $\ln|f(z)|$

$$\ln|f(z)| < (\sigma_f + \varepsilon) r^{\rho(r)},$$

it follows that each circle of the type

$$|z - r| < lr$$

contains a point $z_{1,r}$ at which

$$\ln|\chi(z_{1,r})| > N_\chi^{(2)} r^{\rho(r)} \quad (N_\chi^{(2)} \text{ is a constant}). \tag{2.83}$$

The function $\chi(z)$ satisfies all the conditions of the theorem and *does not have any zeros within the angle* $|\arg z| \leqslant \sigma_2 \leqslant \pi/2$. It remains only to prove the theorem for this particular case.

We shall give an estimate of the function $|\ln|\chi(z)||$ inside the smaller angle $|\arg z| \leqslant \sigma_3$ $(\sigma < \sigma_3 < \sigma_2)$. For this we make use of Lemma 10. From (2.82) and (2.83) it follows that the function

$$\theta(u) = \frac{\chi(r+u)}{\chi(z_{1,r})}$$

satisfies, within the circle $|u| \leqslant r \sin \sigma_2$, the inequality

$$\ln|\theta(u)| < N_\chi^{(3)} r^{\rho(r)} \qquad (N_\chi^{(3)} \text{ is a constant})$$

and at the point $u_0 = z_{1,r} - r$ of this circle, $\theta(u_0) = 1$.

Moreover, the function $\theta(u)$ has no zeros in this circle. From Lemma 10 we see that in the smaller circle $|u| < r \sin \sigma_3$ the following inequality is valid:

$$\big|\ln|\theta(u)|\big| < N_\chi^{(4)} r^{\rho(r)} \qquad \big(N_\chi^{(4)} \text{ is a constant}\big).$$

Finally, returning to the function $\chi(z)$, we find that everywhere within the angle $|\arg z| \leqslant \sigma_3$ the inequality

$$\big|\ln|\chi(z)|\big| < N_\chi r^{\rho(r)} \qquad (N_\chi \text{ is a constant}) \tag{2.84}$$

is valid.

With the aid of the bound (2.84) we establish the equicontinuity of the family of functions of θ when $r > r_0$ and η is arbitrary.

For this purpose we express the harmonic functions $\ln|\chi(r+u)|$ in the circle $|u| \leqslant R$ ($R = r\sin\sigma_2$) by means of the Poisson integral:

$$\ln|\chi(r+u)| = \frac{1}{2\pi}\int_0^{2\pi} \ln|\chi(r+Re^{i\psi})| \operatorname{Re}\left\{\frac{Re^{i\psi}+u}{Re^{i\psi}-u}\right\} d\psi.$$

After we have chosen $k = (\sin\sigma_3)/\sin\sigma_2$ and the numbers u_1 and u_2 in the circle $|u| \leqslant kR$, we have

$$\ln|\chi(r+u_1)| - \ln|\chi(r+u_2)|$$
$$= \frac{1}{2\pi}\int_0^{2\pi} \ln|\chi(r+Re^{i\psi})| \operatorname{Re}\left\{\frac{e^{i\psi}+u_1R^{-1}}{e^{i\psi}-u_1R^{-1}} - \frac{e^{i\psi}+u_2R^{-1}}{e^{i\psi}-u_2R^{-1}}\right\} d\psi,$$

and in view of (2.84) we obtain

$$\left|\ln|\chi(r+u_1)| - \ln|\chi(r+u_2)|\right|$$
$$< N_\chi(2r)^{\rho(2r)} \frac{1}{2\pi}\int_0^{2\pi} \left|\operatorname{Re}\left\{\frac{e^{i\psi}+u_1R^{-1}}{e^{i\psi}-u_1R^{-1}} - \frac{e^{i\psi}+u_2R^{-1}}{e^{i\psi}-u_2R^{-1}}\right\}\right| d\psi.$$

From this inequality it follows that for the given $\varepsilon > 0$, and for some $\delta_1 > 0$, and for $|u_1 - u_2|R^{-1} < \delta_1$,

$$\left|\ln|\chi(r+u_1)| - \ln|\chi(r+u_2)|\right| < \varepsilon r^{\rho(r)}. \qquad (2.85)$$

By choosing $\delta > 0$ sufficiently small, we deduce directly from (2.85) that if $|\theta_1| \leqslant \sigma_3$, $|\theta_2| \leqslant \sigma_3$, and $|\theta_1 - \theta_2| < \delta$, then

$$\left|\ln|\chi(re^{i\theta_1})| - \ln|\chi(re^{i\theta_2})|\right| < \varepsilon r^{\rho(r)} \qquad (r > 0). \qquad (2.86)$$

The function $F(z)$ can be expressed in the form

$$F(z) = \chi(z) f(z).$$

By Theorem 6, the family of functions

$$\frac{\ln|f(re^{i\theta})|}{r^{\rho(r)}}$$

is equicontinuous for r not belonging to some set of arbitrarily small relative measure. Combining this with (2.86), we see that for any $\eta > 0$ the family of functions

$$h_{F,r}(\theta) = \frac{\ln|F(re^{i\theta})|}{r^{\rho(r)}}$$

is equicontinuous in θ if $|\theta| \leqslant \sigma_3 < \pi/2$, and $r \,\overline{\in}\, E_\eta$, where E_η is some set of positive numbers with an upper relative measure less than η. This completes the proof of the theorem.

CHAPTER III

FUNCTIONS OF COMPLETELY REGULAR GROWTH

In this chapter we shall investigate a problem which can be considered as the inverse problem to that of Chapter II. There we established that regularity of the distribution of the zeros of an entire function implies regularity of growth. In this chapter we shall show that regularity of growth implies regularity of the distribution of the zeros of the function.[1]

A function $F(z)$ that is holomorphic and of proximate order $\rho(r)$ within some angle (θ_1, θ_2) will be called a *function of completely regular growth on the ray* $\arg z = \theta$ if the limit

$$h_F(\theta) = \lim_{r\to\infty} \frac{\ln|F(re^{i\theta})|}{r^{\rho(r)}}$$

exists under the condition that r goes to infinity by taking on all positive values except possibly those of a set of zero relative measure (an E^0-set). We shall sometimes write lim* to express this situation.

We shall say that a *function $F(z)$ is of completely regular growth on some set of rays $R_{\mathfrak{M}}$* ($\mathfrak{M}$ is the set of values of θ) if the function

$$h_{F,r}(\theta) = \frac{\ln|F(re^{i\theta})|}{r^{\rho(r)}}$$

converges uniformly to $h_F(\theta)$ for $\theta \in \mathfrak{M}$ when r goes to infinity by taking on all positive values except possibly for a set $E_{\mathfrak{M}}$ of zero relative measure, this set being the same for all rays $R_{\mathfrak{M}}$. The set $E_{\mathfrak{M}}$ will be called the *exceptional set for the given function.*

We shall say that $F(z)$ *is a function of completely regular growth within the angle* (θ_1, θ_2) if this is true for every closed interior angle, and we say simply that $F(z)$ *is of completely regular growth* if it is an entire function and is of regular growth in the entire plane, i.e., for $0 \leqslant \theta \leqslant 2\pi$.

The functions with regularly distributed zeros considered in the preceding chapter are functions of completely regular growth. Indeed, it follows from Theorems 1 and 2 of Chapter II that the function $h_{f,r}(\theta)$ for these functions converges uniformly to $h_f(\theta)$ if r goes to infinity avoiding those values for which the circle $|z| = r$ cuts a circle of some C^0-set. It is obvious that this exceptional set of values of r is an E^0-set.

[1] The basic results of the first three sections of this chapter are due to A. Pfluger [1].

In this chapter we shall show that if $F(z)$ is a function of completely regular growth within the angle (θ_1, θ_2), then the set of its zeros possesses an angular density inside this angle (for an exact formulation see Theorem 2). Furthermore, if $f(z)$ is an entire function of completely regular growth, then its zeros are regularly distributed. Thus, the regular distribution of zeros is a characteristic property of entire functions of completely regular growth. If we combine this with Theorems 1 and 2 of Chapter II, we can conclude that an entire function of completely regular growth satisfies, in the complement of some C^0-set, the asymptotic equation.[2]

$$\ln|f(re^{i\theta})| \approx h_f(\theta) r^{\rho(r)}.$$

In order to obtain these results we shall generalize the formula of Jensen (Chapter I, § 5) by considering a holomorphic function in some sector instead of in a circle. If the sector angle is 2π, this formula takes on the usual form of Jensen.

In § 5 we shall investigate holomorphic functions of some proximate order $\rho(r)$ with a sinusoidal indicator inside some angle. Functions of this type occur in various applications.

The results presented in this section will be used in Chapter VI in the investigation of the zeros of exponential sums.

1. Set of Rays of Completely Regular Growth

In the definition of completely regular growth on some ray, an important role is played by a certain exceptional set E_θ of zero relative measure which depends, in general, on the selected ray. We shall prove, however, that if the holomorphic function $F(z)$ of proximate order $\rho(r)$ within a certain angle has regular growth on each ray of a certain set $R_{\mathfrak{M}}$ then one can construct a set $E_{\mathfrak{M}}$ of zero relative measure in such a way that for $r \bar{\in} E_{\mathfrak{M}}$, $\theta \in \mathfrak{M}$, and $r \to \infty$,

$$\ln|F(re^{i\theta})| \approx h_F(\theta) r^{\rho(r)}.$$

We shall also prove that $F(z)$ is a function of completely regular growth on every ray which is a limiting ray of the set of rays along which $F(z)$ is of completely regular growth.

THEOREM 1. *If a holomorphic function of proximate order $\rho(r)$ within an angle (α, β) is a function of completely regular growth on each ray of some set $R_{\mathfrak{M}}$ belonging to the interior angle ($\alpha < \theta_1 \leqslant \theta \leqslant \theta_2 < \beta$), then it is of completely regular growth on the set of rays $R_{\overline{\mathfrak{M}}}$.*

---

[2] An analogous assertion is true for all functions which are holomorphic and of completely regular growth within some angle $\alpha < \arg z < \beta$. We shall not prove this more general proposition.

1. For the proof we note that if a function $F(z)$ is of completely regular growth on some ray $\arg z = \gamma$ lying within the angle (θ_1, θ_2), then for every $\varepsilon > 0$, and for some positive number $l > 1$, and for all sufficiently large r, each of the circles

$$\left| z - re^{i\frac{\theta_1+\theta_2}{2}} \right| < rl \sin \frac{\theta_2 - \theta_1}{2}$$

contains a point $z = re^{i\gamma}$ for which

$$\ln |F(re^{i\gamma})| > [h_F(\gamma) - \varepsilon] r^{\rho(r)}.$$

Indeed, the ratio of the arc length cut from this circle by the ray $\arg z = \gamma$ to r is a constant quantity, while the ratio of the measure of the part of the exceptional set lying on this arc to r goes to zero.

The conditions of Theorem 7, Chapter II, are thus seen to be satisfied, and therefore for every positive number η there exists a set E_η of positive numbers with upper relative measure less than η such that for $r \,\overline{\in}\, E_\eta$ the functions $h_{F,r}(\theta)$ constitute an equicontinuous family on the interval $\alpha < \theta_1 \leqslant \theta \leqslant \theta_2 < \beta$.

2. For a given $\varepsilon > 0$ we choose $\sigma > 0$ so that for $r \,\overline{\in}\, E_\eta$ and $|\theta' - \theta''| < \sigma$ $(\theta_1 \leqslant \theta' < \theta'' \leqslant \theta_2)$ the inequality

$$|h_{F,r}(\theta') - h_{F,r}(\theta'')| < \frac{\varepsilon}{3} \tag{3.00}$$

is valid. After that, we select from the set $R_{\mathfrak{M}}$ a finite system of rays $\arg z = \theta_j$ $(j = 1, 2, \ldots, n_\sigma)$ so that the σ-neighborhoods of the points θ_j will cover the entire set $\mathfrak{M}$. Next we choose the number r_ε so that for $r > r_\varepsilon$ and $r \,\overline{\in}\, E_{\theta_j}$ the following inequality will hold:

$$|h_{F,r}(\theta_j) - h_F(\theta_j)| < \frac{\varepsilon}{3} \qquad (j = 1, 2, \ldots, n_\sigma).$$

Adding to the set E_η all values of r less than r_ε and all the sets E_{θ_j} $(j = 1, 2, \ldots, n_\sigma)$ we obtain a set $\tilde{E}_\eta$ such that $\overline{m}^*(\tilde{E}_\eta) < \eta$ and

$$|h_{F,r}(\theta) - h_F(\theta)| < \varepsilon \tag{3.01}$$

if $\theta \in \overline{\mathfrak{M}}$, and $r \,\overline{\in}\, \tilde{E}_\eta$.

3. We now select two sequences of positive numbers, $\varepsilon_n \to 0$ and $\eta_n \to 0$ $(n = 1, 2, 3, \ldots)$, and determine the sequence of sets E_n so that $\overline{m}^*(E_n) < \eta_n$ and that for $r \,\overline{\in}\, E_n$ it is true that

$$|h_{F,r}(\theta) - h_F(\theta)| < \varepsilon_n \tag{3.02}$$

for all $\theta \in \overline{\mathfrak{M}}$. By mathematical induction we construct a sequence of numbers $r_1 < r_2 < \ldots < r_n < \ldots$ $(r_n \to \infty)$. We select $r_1 > 0$ arbitrarily. Let us assume that we have chosen the numbers $r_1, r_2, \ldots, r_{n-1}$. Then we pick r_n so that $r \geqslant r_n$, and the following inequality is satisfied

$$\operatorname{mes}(E_k^{r_{n-1}}) < \frac{1}{n}\eta_n r, \quad k \leqslant n-2, \quad \operatorname{mes}(E_{n-1}^r) < \eta_n r, \quad \operatorname{mes}(E_n^r) < \eta_n r \tag{3.03}$$

(E^r denotes, as before, the intersection of the set E with the interval $(0, r)$).

Such a choice is obviously possible. After the sequence $\{r_n\}$ has been selected we determine a set E such that for $r_n < r < r_{n+1}$, $E = E_n$. From (3.03) it follows that for $r_n < r < r_{n+1}$ the inequality mes $E^r < (2\eta_n + \eta_{n-1})r$ will hold and hence that $m^*(E) = 0$.

Furthermore, it follows from (3.02) that for $r > r_n$ and $r \overline{\in} E$, we have the inequality $|h_{F,r}(\theta) - h_F(\theta)| < \varepsilon_n$ $(\theta \in \overline{\mathfrak{M}})$. Thus $F(z)$ is of completely regular growth on $\overline{\mathfrak{M}}$.

In subsection 3 we proved a statement which we shall use below. For this reason we will formulate it as a separate lemma.

LEMMA 1. *If the function $F(z)$ is holomorphic and is of proximate order $\rho(z)$ within the angle (α, β), and if for every pair of numbers $\varepsilon > 0$ and $\eta > 0$ there exists a set $E_{\varepsilon,\eta}$, whose upper relative measure is less than η, such that*

$$|h_{F,r}(\theta) - h_F(\theta)| < \varepsilon$$

for $r \overline{\in} E_{\varepsilon,\eta}$ and $\alpha < \theta < \beta$, then $F(z)$ is a function of completely regular growth within the angle (α, β).[3]

From Theorem 1, one can directly obtain the next result.

COROLLARY. *The set of rays within the angle $\alpha < \theta_1 \leqslant \theta \leqslant \theta_2 < \beta$ on which $F(z)$ is of completely regular growth is closed.*

It follows from this, in particular, that an entire function of completely regular growth on rays that form an everywhere dense set is a function of completely regular growth in the entire plane.[3a]

2. A Generalized Formula of Jensen. Investigation of the Function $J_F^r(\theta)$

For the determination of the number of zeros of a holomorphic function $F(z)$ in the sector $\vartheta \leqslant \arg z \leqslant \theta$, $|z| \leqslant r$, a number which we shall denote by $n(r, \vartheta, \theta)$, we make use of the known principle of the argument. Here we assume that no zeros of $F(z)$ lie on the boundary of the sector.

We have the equation

$$2\pi n(r, \vartheta, \theta) = \int_0^r \frac{d[\arg F(te^{i\theta})]}{dt}\,dt + \int_\vartheta^\theta \frac{d[\arg F(re^{i\varphi})]}{r\,d\varphi}\,r\,d\varphi - \int_0^r \frac{d[\arg F(te^{i\vartheta})]}{dt}\,dt.$$

[3] This lemma remains true if in place of the angle $\alpha < \theta < \beta$ one considers the function on a single ray.

[3a] Azarin [1] showed that every closed set of rays is the set of rays of completely regular growth for some entire function with a given proximate order $\rho(r) \to \rho > 0$. If $\rho(r) \to 0$, then as Gol'dberg and Ostrovskiĭ [2] showed, the set of rays of completely regular growth of an entire function is either empty or covers the whole plane. It was remarked by Grišin [1] that the second case occurs when $\ln M_f(r) = o(\ln^2 r)$.

By the Cauchy-Riemann formula we transform the last equation to

$$2\pi n(r, \vartheta, \theta) = -\int_0^r \frac{d\{\ln|F(te^{i\varphi})|\}_{\varphi=\theta}}{t\,d\varphi}\,dt$$

$$+\int_0^r \frac{d\{\ln|F(te^{i\varphi})|\}_{\varphi=\vartheta}}{t\,d\varphi}\,dt + r\int_\vartheta^\theta \frac{d}{dt}\{\ln|F(te^{i\varphi})|\}_{t=r}\,d\varphi.$$

Assuming that $|F(0)| = 1$, we introduce the function

$$J_F^r(\theta) = \int_0^r \frac{\ln|F(te^{i\theta})|}{t}\,dt.$$

This function will play an important role in the present chapter, and we shall consider some of its properties in this section.

In the proof of theorems in this chapter we shall frequently assume that $|F(0)| = 1$. This assumption will not reduce the generality of our results.

Making use of the function $J_F^r(\theta)$, we may rewrite the formula for the number of zeros of the function $F(z)$ within a sector in the form

$$2\pi n(r, \vartheta, \theta) = -\left\{\frac{d}{d\varphi} J_F^r(\varphi)\right\}_{\varphi=\theta} + \left\{\frac{d}{d\varphi} J_F^r(\varphi)\right\}_{\varphi=\vartheta}$$

$$+ r\int_\vartheta^\theta \left\{\frac{d\ln|F(te^{i\varphi})|}{dt}\right\}_{t=r} d\varphi.$$

Dividing each side of this equation by $2\pi r$, integrating the result from zero to r, and introducing the notation

$$N(r, \vartheta, \theta) = \int_0^r \frac{n(t, \vartheta, \theta)}{t}\,dt,$$

we obtain

$$N(r, \vartheta, \theta) = \frac{1}{2\pi}\left[\frac{d}{d\varphi}\int_0^r J_F^t(\varphi)\frac{dt}{t}\right]_{\varphi=\vartheta} - \frac{1}{2\pi}\left[\frac{d}{d\varphi}\int_0^r J_F(\varphi)\frac{dt}{t}\right]_{\varphi=\theta}$$

$$+\frac{1}{2\pi}\int_\vartheta^\theta \ln|F(re^{i\varphi})|\,d\varphi. \qquad (3.04)$$

This formula yields a generalization of Jensen's formula, and for $\theta = \vartheta + 2\pi$ it goes over into the ordinary formula of Jensen.

We shall use this formula for proving the existence of the angular density of the set of zeros of a function of completely regular growth, and for obtaining asymptotic formulas for the number of zeros of these functions within a sector.

For this purpose we first investigate the asymptotic behavior of the function $J_F^r(\theta)$.

LEMMA 2. *If $F(z)$ is a holomorphic function of proximate order $\rho(r)$ within the angle $\alpha \leqslant \arg z \leqslant \beta$, and if the function is of completely regular growth on the ray* $\arg z = \theta$ $(\alpha < \theta < \beta)$, *then the limit*

$$\lim_{r\to\infty} r^{-\rho(r)} J_F^r(\theta) = \frac{1}{\rho} h_F(\theta) \tag{3.05}$$

will exist.

This assertion can be established quite simply if the function

$$h_{F,r}(\theta) = \frac{\ln|F(re^{i\theta})|}{r^{\rho(r)}}$$

has no exceptional set and hence the limit

$$h_F(\theta) = \lim_{r\to\infty} h_{F,r}(\theta)$$

exists in the ordinary sense. Indeed, in this case the equation (3.05) is a direct consequence of property (5) of a proximate order.

In the general case, when the function has an exceptional set, the proof is considerably more complicated. We note that from the definition of the indicator $h_F(\theta)$ and from property (4) of a proximate order it follows that

$$\overline{\lim_{r\to\infty}}\, r^{-\rho(r)} J_F^r(\theta) \leqslant \frac{1}{\rho} h_F(\theta). \tag{3.06}$$

One needs only to prove that

$$\underline{\lim_{r\to\infty}}\, r^{-\rho(r)} J_F^r(\theta) \geqslant \frac{1}{\rho} h_F(\theta).$$

To establish this inequality one has to estimate the function $\ln|F(te^{i\theta})|$ from below. On the complement of the exceptional set E we have the following asymptotic equation:

$$\ln|F(te^{i\theta})| \approx h_F(\theta)\, t^{\rho(t)}.$$

The problem, therefore, is reduced to the determination of a lower bound on the set E. It is natural to attempt to do this on the basis of the general Theorem 11 of Chapter I on estimation of a holomorphic function from below.

First we shall consider the case when $\theta = 0$ and $h_F(0) = 0$. From the continuity of the indicator $h_F(\theta)$ it is easy to prove, in this case, that for some $\delta > 0$ and $|\theta| < 2\delta$ we have the inequality $|h_F(\theta)| < \varepsilon/4$ and that, in view of the asymptotic inequality

$$\ln|F(re^{i\theta})| < \left(h_F(\theta) + \frac{\varepsilon}{4}\right) r^{\rho(r)},$$

we have the following asymptotic inequality within the angle $|\theta| < \delta$:

$$\ln|F(re^{i\theta})| < \frac{\varepsilon}{2} r^{\rho(r)}. \tag{3.07}$$

Let E_ε be the subset of the positive ray on which

$$\ln|F(t)| < -\frac{\varepsilon}{4} t^{\rho(t)}. \tag{3.08}$$

Since the function $F(z)$ is of completely regular growth on this ray, the relative measure of the set E_ε must be zero. This set can therefore be included in the set $\bar{E}_\varepsilon$ of intervals J_k ($k = 1, 2, 3, \ldots$) of zero relative measure.

Let us denote by d_k the length of the interval J_k and by R_k the abscissa of the right end of this interval. With the point R_k as center we draw a circle of radius $4ed_k$. From the fact that $m^*(\bar{E}_\varepsilon) = 0$ it follows that

$$\lim_{r\to\infty}\left(\frac{1}{r}\sum_{R_k<r} d_k\right) = 0,$$

and hence that

$$\lim_{k\to\infty}\frac{d_k}{R_k} = 0. \tag{3.09}$$

Now we apply Theorem 11 of Chapter I to the function

$$\varphi_k(u) = \frac{F(R_k+u)}{F(R_k)}$$

in the circle C_k: $|u| \leqslant 4ed_k$. For all sufficiently large R_k, in view of the inequality (3.09) and an inequality opposite to (3.08) at the point R_k, the function $\varphi_k(u)$ satisfies in C_k the inequality

$$\ln|\varphi_k(u)| < \frac{\varepsilon}{2}(R_k + 4ed_k)^{\rho(R_k+4ed_k)} + \frac{\varepsilon}{4} R_k^{\rho(R_k)}.$$

Making use of property (2) of a proximate order, we now obtain

$$\ln|\varphi_k(u)| < \varepsilon R_k^{\rho(R_k)}.$$

By Theorem 11 of Chapter I we see that in the circle $|u| \leqslant 2d_k$, but outside the exceptional circles with sum of their radii less than $4\eta d_k$, the following inequality holds:

$$\ln|\varphi_k(u)| > -H(\eta)\varepsilon R_k^{\rho(R_k)}.$$

Returning now to the function $F(z)$ we may conclude that in the circle $|z - R_k| \leqslant 2d_k$, but outside the exceptional circles with sum of their radii less than $4\eta d_k$,

$$\ln|F(z)| > -\left(H(\eta)+\frac{1}{4}\right)\varepsilon R_k^{\rho(R_k)} > -2\varepsilon\left(H(\eta)+\frac{1}{4}\right) r^{\rho(r)}. \tag{3.10}$$

When $\eta < \frac{1}{8}$ we find a circle C_k with center at R_k, and with a radius whose length lies between d_k and $2d_k$. This circle must not intersect the exceptional circles. On the entire circumference which includes the segment d_k inequality (3.10) will be satisfied.

The circles which are bounded by the circumferences C_k may intersect, and they may form "clusters" of regions Ω_p. The diameter δ_p of each region will not exceed the sum of the diameters of the constituent circles. Therefore, if $R_p = \max_{z \in \Omega_p} |z|$,

$$\lim_{r\to\infty}\left(\frac{1}{r}\sum_{R_p\leqslant r}\delta_p\right)=0$$

and

$$\lim_{p\to\infty}\frac{\delta_p}{R_p}=0. \tag{3.11}$$

Let a_n be the zeros of $F(z)$. We set

$$\Phi_p(z)=\prod_{a_n\in\Omega_p}\left(1-\frac{z}{a_n}\right)$$

and

$$\psi_p(z)=F(z)\,\Phi_p^{-1}(z).$$

From (3.11) it follows that if p is greater than some fixed number p_0, then

$$\ln|\Phi_p(z)|<0 \qquad (z\in\Omega_p).$$

Combining this with (3.10) we see that for all sufficiently large values of p we must have on the entire boundary of Ω_p the inequality

$$\ln|\psi_p(z)|>-2\left[H\left(\frac{1}{8}\right)+1\right]\varepsilon R_p^{\rho(R_p)}. \tag{3.12}$$

The function $\ln|\psi_p(z)|$ is harmonic in Ω_p, and hence, by the minimum principle, the inequality (3.12) must hold in this entire region.

Let I_p be the interval which is obtained by cutting the region Ω_p with the positive ray, and let $\tilde{E}_\varepsilon$ be the set of all such intervals. It is obvious that $m^*(\tilde{E}_\varepsilon)=0$, and $\tilde{E}_\varepsilon \supset E_\varepsilon$.

From (3.12) we see that on each interval I_p

$$\ln|F(t)|>\ln|\Phi_p(t)|-m\varepsilon t^{\rho(t)} \qquad (t\in I_p),$$

where $m = 2[H(\frac{1}{8})+1]$. Furthermore, if $t \bar{\in} \tilde{E}_\varepsilon$,

$$\ln|F(t)|\geqslant-\frac{\varepsilon}{4}t^{\rho(t)}.$$

From these inequalities it follows that for sufficiently large values of r_0

$$\int_{r_0}^{r}\frac{\ln|F(t)|}{t}\,dt>-m\varepsilon\int_{r_0}^{r}t^{\rho(t)-1}\,dt + \sum_{R_p-\delta_p<r}\int_{(I_p)}\frac{\ln|\Phi_p(t)|}{t}\,dt. \tag{3.13}$$

Next, from the definition of the function $\Phi_p(z)$ we have

$$\int_{I_p} \frac{\ln|\Phi_p(t)|}{t}\,dt \geqslant \frac{1}{R_p-\delta_p} \sum_{a_k \in \Omega_p} \int_{R_p-\delta_p}^{R_p} \ln\left|\frac{t}{|a_k|}-1\right| dt. \tag{3.14}$$

For the estimation of the right hand side we note that if $R-\delta \leqslant a < R$, then

$$\int_{R-\delta}^{R} \ln\left|\frac{t}{a}-1\right| dt = a \int_{\frac{R-\delta}{a}}^{\frac{R}{a}} \ln|u-1|\,du$$

$$= a \int_{\frac{R-a-\delta}{a}}^{\frac{R-a}{a}} \ln|v|\,dv \geqslant 2a \int_0^{\delta/2a} \ln v\,dv \geqslant 4\delta\left(\ln\frac{\delta}{2R}-1\right).$$

Substitution into (3.14) yields

$$\int_{(I_p)} \frac{\ln|\Phi_p(t)|}{t}\,dt \geqslant \frac{4\delta_p}{R_p-\delta_p}\left(\ln\frac{\delta_p}{2R_p}-1\right) n_p,$$

where n_p is the number of zeros of the function $F(z)$ in the region Ω_p. Returning now to (3.13) we may write

$$\int_0^r \frac{\ln|F(t)|}{t}\,dt > -\frac{m}{\rho}\varepsilon r^{\rho(r)}$$

$$+ \sum_{R_p-\delta_p \leqslant r} \frac{4\delta_p}{R_p-\delta_p}\left(\ln\frac{\delta_p}{R_p}-1\right) n_p + o(r^{\rho(r)}).$$

From (3.11) it follows that for all sufficiently large p

$$\left|\frac{4\delta_p}{R_p-\delta_p}\left(\ln\frac{\delta_p}{R_p}-1\right)\right| < \varepsilon$$

and therefore

$$\int_0^r \frac{\ln|F(t)|}{t}\,dt > -\frac{m}{\rho}\varepsilon r^{\rho(r)} - \Big(\sum_{R_p-\delta_p \leqslant r} n_p\Big)\varepsilon + o(r^{\rho(r)}). \tag{3.15}$$

It is obvious that for every $\sigma > 0$

$$\sum_{R_p-\delta_p \leqslant r} n_p \leqslant n(r, -\sigma, \sigma) + O(1).$$

Furthermore, by Lemma 11 of Chapter II, we have

$$n(r, -\sigma, \sigma) < c_F r^{\rho(r)},$$

where c_F is some constant. Returning to the inequality (3.15), we obtain

$$\int_0^r \frac{\ln|F(t)|}{t}\,dt > -\left(\frac{m}{\rho}+c_F\right)\varepsilon r^{\rho(r)} + o\left(r^{\rho(r)}\right).$$

From this it follows, in view of the arbitrariness of $\varepsilon > 0$, that

$$\varliminf_{r\to\infty}\left\{r^{-\rho(r)}\int_0^r \frac{\ln|F(t)|}{t}\,dt\right\} \geqslant 0.$$

Let us free ourselves now from the restriction that $h_F(0) = 0$. For this purpose we construct an auxiliary function $W(z)$ (see § 17, Chapter I) which has no zeros on the positive ray and is such that $h_W(0) = h_F(0)$.

The function

$$\Phi(z) = \frac{F(z)}{W(z)}$$

has completely regular growth, and $h_\Phi(0) = 0$. We thus find that

$$\varliminf_{r\to\infty} r^{-\rho(r)}\int_0^r \frac{\ln|F(t)|}{t}\,dt$$

$$= \lim_{r\to\infty} r^{-\rho(r)}\int_0^r \frac{\ln|\Phi(t)|}{t}\,dt$$

$$+ \varliminf_{r\to\infty} r^{-\rho(r)}\int_0^r \frac{\ln|W(t)|}{t}\,dt \geqslant \rho^{-1}h_F(0).$$

By a rotation of the plane one can superpose any ray $\arg z = \theta$ of the angle $\alpha \leqslant \theta \leqslant \beta$ onto the positive ray, and thus obtain the inequality

$$\varliminf_{r\to\infty} r^{-\rho(r)}\int_0^r \frac{\ln|F(te^{i\theta})|}{t}\,dt \geqslant \rho^{-1}h_F(\theta).$$

It is sufficient to combine this inequality with the inequality (3.06) to complete the proof of the lemma.

We note the following consequence of the lemma.

COROLLARY. *If $F(z)$ is a holomorphic function of proximate order $\rho(r)$ within the angle $\alpha \leqslant \arg z \leqslant \beta$ and of completely regular growth on the ray $\arg z = \theta$, $\alpha < \theta < \beta$, then the following limit exists:*

$$\lim_{r\to\infty} r^{-\rho(r)}\int_{r_0}^r J_F^t(\theta)\frac{dt}{t} = \rho^{-2}h_F(\theta). \tag{3.05'}$$

The truth of this assertion follows directly from Lemma 2 and property (5) of the definition of proximate order (Chapter I, page 35). Lemma 2 and its corollary have valid converses, in which one does not have to assume that $F(z)$ is holomorphic.

LEMMA 3. *Suppose that the function $F(t)$ is defined for $t \geqslant 0$ and that*

$$h_F(0) = \overline{\lim_{r\to\infty}} \frac{\ln |F(r)|}{r^{\rho(r)}}.$$

Furthermore, if

$$\lim_{r\to\infty} r^{-\rho(r)} J_F^r = \rho^{-1} h_F(0) \tag{3.16}$$

or

$$\lim_{r\to\infty} r^{-\rho(r)} \int_0^r J_F^r \frac{dt}{t} = \rho^{-2} h_F(0), \tag{3.16'}$$

then the limit

$$\lim_{r\to\infty}{}^{*} \frac{\ln |F(r)|}{r^{\rho(r)}}$$

exists.

PROOF. The condition (3.16′) is a consequence of (3.16). Hence it is sufficient to give the proof under the assumption that condition (3.16′) holds. Without loss of generality we shall assume in the sequel that $h_F(0) = 0$. Let γ be some positive number and E be the set of positive numbers on which

$$\ln |F(t)| < -\gamma t^{\rho(t)}. \tag{3.17}$$

Let us denote by $\chi_E(t)$ the characteristic function of the set E, and try to estimate the value of the integral

$$\int_0^r \left\{ \int_0^u \chi_E(t) t^{\rho(t)} \frac{dt}{t} \right\} \frac{du}{u}.$$

We have

$$\int_0^r \left\{ \int_0^u \chi_E(t) t^{\rho(t)} \frac{dt}{t} \right\} \frac{du}{u} \geqslant \int_{\frac{r}{2}}^r \left\{ \int_0^u \chi_E(t) t^{\rho(t)} \frac{dt}{t} \right\} \frac{du}{u}$$
$$\geqslant \ln 2 \int_0^{\frac{r}{2}} \chi_E(t) t^{\rho(t)} \frac{dt}{t},$$

from which it follows that

$$\int_0^r \left\{ \int_0^u \chi_E(t) t^{\rho(t)} \frac{dt}{t} \right\} \frac{du}{u} \geqslant \ln 2 \int_{\frac{r}{4}}^{\frac{r}{2}} \chi_E(t) t^{\rho(t)} \frac{dt}{t} \geqslant \frac{2 \ln 2}{r} \left(\frac{r}{4}\right)^{\rho\left(\frac{r}{4}\right)} \int_{\frac{r}{4}}^{\frac{r}{2}} \chi_E(t)\, dt.$$

Finally, denoting by $m(E^r)$ the measure of the set E on the interval $(0, r)$ we obtain

$$\int_0^r \left\{ \int_0^u \chi_E(t)\, t^{\rho(t)} \frac{dt}{t} \right\} \frac{du}{u} \geqslant \frac{2 \ln 2}{4^{\rho+1}} \frac{m(E^{\frac{r}{2}}) - m(E^{\frac{r}{4}})}{r} r^{\rho(r)}.$$

From this inequality and from inequality (3.17), which is valid on the set E, we can conclude that

$$\int_0^r \left\{ \int_0^u \chi_E(t) \frac{\ln |F(t)|}{t} dt \right\} \frac{du}{u} \leqslant -\gamma \frac{2 \ln 2}{4^{\rho+1}} \frac{m(E^{\frac{r}{2}}) - m(E^{\frac{r}{4}})}{r} r^{\rho(r)}.$$

From this inequality and the asymptotic relation

$$\ln |F(t)| < \varepsilon t^{\rho(t)}$$

it is easy to obtain the inequality

$$\int_0^r \left\{ \int_0^u \frac{\ln |F(t)|}{t} dt \right\} \frac{du}{u} < \frac{\varepsilon}{\rho^2} r^{\rho(r)} - \gamma \frac{2 \ln 2}{4^{\rho+1}} \frac{m(E^{\frac{r}{2}}) - m(E^{\frac{r}{4}})}{r} r^{\rho(r)} + o(r^{\rho(r)}).$$

Dividing each side by $r^{\rho(r)}$, and taking the limit as $r \to \infty$, we obtain from the condition of the lemma

$$\lim_{r\to\infty} \frac{m(E^{\frac{r}{2}}) - m(E^{\frac{r}{4}})}{r} = 0.$$

From this equation we obtain

$$\overline{\lim_{r\to\infty}} \frac{m(E^r)}{r} = \lim_{r\to\infty} \frac{m(E^r) - m(E^{\frac{r}{2}})}{r} + \overline{\lim_{r\to\infty}} \frac{m(E^{\frac{r}{2}})}{r} = \frac{1}{2} \overline{\lim_{r\to\infty}} \frac{m(E^r)}{r},$$

and hence we see that the relative measure of the set E is zero. Thus we obtain the result that for every $\gamma > 0$

$$\left| \frac{\ln |F(r)|}{r^{\rho(r)}} \right| < \gamma$$

everywhere except possibly on a set of zero relative measure. It follows from Lemma 1 that the following limit exists:

$$\lim_{r\to\infty}{}^{*} \frac{\ln|F(r)|}{r^{\rho(r)}} = 0.$$

This completes the proof of the lemma.

Combining this lemma with the corollary of Lemma 2, we obtain the following theorem.

THEOREM 2. *In order that a function $F(z)$ which is holomorphic and of proximate order $\rho(r)$ within the angle $\alpha \leqslant \theta \leqslant \beta$ ($\theta = \arg z$) be of completely regular growth, it is necessary and sufficient that one of the following conditions hold*:

$$\rho^{-1}h_F(\theta) = \lim_{r\to\infty} r^{-\rho(r)} J_F^r(\theta), \quad \rho^{-2}h_F(\theta) = \lim_{r\to\infty} r^{-\rho(r)} \int_0^r J_F^t(\theta)\frac{dt}{t}.$$ [3b]

In conclusion we shall establish one more property of the function $J_F^r(\theta)$ which will be used in Chapter V.

LEMMA 4. *For an entire function $f(z)$ of proximate order $\rho(r)$ the function $J_f^r(\theta)$ satisfies the inequality $|J_f^r(\theta)| < k_f r^{\rho(r)}$, where k_f is a constant depending on the function $f(z)$.*

PROOF. First of all we note that from the asymptotic inequality

$$\ln|f(re^{i\theta})| < [h_f(\theta) + \varepsilon]\, r^{\rho(r)}$$

we have the following asymptotic inequality:

$$J_f^r(\theta) < \rho^{-1} \max_{0 \leqslant \theta < 2\pi} [h_f(\theta) + \varepsilon]\, r^{\rho(r)}. \tag{3.18}$$

In the proof of the estimate from below of the function $J_f^r(\theta)$ one must distinguish the cases of integral and nonintegral order ρ. Here we consider only the more complicated case of integral order.

The function $f(z)$ can be expressed in the form

$$\ln|f(z)| = \operatorname{Re}(P_{\rho-1}(z) + \delta_f(\tau r) z^\rho) + \sum_{|a_n| \leqslant \tau r} \ln\left|G\left(\frac{z}{a_n}; \rho - 1\right)\right| + \sum_{|a_n| > \tau r} \ln\left|G\left(\frac{z}{a_n}; \rho\right)\right|. \tag{3.19}$$

By Lemma 8 of § 17, Chapter I, the last sum ${}_\tau f(z)$ satisfies, for a sufficiently large τ, the inequality

$$\left|\ln|{}_\tau f(z)|\right| < \varepsilon r^{\rho(r)}. \tag{3.20}$$

[3b] A necessary and sufficient condition of a different kind was suggested by Gol'dberg (it is presented in Azarin [11]). He showed that completely regular growth is equivalent to the existence of the limit

$$\lim_{r\to\infty} r^{-\rho(r)}\left\{\int_\alpha^\beta \ln|f(re^{i\vartheta})|\,d\vartheta\right\}$$

for all α and β ($0 \leqslant \alpha < \beta \leqslant 2\pi$). Kondratjuk [3] proved a similar proposition for meromorphic functions. It is clear that either condition can be taken as the definition instead of the usual one. The advantage of these definitions is the absence of the exceptional sets. Azarin [11] and Agranovič [1] have introduced still another equivalent definition of functions of completely regular growth, one that is useful in the theory of functions of several variables. We state it for functions of a singled variable: a function $f(z)$, holomorphic inside an angle and of proximate order $\rho(r)$, is called a *function of completely regular growth* if the limit

$$\lim_{r\to\infty}\left\{\int_K \frac{\ln|f(rz)|}{r^{\rho(r)}}\varphi(z)\,d\sigma(z)\right\}$$

exists, where $\varphi(z)$ is an infinitely differentiable function with support in a compact set K, and $d\sigma(z)$ is the element of area; in other words we require the existence of the weak limit of $r^{-\rho(r)}\ln|f(rz)|$. Evidently this indicator, which is called the *circular indicator*, is equal to $|z|^\rho h(\vartheta)$.

From the inequality $n(t) < ct^{\rho(t)}$ one can quite easily obtain the inequality

$$r^{\lambda} \sum_{|a_n| < \tau r} \frac{1}{|a_n|^{\lambda}} = r^{\lambda} \int_0^{\tau r} \frac{dn(t)}{t^{\lambda}} < c_{\tau} r^{\rho(r)}$$

$$(c_{\tau} \text{ is a constant, } \lambda = 1, 2, \ldots, \rho - 1). \tag{3.21}$$

Furthermore,

$$|\delta(\tau r) z^{\rho}| < c'_{\tau} r^{\rho(r)} \qquad (c'_{\tau} \text{ is a constant}). \tag{3.22}$$

From (3.19), (3.20), (3.21) and (3.22) we find that

$$\ln|f(re^{i\theta})| > -c''_{\tau} r^{\rho(r)} + \sum_{|a_n| \leq \tau r} \ln\left|1 - \frac{z}{|a_n|}\right| \quad (c''_{\tau} \text{ is a constant}). \tag{3.23}$$

We note that

$$\int_0^r \ln\left|1 - \frac{t}{|a_n|}\right| \frac{dt}{t} = \int_0^{\frac{r}{|a_n|}} \ln|1 - u| \frac{du}{u} \geqslant \int_0^2 \ln|1 - u| \frac{du}{u} = -\frac{7}{24}\pi^2.$$

Hence, integrating both sides of the inequality (3.23) we get the asymptotic relation

$$J_f^r(\theta) = \int_0^r \frac{\ln|f(te^{i\theta})|}{t} dt > -\frac{c''_{\tau}}{\rho} r^{\rho(r)} - \pi^2 n(\tau r)$$

$$> -\left(\frac{c_{\tau}}{\rho} + \pi^2 \tau^{2\rho}\right) r^{\rho(r)}. \tag{3.24}$$

The relations (3.18) and (3.24) imply the truth of the lemma.

3. The Basic Theorem on Functions of Completely Regular Order

Starting with Jensen's generalized formula and using the established properties (see § 2) of the function $J_F^r(\theta)$, we can now prove the fundamental theorem on the distribution of the zeros of a function of completely regular growth. We formulate it as follows.

THEOREM 3. *If a holomorphic function $F(z)$ of order $\rho(r)$ has completely regular growth within an angle (θ_1, θ_2), then for all values of ϑ and θ $(\theta_1 < \vartheta < \theta < \theta_2)$ except possibly for a denumerable set the following limit will exist:*

$$\frac{1}{2\pi\rho} s_F(\vartheta, \theta) = \lim_{r \to \infty} \frac{n(r, \vartheta, \theta)}{r^{\rho(r)}}, \tag{3.25}$$

where

$$s_F(\vartheta, \theta) = \left[h'_F(\theta) - h'_F(\vartheta) + \rho^2 \int_{\vartheta}^{\theta} h_F(\varphi)\, d\varphi\right]. \tag{3.26}$$

The exceptional denumerable set can only consist of points for which $h'_F(\theta + 0) \neq h'_F(\theta - 0)$.

PROOF. First of all we transform formula (3.04) so as to eliminate the derivatives with respect to φ. For this purpose we average each side of the equation with respect to ϑ and θ, obtaining

$$\frac{1}{lk}\int_{\theta}^{\theta+k}\int_{\vartheta}^{\vartheta+l} N(r,\vartheta,\theta)\,d\vartheta\,d\theta$$

$$=\frac{1}{2\pi}\left\{\int_0^r \frac{J_F^t(\theta+k)-J_F^t(\theta)}{k}\frac{dt}{t}-\int_0^r \frac{J_F^t(\vartheta+l)-J_F^t(\vartheta)}{l}\frac{dt}{t}\right\}$$

$$+\frac{1}{2\pi kl}\int_{\theta}^{\theta+k}\int_{\vartheta}^{\vartheta+l}\int_{\vartheta}^{\theta}\ln|F(re^{i\varphi})|\,d\varphi\,d\vartheta\,d\theta. \tag{3.27}$$

Next we choose the numbers ϑ and θ in the formula (3.27) so that the derivatives $h_F'(\theta)$ and $h_F'(\vartheta)$ will exist and select numbers l_ε and k_ε in such a way that if $|l| < l_\varepsilon$ and $|k| < k_\varepsilon$, then

$$\left.\begin{aligned}&\left|\frac{h_F(\theta+k)-h_F(\theta)}{k}-h_F'(\theta)\right|<\frac{\varepsilon}{6}\\ \text{and}\qquad&\\ &\left|\frac{h_F(\vartheta+l)-h_F(\vartheta)}{l}-h_F'(\vartheta)\right|<\frac{\varepsilon}{6}.\end{aligned}\right\} \tag{3.28}$$

Furthermore, if k_ε and l_ε are sufficiently small, then

$$\left|\frac{1}{kl}\int_{\theta}^{\theta+k}\int_{\vartheta}^{\vartheta+l}\int_{\vartheta}^{\theta}h_F(\varphi)\,d\varphi\,d\vartheta\,d\theta-\int_{\vartheta}^{\theta}h_F(\varphi)\,d\varphi\right|<\frac{\varepsilon}{6}. \tag{3.29}$$

Since the function $F(z)$ is of completely regular growth, by Lemma 2 we have for $r > r_\varepsilon$ the inequality

$$\left|r^{-\rho(r)}\int_0^r \frac{J_F^t(\theta+k)-J_F^t(\theta)}{k}\frac{dt}{t}-\frac{1}{\rho^2}\frac{h_F(\theta+k)-h_F(\theta)}{k}\right|<\frac{\varepsilon}{6}. \tag{3.30}$$

It is obvious that the following inequality is valid:

$$\left|r^{-\rho(r)}\int_0^r \frac{J_F^t(\vartheta+l)-J_F^t(\vartheta)}{l}\frac{dt}{t}-\frac{1}{\rho^2}\frac{h_F(\vartheta+l)-h_F(\vartheta)}{l}\right|<\frac{\varepsilon}{6}. \tag{3.31}$$

Furthermore, if r does not belong to the exceptional set E for the function $F(z)$, then for a sufficiently large r_ε and for all $r > r_\varepsilon$ we have

$$\left|\int_{\vartheta}^{\theta}\frac{\ln|F(re^{i\varphi})|}{r^{\rho(r)}}\,d\varphi-\int_{\vartheta}^{\theta}h_F(\varphi)\,d\varphi\right|<\frac{\varepsilon}{6}. \tag{3.32}$$

From (3.27), (3.28), (3.29), (3.30), (3.31) and (3.32) it follows that for $r \,\bar{\in}\, E$, $r > r_\varepsilon$ the following inequality must hold:

$$\left| \frac{1}{lkr^{\rho(r)}} \int_{\theta}^{\theta+k} \int_{\vartheta}^{\vartheta+l} N(r, \vartheta, \theta)\, d\vartheta\, d\theta - \frac{1}{2\pi\rho^2} s_F(\vartheta, \theta) \right| < \varepsilon,$$

where $|l| < l_\varepsilon$ and $|k| < k_\varepsilon$.

It is obvious that for $l < 0$ and $k > 0$

$$\frac{N(r, \vartheta, \theta)}{r^{\rho(r)}} \leqslant \frac{1}{lkr^{\rho(r)}} \int_{\theta}^{\theta+k} \int_{\vartheta}^{\vartheta+l} N(r, \vartheta, \theta)\, d\vartheta\, d\theta$$

$$< \frac{1}{2\pi\rho^2} s_F(\vartheta, \theta) + \varepsilon.$$

Taking $l > 0$ and $k < 0$, we obtain in an analogous way the inequality

$$\frac{N(r, \vartheta, \theta)}{r^{\rho(r)}} > \frac{1}{2\pi\rho^2} s_F(\vartheta, \theta) - \varepsilon.$$

From these two inequalities it follows that

$$\lim_{\substack{r \to \infty \\ r \,\bar{\in}\, E}} \frac{N(r, \vartheta, \theta)}{r^{\rho(r)}} = \frac{1}{2\pi\rho^2} s_F(\vartheta, \theta). \tag{3.33}$$

In view of the fact that the relative measure of the set E is equal to zero, every interval $(R, R(1+\delta))$ $(\delta > 0, R > R_\delta)$ will contain points which do not belong to the set E. Taking into account the monotone nature of the function $N(r, \vartheta, \theta)$ we find from the last equation that for every $r > r_{\varepsilon,\delta}$ and $r_1 < r < r_2$, $r_1, r_2 \,\bar{\in}\, E$

$$\left(\frac{r_1}{r_2}\right)^{\rho} \frac{L(r_1)}{L(r_2)} \left((2\pi\rho^2)^{-1} S_F(\vartheta, \theta) - \varepsilon\right) < \frac{N(r_1, \vartheta, \theta)}{r^{\rho(r)}} \leqslant \frac{N(r, \vartheta, \theta)}{r^{\rho(r)}}$$

$$\leqslant \frac{N(r_2, \vartheta, \theta)}{r^{\rho(r)}} < \left(\frac{r_2}{r_1}\right)^{\rho} \frac{L(r_2)}{L(r_1)} \left((2\pi\rho^2)^{-1} S_F(\vartheta, \theta) + \varepsilon\right).$$

The numbers r_1 and r_2 can be chosen in the interval $r(1-\delta) < r_1 < r < r_2 < (1+\delta)r$. Since $L(r)$ is a slowly increasing function we thus have the equation

$$\lim_{r \to \infty} \frac{L(r_2)}{L(r_1)} = 1.$$

Furthermore,

$$\left(\frac{1-\delta}{1+\delta}\right)^{\rho} < \left(\frac{r_2}{r_1}\right)^{\rho} < \left(\frac{1+\delta}{1-\delta}\right)^{\rho}.$$

Having selected $\delta > 0$ sufficiently small, we see that for sufficiently large values of r

$$\frac{1}{2\pi\rho^2} s_F(\vartheta, \theta) - 2\varepsilon < \frac{N(r, \vartheta, \theta)}{r^{\rho(r)}} < \frac{1}{2\pi\rho^2} s_F(\vartheta, \theta) + 2\varepsilon$$

and hence

$$\lim_{r\to\infty} \frac{N(r, \vartheta, \theta)}{r^{\rho(r)}} = \frac{1}{2\pi\rho^2} s_F(\vartheta, \theta). \tag{3.34}$$

In order to pass from the function $N(r, \vartheta, \theta)$ to the function $n(r, \vartheta, \theta)$ we note that, for $k > 1$, the fact that the function $n(r, \vartheta, \theta)$ is monotone implies the inequality

$$\frac{n(r,\vartheta,\theta)}{r^{\rho(r)}} \ln k \leqslant \frac{1}{r^{\rho(r)}} \int_r^{kr} \frac{n(t, \vartheta, \theta)}{t} dt$$
$$= \frac{N(kr, \vartheta, \theta)}{r^{\rho(r)}} - \frac{N(r, \vartheta, \theta)}{r^{\rho(r)}},$$

and in view of (3.34) we obtain

$$\overline{\lim_{r\to\infty}} \frac{n(r, \vartheta, \theta)}{r^{\rho(r)}} \leqslant \frac{k^\rho - 1}{\ln k} \frac{1}{2\pi\rho^2} s_F(\vartheta, \theta).$$

Taking the limit as $k \to 1$, we have

$$\overline{\lim_{r\to\infty}} \frac{n(r, \vartheta, \theta)}{r^{\rho(r)}} \leqslant \frac{1}{2\pi\rho} s_F(\vartheta, \theta).$$

Analogously, making use of the inequality

$$n(r, \vartheta, \theta) \ln \frac{1}{k} \geqslant \int_{kr}^{r} \frac{n(t, \vartheta, \theta)}{t} dt = N(r, \vartheta, \theta) - N(kr, \vartheta, \theta),$$

which holds for $0 < k < 1$, we find that

$$\underline{\lim_{r\to\infty}} \frac{n(r, \vartheta, \theta)}{r^{\rho(r)}} \geqslant \frac{1}{2\pi\rho} s_F(\vartheta, \theta).$$

Thus the limit (3.25) exists for those values of ϑ and θ for which the derivatives $h'(\vartheta)$ and $h'(\theta)$ exist, i.e., for all except possibly a denumerable set. This proves the theorem.

COROLLARY. *In order that the set of zeros of a function of completely regular growth may have a density within some angle (α, β), it is necessary and sufficient that its indicator be sinusoidal within this angle.*

Indeed, if the indicator is sinusoidal, then $s(\alpha, \beta) = 0$, and in view of (3.25) the density of the set of zeros is equal to zero. Conversely, a zero density of the set of zeros implies that $s(\alpha, \beta) = 0$, and by property (g) of the indicator (Chapter I, page 56), it will be sinusoidal.

From the theorem just established, and from Theorem 1, Chapter II, it follows that for nonintegral ρ, the class of entire functions with regularly distributed zeros and the class of functions of regular growth coincide.

We shall prove that this is also the case for entire functions $f(z)$ of integral order. By Theorem 3 the set of zeros of the function $f(z)$ has a density. But in this case, in line with the definition of a regularly distributed set, one must prove also that the following limit exists:

$$\lim_{r\to\infty} \frac{1}{L(r)}\Big\{a+\frac{1}{\rho}\sum_{|a_n|\leqslant r} a_n^{-\rho}\Big\},$$

where a is the coefficient of the term of degree ρ in the polynomial $P(z)$ of the expression (1.38) for the function $f(z)$; the a_n are the zeros of this function, and $L(r) = r^{\rho(r)-\rho}$.

In order to prove the existence of this limit we form the function

$$\Phi_r(z) = \prod_{|a_n|\leqslant r} G\left(\frac{z}{a_n};\ \rho-1\right) \prod_{|a_n|>r} G\left(\frac{z}{a_n};\ \rho\right).$$

By Lemma 5 of Chapter II, for any r not belonging to some E^0-set E_Φ, the following limit exists:

$$\lim_{r\to\infty} \frac{\ln|\Phi_r(re^{i\theta})|}{r^{\rho(r)}} = h_\Phi(\theta).$$

Furthermore, since $f(z)$ is of completely regular growth, we have the following result, if $r \overline{\in} E_f$,

$$\lim_{r\to\infty} \frac{\ln|f(re^{i\theta})|}{r^{\rho(r)}} = h_f(\theta).$$

Noting that

$$\frac{1}{L(r)}\operatorname{Re}\Big\{\Big(a+\frac{1}{\rho}\sum_{|a_n|\leqslant r} a_n^{-\rho}\Big)e^{i\rho\theta}\Big\} = \frac{\ln|f(re^{i\theta})|}{r^{\rho(r)}} - \frac{\ln|\Phi(re^{i\theta})|}{r^{\rho(r)}} + o(1),$$

we find that for arbitrary θ and r not belonging to the set $E_\Phi + E_f$ the limit

$$\lim_{r\to\infty} \frac{1}{L(r)}\operatorname{Re}\Big\{\Big(a+\frac{1}{\rho}\sum_{|a_n|\leqslant r} a_n^{-\rho}\Big)e^{i\rho\theta}\Big\} = h_f(\theta) - h_\Phi(\theta)$$

exists.

Taking first $\theta = 0$, and then $\theta = \pi/2\rho$, we establish the existence of the limit

$$\delta_f = \lim_{r\to\infty} \frac{1}{L(r)}\Big\{a+\frac{1}{\rho}\sum_{|a_n|\leqslant r} a_n^{-\rho}\Big\} \tag{3.35}$$

if $r \overline{\in} E_f + E_\Phi$.

Let us free ourselves now of the requirement that $r \overline{\in} E_f + E_\Phi$. For this purpose we estimate the quantity

$$\frac{1}{L(r)} \sum_{r<|a_n|\leqslant kr} |a_n|^{-\rho}$$

Expressing it as a Stieltjes integral, we obtain

$$\frac{1}{L(r)} \sum_{r<|a_n|\leqslant kr} |a_n|^{-\rho}$$
$$= \frac{1}{L(r)} \int_r^{kr} \frac{dn(t)}{t^\rho} = \frac{1}{L(r)} \left[\frac{n(kr)}{(kr)^\rho} - \frac{n(r)}{r^\rho} \right] + \frac{\rho}{L(r)} \int_r^{kr} \frac{n(t)}{t^{\rho+1}} dt.$$

The first summand on the right hand side may be written in the form

$$\frac{1}{L(r)} \left[\frac{n(kr)}{(kr)^\rho} - \frac{n(r)}{r^\rho} \right] = \frac{n(kr)}{(kr)^{\rho(kr)}} \frac{L(kr)}{L(r)} - \frac{n(r)}{r^{\rho(r)}}.$$

It will tend to zero as $r \to \infty$ if the density Δ of the set $\{a_n\}$ of zeros exists.

The last summand satisfies the asymptotic inequality

$$\frac{\rho}{L(r)} \int_r^{kr} \frac{n(t)}{t^{\rho+1}} dt < \frac{\left(\Delta + \frac{\varepsilon}{2}\right)\rho}{L(r)} \int_r^{kr} t^{\rho(t)-\rho-1} dt$$
$$< \frac{\left(\Delta + \frac{\varepsilon}{2}\right)\rho}{L(r)} (kr)^{\rho(kr)} \int_r^{kr} t^{-\rho-1} dt = \left(\Delta + \frac{\varepsilon}{2}\right) \frac{L(kr)}{L(r)} (k^\rho - 1).$$

Finally, we obtain the asymptotic bound

$$\frac{1}{L(r)} \sum_{r<|a_n|\leqslant kr} |a_n|^{-\rho} < (\Delta + \varepsilon)(k^\rho - 1). \tag{3.36}$$

For an arbitrary $k > 1$, and for sufficiently large r, every interval (r, kr) contains a point r_1 that does not belong to the set $E_f + E_\Phi$. In view of (3.35) we have the asymptotic inequality

$$\left| \frac{1}{L(r_1)} \left(a + \frac{1}{\rho} \sum_{|a_n|\leqslant r_1} a_n^{-\rho} \right) - \delta_f \right| < \frac{\varepsilon}{4},$$

or

$$\left| \frac{1}{L(r)} \left(a + \frac{1}{\rho} \sum_{|a_n|\leqslant r_1} a_n^{-\rho} \right) - \delta_f \right| < \frac{\varepsilon}{2},$$

since $\lim r \to \infty\ L(r)/L(r_1) = 1$. Furthermore, from (3.36) we see that for k sufficiently near unity, the following asymptotic inequality holds:

$$\left| \frac{1}{\rho L(r)} \sum_{r<|a_n|\leqslant r_1} a_n^{-\rho} \right| < \frac{\varepsilon}{2}.$$

Therefore

$$\left|\frac{1}{L(r)}\left(a+\frac{1}{\rho}\sum_{|a_n|\leqslant r} a_n^{-\rho}\right)-\delta_f\right|<\varepsilon.$$

It follows that the limit

$$\text{(B)} \qquad \delta_f=\lim_{r\to\infty}\frac{1}{L(r)}\left(a+\frac{1}{\rho}\sum_{|a_k|\leqslant r}\frac{1}{a_k^{\rho}}\right)$$

exists if r goes to infinity in an arbitrary way. Combining this result with Theorem 2 of Chapter II, we obtain the next theorem, which may be considered to be the fundamental theorem of the theory of functions of completely regular growth.

THEOREM 4. *In order that an entire function $F(x)$ of proximate order $\rho(r)$ be a function of completely regular growth, it is necessary and sufficient that its zeros be regularly distributed.*

In Chapter II there was given (in Theorem 4) an asymptotic representation of entire functions with regularly distributed zeros.

From this representation, and from the theorem just proved for functions of completely regular growth, it follows that for all but a denumerable set of values of θ the limit

$$k_F(\theta)=\lim_{r\to\infty}\frac{\arg f(re^{i\theta})}{r^{\rho(r)}}$$

exists.

The same result can be obtained for functions of proximate order $\rho(r)$ and of completely regular growth within some angle; however, we shall not dwell upon this matter.[4]

REMARK. *From condition* (B) *it follows easily that the canonical representation* (1.38) *of an entire function $f(z)$ of completely regular growth of integral order ρ and of the normal type can be written in the form*

$$f(z)=z^m e^{P_\rho(z)}\lim_{R\to\infty}\prod_{|a_k|<R}\left(1-\frac{z}{a_k}\right)e^{\frac{z}{a_k}+\frac{z^2}{2a_k^2}+\cdots+\frac{z^{\rho-1}}{(\rho-1)a_k^{\rho-1}}},$$

where m is a non-negative integer and $P_\rho(z)$ is a polynomial of degree not greater than ρ. In particular, an entire function of exponential type and of completely regular growth admits the representation

$$f(z)=z^m e^{az+b}\lim_{R\to\infty}\prod_{|a_k|<R}\left(1-\frac{z}{a_k}\right).$$

[4] See A. Pfluger [3].

Indeed, an entire function is representable in the form

$$f(z) = z^m e^{P(z)} \lim_{R\to\infty} \prod_{|a_k|<R} \left(1 - \frac{z}{a_k}\right) e^{\frac{z}{a_k} + \dots + \frac{z^{\rho-1}}{(\rho-1)a_k^{\rho-1}} + \frac{z^\rho}{\rho a_k^\rho}},$$

that is, as

$$f(z) = z^m e^{P(z)} \lim_{R\to\infty} \exp\left\{\left(\sum_{|a_k|<R} a_k^{-\rho}\right) z^\rho\right\}$$

$$\times \lim_{R\to\infty} \prod_{|a_k|<R} \left(1 - \frac{z}{a_k}\right) e^{\frac{z}{a_k} + \dots + \frac{z^{\rho-1}}{(\rho-1)a_k^{\rho-1}}}$$

In order to obtain from this the required representation of the function $f(z)$ one needs only to use condition (B).

4. Indicator of the Product of Two Functions

In § 15 of Chapter I we mentioned that the indicator of the product of two functions $f(z)$ and $\varphi(z)$ holomorphic within an angle (θ_1, θ_2) does not exceed the sum of the indicators of the factors, i.e.,

$$h_{f\varphi}(\theta) \leqslant h_f(\theta) + h_\varphi(\theta) \qquad (\theta_1 \leqslant \theta \leqslant \theta_2). \tag{3.37}$$

The same relation obviously exists for the generalized indicators if one assumes that all three indicators have been computed for the same proximate order.

We shall show that if one of the factors has completely regular growth, then the equality holds in (3.37).

Indeed, from V. Bernstein's theorem (Theorem 31, Chapter I) it follows that the inequality

$$\ln|f(re^{i\theta})| > [h_f(\theta) - \varepsilon]\, r^{\rho(r)}$$

holds for arbitrary fixed θ $(\theta_1 < \theta \leqslant \theta_2)$ on a set E' of positive upper relative measure. If one of the factors (for example, $\varphi(z)$) is a function of completely regular growth, then for all values of $r > 0$, except possibly for a set E^0 of zero relative measure, we have the inequality $\ln|\varphi(re^{i\theta})| > [h_\varphi(\theta) - \varepsilon]\, r^{\rho(r)}$.

Thus for all values of r belonging to the set $E' - E^0 \cdot E'$ of positive upper relative measure we have

$$\ln|\varphi(re^{i\theta}) f(re^{i\theta})| > [h_f(\theta) + h_\varphi(\theta) - 2\varepsilon]\, r^{\rho(r)}. \tag{3.38}$$

Combining (3.37) and (3.38) we obtain the following theorem.

THEOREM 5. *If the functions $f(z)$ and $\varphi(z)$ are holomorphic within the angle (θ_1, θ_2) and if one of them is of completely regular growth, then the indicator of the product of these functions is equal to the sum of their indicators.*[4a]

[4a] Azarin [2] has proved the following converse theorem. *If $f(z)$ is holomorphic and has proximate order $\rho(r)$ inside an angle (θ_1, θ_2), and if for every holomorphic function $\varphi(z)$ of proximate order $\rho(r)$ the indicator of the product $f(z)\varphi(z)$ equals the sum of the indicators, then $f(z)$ is of completely regular growth.*

Let us also note that if both factors are of completely regular growth on some ray, then the product is of completely regular growth on that ray.

If the entire function $f(z)$ satisfies the condition

$$\lim_{r\to\infty} \frac{\ln M_f(r)}{r^{\rho(r)}} = 0,$$

then it follows from Lemma 4, Chapter I, that

$$\lim_{r\to\infty} \frac{n_f(r)}{r^{\rho(r)}} = 0,$$

and hence the set of zeros of this function will have zero angular density. If $\rho = \lim_{r\to\infty} \rho(r)$ is an integer, then it follows from Theorem 18 of Chapter I that

$$\delta_f = \lim_{r\to\infty} \frac{\delta_f(r)}{L(r)} = 0.$$

Hence, by Theorems 1 and 2 of Chapter II, the function $f(z)$ has completely regular growth and its indicator is equal to zero. Thus by Theorem 5 we obtain the following result.

COROLLARY 1. *If $F(z)$ is a holomorphic function of proximate order $\rho(r)$ within an angle (θ_1, θ_2), while $\varphi(z)$ is an entire function such that*

$$\lim_{r\to\infty} \frac{M_\varphi(r)}{r^{\rho(r)}} = 0,$$

then the indicator of the product of these functions within the angle (θ_1, θ_2) is equal to the indicator of the function $F(z)$.

From Theorem 4 it is easy to obtain the following corollary.

COROLLARY 2. *If the functions $f(z)$ and $\varphi(z)$ are holomorphic within an angle (θ_1, θ_2), if $\varphi(z)$ is of completely regular growth for some $\rho(r)$, and if*

$$\Psi(z) = \frac{f(z)}{\varphi(z)}$$

is a holomorphic function within the same angle, then

$$h_\Psi(\theta) = h_f(\theta) - h_\varphi(\theta) \qquad (\theta_1 < \theta < \theta_2).$$

To derive this equation one considers the product $\varphi(z)\Psi(z)$ and makes use of Theorem 5.

We call attention to the fact that here, as well as in Theorem 5, we have not excluded the case that one or both of the indicators are zero.

5. Some Consequences of the Generalized Formula of Jensen. Case of a Sinusoidal Indicator

In § 1 we showed that if an entire function has completely regular growth on every ray of some dense set of rays, then the function is of completely regular growth in the whole plane.

It is not difficult to construct an example of an entire function which has completely regular growth on a finite set of rays ($\arg z = k\pi/n$, $k = 1, 2, \ldots, 2n$; n arbitrary) but which is not of completely regular growth in any angle greater than π/n.

Indeed, let $\rho(r)$ be an arbitrary proximate order, and let r_k be a root of the equation

$$r^{\rho(r)} = k \qquad (k = 1, 2, \ldots).$$

Let n be an integer, $n > \rho$. The set of zeros of the entire function

$$\Phi(z) = \prod_{k=1}^{\infty}\left(1 - \frac{z^{2n}}{r_k^{2n}}\right)$$

has an angular density. Furthermore, $\delta_\Phi = 0$. Thus $\Phi(z)$ is of completely regular growth.

Let us mark the points r_k lying in the intervals $(2^{2m-1}, 2^{2m})$ $(m = 1, 2, \ldots)$. Next we renumber them in the order of increasing moduli, and denote them by $\{a_p\}$. Analogously, we denote by $\{b_p\}$ the set of points r_k lying in the intervals $(2^{2m}, 2^{2m+1})$, and we set

$$\Psi(z) = \prod_{p=1}^{\infty}\left(1 - \frac{z^n}{a_p^n}\right)^2 \prod_{p=1}^{\infty}\left(1 + \frac{z^n}{b_p^n}\right)^2.$$

The set of zeros of the function $\Psi(z)$ does not have an angular density in any angle containing a ray of the set $\arg z = k\pi/n$ $(k = 0, 1, \ldots, 2n - 1)$, and therefore $\Psi(z)$ is not a function of completely regular growth in any one of these angles. However, on each ray of the type $\arg z = (2k + 1)\pi/2n$ $(k = 0, 1, \ldots, n - 1)$ we have

$$|\Psi(z)| = |\Phi(z)|,$$

and therefore the function $\Psi(z)$ is of completely regular growth on each one of these rays. The function $\Psi(z)$ thus serves as the required example.

The situation is quite different if one requires additionally that the indicator of the growth of the function $f(z)$ be sinusoidal within the angle $[\alpha, \beta]$, i.e., if

$$h_f(\theta) = A\cos\theta + B\sin\theta \qquad (\alpha \leqslant \theta \leqslant \beta).$$

In this case, for the investigation of the behavior of the function within the angle $[\alpha, \beta]$, we transform formula (3.04) by integrating both sides of the equation with respect to ϑ from φ_1 to φ_2, and with respect to θ from φ_2 to φ_3 ($\alpha < \varphi_1 < \varphi_2 < \varphi_3 < \beta$). Thus we obtain the equation

$$2\pi \int_{\varphi_2}^{\varphi_3} \int_{\varphi_1}^{\varphi_2} N_F(r, \vartheta, \theta)\, d\vartheta\, d\theta$$
$$= \int_0^r [J_F^t(\varphi_3)(\varphi_2 - \varphi_1) + J_F^t(\varphi_2)(\varphi_1 - \varphi_3) + J_F^t(\varphi_1)(\varphi_3 - \varphi_2)] \frac{dt}{t}$$
$$+ \int_{\varphi_2}^{\varphi_3} \int_{\varphi_1}^{\varphi_2} \int_{\vartheta}^{\theta} \ln|F(re^{i\varphi})|\, d\varphi\, d\vartheta\, d\theta. \qquad (3.39)$$

For the purpose of estimating the value of the right hand side of (3.39) we note the asymptotic relation

$$\ln|F(te^{i\varphi})| < \left[h_F(\varphi) + \frac{\varepsilon}{2}\right] t^{\rho(t)} \qquad (\varepsilon > 0).$$

This implies that for $\rho > 0$ we have the asymptotic inequality

$$J_F^r(\varphi) = \int_0^r \frac{\ln|F(te^{i\varphi})|}{t}\, dt < o(r^{\rho(r)}) + \left[h_F(\varphi) + \frac{\varepsilon}{2}\right] \int_0^r t^{\rho(t)-1}\, dt$$
$$< \frac{1}{\rho}[h_F(\varphi) + \varepsilon]\, r^{\rho(r)}.$$

From this it is easy to see that

$$\overline{\lim_{r \to \infty}}\, r^{-\rho(r)} \int_0^r J_F^t(\varphi) \frac{dt}{t} \leqslant \rho^{-2} h_F(\varphi). \qquad (3.40)$$

Furthermore, Lemma 2 implies that if $F(z)$ is of completely regular growth on the ray $\arg z = \varphi_2$, then

$$\lim_{r \to \infty} r^{-\rho(r)} \int_0^r J_f^t(\varphi_2) \frac{dt}{t} = \rho^{-2} h_f(\varphi_2). \qquad (3.41)$$

From (3.39), (3.40) and (3.41) we find that for every given $\varepsilon > 0$

$$\overline{\lim_{r \to \infty}}\, 2\pi r^{-\rho(r)} \int_{\varphi_2}^{\varphi_3} \int_{\varphi_1}^{\varphi_2} N_F(r, \vartheta, \theta)\, d\vartheta\, d\theta$$
$$\leqslant \rho^{-2}[h_F(\varphi_3)(\varphi_2 - \varphi_1) + h_F(\varphi_2)(\varphi_1 - \varphi_3) + h_F(\varphi_1)(\varphi_3 - \varphi_2)]$$
$$+ \int_{\varphi_2}^{\varphi_3} \int_{\varphi_1}^{\varphi_2} \int_{\vartheta}^{\theta} h_F(\varphi)\, d\varphi\, d\vartheta\, d\theta. \qquad (3.42)$$

If in the left hand side of the equation we replace the upper limit by the lower one, then the sign of equality in (3.42) will apply only for functions of completely regular growth. This we shall show later.

LEMMA 5. *If the function $F(z)$ is holomorphic and of order $\rho(r)$ when $\alpha \leqslant \arg z \leqslant \beta$, and is of completely regular growth on some ray $\arg z = \varphi_2$ $(\alpha < \varphi_2 < \beta)$, and if for $\alpha < \varphi_1 < \varphi < \varphi_3 < \beta$*

$$2\pi \varliminf_{r\to\infty} r^{-\rho(r)} \int_{\varphi_2}^{\varphi_3} \int_{\varphi_1}^{\varphi_2} N_F(r, \theta, \vartheta)\, d\vartheta\, d\theta$$

$$= \rho^{-2}\{h_F(\varphi_3)(\varphi_2-\varphi_1)+h_F(\varphi_2)(\varphi_1-\varphi_3)+h_F(\varphi_1)(\varphi_3-\varphi_2)\}$$

$$+\int_{\varphi_2}^{\varphi_3}\int_{\varphi_1}^{\varphi_2}\int_{\vartheta}^{\theta} h_F(\varphi)\, d\varphi\, d\vartheta\, d\theta,$$

then $F(z)$ is of completely regular growth within the angle $\alpha \leqslant \varphi \leqslant \beta$.

PROOF. First of all we note that in view of (3.42) one must replace the lower limit simply by the limit in the formulation of the lemma. From equation (3.39) and the hypotheses of the lemma, we obtain

$$\lim_{r\to\infty}\left\{\left[r^{-\rho(r)}\int_0^r J_F^t(\varphi_3)\frac{dt}{t}-\rho^{-2}h_F(\varphi_3)\right](\varphi_2-\varphi_1)\right.$$

$$+\left[r^{-\rho(r)}\int_0^r J_F^t(\varphi_2)\frac{dt}{t}-\rho^{-2}h_F(\varphi_2)\right](\varphi_1-\varphi_3)$$

$$+\left[r^{-\rho(r)}\int_0^r J_F^t(\varphi_1)\frac{dt}{t}-\rho^{-2}h_F(\varphi_1)\right](\varphi_3-\varphi_2)$$

$$\left.+\int_{\varphi_2}^{\varphi_3}\int_{\varphi_1}^{\varphi_2}\int_{\vartheta}^{\theta}[h_{F,r}(\varphi)-h_F(\varphi)]\, d\varphi\, d\vartheta\, d\theta\right\}=0.$$

From (3.40), (3.41) and the asymptotic inequality

$$h_{F,r}(\theta) < h_F(\theta)+\varepsilon \qquad (\varepsilon > 0)$$

it easily follows that the upper limits of all four summands are nonpositive. Transposing all but one of the summands to the right hand side, we see that the lower limit of this summand is non-negative.

Hence each of the summands must have the limit zero, i.e.,

$$\lim_{r\to\infty} r^{-\rho(r)} \int_0^r J_F^t(\varphi)\frac{dt}{t} = \rho^{-2} h_F(\varphi),$$

$$\lim_{r\to\infty} \int_{\varphi_2}^{\varphi_3} \int_{\varphi_1}^{\varphi_2} \int_{\vartheta}^{\theta} [h_{F,r}(\varphi) - h_F(\varphi)]\, d\varphi\, d\vartheta\, d\theta = 0 \qquad (\alpha < \varphi < \beta).$$

By Lemma 3, it follows from the first equation that $F(z)$ is of completely regular growth within the angle $\alpha < \arg z < \beta$.

This proves the lemma. We now state a theorem on functions with sinusoidal indicator.

THEOREM 6. *If the function $F(z)$ is holomorphic and of proximate order $\rho(r)$, if it has within the angle $\alpha < \arg z < \beta$ a sinusoidal indicator, and if the function is of completely regular growth on some ray within this angle, then the function is of completely regular growth within the whole angle.*

PROOF. We begin with the inequality (3.42) The right hand side of this inequality can be written in the form

$$\rho^{-2} \int_{\varphi_2}^{\varphi_3} \int_{\varphi_1}^{\varphi_2} [s_F(\theta) - s_F(\vartheta)]\, d\vartheta\, d\theta,$$

where

$$s_F(\varphi) = h_F'(\varphi) + \rho^2 \int^{\varphi} h(\psi)\, d\psi.$$

For a sinusoidal indicator, $S_F(\varphi) = \text{const}$ and the right hand side of (3.42) is equal to zero. Since $N(r, \vartheta, \theta)$ is a non-negative number it is obvious that

$$\lim_{r\to\infty} r^{-\rho(r)} \int_{\varphi_2}^{\varphi_3} \int_{\varphi_1}^{\varphi_2} N(r, \vartheta, \theta)\, d\vartheta\, d\theta = 0.$$

The conditions of Lemma 5 are thus fulfilled and $F(z)$ is therefore of completely regular growth within the angle (α, β).

We note that for a sinusoidal indicator, (3.25) and (3.26) imply that

$$\lim_{r\to\infty} \frac{n(r, \vartheta, \theta)}{r^{\rho(r)}} = 0.$$

This completes the proof of the theorem.

The requirement that the function be of completely regular growth on some ray within an angle in which the indicator is sinusoidal may be relaxed. One has the following theorem.

THEOREM 7. *If for $\alpha < \theta < \beta$ a holomorphic function of proximate order $\rho(r)$ has a sinusoidal indicator within the angle $\alpha < \theta_1 \leqslant \theta \leqslant \theta_2 < \beta$, and if in addition the function is of completely regular growth on the ray* $\arg z = \theta_1$, *where $h_F'(\theta_1 - 0) = h_F'(\theta_1 + 0)$, then it is of completely regular growth within the angle $\theta_1 \leqslant \theta \leqslant \theta_2$.*

PROOF. From formula (3.39) it follows directly that

$$\int_0^r [J_F^t(\varphi_3)(\varphi_2-\varphi_1)+J_F^t(\varphi_2)(\varphi_1-\varphi_3)+J_F^t(\varphi_1)(\varphi_3-\varphi_2)]\frac{dt}{t}$$

$$+\int_{\varphi_2}^{\varphi_3}\int_{\varphi_1}^{\varphi_2}\int_{\vartheta}^{\theta}\ln|F(re^{i\varphi})|\,d\varphi\,d\vartheta\,d\theta\geqslant 0\qquad(\varphi_1<\varphi_2<\varphi_3).$$

We rewrite this inequality in the form

$$(\varphi_2-\varphi_1)\int_0^r J_F^t(\varphi_3)\frac{dt}{t}\geqslant(\varphi_3-\varphi_1)\int_0^r J_F^t(\varphi_2)\frac{dt}{t}$$

$$-(\varphi_3-\varphi_2)\int_0^r J_F^t(\varphi_1)\frac{dt}{t}-\int_{\varphi_2}^{\varphi_3}\int_{\varphi_1}^{\varphi_2}\int_{\vartheta}^{\theta}\ln|F(re^{i\varphi})|\,d\varphi\,d\vartheta\,d\theta.$$

Let us set $\varphi_2=\theta_1$, $\theta_1<\varphi_3<\theta_2$, and $\alpha<\varphi_1<\theta_1$. Since the function $F(z)$ is of completely regular growth on the ray $\arg z=\varphi_2$, we may apply the corollary to Lemma 2, obtaining

$$(\varphi_2-\varphi_1)\varliminf_{r\to\infty} r^{-\rho(r)}\int_0^r J_F^t(\varphi_3)\frac{dt}{t}$$

$$\geqslant\frac{1}{\rho^2}h_F(\varphi_2)(\varphi_3-\varphi_1)-(\varphi_3-\varphi_2)\varlimsup_{r\to\infty} r^{-\rho(r)}\int_0^r J_F^t(\varphi_1)\frac{dt}{t}$$

$$-\varlimsup_{r\to\infty}\int_{\varphi_2}^{\varphi_3}\int_{\varphi_1}^{\varphi_2}\int_{\vartheta}^{\theta}h_{F,r}(\varphi)\,d\varphi\,d\vartheta\,d\theta.$$

Together with the asymptotic inequality

$$r^{-\rho(r)}\int_0^r J_F^t(\varphi_1)\frac{dt}{t}<\frac{1}{\rho^2}h_F(\varphi_1)+\varepsilon,\qquad h_{F,r}(\varphi)<h_F(\varphi)+\varepsilon$$

this implies that

$$(\varphi_2-\varphi_1)\varliminf_{r\to\infty} r^{-\rho(r)}\int_0^r J_F^t(\varphi_3)\frac{dt}{t}$$

$$\geqslant\frac{1}{\rho^2}[h_F(\varphi_2)(\varphi_3-\varphi_1)-h_F(\varphi_1)(\varphi_3-\varphi_2)]$$

$$-\int_{\varphi_2}^{\varphi_3}\int_{\varphi_1}^{\varphi_2}\int_{\vartheta}^{\theta}h_F(\varphi)\,d\varphi\,d\vartheta\,d\theta.$$

This last inequaltiy is valid for arbitrary $\varphi_1 < \varphi_2$, and $\varphi_2 < \varphi_3 < \theta_2$. Dividing it by $\varphi_2 - \varphi_1$, and taking the limit as $\varphi_1 \to \varphi_2$, we obtain

$$\varliminf_{r\to\infty} r^{-\rho(r)} \int_0^r J_F^t(\varphi_3) \frac{dt}{t}$$

$$\geqslant \frac{1}{\rho^2}\left\{h_F(\varphi_2) + h'_F(\varphi_2 - 0)(\varphi_3 - \varphi_2) - \rho^2 \int_{\varphi_2}^{\varphi_3}\int_{\varphi_2}^{\theta} h_F(\varphi)\, d\varphi\, d\theta\right\}.$$

From the condition $h'_F(\varphi_2 + 0) = h'_F(\varphi_2 - 0)$, we see that

$$\varliminf_{r\to\infty} r^{-\rho(r)} \int_0^r J_F^t(\varphi_3) \frac{dt}{t}$$

$$\geqslant \frac{1}{\rho^2}\left\{h_F(\varphi_2) + h'_F(\varphi_2 + 0)(\varphi_3 - \varphi_2) - \rho^2 \int_{\varphi_2}^{\varphi_3}\int_{\varphi_2}^{\theta} h_F(\varphi)\, d\varphi\, d\theta\right\}.$$

Since by hypothesis $h_F(\varphi) = A\cos\rho(\varphi - \varphi_0)$ when $\varphi_2 \leqslant \varphi \leqslant \theta_2$, this inequality yields

$$\varliminf_{r\to\infty} r^{-\rho(r)} \int_0^r J_F^t(\varphi_3) \frac{dt}{t} \geqslant \frac{1}{\rho^2} h_F(\varphi_3).$$

Combining this with the inequality (3.06) we easily see that for $\theta_1 \leqslant \varphi \leqslant \theta_2$ the following limit exists:

$$\lim_{r\to\infty} r^{-\rho(r)} \int_0^r J_F^t(\varphi) \frac{dt}{t} = \frac{1}{\rho^2} h_F(\varphi).$$

It follows now from Theorem 2 that $F(z)$ is of completely regular growth within the angle $\theta_1 \leqslant \arg z \leqslant \theta_2$. The theorem is thus proved.

Formula (3.39) permits us to draw some other conclusions about the distribution of zeros on rays of completely regular growth. Let us denote by $p(t, \varphi)$ the number of zeros of the function $F(z)$ on the ray $\arg z = \varphi$ whose moduli (absolute values) do not exceed a number $t > 0$, and let us consider the integral

$$P(r, \varphi) = \int_0^r \frac{p(t, \varphi)}{t}\, dt.$$

It is obvious that for $\vartheta < \varphi < \theta$

$$N(r, \vartheta, \theta) \geqslant P(r, \varphi).$$

If the function $F(z)$ is of completely regular growth on the ray $\arg z = \varphi$, then setting $\varphi_1 = \varphi - k$, $\varphi_2 = \varphi$, $\varphi_3 = \varphi + k$ $(k > 0)$ in formula (3.42) we obtain the inequality

$$k^2 \overline{\lim_{r\to\infty}} \frac{P(r, \varphi)}{r^{\rho(r)}} \leqslant \rho^{-2} k [h(\varphi + k) - 2h(\varphi) + h(\varphi + k)] + \int_{\varphi}^{\varphi+k} \int_{\varphi-k}^{\varphi} \int_{\vartheta}^{\theta} h(\tau)\, d\tau\, d\vartheta\, d\theta.$$

This yields directly the next theorem.

THEOREM 8. *If $F(z)$ is a holomorphic function of proximate order $\rho(r)$ within an angle containing the ray* $\arg z = \varphi$, *and if the function $F(z)$ is of regular growth on this ray, then*

$$\overline{\lim_{r\to\infty}} \frac{P(r, \varphi)}{r^{\rho(r)}} \leqslant \frac{1}{2\pi\rho^2} [h'_F(\varphi + 0) - h'_F(\varphi - 0)].$$

We might recall that for functions of the first order the quantity $h'_F(\varphi + 0) - h'_F(\varphi - 0)$ has a simple geometric meaning. This quantity is equal to the length of a straight line segment of the boundary of the indicator diagram whose outer normal has the direction of the ray $\arg z = \varphi$.

We note also that the theorem remains true if we understand by $P(r, \varphi)$ the number of zeros in the intersection of the circle $|z| \leqslant r$ with some region which lies almost entirely[5] within an arbitrary angle containing in its interior the ray $\arg z = \varphi$. For example, such a region may be a parabola with axis directed along this ray.

The requirement that the function $F(z)$ be of completely regular growth on the ray $\arg z = \varphi$ is essential. In Chapter V we shall construct a function of exponential type and of nonregular growth for which

$$\overline{\lim_{r\to\infty}} \frac{P(r, 0)}{r} = 1$$

and for which the boundary of the indicator diagram does not contain a segment of a straight line parallel to the imaginary axis (Chapter V, §5).

[5] The term "almost entirely" is to be understood here in the sense: entirely except for a finite region.

CHAPTER IV

UNIQUENESS, INTERPOLATION AND COMPLETENESS

In this chapter we apply the apparatus developed in the preceding chapters to problems about the uniqueness of entire functions, the interpolation of entire functions and the completeness of systems of functions in the complex domain. These questions are, as we shall see, closely connected with each other. There is also a connection between interpolation and the problem of the dependence of the asymptotic behavior of an entire function in the whole plane on its asymptotic behavior on sequences of points tending to infinity.

The first three sections of this chapter are devoted to questions of uniqueness. The general problem of uniqueness can be formulated in the following way.

Let the holomorphic function $f(z)$ belong to a linear class (for example, let it be at most of order ρ and normal type). For which systems of linear functionals Λ_n $(n = 1, 2, \ldots)$ will the values $a_n = \Lambda_n(f)$ determine $f(z)$ uniquely?[1] As functionals we may take, in particular, the values of the function at certain points, i.e., $\Lambda_n(f) = f(z_n)$. A set $\{z_n\}$ such that when two functions of a class coincide on the set they are identically equal will be called a *set of uniqueness* for the given class.

By a familiar theorem of the theory of functions, a set which has a limit point in a region G is a set of uniqueness for all functions that are holomorphic in the region. Hence for entire functions the interesting sets of uniqueness are those with a limit point at infinity. The corresponding uniqueness theorems state that a function of "not too rapid growth" cannot vanish on a "sufficiently dense" set unless it is identically zero. Thus these theorems can be considered as generalizations of a familiar theorem on polynomials: a polynomial of degree at most n which is not identically zero can not have more than n zeros. Theorems of the same character also hold for functions holomorphic in an angle.

A typical uniqueness theorem of this kind is the following theorem of Carlson [1], which plays a prominent role in interpolatory questions.

If a holomorphic function $f(z)$ of exponential type in the half-plane $\operatorname{Re} z > 0$ *vanishes at the integral points $n = 1, 2, 3, \ldots$ and the width $h_f(\pi/2) + h_f(-\pi/2)$ of its indicator diagram is less than 2π then $f(z) \equiv 0$.*

---

[1] This formulation of the uniqueness problem was given by A. O. Gel'fond [2] and V. L. Gončarov [1].

This theorem is sharp, since sin πz vanishes at all the integers and the width of its indicator diagram is exactly 2π.[2]

In §§ 2 and 3 we shall prove considerably more general theorems of the same type.

In most of these theorems the class of functions of completely regular growth plays a fundamental role. They represent the class of "boundary functions," which have the greatest density of zeros for a given rate of growth. To clarify this fundamental role we mention Theorem 3 of § 2.

The set of zeros of every entire function, not identically zero, of proximate order $\rho(r)$ *satisfies the inequality*

$$\underline{\lim}_{r\to\infty} \frac{n_f(r)}{r^{\rho(r)}} \leqslant \frac{\rho}{2\pi} \int_0^{2\pi} h_f(\theta)\, d\theta.$$

There is equality for entire functions of completely regular growth, and only for such functions.

This characteristic extremal property of entire functions of completely regular growth could serve as their definition. Therefore entire functions of completely regular growth are the functions with given proximate order and given indicator which have the greatest lower density for their sets of zeros.

We shall prove analogous theorems for functions of finite order which are holomorphic in an angle.

In § 3 we present theorems in which the set of uniqueness is either a linear set or a set of points with an angular density in an angle.

Along with the question of the unique determination of an entire function by its values on a set of points, it is natural to consider the question of reconstructing the function from these values. This problem is solved by interpolation series. Here we shall study the question of the representation of an entire function by a Lagrange interpolation series, where we choose the nodes to be an R-set (Chapter II, § 1). Under this restriction we shall obtain definitive results.

The problem of the uniqueness of entire functions is connected with the problem of completeness of systems of holomorphic functions in a region. This

---

[2] To prove this theorem it is enough to remark that if $f(z)$ satisfies the hypotheses and is not identically zero then the quotient

$$\varphi(z) = \frac{zf(z)}{\sin \pi z}$$

is a function of exponential type in the half-plane Re $z > 0$ with indicator

$$h_\varphi(\theta) = h_f(\theta) - \pi\,|\sin\theta|.$$

Therefore

$$h_\varphi\left(\frac{\pi}{2}\right) + h_\varphi\left(-\frac{\pi}{2}\right) < 0,$$

which contradicts property (h) of the indicator (cf., § 16, Chapter I).

connection is investigated very fully in a paper of A. I. Markuševič [2]. He established that every theorem on uniqueness implies a theorem on completeness, and conversely.

In § 6 we present theorems on the connection between completeness and uniqueness, and in § 7, by using this connection and the uniqueness theorems of the first few sections, we obtain some theorems on the completeness of systems of functions.

1. Uniqueness Theorems for Entire Functions of Finite Order

Uniqueness theorems can be formulated as indicated above, i.e., under restrictions on the growth of a function that is entire (or holomorphic in an angle) the vanishing of the function on a set of large density (ordinary, upper or lower) implies that it vanishes identically.

Evidently we can give an equivalent formulation of each such theorem, namely that for a function that is not identically zero the density of the set of zeros can not exceed a number that depends on its growth. It is natural to obtain such estimates for the density of zeros of an entire function of finite order from Jensen's theorem (cf. Lemma 4 of Chapter I).

THEOREM 1. *Let $f(z)$ be an entire function of proximate order $\rho(r)$ ($\rho > 0$), not identically zero, $n_f(r)$ the number of zeros (not at the origin) of $f(z)$ in a disk of radius r and*

$$N_f(r) = \int_0^r \frac{n_f(t)}{t}\,dt. \tag{4.00}$$

Then

$$\overline{\lim_{r\to\infty}} \frac{N_f(r)}{r^{\rho(r)}} \leq \frac{1}{2\pi}\int_0^{2\pi} h_f(\theta)\,d\theta. \tag{4.01}$$

PROOF. By Jensen's theorem[3] we have the equation

$$\frac{N_f(r)}{r^{\rho(r)}} = \frac{1}{2\pi}\int_0^{2\pi} h_{f,r}(\theta)\,d\theta - \frac{\ln|f(0)|}{r^{\rho(r)}}. \tag{4.02}$$

The right side of this equation is easily estimated if we use the asymptotic inequality

$$h_{f,r}(\theta) < h_f(\theta) + \varepsilon, \tag{4.03}$$

which was proved in Theorem 28 of Chapter I. From (4.02) and (4.03) the conclusion of the theorem follows at once.

---

[3] If $f(0) = 0$, we apply Jensen's theorem to $f(z)/z^m$, where m is the multiplicity of the zero at 0.

THEOREM 2. *Under the hypotheses of Theorem* 1 *we have the inequality*

$$\overline{\lim_{r\to\infty}} \frac{n_f(r)}{r^{\rho(r)}} \leqslant \frac{e\rho}{2\pi} \int_0^{2\pi} h_f(\theta)\, d\theta. \tag{4.04}$$

PROOF. From the monotonicity of $n_f(t)$ we easily get the inequality

$$\frac{n_f(r)}{r^{\rho(r)}} \leqslant \frac{\rho}{r^{\rho(r)}} \int_r^{e^{\frac{1}{\rho}}r} \frac{n_f(t)}{t}\, dt \leqslant \rho \frac{N_f(e^{\frac{1}{\rho}} r)}{r^{\rho(r)}}, \tag{4.05}$$

from which we obtain

$$\overline{\lim_{r\to\infty}} \frac{n_f(r)}{r^{\rho(r)}} \leqslant e\rho \overline{\lim_{r\to\infty}} \frac{N_f(r)}{r^{\rho(r)}}. \tag{4.06}$$

Inequality (4.04) follows directly from (4.06) and Theorem 1.

We now show that (4.04) is sharp; indeed, we show that there is an entire function for which there is equality in (4.04).

In constructing such a function it is helpful to remark that there can be equality in (4.04) only for functions of very irregular growth. In fact, there is equality on the left of (4.05) only when $f(z)$ has no zeros in the annulus $r < |z| < e^{1/\rho}r$. There is equality on the right of (4.05) only when $f(z)$ has no zeros in the disk $|z| < r$. Hence equality in (4.04) is possible only when $n(r)$ increases only on short intervals which are sparsely distributed along the positive real axis.

After these preliminary remarks we turn to the construction of a function for which there is equality in (4.04). We consider only the case of entire functions of normal type of the ordinary order ρ.

EXAMPLE. We consider the sequence of integers $n_p = 2^{2^p}$ and construct the function

$$f(z) = \prod_{p=1}^{\infty} \left[1 - \left(\frac{z}{n_p^{\frac{1}{\rho}}}\right)^{n_p}\right]. \tag{4.07}$$

For $n_k^{1/\rho} < r < n_{k+1}^{1/\rho}$ we have

$$n(r) = 2^2 + 2^4 + \dots + 2^{2^k} = n_k(1 + \varepsilon_k),$$

where $\varepsilon_k < (k - 1)2^{-2^{k-1}}$.

Hence

$$\overline{\lim_{r\to\infty}} \frac{n(r)}{r^{\rho}} = 1. \tag{4.08}$$

On the other hand

$$\ln|f(z)| = \sum_{p=1}^{k-1} \ln\left|1-\left(\frac{z}{n_p^{\frac{1}{\rho}}}\right)^{n_p}\right| + \ln\left|1-\left(\frac{z^\rho}{n_k}\right)^{\frac{n_k}{\rho}}\right|$$

$$+\sum_{p=k+1}^{\infty} \ln\left|1-\left(\frac{z}{n_p^{\frac{1}{\rho}}}\right)^{n_p}\right|. \tag{4.09}$$

For $n_k^{1/\rho} < r < n_{k+1}^{1/\rho}$ we obtain

$$\sum_{p=1}^{k-1} \ln\left|1-\left(\frac{z^\rho}{n_p}\right)^{\frac{n_p}{\rho}}\right| < \left(\sum_{p=1}^{k-1} n_p\right) \ln 2r$$

$$< 2n_{k-1} \ln 2r < 2r^{\frac{\rho}{2}} \ln 2r = o(r^\rho). \tag{4.10}$$

If $p \geqslant k+2$, we have asymptotically $rn^{-1/\rho} < \frac{1}{2}$ and from the inequality

$$|\ln(1-u)| < \frac{|u|}{1-|u|} < 2|u|,$$

which holds for $|u| < \frac{1}{2}$, it follows that

$$\sum_{p=k+1}^{\infty} \ln\left|1-\left(\frac{z^\rho}{n_p}\right)^{\frac{n_p}{\rho}}\right| < \ln 2 + 2\sum_{p=k+2}^{\infty}\left(\frac{r^\rho}{n_p}\right)^{\frac{n_p}{\rho}} < \ln 2 + 1. \tag{4.11}$$

Combining (4.09), (4.10) and (4.11), we see that asymptotically for $n_k^{1/\rho} < r < n_{k+1}^{1/\rho}$

$$\ln|f(z)| = \ln\left|1-\left(\frac{z^\rho}{n_k}\right)^{\frac{n_k}{\rho}}\right| + o(r^\rho), \tag{4.12}$$

or for $n_k(1+\varepsilon) < r^\rho < n_{k+1}$ $(\varepsilon > 0)$

$$\frac{\ln|f(z)|}{r^\rho} = \frac{n_k}{\rho r^\rho} \ln \frac{r^\rho}{n_k} + o(1). \tag{4.13}$$

The maximum of the first term on the right side of this equation is attained for $r^\rho = en_k$ and consequently

$$h_f(\vartheta) = \overline{\lim_{r\to\infty}} \frac{\ln|f(re^{i\vartheta})|}{r^\rho} = \frac{1}{\rho e}. \tag{4.14}$$

Combining this with (4.08), we obtain

$$\overline{\lim_{r\to\infty}} \frac{n(r)}{r^\rho} = \frac{\rho e}{2\pi} \int_0^{2\pi} h_f(\vartheta)\, d\vartheta, \tag{4.15}$$

i.e., (4.04) becomes an equality for the function we have constructed.

In estimating the *lower* density of the set of zeros of an entire function a fundamental role is played by functions of completely regular growth. We have the following theorem.

THEOREM 3. *The lower density of the set of zeros of every entire function $f(z)$, not identically zero, of proximate order $\rho(r)$ ($\rho > 0$) satisfies the inequality*

$$\underline{\lim}_{r\to\infty} \frac{n_f(r)}{r^{\rho(r)}} \leqslant \frac{\rho}{2\pi} \int_0^{2\pi} h_f(\theta)\, d\theta. \tag{4.16}$$

There is equality for functions of completely regular growth and only for such functions.[4]

PROOF. The first part of the theorem follows immediately from (4.01) and the inequality

$$\underline{\lim}_{r\to\infty} \frac{n_f(r)}{r^{\rho(r)}} \leqslant \rho \underline{\lim}_{r\to\infty} \frac{N_f(r)}{r^{\rho(r)}}. \tag{4.17}$$

The latter inequality follows from property (4) (cf. Chapter I, § 12) of proximate orders. The essential part is the second statement of the theorem.

In Chapter III (Theorem 2) it was proved that the set of zeros of a holomorphic function $f(z)$ of finite order and completely regular growth in an angle has an angular density, and that

$$\lim_{r\to\infty} \frac{n_f(r, \vartheta, \theta)}{r^{\rho(r)}} = \frac{1}{2\pi\rho}\left[h_f'(\theta) - h_f'(\vartheta) + \rho^2 \int_\vartheta^\theta h_f(\varphi)\, d\varphi\right].$$

Taking $\vartheta = 0$, $\theta = 2\pi$, we have

$$\lim_{r\to\infty} \frac{n_f(r)}{r^{\rho(r)}} = \frac{\rho}{2\pi} \int_0^{2\pi} h_f(\theta)\, d\theta.$$

It remains to show that there is equality in (4.16) *only* for entire functions of completely regular growth. It follows from (4.17) that when there is equality

$$\underline{\lim}_{r\to\infty} \frac{N_f(r)}{r^{\rho(r)}} \geqslant \frac{1}{2\pi} \int_0^{2\pi} h_f(\theta)\, d\theta,$$

[4] The estimates (4.04) and (4.16) were essentially given by R. P. Boas. Cf., the book [1]. In a recent paper R. C. Buck [1] gave the following estimate for the distribution of the zeros of a function of order ρ and normal type σ:

$$e \overline{\lim}_{r\to\infty} \frac{n(r)}{r^\rho} + \underline{\lim}_{r\to\infty} \frac{n(r)}{r^\rho} \leqslant e^2\rho\sigma.$$

A. Pfluger [1] gave a theorem closely related to Theorem 3.

and since by Theorem 1

$$\overline{\lim_{r\to\infty}} \frac{N_f(r)}{r^{\rho(r)}} \leqslant \frac{1}{2\pi} \int_0^{2\pi} h_f(\theta)\, d\theta,$$

we obtain

$$\lim_{r\to\infty} \frac{N_f(r)}{r^{\rho(r)}} = \frac{1}{2\pi} \int_0^{2\pi} h_f(\theta)\, d\theta.$$

Combining this equation with Jensen's theorem (4.02) we obtain

$$\lim_{r\to\infty} \int_0^{2\pi} [h_f(\theta) - h_{f,r}(\theta)]\, d\theta = 0$$

and by using the asymptotic inequality $h_{f,r}(\theta) < h_f(\theta) + \varepsilon$ ($\varepsilon > 0$) we infer that

$$\lim_{r\to\infty} \int_0^{2\pi} |h_f(\theta) - h_{f,r}(\theta)|\, d\theta = 0. \tag{4.18}$$

To deduce from this that $f(z)$ is of completely regular growth we first use Theorem 6 of Chapter II, according to which the family of functions $h_{f,r}(\theta)$ is equicontinuous for all positive r except possibly for a set E of values of arbitrarily small upper relative measure.

We shall show, by using this theorem and (4.18), that for $r \overline{\in} E$ and $r \to \infty$ the function $h_{f,r}(\theta)$ tends uniformly to $h_f(\theta)$. In fact, for sufficiently small δ, $|k| < \delta$ and $r \overline{\in} E$

$$\begin{aligned} |h_{f,r}(\theta) - h_f(\theta)| &< \frac{\varepsilon}{2} + \frac{1}{k} \int_\theta^{\theta+k} |h_{f,r}(\varphi) - h_f(\varphi)|\, d\varphi \\ &< \frac{\varepsilon}{2} + \frac{1}{|k|} \int_0^{2\pi} |h_{f,r}(\varphi) - h_f(\varphi)|\, d\varphi \end{aligned}$$

and for $r > r_\varepsilon$

$$|h_{f,r}(\theta) - h_f(\theta)| < \varepsilon. \tag{4.19}$$

Hence we have shown that for arbitrary $\varepsilon > 0$ and $\eta > 0$ the inequality (4.19) is satisfied for all positive r except perhaps for a set of upper relative measure less than η. By Lemma 1 of Chapter III it follows from this that $f(z)$ is of completely regular growth. The theorem is established.

REMARK. For $\rho = 1$ the integral on the right of (4.16) is equal to the length of the boundary of the indicator diagram of $f(z)$.

2. Uniqueness Theorems for Functions of Finite Order, Holomorphic in an Angle

In this section we obtain uniqueness theorems for functions that are holomorphic and of finite order in an angle. These are analogs of the theorems for entire functions that we established in the preceding section. In Chapter III, by using the generalized Jensen theorem, we obtained estimates for the density of the set of zeros of a function $F(z)$ in an angle (inequality (3.42) and Lemma 5) analogous to Theorems 1 and 3 of the preceding section. However, these estimates were obtained under the very restrictive supplementary hypothesis that $F(z)$ has completely regular growth on a ray inside the angle.

In this section we shall prove further uniqueness theorems for holomorphic functions of finite order in an angle without making any assumptions about the regularity of their growth. For this purpose we shall develop a new formula analogous to Jensen's. We note that without loss of generality we may suppose that the angle in which $F(z)$ is given is the half-plane $\operatorname{Re} z \geqslant 0$, since the general case can be reduced to this by the transformation $w = e^{i\alpha}z^{\gamma}$ ($\gamma > 0, 0 \leqslant \alpha < 2\pi$).

We now establish the analog of Jensen's formula for a half-plane.

Let $F(z)$ be holomorphic in the half-plane $\operatorname{Re} z \geqslant 0$ and let $F(0) \neq 0$. Take a number d such that $F(z)$ has no zeros in the closed disk $|z| \leqslant d$. We construct a contour $\Gamma_{d,R}$ consisting of arcs of the circumferences $|z| = d$ and $|z - R/2| = R/2$ ($R > d$). With each zero of $F(z)$ as center, draw the circumference C_j of radius δ, where δ is chosen smaller than the distance from $\Gamma_{d,R}$ to the set of zeros that lie inside this contour, and smaller than the distance between the zeros. In the region bounded by the contour $\Gamma_{d,R}$ and the circumferences C_j, consider the two harmonic functions $u(z) = \ln|F(z)|$ and $v(z) = 1/R - \cos\theta/r$ ($z = re^{i\theta}$) and apply Green's theorem to them. We obtain

$$\int_{(\Gamma_{d,R})} \left(u\frac{dv}{dn} - v\frac{du}{dn}\right) ds = \sum_j \int_{(C_j)} \left(u\frac{dv}{dn} - v\frac{du}{dn}\right) ds.$$

Taking the limit as $\delta \to 0$ and noting that $v = 0$ on the circumference $|z - R/2| = R/2$, which we denote by Γ_R, we obtain for large R

$$\sum_{r_j < R\cos\theta_j} \left(\frac{1}{R} - \frac{\cos\theta_j}{r_j}\right) = \frac{1}{2\pi} \int_{-\frac{\pi}{2}+\alpha(R)}^{\frac{\pi}{2}-\alpha(R)} \ln|F(R\cos\theta e^{i\theta})| \frac{dv}{dn} R\, d\theta + O(1),$$

where $z_j = r_j e^{i\theta_j}$ are the zeros of $F(z)$, $\alpha(R) = \arcsin d/R$, and $O(1)$ is an integral of a bounded function along an arc of $|z| = d$.

Noting that $r_j \cos^{-1}\theta_j$ is equal to the diameter R_j of the circle Γ_{R_j} about the zero $r_j e^{i\theta_j}$, we can write the left side of the equation in the form

$$\sum_{r_j < R\cos\theta_j} \left(\frac{1}{R} - \frac{\cos\theta_j}{r_j}\right) = \int_0^R \left(\frac{1}{R} - \frac{1}{t}\right) d\nu_F(t),$$

where $\nu_F(t)$ is the number of zeros of $F(z)$ inside Γ_t. Transforming the right side of the last equation by integration by parts, we obtain

$$-\sum_{r_j<R\cos\theta_j}\left(\frac{1}{R}-\frac{\cos\theta_j}{r_j}\right)=\int_0^R\frac{\nu_F(t)}{t^2}\,dt.$$

On the other hand, noting that $dv/dn = 1/R^2\cos^2\theta$ on Γ_R, we obtain[5]

$$\int_0^R\frac{\nu_F(t)}{t^2}\,dt=\frac{1}{2\pi}\int_{-\frac{\pi}{2}+\alpha(R)}^{\frac{\pi}{2}-\alpha(R)}\frac{\ln|F(R\cos\theta e^{i\theta})|}{R}\,\frac{d\theta}{\cos^2\theta}+O(1). \tag{4.20}$$

Equation (4.20) is an analog of Jensen's formula and provides a means of establishing uniqueness theorems for functions holomorphic in a half-plane.

We shall give another derivation of this formula based on the same principle as the derivation of Jensen's formula in Chapter I, or of the generalized Jensen formula in Chapter III.

Let Γ_t denote the circumference $|z - t/2| = t/2$, and choose t so that this circumference does not pass through any zeros of $F(z)$. We have

$$\nu_F(t)=\frac{1}{2\pi}\int_{(\Gamma_t)}d\,[\arg F(z)].$$

The transformation

$$w=\frac{1}{z}$$

takes Γ_t into the line $u = 1/t$ of the $w = u + iv$ plane, and we have

$$\nu_F(t)=\frac{1}{2\pi}\int_{-\infty}^{\infty}\frac{d}{dv}\left[\arg F\left(\frac{1}{u+iv}\right)\right]_{u=t^{-1}}dv$$

$$=\frac{1}{2\pi}\int_{-\infty}^{\infty}\frac{d}{du}\left[\ln\left|F\left(\frac{1}{u+iv}\right)\right|\right]_{u=t^{-1}}dv.$$

Integrating both sides of this equation with respect to u between the limits ε^{-1} and R^{-1} we obtain

$$-\int_\varepsilon^R\frac{\nu_F(t)}{t^2}\,dt=\frac{1}{2\pi}\int_{-\infty}^{\infty}\ln\left|F\left(\frac{1}{R^{-1}+iv}\right)\right|dv-\frac{1}{2\pi}\int_{-\infty}^{\infty}\ln\left|F\left(\frac{1}{\varepsilon^{-1}+iv}\right)\right|dv.$$

[5] If we analyze the term $O(1)$ when $F(0) = 1$, it is easy to see that as $d \to 0$ the term $O(1)$ tends to $\frac{1}{2}$ Re $F'(0)$, and taking the limit in (4.20) we obtain the equation

$$\int_0^R\frac{\nu_F(t)}{t^2}\,dt=\frac{1}{2\pi}\int_{-\frac{\pi}{2}}^{\frac{\pi}{2}}\frac{\ln|F(R\cos\theta e^{i\theta})|}{R}\,\frac{d\theta}{\cos^2\theta}-\frac{1}{2}\operatorname{Re}F'(0).$$

Here the integral on the right side of the equation is to be taken as a principal value (see Levin [**14**]).

Noting that on Γ_t we have

$$z = t\cos\theta e^{i\theta}, \qquad v = -t^{-1}\operatorname{tg}\theta, \qquad dv = -\frac{d\theta}{t\cos^2\theta},$$

we find that

$$\int_\varepsilon^R \frac{\nu_F(t)}{t^2}\,dt$$

$$= \frac{1}{2\pi}\int_{-\frac{\pi}{2}}^{\frac{\pi}{2}} (R^{-1}\ln|F(R\cos\theta e^{i\theta})| - \varepsilon^{-1}\ln|F(\varepsilon\cos\theta e^{i\theta})|)\frac{d\theta}{\cos^2\theta}.$$

Furthermore, it is easily seen from the expansion

$$F(z) = 1 + c_1 z + c_2 z^2 + \ldots$$

that for given $d > 0$, $R\cos\theta < d$ and $\varepsilon < R$ we may choose M so that

$$\left|R^{-1}\ln|F(R\cos\theta e^{i\theta})| - \varepsilon^{-1}\ln|F(\varepsilon\cos\theta e^{i\theta})|\right| < MR\cos^2\theta.$$

Using this, we obtain

$$\int_\varepsilon^R \frac{\nu_F(t)}{t^2}\,dt$$

$$= \frac{1}{2\pi}\int_{-\frac{\pi}{2}+\alpha(R)}^{\frac{\pi}{2}-\alpha(R)} \left(R^{-1}\ln|F(R\cos\theta e^{i\theta})| - \varepsilon^{-1}\ln|F(\varepsilon\cos\theta e^{i\theta})|\right)\frac{d\theta}{\cos^2\theta} + O(1)$$

and finally

$$\int_0^R \frac{\nu_F(t)}{t^2}\,dt = \frac{1}{2\pi}\int_{-\frac{\pi}{2}+\alpha(R)}^{\frac{\pi}{2}-\alpha(R)} R^{-1}\ln|F(R\cos\theta e^{i\theta})|\frac{d\theta}{\cos^2\theta} + O(1).$$

To explain the fundamental role which the class of functions of completely regular growth in an angle plays in uniqueness theorems, we need a lemma giving a criterion for a function that is holomorphic in an angle to be a function of completely regular growth.

LEMMA 1. *For a function $F(z)$ holomorphic in the angle (α, β) and of proximate order $\rho(r)$ to be of completely regular growth, it is necessary and sufficient that the areal mean*

$$\psi(R, \alpha, \beta) = \frac{2}{3(\beta-\alpha)R^2}\int_\alpha^\beta \int_R^{2R} [h_F(\theta) - h_{F,R}(\theta)]\,r\,dr\,d\theta$$

tends to zero as $R \to \infty$.

PROOF OF NECESSITY. For a function of completely regular growth we have

$$\lim_{r\to\infty}\frac{N_F(r, \vartheta, \theta)}{r^{\rho(r)}} = \frac{1}{2\pi\rho^2} s_F(\vartheta, \theta),$$

where

$$s_F(\vartheta, \theta) = h'_F(\theta) - h'_F(\vartheta) + \rho^2\int_\vartheta^\theta h_F(\varphi)\, d\varphi$$

(cf., (3.34) of § 3 of Chapter III). From this equation it follows immediately that

$$\begin{aligned}
\lim_{r\to\infty}\frac{r^{-\rho(r)}}{lk}\int_\theta^{\theta+k}\int_\vartheta^{\vartheta+l} N_F(r, \vartheta, \theta)\, d\vartheta\, d\theta
&= \frac{1}{2\pi lk\rho^2}\int_\theta^{\theta+k}\int_\vartheta^{\vartheta+l} s_F(\vartheta, \theta)\, d\vartheta\, d\theta \\
&= \frac{1}{2\pi\rho^2}\Big\{\frac{h_F(\theta+k)-h_F(\theta)}{k} - \frac{h_F(\vartheta+l)-h_F(\vartheta)}{l} \\
&\quad + \frac{\rho^2}{kl}\int_\theta^{\theta+k}\int_\vartheta^{\vartheta+l}\int_\vartheta^\theta h_F(\varphi)\, d\varphi\, d\vartheta\, d\theta\Big\}. \qquad (4.21)
\end{aligned}$$

In Chapter III we proved the following inequality (cf. (3.27)) by using the generalized Jensen formula:

$$\begin{aligned}
\frac{1}{lk}\int_\theta^{\theta+k}\int_\vartheta^{\vartheta+l} N_F(r, \vartheta, \theta)\, d\vartheta\, d\theta
&= \frac{1}{2\pi}\int_0^r\Big[\frac{J_F^t(\theta+k)-J_F^t(\theta)}{k} - \frac{J_F^t(\vartheta+l)-J_F^t(\vartheta)}{l}\Big]\frac{dt}{t} \\
&\quad + \frac{1}{2\pi lk}\int_\theta^{\theta+k}\int_\vartheta^{\vartheta+l}\int_\vartheta^\theta h_{F,r}(\varphi)\, d\varphi\, d\vartheta\, d\theta. \qquad (4.22)
\end{aligned}$$

As a corollary of Lemma 2 of Chapter III we have for functions of completely regular growth

$$\lim_{r\to\infty} r^{-\rho(r)}\int_0^r J_F^t(\varphi)\frac{dt}{t} = \frac{1}{\rho^2}h_F(\varphi).$$

By using this, we can deduce from (4.21) and (4.22) that

$$\lim_{r\to\infty}\frac{1}{kl}\int_\theta^{\theta+k}\int_\vartheta^{\vartheta+l}\int_\vartheta^\theta [h_F(\varphi) - h_{F,r}(\varphi)]\, d\varphi\, d\vartheta\, d\theta = 0. \qquad (4.23)$$

It follows from the asymptotic inequality

$$h_{F,r}(\varphi) < h_F(\varphi) + \varepsilon \qquad (\varepsilon > 0,\ \vartheta < \varphi < \vartheta + l)$$

that in (4.23) we may replace the quantity $h_{F,r}(\theta) - h_F(\theta)$ by its absolute value and obtain the equation

$$\lim_{r\to\infty} \int_{\vartheta+l}^{\theta} | h_F(\varphi) - h_{F,r}(\varphi) | \, d\varphi = 0 \qquad (0 < l_1 < l).$$

Integrating this equation with respect to r from R to $2R$ we obtain

$$\lim_{R\to\infty} \psi(R, \alpha, \beta) = 0. \tag{4.24}$$

The necessity is established.

Proof of sufficiency. Let (4.24) hold. If we write

$$K_F(r) = \int_{\alpha}^{\beta} | h_F(\theta) - h_{F,r}(\theta) | \, d\theta,$$

then we easily obtain from (4.24)

$$\lim_{R\to\infty} \frac{1}{R} \int_{R}^{2R} K_F(r)\, dr = 0. \tag{4.25}$$

Let E_γ be the set of values of r for which

$$K_F(r) \geqslant \gamma > 0.$$

Then evidently

$$\frac{1}{R} \int_{R}^{2R} K_F(r)\, dr \geqslant \gamma \frac{\operatorname{mes}(E_\gamma^{2R}) - \operatorname{mes}(E_\gamma^{R})}{R}$$

and by (4.25)

$$\overline{\lim_{R\to\infty}} \frac{m(E_\gamma^{2R}) - m(E_\gamma^{R})}{R} = 0.$$

It follows from this that the relative measure of E_γ is zero. Thus if $r \to \infty$ outside a set E_1 of arbitrarily small upper relative measure, we have

$$\lim_{\substack{r\to\infty \\ r\,\bar{\in}\, E_1}} \int_{\alpha}^{\beta} | h_F(\theta) - h_{F,r}(\theta) | \, d\theta = 0. \tag{4.26}$$

We now show that the functions $|h_F(\theta) - h_{F,r}(\theta)|$ form an equicontinuous family in the angle $\alpha < \alpha' \leqslant \theta \leqslant \beta' < \beta$ for all values of the parameter r, except perhaps for a set E_1 of arbitrarily small upper relative measure $(\overline{m}^*(E_1) < \eta/2)$. To do this we choose a positive number l so small that every disk $C(R, l)$ of the form

$$\left| z - Re^{i\frac{\alpha'+\beta'}{2}} \right| \leqslant lR$$

is contained in the region $\frac{2}{3}R \leqslant |z| \leqslant \frac{4}{3}R$, $\alpha \leqslant \arg z \leqslant \beta$, over which the integral is taken in the definition of $\psi(\frac{2}{3}R, \alpha, \beta)$. It is enough to take $l = \frac{1}{3}\sin(\beta - \alpha)/2$. It follows from (4.24) that the average of $|h_F(\theta) - h_{F,r}(\theta)|$ over the disk $C(R, l)$ tends to zero as $R \to \infty$. Hence for sufficiently large R, every disk $C(R, l)$ contains a point $z_R = R_1 e^{i\theta_1}$ at which

$$|h_F(\theta_1) - h_{F,R_1}(\theta_1)| < \varepsilon.$$

This inequality implies

$$\ln|F(z_R)| > -NR_1^{\rho(R_1)},$$

where

$$N = -[\min_{\alpha \leqslant \theta \leqslant \beta} h_F(\theta) - \varepsilon].$$

We now have the hypotheses of Theorem 7 of Chapter II, and consequently for $r \overline{\in} E_2(\overline{m}^*(E_2) < \eta/2)$ the family of functions $|h_F(\theta) - h_{F,r}(\theta)|$ is equicontinuous on the interval $(\alpha < \alpha' \leqslant \theta \leqslant \beta' < \beta)$.

It follows from equicontinuity and (4.26) that for sufficiently small $\delta > 0$, $|k| < \delta$, $\alpha' \leqslant \theta \leqslant \beta'$, $r \overline{\in} E_1 + E_2$, we have

$$|h_F(\theta) - h_{F,r}(\theta)| < \frac{\varepsilon}{2} + \frac{1}{k}\int_\theta^{\theta+k} |h_F(\varphi) - h_{F,r}(\varphi)|\, d\varphi$$

$$\leqslant \frac{\varepsilon}{2} + \frac{1}{|k|}\int_\alpha^\beta |h_F(\varphi) - h_{F,r}(\varphi)|\, d\varphi$$

and for $r > r_\varepsilon$

$$|h_F(\theta) - h_{F,r}(\theta)| < \varepsilon. \tag{4.27}$$

Thus for arbitrary $\varepsilon > 0$ and $\eta > 0$, and for all r not belonging to a certain set of upper relative measure less than η, (4.27) holds for $\alpha' \leqslant \theta \leqslant \beta'$. Then Lemma 1 of Chapter III establishes completely regular growth in every angle $\alpha < \theta < \beta$. The lemma is established.

The following lemma also plays an essential role in proving uniqueness theorems by means of (4.20).

LEMMA 2. *If $F(z)$ is holomorphic and of proximate order $\rho(r)(\rho > 1)$ inside an angle $|\arg z| < \pi/2 + \varepsilon$, $\varepsilon > 0$, and of completely regular growth, its zeros satisfy*

$$\frac{1}{\rho - 1}\lim_{r\to\infty}\frac{\nu_F(r)}{r^{\rho(r)}} = \lim_{r\to\infty}\frac{\mathrm{N}_F(r)}{r^{\rho(r)-1}}$$

$$= \frac{1}{2\pi}\int_{-\frac{\pi}{2}}^{\frac{\pi}{2}} h_F(\theta)\cos^{\rho-2}\theta\, d\theta, \tag{4.28}$$

where

$$\mathrm{N}_F(r) = \int_0^r \frac{\nu_F(t)}{t^2}\, dt. \tag{4.29}$$

PROOF. In proving this lemma it is natural to use Theorem 3 of Chapter III. Choose $\delta > 0$ and points $\theta_0 < \theta_1 < \ldots < \theta_n$ ($\theta_0 = -\pi/2 + \delta$, $\theta_n = \pi/2 - \delta$) at which $h_F(\theta)$ is differentiable. Let θ_j' ($j = 0, 1, \ldots, n-1$) be the point of the interval $[\theta_j, \theta_{j+1}]$ at which $\cos\theta$ takes its maximum value. Then we have

$$\nu_F(r) < \sum_{j=0}^{n-1} n(r\cos\theta_j', \theta_j, \theta_{j+1}) + n\left(r\sin\delta, \frac{\pi}{2}-\delta, \frac{\pi}{2}\right) + n\left(r\sin\delta, -\frac{\pi}{2}, -\frac{\pi}{2}+\delta\right),$$

since the right side is the number of zeros of $F(z)$ in a region containing the

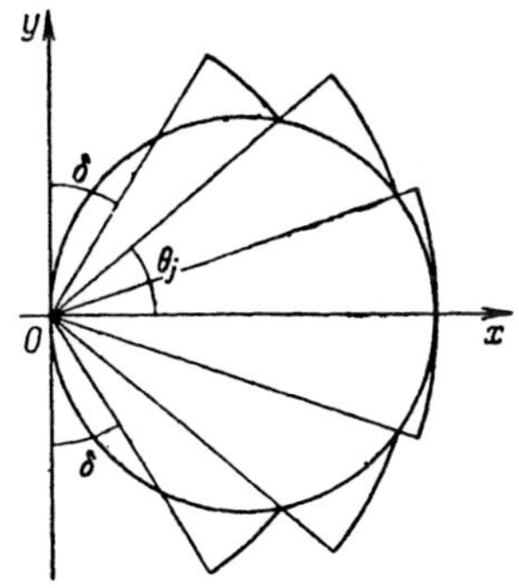

Fig. 7

disk $|z - r/2| \leqslant r/2$ (Fig. 7). From this inequality we have by Theorem 3 of Chapter III

$$\overline{\lim_{r\to\infty}} \frac{\nu_F(r)}{r^{\rho(r)}} \leqslant \frac{1}{2\pi\rho} \sum_{j=0}^{n-1} [s_F(\theta_{j+1}) - s_F(\theta_j)] \cos^\rho \theta_j' + \frac{1}{2\pi\rho}(\sin\delta)^\rho\left[s_F\left(\frac{\pi}{2}\right) - s_F\left(\frac{\pi}{2}-\delta\right) + s_F\left(-\frac{\pi}{2}+\delta\right) - s_F\left(-\frac{\pi}{2}\right)\right].$$

Taking limits as $\max_j |\theta_{j+1} - \theta_j| \to 0$ and then as $\delta \to 0$, we obtain

$$\overline{\lim_{r\to\infty}} \frac{\nu_F(r)}{r^{\rho(r)}} \leqslant \frac{1}{2\pi\rho} \int_{-\frac{\pi}{2}}^{\frac{\pi}{2}} \cos^\rho\theta \, ds_F(\theta). \tag{4.30}$$

In a similar way we can obtain a lower estimate for the lower limit. For this we let θ_j'' be the point of $[\theta_j, \theta_{j+1}]$ at which $\cos\theta$ takes its minimum value, obtaining

$$\nu_F(r) \geqslant \sum_{j=0}^{n-1} n(r\cos\theta_j'', \theta_j, \theta_{j+1}).$$

From this it follows that

$$\varliminf_{r\to\infty} \frac{\nu_F(r)}{r^{\rho(r)}} \geqslant \frac{1}{2\pi\rho} \sum_{j=0}^{n-1} [s_F(\theta_{j+1}) - s_F(\theta_j)] \cos^\rho \theta_j.$$

Letting $\max_j |\theta_{j+1} - \theta_j| \to 0$ and then $\delta \to 0$ on the right hand side, we obtain

$$\varliminf_{r\to\infty} \frac{\nu_F(r)}{r^{\rho(r)}} \geqslant \frac{1}{2\pi\rho} \int_{-\frac{\pi}{2}}^{\frac{\pi}{2}} \cos^\rho \theta \, ds_F(\theta).$$

Combining this with (4.30), we obtain

$$\lim_{r\to\infty} \frac{\nu_F(r)}{r^{\rho(r)}} = \frac{1}{2\pi\rho} \int_{-\frac{\pi}{2}}^{\frac{\pi}{2}} \cos^\rho \theta \, ds_F(\theta). \tag{4.28'}$$

Moreover, if $\rho \neq 1$, (4.29) implies, by means of property (5) of proximate orders,

$$\lim_{r\to\infty} \frac{N_F(r)}{r^{\rho(r)-1}} = \frac{1}{\rho-1} \lim_{r\to\infty} \frac{\nu_F(r)}{r^{\rho(r)}} = \frac{1}{2\pi\rho(\rho-1)} \int_{-\frac{\pi}{2}}^{\frac{\pi}{2}} \cos^\rho \theta \, ds_F(\theta).$$

We now transform the right hand side of this equation. The integral on the right can be written in the form

$$\int_{-\frac{\pi}{2}}^{\frac{\pi}{2}} \cos^\rho \theta \, ds_F(\theta) = \int_{-\frac{\pi}{2}}^{\frac{\pi}{2}} \cos^\rho \theta \, d\left[h_F'(\theta) + \rho^2 \int^{\theta} h_F(\varphi)\, d\varphi\right]$$

$$= \int_{-\frac{\pi}{2}}^{\frac{\pi}{2}} \cos^\rho \theta \, dh_F'(\theta) + \rho^2 \int_{-\frac{\pi}{2}}^{\frac{\pi}{2}} h_F(\theta) \cos^\rho \theta \, d\theta.$$

When $\rho > 1$ the first integral can be integrated twice by parts We obtain

$$\int_{-\frac{\pi}{2}}^{\frac{\pi}{2}} \cos^\rho \theta \, ds_F(\theta) = \rho(\rho-1) \int_{-\frac{\pi}{2}}^{\frac{\pi}{2}} h_F(\theta) \cos^{\rho-2} \theta \, d\theta \tag{4.31}$$

and substituting this into (4.28′) we obtain the conclusion of the lemma.

We now use (4.20) to prove a uniqueness theorem for functions holomorphic in the half-plane $\operatorname{Re} z > 0$. This is an analog of Theorem 1 of the preceding section for entire functions.

THEOREM 4. *If $F(z)$ $(F(0) \neq 0)$ is holomorphic in the right half-plane, of proximate order $\rho(r)$ $(\rho > 1)$, and*

$$N_F(r) = \int_0^r \frac{\nu_F(t)}{t^2}\, dt,$$

then

$$\overline{\lim_{r\to\infty}} \frac{N_F(r)}{r^{\rho(r)-1}} \leqslant \frac{1}{2\pi} \int_{-\frac{\pi}{2}}^{\frac{\pi}{2}} h_F(\theta) \cos^{\rho-2}\theta\, d\theta. \tag{4.32}$$

PROOF. Suppose first that $h_F(\pi/2) < 0$ and $h_F(-\pi/2) < 0$. Then if $\delta > 0$ is sufficiently small, the indicator $h_F(\theta)$ remains negative in the angle $\pi/2 - \delta \leqslant |\theta| \leqslant \pi/2$, and for some $d > 0$ and all $r \geqslant d$

$$\ln |F(re^{i\theta})| < 0 \qquad \left(\frac{\pi}{2} - \delta \leqslant |\theta| \leqslant \frac{\pi}{2}\right). \tag{4.33}$$

By (4.20) we have

$$\frac{N_F(r)}{r^{\rho(r)-1}} = \frac{1}{2\pi} \int_{-\frac{\pi}{2}+\alpha(r)}^{\frac{\pi}{2}-\alpha(r)} \frac{\ln|F(r\cos\theta\, e^{i\theta})|}{r^{\rho(r)}} \frac{d\theta}{\cos^2\theta} + \frac{O(1)}{r^{\rho(r)-1}}$$

and using (4.33) we obtain the asymptotic inequality

$$\frac{N_F(r)}{r^{\rho(r)-1}} < \frac{1}{2\pi} \int_{-\frac{\pi}{2}+\delta}^{\frac{\pi}{2}-\delta} \frac{\ln|F(r\cos\theta\, e^{i\theta})|}{r^{\rho(r)}} \frac{d\theta}{\cos^2\theta} + \frac{O(1)}{r^{\rho(r)-1}}.$$

We transform this inequality by using the fact that for arbitrary ε and all sufficiently large r

$$\ln|F(r\cos\theta\, e^{i\theta})| < [h_F(\theta) + \varepsilon](r\cos\theta)^{\rho(r\cos\theta)}$$
$$\left(|\theta| \leqslant \frac{\pi}{2} - \delta\right)$$

(cf., Chapter I, Theorem 28), and also the fact that

$$\lim_{r\to\infty} \frac{(r\cos\theta)^{\rho} L(r\cos\theta)}{r^{\rho} L(r)} = \cos^{\rho}\theta$$

uniformly for $|\theta| \leqslant \pi/2 - \delta$. After making the transformation we obtain the asymptotic inequality

$$\frac{N_F(r)}{r^{\rho(r)-1}} < \frac{1}{2\pi} \int_{-\frac{\pi}{2}+\delta}^{\frac{\pi}{2}-\delta} [h_F(\theta)+\varepsilon] \cos^{\rho-2}\theta \, d\theta + \frac{O(1)}{r^{\rho(r)-1}}$$

and since $\varepsilon > 0$ is arbitrary

$$\overline{\lim_{r\to\infty}} \frac{N_F(r)}{r^{\rho(r)-1}} \leqslant \frac{1}{2\pi} \int_{-\frac{\pi}{2}+\delta}^{\frac{\pi}{2}-\delta} h_F(\theta) \cos^{\rho-2}\theta \, d\theta.$$

Finally, since $\delta > 0$ is arbitrary

$$\overline{\lim_{r\to\infty}} \frac{N_F(r)}{r^{\rho(r)-1}} \leqslant \frac{1}{2\pi} \int_{-\frac{\pi}{2}}^{\frac{\pi}{2}} h_F(\theta) \cos^{\rho-2}\theta \, d\theta.$$

We now dispose of the other possibilities $h_F(-\pi/2) < 0$ and $h_F(\pi/2) < 0$. To do this we construct an entire function $\Phi(z)$ of proximate order $\rho(r)$ which is of completely regular growth and has an indicator $h_\Phi(\theta)$ such that $h_\Phi(\pi/2) < -h_F(\pi/2)$ and $h_\Phi(-\pi/2) < -h_F(-\pi/2)$. The possibility of such a construction follows from Theorem 3 of Chapter II when $\rho > 1$.

The function

$$\Psi(z) = F(z)\Phi(z)$$

has the indicator

$$h_\Psi\left(-\frac{\pi}{2}\right) < 0 \text{ и } h_\Psi\left(\frac{\pi}{2}\right) < 0$$

and consequently $h_\Psi(-\pi/2) < 0$ and $h_\Psi(\pi/2) < 0$.

We now have the inequality

$$\overline{\lim_{r\to\infty}} \frac{N_\Psi(r)}{r^{\rho(r)-1}} \leqslant \frac{1}{2\pi} \int_{-\frac{\pi}{2}}^{\frac{\pi}{2}} h_\Psi(\theta) \cos^{\rho-2}\theta \, d\theta,$$

which can be written in the form

$$\overline{\lim_{r\to\infty}} \frac{N_F(r)}{r^{\rho(r)-1}} + \lim_{r\to\infty} \frac{N_\Phi(r)}{r^{\rho(r)-1}}$$
$$\leqslant \frac{1}{2\pi} \int_{-\frac{\pi}{2}}^{\frac{\pi}{2}} h_F(\theta) \cos^{\rho-2}\theta \, d\theta + \frac{1}{2\pi} \int_{-\frac{\pi}{2}}^{\frac{\pi}{2}} h_\Phi(\theta) \cos^{\rho-2}\theta \, d\theta.$$

The limit

$$\lim_{r\to\infty} \frac{N_\Phi(r)}{r^{\rho(r)-1}}$$

exists by Lemma 2 and is equal to the second term on the right. Hence we obtain

$$\overline{\lim_{r\to\infty}} \frac{N_F(r)}{r^{\rho(r)-1}} \leqslant \frac{1}{2\pi} \int_{-\frac{\pi}{2}}^{\frac{\pi}{2}} h_F(\theta)\cos^{\rho-2}\theta\, d\theta.$$

The theorem is established.

In the preceding section we derived from Theorem 1 an estimate for the upper density $\bar{\Delta}$ of the set of zeros of an entire function. In an analogous way we can derive from Theorem 4 an estimate for the quantity

$$\overline{\lim_{r\to\infty}} \frac{\nu_F(r)}{r^{\rho(r)}}.$$

COROLLARY. *Under the hypotheses of Theorem 4*

$$\overline{\lim_{r\to\infty}} \frac{\nu_F(r)}{r^{\rho(r)}} \leqslant \frac{\left(1+\frac{1}{\rho}\right)^{\rho}}{2\pi(\rho-1)} \int_{-\frac{\pi}{2}}^{\frac{\pi}{2}} \cos^{\rho}\theta\, ds_F(\theta). \tag{4.34}$$

PROOF. For $k > 1$ it follows from the monotonicity of $\nu_F(r)$ that

$$\frac{\nu_F(r)}{r}\left(1-\frac{1}{k}\right) \leqslant \int_{r}^{kr} \frac{\nu_F(t)}{t^2}\, dt \leqslant N_F(kr).$$

Hence by using (4.32) and (4.31) we obtain

$$\overline{\lim_{r\to\infty}} \frac{\nu_F(r)}{r^{\rho(r)}} \leqslant \frac{k^{\rho}}{2\pi(\rho-1)\rho(k-1)} \int_{-\frac{\pi}{2}}^{\frac{\pi}{2}} \cos^{\rho}\theta\, ds_F(\theta).$$

The right hand side of this inequality takes its smallest value for $k = 1 + 1/\rho$. Putting in this value of k, we obtain (4.34). Just as in § 1, we can construct a function for which there is equality in (4.34); but we shall not go into this.

An estimate for the lower limit

$$\underline{\lim_{r\to\infty}} \frac{\nu_F(r)}{r^{\rho(r)}}$$

leads us to a theorem analogous to Theorem 2 of the preceding section.

THEOREM 5. *If $F(z)$ is a holomorphic function, not identically zero, in the half-plane* Re $z \geqslant 0$, *of proximate order $\rho(r)$ with $\rho > 1$, then*

$$\underline{\lim_{r\to\infty}} \frac{\nu_F(r)}{r^{\rho(r)}} \leqslant \frac{1}{2\pi\rho} \int_{-\frac{\pi}{2}}^{\frac{\pi}{2}} \cos^{\rho}\theta\, ds_F(\theta). \tag{4.35}$$

If equality holds, $F(z)$ is a function of completely regular growth for $|\arg z| < \pi/2$.

PROOF. The proof of the first conclusion presents no difficulty. In fact, letting

$$\mu = \varliminf_{r\to\infty} \frac{\nu_F(r)}{r^{\rho(r)}},$$

we can obtain the asymptotic inequality

$$\nu_F(r) > (\mu - \varepsilon)\, r^{\rho(r)}.$$

Without loss of generality we may suppose that $F(0) \neq 0$ and obtain the asymptotic equation

$$N_F(r) = \int_0^r \frac{\nu_F(t)}{t^2}\,dt > (\mu - \varepsilon)\int_0^r t^{\rho(t)-2}\,dt$$

$$\approx \frac{\mu-\varepsilon}{\rho - 1} r^{\rho(r)-1} + o(r^{\rho(r)-1}).$$

It follows immediately from this that

$$\mu \leqslant (\rho - 1) \varliminf_{r\to\infty} \frac{N_F(r)}{r^{\rho(r)-1}}, \tag{4.35'}$$

and (4.35) follows immediately from (4.32) and (4.31).

It remains to show that there is equality *only* for functions of completely regular growth. If we assume that there is equality in (4.35), then by using (4.20) and (4.35′) we obtain

$$\lim_{r\to\infty} \int_{-\frac{\pi}{2}+\alpha(r)}^{\frac{\pi}{2}-\alpha(r)} \frac{\ln|F(r\cos\theta e^{i\theta})|}{r^{\rho(r)}} \frac{d\theta}{\cos^2\theta} = \int_{-\frac{\pi}{2}}^{\frac{\pi}{2}} h_F(\theta)\cos^{\rho-2}\theta\, d\theta.$$

We have already noticed in the proof of Theorem 4 that, without loss of generality, we may suppose that the following condition is satisfied: for sufficiently small $\delta > 0$, some $d > 0$, and all $r > d$,

$$\ln|F(re^{i\theta})| < 0 \qquad \left(\frac{\pi}{2} - \delta \leqslant |\theta| \leqslant \frac{\pi}{2}\right),$$

and consequently we may suppose that

$$\varlimsup_{r\to\infty}\left[\left(\int_{-\frac{\pi}{2}+\alpha(r)}^{-\frac{\pi}{2}+\delta} + \int_{\frac{\pi}{2}-\delta}^{\frac{\pi}{2}-\alpha(r)}\right) \frac{\ln|F(r\cos\theta e^{i\theta})|}{r^{\rho(r)}} \frac{d\theta}{\cos^2\theta}\right] \leqslant 0.$$

Using this we obtain

$$\lim_{r\to\infty}\int_{-\frac{\pi}{2}+\delta}^{\frac{\pi}{2}-\delta}\left[\frac{\ln|F(r\cos\theta e^{i\theta})|}{r^{\rho(r)}}-h_F(\theta)\cos^{\rho}\theta\right]\frac{d\theta}{\cos^2\theta}$$

$$\geqslant\left(\int_{-\frac{\pi}{2}}^{-\frac{\pi}{2}+\delta}+\int_{\frac{\pi}{2}-\delta}^{\frac{\pi}{2}}\right)h_F(\theta)\cos^{\rho-2}\theta\,d\theta=O(\delta^{\rho-1}).$$

It follows from this and from the asymptotic inequality

$$\frac{\ln|F(r\cos\theta e^{i\theta})|}{r^{\rho(r)}}<(h_F(\theta)+\varepsilon)\cos^{\rho}\theta,$$

which holds for arbitrary $\varepsilon>0$ and $|\theta|\leqslant\pi/2-\delta$, that asymptotically

$$\overline{\lim_{r\to\infty}}\int_{-\frac{\pi}{2}+\delta}^{\frac{\pi}{2}-\delta}\left|\frac{\ln|F(r\cos\theta e^{i\theta})|}{r^{\rho(r)}}-h_F(\theta)\cos^{\rho}\theta\right|\frac{d\theta}{\cos^2\theta}\leqslant O(\delta^{\rho-1}),$$

and since $\delta>0$ is arbitrarily small,

$$\lim_{\delta\to0}\left\{\overline{\lim_{r\to\infty}}\int_{-\frac{\pi}{2}+\delta}^{\frac{\pi}{2}-\delta}\left|\frac{\ln|F(r\cos\theta e^{i\theta})|}{r^{\rho(r)}}-h_F(\theta)\cos^{\rho}\theta\right|\frac{d\theta}{\cos^2\theta}\right\}=0.$$

The quantity after the limit sign on the left does not decrease as δ decreases, and hence

$$\overline{\lim_{r\to\infty}}\int_{-\frac{\pi}{2}+\delta}^{\frac{\pi}{2}-\delta}\left|\frac{\ln|F(r\cos\theta e^{i\theta})|}{r^{\rho(r)}}-h_F(\theta)\cos^{\rho}\theta\right|\frac{d\theta}{\cos^2\theta}\leqslant0.$$

The quantity after the upper limit sign on the left is non-negative and consequently the following limit exists:

$$\lim_{r\to\infty}\int_{-\frac{\pi}{2}+\delta}^{\frac{\pi}{2}-\delta}\left|\frac{\ln|F(r\cos\theta e^{i\theta})|}{r^{\rho(r)}}-h_F(\theta)\cos^{\rho}\theta\right|\frac{d\theta}{\cos^2\theta}=0.$$

It follows from this equation that for arbitrary γ

$$\lim_{r\to\infty}\int_{-\gamma}^{\gamma}\left|\frac{\ln|F(r\cos\theta e^{i\theta})|}{r^{\rho(r)}}-h_F(\theta)\cos^{\rho}\theta\right|\frac{d\theta}{\cos^2\theta}=0.$$

We now apply Lemma 1. To do this, we multiply the integral on the left of the last equation by r/R^2 and integrate from R to $2R \sec \gamma$. We obtain

$$\lim_{R\to\infty} \frac{1}{R^2} \int_R^{2R\sec\gamma} \int_{-\gamma}^{\gamma} \left| \frac{\ln |F(r\cos\theta e^{i\theta})|}{r^{\rho(r)}} - h_F(\theta)\cos^\rho\theta \right| \frac{r\,d\theta\,dr}{\cos^2\theta} = 0.$$

Denoting $r \cos \theta$ by r_1, we have

$$\lim_{R\to\infty} \frac{1}{R^2} \int_{-\gamma}^{\gamma} \int_R^{2R\frac{\cos\theta}{\cos\gamma}} \left| \frac{\ln |F(r_1 e^{i\theta})|}{r_1^{\rho(r_1)}} - h_F(\theta) \right| \cos^{\rho-4}\theta\, r_1\, dr_1\, d\theta = 0$$

and finally

$$\lim_{R\to\infty} \frac{1}{R^2} \int_{-\gamma}^{\gamma} \int_R^{2R} | h_{F,r}(\theta) - h_F(\theta) |\, r\, dr\, d\theta = 0.$$

From this it follows by Lemma 1 that $F(z)$ is of completely regular growth in the half-plane $\operatorname{Re} z \geqslant 0$. The theorem is established.

As a corollary of this theorem we obtain the following fundamental theorem of M. L. Cartwright [3].

THEOREM 6. *If $F(z)$ is holomorphic and of proximate order $\rho(r)$ in the angle (α, β) and has a sinusoidal indicator for $\alpha < \theta < \beta$, and the opening of the angle (α, β) is greater than π/ρ, then it is of completely regular growth and the density of its zeros in the angle is zero.*

PROOF. Without loss of generality we may suppose that the angle (α, β) is the half-plane $\operatorname{Re} z > 0$ and that $\rho > 1$. For a sinusoidal indicator, $ds_F(\theta) = 0$, and hence there is equality in (4.35).

It follows from Theorem 5 that $F(z)$ is of completely regular growth. In addition, the density of its set of zeros in the half-plane $\operatorname{Re} z > 0$ is $1/2\pi\rho[s_F(\pi/2) - s_F(-\pi/2)] = 0$. The theorem is established.

It is interesting to compare this theorem with Theorem 5 of § 3 of Chapter III. That theorem requires that the indicator be sinusoidal and that the angle contains a ray on which $F(z)$ is of completely regular growth. It follows from Cartwright's theorem that if the opening of the angle exceeds π/ρ, the requirement of the existence of such a ray is superfluous.

All the theorems that we have proved on estimates for the density of the set of zeros in a half-plane concern the case $\rho > 1$. The case $\rho = 1$ is exceptional. In this case there is no estimate of the kind given in Theorem 4.

We consider only functions of exponential type and prove a theorem which contains, in particular, the theorem of Cartwright that we stated at the beginning of this chapter.

THEOREM 7. *If $F(z)$ is holomorphic, not identically zero, and of exponential type in the half-plane* $\operatorname{Re} z \geqslant 0$, *then*

$$\overline{\lim_{R\to\infty}} \frac{\nu_F(r)}{r} \leqslant \frac{1}{2\pi}\left[h_F\left(\frac{\pi}{2}\right)+h_F\left(-\frac{\pi}{2}\right)\right]. \tag{4.36}$$

PROOF. We start from formula (4.20). For a given $\varepsilon > 0$ such that

$$\ln|F(re^{i\theta})| < [h_F(\theta)+\varepsilon]\, r \qquad \left(|\theta| \leqslant \frac{\pi}{2}\right);$$

we can choose $d > 0$ so that

$$N_F(r) < \frac{1}{2\pi}\int_{-\frac{\pi}{2}+\alpha(r)}^{\frac{\pi}{2}-\alpha(r)} [h_F(\theta)+\varepsilon]\frac{d\theta}{\cos\theta} + O(1)$$

$$\left(\alpha(r) = \arcsin\frac{d}{r}\right).$$

We shall show that the first term on the right is the principal term. To do this, integrate by parts. We obtain

$$N_F(r) < \frac{1}{2\pi}[h_F(\theta)+\varepsilon]\ln\left|\operatorname{tg}\left(\frac{\theta}{2}+\frac{\pi}{4}\right)\right| \Bigg|_{-\frac{\pi}{2}+\alpha(r)}^{\frac{\pi}{2}-\alpha(r)}$$

$$-\frac{1}{2\pi}\int_{-\frac{\pi}{2}+\alpha(r)}^{\frac{\pi}{2}-\alpha(r)} \ln\left|\operatorname{tg}\left(\frac{\theta}{2}+\frac{\pi}{4}\right)\right| h_F'(\theta)\,d\theta = \frac{1}{2\pi}\left[h_F\left(\frac{\pi}{2}-\alpha(r)\right)\right.$$

$$\left.+h_F\left(-\frac{\pi}{2}+\alpha(r)\right)+2\varepsilon\right]\ln\left|\operatorname{ctg}\frac{\alpha(r)}{2}\right| + O(1).$$

Since $\ln \operatorname{ctg}\dfrac{\alpha(r)}{2}\Big/ \ln r \to 1$ as $r \to \infty$, it follows from this that

$$N_F(r) < \frac{1}{2\pi}\left[h_F\left(\frac{\pi}{2}\right)+h_F\left(-\frac{\pi}{2}\right)+2\varepsilon\right]\ln r + o(\ln r),$$

and since $\varepsilon > 0$ is arbitrary

$$\overline{\lim_{r\to\infty}} \frac{N_F(r)}{\ln r} \leqslant \frac{1}{2\pi}\left[h_F\left(\frac{\pi}{2}\right)+h_F\left(-\frac{\pi}{2}\right)\right]. \tag{4.37}$$

We turn now to the quantities $N_F(r)$ and $\nu_F(r)$. Put

$$\mu = \overline{\lim_{r\to\infty}} \frac{\nu_F(r)}{r}.$$

Then for every $\varepsilon > 0$ we have the asymptotic inequality

$$\nu_F(r) > (\mu - \varepsilon) r.$$

Using this, we find from the definition of $N_F(r)$ that

$$N_F(r) > O(1) + (\mu - \varepsilon) \ln r.$$

Finally, since $\varepsilon > 0$ is arbitrary this implies

$$\mu \leqslant \varliminf_{r \to \infty} \frac{N_F(r)}{\ln r}.$$

Substituting this into (4.37), we obtain

$$\varliminf_{r \to \infty} \frac{\nu_F(r)}{r} \leqslant \frac{1}{2\pi} \left[h_F\left(\frac{\pi}{2}\right) + h_F\left(-\frac{\pi}{2}\right) \right].$$

The theorem is established. We note in addition that the width $h_F(\pi/2) + h_F(-\pi/2)$ of the indicator diagram can be represented in the form

$$h_F\left(\frac{\pi}{2}\right) + h_F\left(-\frac{\pi}{2}\right) = \int_{-\frac{\pi}{2}}^{\frac{\pi}{2}} \cos\theta \, ds_F(\theta).$$

In fact, it is easily seen that the integral on the right side of this equation gives the projection on the imaginary axis of the arc of the boundary of the indicator diagram of $F(z)$ between the supporting lines $l_{-\pi/2}$ and $l_{\pi/2}$ and consequently is equal to the width of the indicator diagram in the direction of the imaginary axis. Therefore (4.36) follows from (4.35) if we put $\rho = 1$.[6]

3. Functions Vanishing on a Set with an Angular Density[7]

In this section we give a different generalization of Carlson's theorem for holomorphic functions of finite order in an angle. We consider functions which vanish on a set of points that has an angular density, and show that under certain restrictions on the growth of the function, the function must be identically zero.

THEOREM 8. *Let $F(z)$ be holomorphic and at most of normal type with respect to the proximate order $\rho(r)$ in the angle $\alpha \leqslant \arg z \leqslant \alpha + \pi/\rho$, and vanish on a set $N = \{a_k\}$ in this angle, with angular density $\Delta_N(\psi)$, and let*

$$H_N(\theta) = \frac{\pi}{\sin \pi\rho} \int_{\alpha}^{\alpha + \frac{\pi}{\rho}} \cos \rho(|\theta - \psi| - \pi) \, d\Delta_N(\psi)$$

[6] We note that in this case it is an open question whether there is equality in (4.36) only for functions of completely regular growth.

[7] B. Ja. Levin [1].

when ρ is nonintegral and

$$H_N(\theta) = \pi \int_{\alpha}^{\alpha+\frac{\pi}{\rho}} \sin \rho \,|\psi - \theta|\, d\Delta_N(\psi)$$

when ρ is integral. Then if $F(z)$ is not identically zero,

$$h_F(\alpha) + h_F\left(\alpha + \frac{\pi}{\rho}\right) \geqslant H_N(\alpha) + H_N\left(\alpha + \frac{\pi}{\rho}\right).$$

PROOF. Consider the holomorphic function

$$\Psi(z) = \frac{F(z)}{\Phi_N(z)}$$

in the angle $(\alpha, \alpha + \pi/\rho)$, where

$$\Phi_N(z) = \prod_{k=1}^{\infty} G\left(\frac{z}{a_k}; p\right)$$

when ρ is not integral and

$$\Phi_N(z) = \prod_{k=1}^{\infty} \left(1 - \frac{z^{2\rho}}{a_k^{2\rho}}\right)$$

when ρ is integral.

Since $\Phi(z)$ is a function of completely regular growth with indicator $H_N(\theta)$ (cf., Chapter II, § 4) we have

$$h_\Psi(\theta) = h_F(\theta) - h_N(\theta).$$

By property (h) of the indicator (cf., Chapter I, § 16)

$$h_\Psi\left(\frac{\pi}{\rho}\right) + h_\Psi\left(\frac{\pi}{\rho} + \alpha\right) \geqslant 0$$

and consequently

$$h_F\left(\alpha + \frac{\pi}{\rho}\right) + h_F(\alpha) \geqslant H_N\left(\alpha + \frac{\pi}{\rho}\right) + H_N(\alpha).$$

The theorem is established. We note that when $\rho = 1$ the quantity $h_F(\alpha + \pi) + h_F(\alpha)$ is the width of the indicator diagram of the function in the direction $\arg z = \alpha$.

If $\rho(r) = 1$ and the set N is the sequence of integers $a_k = 1, 2, \ldots,$ then $H_N(\theta) = \pi\,|\sin \theta|$, and by the theorem just established

$$h_F\left(\frac{\pi}{2}\right) + h_F\left(-\frac{\pi}{2}\right) \geqslant 2\pi.$$

Hence Carlson's theorem is a special case of Theorem 8.

We obtain a different uniqueness theorem by using property (g) of the indicator instead of property (h). Before stating this theorem we introduce the concept of an indicator of uniqueness. We call a trigonometrically convex function $h(\theta)$ the indicator of uniqueness of the regular set N with angular density $\Delta_N(\theta)$ and indicator $H_N(\theta)$ if there is no entire function $F(z)$, not identically zero, of proximate order $\rho(r)$ with indicator $h(\theta)$, which vanishes at all points of N.

THEOREM 9. *In order that a trigonometrically convex function $h(\theta)$ should not be the indicator of uniqueness of a given regular set N, it is necessary and sufficient that the following inequality holds for arbitrary α and β ($\alpha < \beta$):*

$$s(\alpha \pm 0, \beta \pm 0) \geqslant 2\pi\rho[\Delta_N(\beta \pm 0) - \Delta_N(\alpha \pm 0)], \tag{4.38}$$

where

$$s(\alpha, \beta) = h'(\beta) - h'(\alpha) + \rho^2 \int_\alpha^\beta h(\varphi)\, d\varphi.$$

PROOF. We note first that $h(\theta)$ fails to be an indicator of uniqueness if and only if it can be represented in the form

$$h(\theta) = H_N(\theta) + h_1(\theta), \tag{4.39}$$

where $h_1(\theta)$ is a trigonometrically convex function of period 2π. In fact, if $h(\theta)$ is not an indicator of uniqueness, there is an entire function $F(z)$ of proximate order $\rho(r)$ with indicator $h(\theta)$, vanishing on N. Let $\Phi_N(z)$ be the canonical function of the regular set (Chapter II, § 1). Then evidently the function

$$\Psi(z) = \frac{F(z)}{\Phi_N(z)}$$

is entire and its indicator

$$h_\Psi(\theta) = h(\theta) - H_N(\theta).$$

Conversely, if (4.39) is satisfied then, constructing an entire function $\Psi(z)$ with indicator $h_1(\theta)$ (Chapter II, § 4), we see that the function

$$F(z) = \Phi_N(z)\Psi(z)$$

has the indicator

$$h(\theta) = H_N(\theta) + h_1(\theta)$$

and vanishes on N.

To establish the equivalence of (4.39) and (4.38) for arbitrary α and β ($\alpha < \beta$) it is enough to show that it is equivalent for a trigonometrically convex function $h_1(\theta)$ to satisfy the inequality

$$h_1'(\beta) - h_1'(\alpha) + \rho^2 \int_\alpha^\beta h_1(\theta)\, d\theta \geqslant 0$$

for $\alpha < \beta$ (Chapter I, § 16) or the equation

$$H_N'(\beta) - H_N'(\alpha) + \rho^2 \int_\alpha^\beta H_N(\varphi)\, d\varphi = 2\pi\rho[\Delta_N(\beta) - \Delta_N(\alpha)],$$

established in § 3 of Chapter III.

We recall that for $\rho = 1$ the quantity $s(\alpha, \beta)$ is the length of the boundary of the region of which $h(\theta)$ is the supporting function, between the points of contact of the supporting lines perpendicular to the rays $\arg z = \alpha$, $\arg z = \beta$.

Hence for $\rho = 1$ we have a geometrical criterion for the supporting function $h(\theta)$ of a convex region to be an indicator of uniqueness. In fact, the length of the arc in question must be less than 2π times the density of the set N in the angle.

We also remark that the function $H_N(\theta)$ minimizes the functional

$$s(\alpha, \beta, h) = h'(\beta) - h'(\alpha) + \rho^2 \int_\alpha^\beta h(\theta)\, d\theta,$$

with arbitrary α and β ($\alpha < \beta$), among all trigonometrically convex functions that are not indicators of uniqueness for a given regular set.

In fact,

$$s(\alpha, \beta, H_N) = 2\pi\rho\,[\Delta(\beta) - \Delta(\alpha)],$$

and in the general case it follows from (4.39) that

$$s(\alpha, \beta, h) = 2\pi\rho\,[\Delta(\beta) - \Delta(\alpha)] + s(\alpha, \beta, h_1).$$

But by property (g) of trigonometrically convex functions (Chapter I, § 16), $s(\alpha, \beta, h_1) \geqslant 0$ and consequently

$$s(\alpha, \beta, h) \geqslant s(\alpha, \beta, H_N).$$

We note (Chapter I, § 16) that the equation $s(\alpha, \beta, h_1) = 0$ is possible only when $h_1(\theta) = A \cos \rho\theta + B \sin \rho\theta$.

It follows from this that when ρ is not integral $H_N(\theta)$ is the unique function minimizing the functional $s(0, 2\pi, h)$, and when ρ is integral the general form of minimizing functions is $h(\theta) = H_N(\theta) + A \cos \rho\theta + B \sin \rho\theta$.

In an analogous way we can define an indicator of uniqueness for holomorphic functions of proximate order $\rho(r)$ in an angle (α, β).

We shall call a trigonometrically convex function $h(\theta)$ ($\alpha \leqslant \theta \leqslant \beta$) *an indicator of uniqueness for the set N which is inside the angle* (α, β) *and has angular density* $\Delta_N(\theta)$ provided that there is no function that is holomorphic in that angle, of proximate order $\rho(r)$, with indicator $h(\theta)$, vanishing at all points of N and not identically zero. We can prove the following theorem about indicators of uniqueness for an angle.

THEOREM 10. *If the trigonometrically convex function* $h(\theta)$ ($\alpha \leqslant \theta \leqslant \beta$) *is not an indicator of uniqueness for the set N, then the following inequality holds for arbitrary* α' *and* β' ($\alpha \leqslant \alpha' < \beta' \leqslant \beta$):

$$s(\alpha', \beta', h) \geqslant 2\pi\rho\,[\Delta_N(\beta') - \Delta_N(\alpha')]. \tag{4.40}$$

PROOF. Suppose first that ρ is not integral, and construct the function

$$\Phi(z)=\prod_{k=1}^{\infty} G\left(\frac{z}{a_k};\ p\right),$$

where a_k are the points of N that are in the angle (α', β'). Let $F(z)$ be a holomorphic function of proximate order $\rho(r)$ in (α, β) with indicator $h(\theta)$. Then

$$\Psi(z)=\frac{F(z)}{\Phi(z)}$$

is a holomorphic function in (α', β') of proximate order $\rho(r)$, with indicator

$$h_\Psi(\theta)=h(\theta)-H_N(\theta).$$

It follows that

$$s(\alpha', \beta', h)-s(\alpha', \beta', H_N)\geqslant 0,$$

and since $\Phi(z)$ is of completely regular growth, we have

$$s(\alpha', \beta', H_N)=2\pi\rho\left[\Delta_N(\beta')-\Delta_N(\alpha')\right].$$

To treat the case of integral ρ, we first note that it is enough to prove (4.40) when $\beta'-\alpha'<\pi/\rho$. In fact, if we have proved this, then to prove (4.40) in general we have only to divide (α', β') into parts and add the corresponding inequalities. Put

$$\Phi(z)=\prod_{k=1}^{\infty}\left(1-\frac{z^{2\rho}}{a_k^{2\rho}}\right),$$

where a_k are the points of N lying in the angle (α', β') $(\beta'-\alpha'<\pi/\rho)$. The function $\Phi(z)$ is holomorphic, of proximate order $\rho(r)$, and of completely regular growth. The rest of the proof is just the same as for nonintegral ρ.

4. Representation of Entire Functions by Lagrange Interpolation Series

A problem closely related to that of the unique determination of an entire function by its values on a set of points is the problem of interpolation of entire functions, i.e., the problem of constructing an entire function that belongs to a prescribed class and takes given values $\{b_n\}$ at the points of a given set $\{a_n\}$ (nodes).

For this construction we use *interpolation formulas* which generalize the corresponding formulas for interpolation by means of polynomials.

A *Newton series* is an interpolation series of the form

$$\sum_{n=0}^{\infty} a_n(z-\alpha_1)(z-\alpha_2)\ \ldots\ (z-\alpha_n).$$

It is known[8] that when $a_n = n$ ($n = 0, 1, 2, \ldots$) the region of convergence of this series is a half-plane $\operatorname{Re} z \geqslant \gamma$ (γ is a real number). In addition, the sum of a Newton series is a function of exponential type in this half-plane and its indicator satisfies the inequality

$$h_F(\theta) \leqslant \cos\theta \ln(2\cos\theta) + \theta \sin\theta \qquad \left(|\theta| < \frac{\pi}{2}\right).$$

On the other hand, by Carlson's theorem a function of exponential type in the half-plane $\operatorname{Re} z \geqslant \gamma$ is determined by its values at the integral points $n = 0, 1, \ldots$ if $h_F(\pi/2) + h_F(-\pi/2) < 2\pi$.

Hence there is a class of functions of exponential type which are determined by their values at the points $0, 1, 2, \ldots$ but cannot be constructed from these values by means of a Newton series.[9] In particular, this class contains all functions of exponential type for which $\pi < h_F(\pi/2) + h_F(-\pi/2) < 2\pi$. Hence for the set $0, 1, 2, \ldots$ the "convergence class" is narrower than the "uniqueness class" defined by $h_F(\pi/2) + h_F(-\pi/2) < 2\pi$.

In many cases it is more convenient to represent entire functions by a different interpolation formula, which represents an entire function, not as a series of polynomials, but as a series of special entire functions.

Let $f(z)$ be an entire function which has simple zeros at all the nodes $\{a_n\}$ ($|a_n| \to \infty$), and let $\{b_n\}$ be values assigned at the nodes. If the series

$$\sum_{n=1}^{\infty} \frac{b_n f(z)}{f'(a_n)(z - a_n)} \tag{4.41}$$

converges uniformly in every bounded region, it represents an entire function which solves the interpolation problem. This series is called a *Lagrange interpolation series*.

We shall investigate the question of the representation of entire functions by Lagrange interpolation series in the case when the nodes form an R-set with index[10] $\rho(r)$ (for the definition see Chapter II, § 1). We denote the set of nodes by $\mathfrak{A}$, the canonical function $f(z, \mathfrak{A})$ of this set by $f(z)$, and the indicator $H_{\mathfrak{A}}(\theta)$ of the set by $H(\theta)$. The following theorem about the convergence of the Lagrange interpolation series then holds.

THEOREM 11. *If the set $\mathfrak{A}$ of nodes is an R-set* (*Chapter* II, § 1) *with index $\rho(r)$ and if for some $\varepsilon > 0$ and $n > n(\varepsilon)$ all the numbers b_n satisfy*

$$\ln|b_n| < [H(\psi_n) - \varepsilon]\,|a_n|^{\rho(|a_n|)} \qquad (\psi_n = \arg a_n), \tag{4.42}$$

[8] Cf., A. O. Gel'fond [**1; 2**].

[9] The question of the representability of entire functions by Newton series has been investigated in a number of papers. The most complete results are obtained in the papers by A. O. Gel'fond [**1**] and I. I. Ibragimov and M. V. Keldyš [**1**]. Gel'fond gave a very general method for constructing an entire function from its values on a sequence of functionals [**1; 2**].

[10] Part of the results of §§ 4 and 5 are given in the author's paper [**2**]. See also A. Pfluger [**2**].

then the Lagrange series (4.41) *converges uniformly in every bounded region and the limit is an entire function with indicator not exceeding* $H(\theta)$.

PROOF. By Theorem 5 of Chapter II, $f(z)$ satisfies the asymptotic equation

$$\ln|f(re^{i\theta})| \approx H(\theta)\, r^{\rho(r)}$$

everywhere outside the excluded C_R-disks. We now estimate $|f'(a_n)|$. Consider the function

$$\ln\left|\frac{f(z)}{z-a_n}\right|.$$

It is harmonic outside the excluded C_R-disk with center at a_n and satisfies the following asymptotic equation (as $n \to \infty$) on the boundary of this disk:

$$\ln\left|\frac{f(z)}{z-a_n}\right| \approx H(\psi_n)\,|a_n|^{\rho(|a_n|)}.$$

By the maximum and minimum principle for harmonic functions we obtain

$$\ln|f'(a_n)| \approx H(\psi_n)\,|a_n|^{\rho(|a_n|)}. \tag{4.43}$$

From this and (4.42) it follows that when z lies outside the excluded C_R-disks, and $n > n'(\varepsilon)$,

$$\ln\left|\frac{b_n}{f'(a_n)(z-a_n)}\right| < -\frac{\varepsilon}{2}\,|a_n|^{\rho(|a_n|)}.$$

On the other hand, we have

$$n_f(r) \approx \Delta(2\pi)\, r^{\rho(r)},$$

and consequently

$$|a_n|^{\rho(|a_n|)} \approx \frac{n}{\Delta(2\pi)}.$$

This relationship makes it possible to write the preceding inequality in the form

$$\left|\frac{b_n}{f'(a_n)(z-a_n)}\right| < e^{-cn} \qquad \left(c = \frac{\varepsilon}{2\Delta(2\pi)}\right). \tag{4.44}$$

Hence the series

$$\sum_{n=1}^{\infty} \frac{b_n}{f'(a_n)(z-a_n)}$$

converges outside the excluded C_R-disks to a function $\varphi(z)$.

Applying the maximum principle in each excluded disk to the sections of the series (4.41) of entire functions, we find that the series converges uniformly in each bounded region to an entire function $F(z) = f(z)\varphi(z)$.

We show further that $h_F(\theta) \leqslant H(\theta)$. To do this we first show that $\varphi(z)$

tends to zero as $|z| \to \infty$ outside the set of C_R-disks. In fact, we have from (4.44) for $z \bar{\in} C_R$

$$|\varphi(z)| < \sum_{n=1}^{N} \frac{|b_n|}{|f'(a_n)||z-a_n|} + \frac{e^{-cN}}{1-e^{-c}}.$$

For any $\varepsilon > 0$ we can choose N, and then r_ε, so that

$$|\varphi(z)| < \varepsilon$$

for $|z| > r_\varepsilon$ and $z \in C_\varepsilon$.

Therefore

$$|F(z)| < |f(z)|$$

outside the excluded disks and for $|z| > r_\varepsilon$, and consequently $h_F(\theta) \leqslant H(\theta)$. The theorem is established.

REMARK 1. *The function $F(z)$ represented by the series* (4.41) *is not the unique solution of the interpolation problem even under the condition*

$$h_F(\theta) \leqslant H(\theta). \tag{4.45}$$

We shall show that all solutions that satisfy this condition are represented by the formula

$$\Phi(z) = F(z) + \omega(z) f(z), \tag{4.46}$$

where $\omega(z)$ is an entire function of at most minimal type of the proximate order $\rho(r)$.

In fact, if $\Phi(z)$ is any solution, then

$$\omega(z) = \frac{\Phi(z) - F(z)}{f(z)}$$

is an entire function. In addition, if $\Phi(z)$ satisfies (4.45) the indicator of the numerator does not exceed $H(\theta)$. The denominator is an entire function of completely regular growth with indicator $H(\theta)$. Hence the indicator of $\omega(z)$ is at most zero and consequently $\omega(z)$ is at most of minimal type of the proximate order $\rho(r)$.

Conversely, every function $\Phi(z)$ represented by (4.46) is evidently a solution of the interpolation problem under the supplementary condition (4.45).

REMARK 2. If we replace the supplementary condition (4.45) by the stronger condition

$$\lim_{\substack{|z| \to \infty \\ z \bar{\in} C}} \frac{|\Phi(z)|}{|f(z)|} = 0, \tag{4.47}$$

which, as has been shown, $F(z)$ satisfies, we find that the entire function $\omega(z)$ tends to zero when $|z| \to \infty$ ($z \bar{\in} C$), and by the maximum principle it follows that $\omega(z) \equiv 0$.

Hence $F(z)$ *is the only solution of the problem under the supplementary condition* (4.47). In the following theorem we give a sufficient condition for the representability of a given entire function by a Lagrange interpolation series.

THEOREM 12. *If the points* $\{a_k\}$ *form an R-set* $\mathfrak{A}$ *with index* $\rho(r)$, *then every entire function* $F(z)$ *that satisfies*

$$h_F(\theta) < H(\theta) \qquad (0 \leqslant \theta < 2\pi) \tag{4.48}$$

is represented by the series

$$F(z) = \sum_{n=1}^{\infty} \frac{F(a_n) f(z)}{f'(a_n)(z - a_n)}, \tag{4.49}$$

with uniform convergence in each bounded region.

PROOF. We shall show that when (4.48) is satisfied the sequence $\{b_n = F(a_n)\}$ satisfies (4.42) of Theorem 11. In fact, it follows from (4.48) that for some $\varepsilon > 0$

$$h_F(\theta) < H(\theta) - 2\varepsilon. \tag{4.50}$$

In addition, by Theorem 28 of Chapter I we have the asymptotic inequality

$$\ln |F(re^{i\theta})| < [h_F(\theta) + \varepsilon]\, r^{\rho(r)} < [H(\theta) - \varepsilon]\, r^{\rho(r)}$$

and consequently for $n > n(\varepsilon)$

$$\ln |F(a_n)| < [H(\psi_n) - \varepsilon]\, |a_n|^{\rho(|a_n|)}.$$

By Theorem 11 this inequality implies the convergence of (4.49). We still have to show that the sum of this series is equal to the given function $F(z)$. To do this we use Remark 2 after Theorem 11. As we see from (4.48), $F(z)$ satisfies (4.47). The sum of the series satisfies the same condition. Then (4.49) follows from the uniqueness of the solution of the interpolation problem under the supplementary condition (4.47). The theorem is established.

REMARK. Theorem 12 remains true if the requirement that the set of nodes is an R-set is replaced by the much weaker requirement that it is a regular set. In this case, however, the sum of the series must be taken in a special sense. Indeed, we can say that

$$F(z) = \lim_{\substack{R \to \infty \\ R \overline{\in} E}} \sum_{|a_n| < R} \frac{F(a_n) f(z)}{f'(a_n)(z - a_n)}, \tag{4.51}$$

where E is an E^0-set.

In fact, from (4.48) and from the asymptotic equation $\ln|f(re^{i\theta})| \approx H(\theta) r^{\rho(r)}$, which holds for r not belonging to an E^0-set E, it follows that

$$\ln \left| \frac{F(re^{i\theta})}{f(re^{i\theta})} \right| < -\varepsilon r^{\rho(r)} \qquad (r \overline{\in} E).$$

This inequality implies that the integral

$$J_r(z) = \frac{1}{2\pi i} \int\limits_{|\zeta|=r} \frac{F(\zeta) f(z)}{f(\zeta)(\zeta - z)} d\zeta$$

tends to zero when r tends to infinity outside the set E. On the other hand,

$$J_r(z) = F(z) - \sum_{|a_n|<r} \frac{F(a_n) f(z)}{f'(a_n)(z - a_n)}$$

and consequently (4.51) holds.

We call a trigonometrically convex function $h(\theta)$ a *convergence indicator* (for a given $\rho(r)$) if every entire function of proximate order $\rho(r)$ with indicator not exceeding $h(\theta)$ is represented by the series (4.49). It is clear that a convergence indicator is also a uniqueness indicator. It follows from Theorem 12 that a convergence indicator can be as close as we please to a set indicator $H(\theta)$ which is not a uniqueness indicator. We recall that the situation is different for the Newton interpolation series.[11]

It was shown in Theorem 11 that if an R-set $\{a_n\}$ is given and the sequence $\{b_n\}$ satisfies (4.42), there exists an entire function $F(z)$ whose indicator does not exceed the set indicator $H(\theta)$ and which takes the values $\{a_n\}$ at the points $\{b_n\}$.

In the following theorem we give a necessary and sufficient condition for a given sequence of numbers $\{b_n\}$ to be the values, at the points $\{a_n\}$ of a given R-set, of an entire function whose indicator does not exceed the set indicator $H(\theta)$.

THEOREM 13. *Let $\mathfrak{A}$ be an R-set of points $\{a_n\}$ with index $\rho(r)$ and $\{b_n\}$ a sequence of numbers. A necessary and sufficient condition for the existence of an entire function $F(z)$ satisfying the conditions*

$$F(a_n) = b_n, \tag{α}$$

$$h_F(\theta) \leqslant H(\theta), \tag{β}$$

is that

$$\varlimsup_{n\to\infty} \left\{ \frac{\ln|b_n|}{|a^n|^{\rho(|a_n|)}} - H_{\mathfrak{A}}(\psi_n) \right\} < 0 \qquad (\psi_n = \arg a_n). \tag{4.52}$$

PROOF. The necessity of (4.52) follows immediately from the asymptotic inequality

$$\ln|F(re^{i\theta})| < [h_F(\theta) + \varepsilon]\, r^{\rho(r)},$$

which the interpolating function $F(z)$ must satisfy, and (β).

---

[11] We may, however, remark that the solutions of many problems of the theory of entire functions require the expansion of a function in a Newton series rather than a Lagrange series. This is basically connected with the fact that a Newton series is a series of polynomials.

To establish the sufficiency we note first that if

$$\overline{\lim_{n\to\infty}} \left\{ \frac{\ln|b_n|}{|a^n|^{\rho(|a_n|)}} - H(\psi_n) \right\} < 0,$$

the existence of an interpolating function satisfying (β) follows directly from Theorem 11. However, if there is equality in (4.52), the Lagrange series (4.41) does not necessarily converge, and hence we cannot construct the function $F(z)$ in the same way as in Theorem 11. To obtain a convergent series in this case, we adjoin new nodes to the set $\mathfrak{A}$ in such a way that the new set is an R-set and the canonical function $f_1(z)$ of the new set grows faster than $f(z)$. When $f_1(z)$ grows sufficiently rapidly the series

$$\sum_{n=1}^{\infty} \frac{|b_n|}{|f_1'(a_n)|}$$

converges, and the interpolating function $F(z)$ can be constructed with the aid of the Lagrange formula. At the additional nodes α_n we may put $F(\alpha_n) = 0$. We also require that the indicator $h_F(\theta)$ of the function we have constructed does not exceed $H(\theta)$. This will be true if we extend the set $\mathfrak{A}$ in such a way that its indicator is not changed. We make this extension in the following way. When there is equality in (4.52) we have the equation

$$\overline{\lim_{n\to\infty}} \left(|a_n|^{-\rho(|a_n|)} \ln \left| \frac{b_n}{f'(a_n)} \right| \right) = 0.$$

By Theorem 16 of Chapter I there is a proximate order $\rho(r)$ such that[12]

$$\overline{\lim_{n\to\infty}} \left(|a_n|^{-\rho_1(|a_n|)} \ln \left| \frac{b_n}{f'(a_n)} \right| \right) = 1,$$

and

$$\lim_{r\to\infty} r^{\rho_1(r)-\rho(r)} = 0.$$

We introduce the proximate order

$$\rho_2(r) = \max\left(\rho_1(r), \frac{\rho(r)}{2}\right)$$

and construct an R-set $\mathfrak{A}_1$ with index $\rho_2(r)$ satisfying condition (C′) (Chapter II, § 1) and such that the indicator $H_{\Delta_1}(\theta)$ satisfies

$$H_{\mathfrak{A}_1}(\theta) > 2 \qquad (0 \leqslant \theta < 2\pi).$$

---

[12] In Chapter I we considered not sequences but functions defined on the positive real axis; however, in the present problem the difference is not important.

Here we may suppose that the excluded disks for the sets $\mathfrak{A}$ and $\mathfrak{A}_1$ do not intersect.[13]

Let $\omega(z)$ be the canonical function of $\mathfrak{A}_1$. Outside the excluded disks for $\mathfrak{A}_1$ it satisfies the asymptotic inequality

$$\ln|\omega(z)| > 2r^{\rho_2(r)}$$

and consequently

$$\ln\left|\frac{b_n}{|f'(a_n)\,\omega(a_n)|}\right| < -(1-\varepsilon)|a_n|^{\rho_2(|a_n|)}.$$

It follows[14] from this inequality that the series

$$F(z) = \sum_{n=1}^{\infty} \frac{f(z)\,\omega(z)\,b_n}{f'(a_n)\,\omega(a_n)\,(z-a_n)} \qquad (a_n \in \mathfrak{A})$$

converges uniformly in every finite region and solves the interpolation problem, and the indicator of the function represented by this series does not exceed $(H_{\mathfrak{A}}\theta)$.

---

[13] We outline the construction of $\mathfrak{A}_1$: on the rays $\arg z = 2\pi k/n + \alpha$ $(k = 0, 1, 2, \ldots, n-1;$ n integral and $n \geqslant 2\rho_2;$ α arbitrary) we select the points where the function $\Delta r^{\rho_2(r)}$ takes integral values. The set so constructed has an angular density. When ρ_2 is integral we take n to be the even number $2\rho_2$ and condition (B) (Chapter II, § 1) is satisfied. In addition, for the points on a single ray we have by construction

$$\begin{aligned} 1 = \Delta|a_{n+1}|^{\rho_2(|a_{n+1}|)} &- \Delta|a_n|^{\rho_2(|a_n|)} \\ &= \Delta(|a_{n+1}| - |a_n|)\,[\rho_2(c)\,c^{\rho_2(c)-1} + c^{\rho_2(c)}\rho_2'(c)\ln c] \\ &\qquad (|a_n| < c < |a_{n+1}|), \end{aligned}$$

or asymptotically

$$|a_{n+1}| - |a_n| > \frac{1}{2}\Delta^{-1}\rho_2^{-1}|a_n|^{1-\rho_2(|a_n|)},$$

and consequently (C′) is satisfied. Using the asymptotic formulas (2.05) and (2.07), we find that $H_1(\theta) > 2$ for sufficiently large Δ.

We also need to know that the excluded disks for $\mathfrak{A}$ and $\mathfrak{A}'$ do not intersect. Note that the diameters of the excluded disks for $\mathfrak{A}_1$ do not exceed

$$d|a_n|^{1-\rho_2(|a_n|)} \leqslant d|a_n|^{1-\frac{\rho(|a_n|)}{2}}. \qquad (*)$$

If $\mathfrak{A}$ satisfies (C′) it is enough to take the rays containing points of $\mathfrak{A}_1$ different from the rays containing points of $\mathfrak{A}$, and choose d sufficiently small; then the excluded disks of the two sets cannot intersect.

If, however, $\mathfrak{A}$ satisfies (C), the rays containing the points of $\mathfrak{A}_1$ can be chosen arbitrarily. If some point of $\mathfrak{A}_1$ falls inside an excluded disk $C_{\mathfrak{A}}$, we may move this point to the circumference of the disk. After this transformation $\mathfrak{A}_1$ is still an R-set and its indicator is not changed. If we halve the radii of the excluded disks, then, as we see from (*), the disks for $\mathfrak{A}$ and $\mathfrak{A}_1$ do not intersect.

[14] Theorem 13 is analogous to theorems which A. F. Leont'ev [1; 2; 3; 5; 6] proved by means of Lagrange series. In Leont'ev's theorems the set of nodes is more general, but the characterization of the rate of growth is less precise than in Theorem 13. Leont'ev's investigations have been extended by Lapin [1; 2], Trošin [1–3] and Firsakov [1].

REMARK. *The hypothesis that the set of nodes is an R-set can be weakened to*

$$\lim_{n\to\infty}\left\{\frac{\ln|f'(a_n)|}{|a_n|^{\rho(|a_n|)}} - H(\psi_n)\right\} = 0.$$

We shall not discuss the proof of this.[14a]

5. Some Applications of Lagrange Interpolation Series

By using Lagrange interpolation series we can investigate the question of how the behavior of an entire function in the whole plane depends on its behavior on a set of points with infinity as a limit. We shall discuss several results from this range of ideas.

THEOREM 14. *If the entire function $\Phi(z)$ is (1) bounded on an R-set with index $\rho(r)$ and positive indicator ($H(\theta) > 0$ for $0 \leqslant \theta < 2\pi$), (2) of at most normal type of the proximate order $\rho(r)$, and its indicator satisfies the inequality*

$$h_\Phi(\theta_j) < H(\theta_j) \qquad (j = 1, 2, \ldots, n), \tag{4.53}$$

where

$$|\theta_{j+1} - \theta_j| \leqslant \frac{\pi}{\rho}, \qquad |\theta_1 - \theta_n + 2\pi| \leqslant \frac{\pi}{\rho}$$

then $\Phi(z) \equiv \text{const.}$

PROOF. We show first that the hypothesis of the theorem implies

$$h_\Phi(\theta) < H(\theta) \qquad (0 \leqslant \theta < 2\pi). \tag{4.54}$$

To do this we construct a function $\Psi(z)$ of completely regular growth of proximate order $\rho(r)$, for which the indicator $h_\Psi(\theta)$ is a constant h less than

$$\min\,[H(\theta_j) - h_\Phi(\theta_j)] \qquad (j = 1, 2, \ldots, n), \qquad \text{and} \qquad h < \min_{0\leqslant\theta<2\pi} H(\theta).$$

The indicator of $\Phi(z)\Psi(z)$ is equal to the sum $h_\Phi(\theta) + h$ and consequently satisfies the condition (4.53). Using this, we shall show that $\Phi(z)\Psi(z)$ can be represented by a Lagrange interpolation series.

In fact, since the numbers $\Phi(a_n)$ are bounded and $h_\Psi(\theta) = h$, it follows that

$$\overline{\lim_{n\to\infty}}\, \frac{\ln|\Phi(a_n)\Psi(a_n)|}{h\,|a_n|^{\rho(|a_n|)}} \leqslant 1.$$

[14a] Firsakov [1] proved the following stronger theorem: *Let the set of interpolation nodes $\{\lambda_n\}$ be regularly distributed with respect to the proximate order $\rho(r)$ and have indicator $H(\theta)$. In order that to a sequence $\{a_n\}$ satisfying*

$$\overline{\lim_{n\to\infty}}\left\{\frac{\ln|a_n|}{|\lambda_n|^{\rho(|\lambda_n|)}} - H(\psi_n)\right\} < 0 \qquad (\psi_n = \arg\lambda_n),$$

there corresponds at least one entire function, with indicator $h_\varphi(\theta) < H(\theta)$, that solves the interpolation problem, it is necessary and sufficient that

$$\overline{\lim_{n\to\infty}}\left\{\frac{\ln\left|\dfrac{1}{f'(\lambda_n)}\right|}{|\lambda_n|^{\rho(|\lambda_n|)}} + H(\psi_n)\right\} < 0,$$

where $f(z)$ is the canonical function of the set $\{\lambda_n\}$.

Because $h < \min_{0\leqslant\theta<2\pi} H(\theta)$ we obtain

$$\varlimsup_{n\to\infty} \frac{\ln|\Phi(a_n)\Psi(a_n)|}{H(\psi_n)|a_n|^{\rho(|a_n|)}} < 1,$$

and by Theorem 11 the series

$$\theta(z) = \sum_{n=1}^{\infty} \frac{\Phi(a_n)\Psi(a_n)}{f'(a_n)(z-a_n)}$$

converges and, moreover, as has already been shown, tends to zero as $|z| \to \infty$ with z outside the excluded set of C_R-disks. The entire function $\theta(z)f(z)$ takes the values $\Phi(a_n)\Psi(a_n)$ at the points $\{a_n\}$. Hence the difference

$$\zeta(z) = \Phi(z)\Psi(z) - \theta(z)f(z)$$

is an entire function vanishing on the points $\{a_n\}$ and consequently

$$\chi(z) = \frac{\zeta(z)}{f(z)} = \frac{\Phi(z)\Psi(z)}{f(z)} - \theta(z)$$

is also an entire function. It is easily seen that $\chi(z)$ is at most of normal type of the proximate order $\rho(r)$. We shall show that $\chi(z) \equiv 0$. To do this, we remark that the indicator $h_\Phi(\theta) + h$ of the function $\Phi(z)\Psi(z)$ is less than $H(\theta)$ at the points θ_j ($j = 1, 2, \ldots, n$), and by continuity we have the inequality

$$h_\Phi(\theta) + h < H(\theta)$$

inside the angles $|\theta - \theta_j| \leqslant \delta$ ($j = 1, 2, \ldots, n$) for sufficiently small $\delta > 0$. From this it follows that inside the angles $|\theta - \theta_j| \leqslant \delta/2$ ($j = 1, 2, \ldots, n$), but outside the excluded C_R-disks, the function

$$\frac{\Phi(z)\Psi(z)}{f(z)}$$

tends uniformly to zero as $|z| \to \infty$.

Since $\theta(z)$ also tends uniformly to zero outside the excluded disks, it follows that the entire function $\chi(z)$ also tends uniformly to zero in the same angles, outside the excluded disks. Applying the maximum principle to $|\chi(z)|$ in each of these disks, we infer that $\chi(re^{i\theta_j}) \to 0$ as $r \to \infty$. Applying the Phragmén-Lindelöf theorem to this function in each angle $\theta_j \leqslant \theta \leqslant \theta_{j+1}$ ($j = 1, 2, \ldots, n$ and $\theta_{n+1} = \theta_1 + 2\pi$), we find that $\chi(z)$ is bounded in the whole plane and consequently $\chi(z) \equiv \text{const}$, but since $\chi(re^{i\theta_j}) \to 0$ as $r \to \infty$, we have $\chi(z) \equiv 0$.

Therefore

$$\Phi(z)\Psi(z) \equiv f(z)\theta(z),$$

consequently

$$h_\Phi(\theta) + h \leqslant H(\theta) \quad \text{and} \quad h_\Phi(\theta) < H(\theta).$$

We now show that $\Phi(z)$ is a function of minimal type of the proximate order $\rho(r)$. We construct the auxiliary function $V_C(z)$ (Chapter I, § 17), whose indicator in the angle $|\arg z| < \pi/2\rho$ is equal to $C \cos \rho\theta$.

Then we may suppose that

$$C < \min_{|\theta| \leqslant \frac{\pi}{2\rho}} \frac{H(\theta) - h_\Phi(\theta)}{\cos \rho\theta}, \tag{4.55}$$

since

$$C \cos \rho\theta < H(\theta) - h_\Phi(\theta) \qquad \left(|\theta| \leqslant \frac{\pi}{2\rho}\right).$$

Then the numbers $b_n = \Phi(a_n)V_C(a_n)$ satisfy (4.42), and by Theorem 11 the series

$$\mathfrak{H}(z) = \sum_{|\arg a_n| \leqslant \frac{\pi}{2\rho}} \frac{\Phi(a_n) V_C(a_n)}{f'(a_n)(z - a_n)}$$

converges uniformly outside the excluded C_R-disks and tends uniformly to zero outside these disks as $|z| \to \infty$.

As above, we see that the function

$$\varphi(z) = \frac{\Phi(z) V_C(z)}{f(z)} - \mathfrak{H}(z)$$

is holomorphic inside the angle $|\theta| \leqslant \pi/2\rho$ and at most of normal type of the proximate order $\rho(r)$. In addition, it follows from (4.54) that for $\theta = \pi/2\rho$ we have $h_\Phi(\theta) + C \cos \rho\theta < H(\theta)$ and consequently for sufficiently small $\delta > 0$ the function $\varphi(z)$ tends uniformly to zero as $z \to \infty$ outside the angles $\pi/2\rho - \delta < \theta \leqslant \pi/2\rho$ and $-\pi/2\rho \leqslant \theta < -\pi/2\rho + \delta$. By the Phragmén-Lindelöf theorem it follows from this that $\varphi(z)$ is bounded in the entire angle $|\theta| \leqslant \pi/2\rho$, and consequently

$$h_\varphi(\theta) \leqslant 0 \qquad \left(|\theta| \leqslant \frac{\pi}{2\rho}\right),$$

or in other words

$$h_\Phi(\theta) \leqslant H(\theta) - C \cos \rho\theta \qquad \left(|\theta| \leqslant \frac{\pi}{2\rho}\right).$$

Since C is any number satisfying (4.55), we must have, at the point θ_0 at which the right side of (4.55) attains its minimum,

$$h_\Phi(\theta_0) \leqslant 0.$$

By rotating the plane, we can make any angle of opening π/ρ coincide with the angle $|\theta| \leqslant \pi/2\rho$, and accordingly every angle of opening π/ρ contains a ray on which

$$h_\Phi(\theta) \leqslant 0.$$

From this and the fundamental properties of the indicator (Chapter I, § 16) it follows easily that $h_\Phi(\theta) \leqslant 0$ everywhere, i.e., that $\Phi(z)$ is at most of minimal type of the proximate order $\rho(r)$.

It remains to establish the conclusions of the theorem when $\Phi(z)$ is an entire function of at most minimal type of the proximate order $\rho(r)$. To do this we use a device of V. G. Iyer's [**1**].

For any positive integer m the function $[\Phi(z)]^m$ satisfies all the hypotheses of Theorem 12 and consequently can be represented by the series

$$[\Phi(z)]^m = f(z) \sum_{n=1}^{\infty} \frac{[\Phi(a_n)]^m}{f'(a_n)(z-a_n)}.$$

Under the hypotheses of the theorem $|\Phi(a_n)| \leqslant M$ $(n = 1, 2, \ldots)$ and hence, with the notation

$$X(z) = \sum_{n=1}^{\infty} \frac{1}{|f'(a_n)|\,|z-a_n|},$$

we can assert that

$$|\Phi(z)|^m \leqslant |f(z)|\, M^m X(z),$$

or

$$|\Phi(z)| \leqslant M\,|f(z)\,X(z)|^{\frac{1}{m}}.$$

Finally, since m is arbitrary,

$$|\Phi(z)| \leqslant M.$$

The conclusion of the theorem follows.

We note that the hypothesis (4.53) on the indicator of $\Phi(z)$ is essential. In fact, the indicator of $f(z)$ is equal to $H(\theta)$.

Theorem 14 is a generalization of the following theorem of N. Levinson [**1**].

Let $\{\lambda_n\}$ and $\{\mu_n\}$ be two sequences of positive numbers having densities D_λ and D_μ, i.e.,

$$\lim_{n\to\infty} \frac{n}{\lambda_n} = D_\lambda, \qquad \lim_{n\to\infty} \frac{n}{\mu_n} = D_\mu$$

and

$$\inf(\lambda_{n+1} - \lambda_n) > 0, \qquad \inf(\mu_{n+1} - \mu_n) > 0.$$

If an entire function $\Phi(z)$ of exponential type is bounded at the points $\pm\lambda_n$, $\pm i\mu_n$ and

$$\sigma_\Phi < \pi\sqrt{D_\lambda^2 + D_\mu^2},$$

then $\Phi(z) \equiv \text{const.}$[15]

PROOF. The set of all points of the form $\pm\lambda_n$, $\pm i\mu_n$ evidently forms an R-set with index $\rho(r) = 1$.

The canonical function $f(z)$ of this set can be represented in the form

$$f(z) = \varphi(z)\,\psi(z),$$

[15] This theorem in turn is a strengthening of an analogous theorem of V. G. Iyer [**1**], which involves the stronger hypothesis $\sigma_\Phi < \pi \min(D_\lambda, D_\mu)$ instead of $\sigma_\Phi < \pi\sqrt{D_\lambda^2 + D_\mu^2}$.

where

$$\varphi(z)=\prod_{n=1}^{\infty}\left(1-\frac{z^2}{\lambda_n^2}\right),\qquad \psi(z)=\prod_{n=1}^{\infty}\left(1+\frac{z^2}{\mu_n^2}\right).$$

The functions $\varphi(z)$ and $\psi(z)$ are of completely regular growth, $h_\varphi(\theta) = \pi D_\lambda |\sin\theta|$ and $h_\psi(\theta) = \pi D_\mu |\cos\theta|$, and the indicator diagrams of these functions are, respectively, the segments $x = 0$, $|y| \leqslant \pi D_\lambda$ and $y = 0$, $|x| \leqslant \pi D_\mu$. When functions of completely regular growth are multiplied, their indicator diagrams are added, and consequently the indicator diagram of $f(z)$ is the rectangle I_f: $|x| \leqslant \pi D_\mu$, $|y| \leqslant \pi D_\lambda$. It follows from the condition $\sigma_\Phi < \pi\sqrt{D_\lambda^2 + D_\mu^2}$ that the indicator diagram of $\Phi(z)$ lies in the disk with center at the origin and diameter less than the diagonal of the rectangle I_f.

Therefore for $\theta_j = \pm \operatorname{arc\,tg} D_\mu / D_\lambda$ we have

$$h_\Phi(\theta_j) < H(\theta_j).$$

Condition (4.53) is satisfied and consequently $\Phi(z) \equiv \text{const}$. The theorem is established.

The case when all points of the R-set lie on the real axis and $\rho = 1$ is not covered by Theorem 14, since in this case $H(0) = H(\pi) = 0$.

This important case has been discussed in a number of papers. The following theorem of M. Cartwright may be considered fundamental for this problem.

THEOREM 15. *If an entire function of exponential type whose indicator diagram has width $2k < 2\pi$ is bounded at the real integral points, i.e.,*

$$|f(n)| \leqslant M \qquad (\pm n = 0,\ 1,\ 2,\ \ldots),$$

then it is bounded on the whole real axis, and

$$|f(x)| \leqslant C(k)\, M,$$

where $C(k)$ depends only on k.[16]

PROOF. We may assume without loss of generality that $h_f(\pi/2) = h_f(-\pi/2) = k$. The proof of the theorem is based on the formula

$$f(n+z) = \frac{\sin \pi z}{\pi\omega} \sum_{\nu=-\infty}^{\infty} (-1)^\nu \frac{f(n+\nu)\sin\omega(\nu - z)}{(\nu - z)^2}, \tag{4.56}$$

which holds for entire functions of exponential type that are bounded at the integral points provided that $k < \pi$ ($k = h_f(\pi/2) = h_f - (\pi/2)$) and $0 < \omega < \pi - k$.[17]

---

[16] It follows from this theorem that a function of degree zero that is bounded at the integers is a constant. In this formulation the theorem was first given by G. Pólya.

[17] This interpolation formula was introduced by Boas and Schaeffer [1] and S. N. Bernšteĭn [2] (see also Boas [1]).

To derive this formula we fix an integer n and consider the function of two variables

$$\varphi(\zeta, z) = f(n+\zeta)\frac{\sin\omega(\zeta-z)}{\omega(\zeta-z)} - \frac{\sin\pi\zeta}{\pi\omega}\sum_{\nu=-\infty}^{\infty}(-1)^{\nu}\frac{f(n+\nu)\sin\omega(\nu-z)}{(\nu-z)(\zeta-\nu)}.$$

When z is fixed this is an entire function of finite degree in ζ, vanishing at all integral points.

Hence after dividing this function by $\sin\pi\zeta$ we again obtain an entire function of exponential type in ζ:

$$\psi(\zeta, z) = \frac{f(n+\zeta)}{\sin\pi\zeta}\frac{\sin\omega(\zeta-z)}{\omega(\zeta-z)} + \frac{1}{\pi\omega}\sum_{\nu=-\infty}^{\infty}(-1)^{\nu}\frac{f(n+\nu)\sin\omega(\nu-z)}{(\nu-\zeta)(\nu-z)}.$$

The indicator of the first term is, for $\theta = \pm\pi/2$,

$$h\left(\pm\frac{\pi}{2}\right) = k+\omega-\pi < 0.$$

Since the indicator is continuous it is negative in some neighborhood of the points $\pm\pi/2$.

The second term tends to zero on any ray $\arg\zeta = \theta$ for $\theta \neq 0, \pi$. Hence $\psi(\zeta, z)$ tends to zero on any ray in a neighborhood of the rays $\arg\zeta = \pm\pi/2$. By the Phragmén-Lindelöf theorem it follows that $\psi(\zeta, z) \equiv 0$.

Hence we obtain the identity

$$f(n+\zeta)\frac{\sin\omega(\zeta-z)}{\omega(\zeta-z)} = \frac{\sin\pi\zeta}{\pi\omega}\sum_{\nu=-\infty}^{\infty}(-1)^{\nu}\frac{f(n+\nu)\sin\omega(\nu-z)}{(\nu-z)(\zeta-\nu)},$$

from which (4.56) follows by taking the limit as $\zeta \to z$.

Now put $z = x \pm i\ (-\frac{1}{2} \leqslant x \leqslant \frac{1}{2})$ and note that $|f(n+\nu)| \leqslant M\ (\pm\nu = 0, 1, 2, \ldots)$; from (4.56) we easily obtain the estimate

$$|f(n+x\pm i)| < \frac{M}{\pi\omega}\operatorname{ch}\pi\operatorname{ch}\omega\sum_{-\infty}^{\infty}\frac{1}{|\nu-x\pm i|^2}$$

$$< \frac{M\pi}{\omega}\operatorname{ch}\pi\operatorname{ch}\omega.$$

Since n is arbitrary, it follows from this estimate that for $-\infty < x < \infty$

$$|f(x\pm i)| < P(\omega)M$$

$$\left(P(\omega) = \pi\operatorname{ch}\pi\frac{\operatorname{ch}\omega}{\omega}, \quad \omega < \pi - k\right).$$

By the Phragmén-Lindelöf theorem for a strip (Chapter I, § 14) we find that $f(z)$ satisfies the same estimate in the strip $|y| \leqslant 1$ and, in particular, on the real axis. The theorem is established.

Our estimate for $C(k)$,

$$C(k) \leqslant \min_{\omega < \pi - k} P(\omega) = \pi \operatorname{ch} \pi \frac{\operatorname{ch}(\pi - k)}{\pi - k},$$

is not exact.[18]

Results analogous to Cartwright's theorem have been obtained not only for functions bounded on a sequence of points but also for functions that increase on a sequence.[19] In particular, if the points of an R-set with $\rho = 1$ are sufficiently densely distributed on the positive real axis, the growth of the function on this sequence is the same as its growth on the positive real axis.[20]

THEOREM 16. *Let the sequence of points* $\{a_n\}$ *on the positive real axis* $(a_1 < a_2 < \ldots)$ *have density* Δ *with index* $\rho(r)$ $(\rho = 1)$, i.e.,

$$\lim_{r \to \infty} \frac{n(r)}{r^{\rho(r)}} = \Delta,$$

and for some $d > 0$

$$a_{n+1} - a_n > d a_n^{1 - \rho(a_n)},$$

and let $F(z)$ *be a holomorphic function for* $\operatorname{Re} z \geqslant 0$ *such that*

$$\overline{\lim_{r \to \infty}} \frac{\ln |F(z)|}{r^{\rho(r)}} < \infty \qquad (\operatorname{Re} z \geqslant 0)$$

and

$$h_F\left(\frac{\pi}{2}\right) + h_F\left(-\frac{\pi}{2}\right) < 2\pi\Delta;$$

then

$$h_F(0) = \overline{\lim_{n \to \infty}} \frac{\ln |F(a_n)|}{a_n^{\rho(a_n)}}.$$

[18] Estimates for $C(k)$ have been given in several papers. The most precise estimates are given in S. N. Bernšteĭn's paper [2]. For the case $k = \pi - 1/n$, where n is an integer, Bernšteĭn gives an exact estimate. As $\sigma \to \pi$ Bernšteĭn [2] gives the asymptotic estimate $C(k) \approx 2/\pi \ln 1/(\pi - k)$.

A number of investigations have been devoted to various generalizations of Cartwright's theorem: R. Boas and Schaeffer [1], Duffin and Schaeffer [1], S.N.Bernšteĭn [2], N. I. Ahiezer and B. Ja. Levin [1], S. Agmon [1]. The paper of Plancherel and G. Pólya [1] belongs to the same circle of ideas.

[19] Cf. S. Agmon [1], N. I. Ahiezer [4] and B. Ja. Levin [16].

[20] The question of the connection between the upper limits

$$\overline{\lim_{r \to \infty}} \frac{\ln |F(r)|}{r^{\rho(r)}} \quad \text{and} \quad \overline{\lim_{n \to \infty}} \frac{\ln |F(a_n)|}{|a_n|^{\rho(|a_n|)}}$$

has been the subject of a number of investigations. We mention the work of V. Bernstein [1], A. Pfluger [3], N. Levinson [1] and R. P. Boas [2]. The most complete results were obtained recently by W. H. Fuchs [1].

PROOF. The set of points $\pm a_n$ $(n = 1, 2, \ldots)$ is an R-set with canonical function

$$f(z) = \prod_{n=1}^{\infty}\left(1 - \frac{z^2}{a_n^2}\right)$$

and indicator

$$H(\theta) = \pi\Delta\,|\sin\theta|. \tag{4.57}$$

Let

$$c = \overline{\lim_{n\to\infty}}\, \frac{\ln|F(a_n)|}{a_n^{\rho(a_n)}}. \tag{4.58}$$

Take a sequence of positive numbers $\mu_1 < \mu_2 < \ldots$ with density $\Delta_1 > c\pi^{-1}$ and construct the auxiliary function

$$V(z) = \prod_{n=1}^{\infty}\left(1 + \frac{z^2}{\mu_n^2}\right)$$

with indicator

$$h_V(\theta) = \pi\Delta_1\,|\cos\theta|. \tag{4.59}$$

From (4.58) and (4.59) we easily obtain that for some $\varepsilon > 0$ and all sufficiently large n

$$\ln|F(a_n)V^{-1}(a_n)| < (c - \pi\Delta_1 + \varepsilon)\,a_n^{\rho(a_n)}$$

and consequently for some $\eta > 0$ and all $n > n_\eta$

$$\ln|F(a_n)V^{-1}(a_n)| < -\eta a_n^{\rho(a_n)}.$$

On the other hand, it follows from (4.57) and (4.43) that

$$\lim_{n\to\infty} \frac{\ln|f'(a_n)|}{a_n^{\rho(a_n)}} = 0.$$

The last two relations imply the uniform convergence of the series

$$\psi(z) = \sum_{n=1}^{\infty} \frac{F(a_n)V^{-1}(a_n)}{f'(a_n)(z - a_n)}$$

outside the excluded C_R-disks, and the sum $\psi(z)$ of the series tends uniformly to zero as $|z| \to \infty$ outside these disks. The function

$$\chi(z) = \frac{F(z)}{V(z)f(z)} - \psi(z)$$

is holomorphic in the angle $|\arg z| < \pi/2$ and at most of normal type of the proximate order $\rho(r)$.

We shall show that this function is bounded in the angle $|\arg z| \leqslant q < \pi/2$. To do this we first remark that without loss of generality we may suppose $h_F(-\pi/2) = h_F(\pi/2)$ and that by hypothesis $h_F(\pi/2) < \pi\Delta$.

The indicator of the function

$$\varphi(z)=\frac{F(z)}{V(z)f(z)}$$

is equal to $h_F(\theta) - \pi\Delta\,|\sin\theta| - \pi\Delta_1\,|\cos\theta|$, and consequently for θ close to $\pi/2$ or $-\pi/2$ it takes negative values. Hence for sufficiently small $\delta > 0$ the function $\varphi(re^{i\theta})$ tends to zero as $r \to \infty$ and $\theta = \pm(\pi/2 - \delta)$. Since $\psi(z)$ also tends to zero on these rays, the same is true of $\chi(z)$. Then, applying the Phragmén-Lindelöf theorem for angles to this function, we obtain

$$|\chi(z)| \leqslant M \qquad \left(|\arg z| \leqslant \frac{\pi}{2}-\delta\right).$$

To estimate the indicator of $F(z)$ we may now use the equation

$$F(z)=[\chi(z)+\psi(z)]\,V(z)f(z).$$

We have shown that $|\chi(z)|$ is bounded, and $|\psi(z)|$ is bounded outside the excluded disks. It follows at once that

$$h_F(\theta) \leqslant \pi\Delta\,|\sin\theta| + \pi\Delta_1\,|\cos\theta|.$$

Putting $\theta = 0$, we obtain

$$h_F(0) \leqslant \pi\Delta_1.$$

In constructing $V(z)$ we could have taken Δ_1 to be any number larger than $c\pi^{-1}$. Hence we obtain

$$h_F(0) \leqslant c.$$

The opposite inequality follows directly from the definition of the indicator.

6. Completeness of Systems of Functions. The Connection between Completeness and Uniqueness

A sequence $\{f_n(z)\}$ of functions holomorphic in a region G will be called *uniformly convergent in this region* if it converges uniformly on every bounded closed set contained in G.

A system $\{\varphi_n(z)\}$ of holomorphic functions is called *complete in the region* G if for each function $f(z)$ holomorphic in G there is a sequence of linear combinations

$$\sum_{n=1}^{p_k} c_{n,k}\varphi_n(z) \tag{4.60}$$

with constant coefficients which converges to this function uniformly in G. In other words, the closed linear span of the system $\{\varphi_n(z)\}$ coincides with the set of all holomorphic functions. If this condition is not satisfied, the system is called incomplete. By Runge's theorem[21] the system of powers $1, z, z^2, \ldots$ is complete in every simply-connected region.

We remark that if $\{\varphi_n(z)\}$ is complete in G, then the linear combinations (4.60) will approximate uniformly any function that is analytic in an arbitrary simply-connected bounded closed region $\bar{D}$ that belongs entirely to G. In fact, it follows from the completeness of the system $\{\varphi_n(z)\}$ in G that each power z^m can be approximated arbitrarily closely on $\bar{D}$ by linear combinations of the form (4.60), and consequently any holomorphic function can also be approximated by these combinations.

We now establish the converse, i.e., that if every function holomorphic in an arbitrary bounded simply-connected closed region $\bar{D}$ belonging to the simply-connected region G can be uniformly approximated by linear combinations of the form (4.60) then the system $\{\varphi_n(z)\}$ is complete in G.

In fact, let $f(z)$ be any function holomorphic in G. Consider a sequence of bounded closed simply-connected regions

$$\bar{D}_1 \subset \bar{D}_2 \subset \ldots \subset \bar{D}_n \subset \ldots \sum_{n=1}^{\infty} \bar{D}_n = G$$

and positive numbers $\varepsilon_1 > \varepsilon_2 > \ldots > \varepsilon_n > \ldots$, $\lim_{n\to\infty} \varepsilon_n = 0$.

By hypothesis there are linear combinations $P_n(z)$ of the form (4.60) satisfying the inequalities

$$|f(z) - P_n(z)| < \varepsilon_n \qquad (z \in \bar{D}_n,\ n = 1, 2, \ldots),$$

and consequently

$$f(z) = P_1(z) + \sum_{n=1}^{\infty} [P_{n+1}(z) - P_n(z)],$$

where the series converges uniformly in G.

Let G be simply-connected and let G not contain the point at infinity. Let $\mathfrak{M}_G$ denote the set of functions each of which is holomorphic in some simply-connected region containing the complement of G and vanishes at infinity. Each element of $\mathfrak{M}_G$ is holomorphic in a region containing the boundary of G and infinity. Hence to each function $f(z)$ in $\mathfrak{M}_G$ there is a rectifiable contour L_f, belonging to G, such that $f(z)$ is holomorphic outside and on it.

We say that a function $f(z)$ belonging to $\mathfrak{M}_G$ is orthogonal to a function $\varphi(z)$ holomorphic in G if

$$\int_{(L_f)} f(z)\varphi(z)\,dz = 0. \tag{4.61}$$

THEOREM 17. *A system of functions $\{\varphi_n(z)\}$ is complete in G if and only if the orthogonality of a function $f(z)$ of $\mathfrak{M}_G$ to every function of the system $\{\varphi_n(z)\}$ implies $f(z) \equiv 0$.*

---

[21] Cf. for example, A. I. Markuševič [1].

PROOF. Suppose first that G is the unit disk. Let $\{\varphi_n(z)\}$ be incomplete in this disk. Then there exist a disk $|z| \leqslant r < 1$ and a power z^m which cannot be approximated in this disk by linear combinations of the form (4.60). But then this power cannot be approximated in mean square by linear combinations (4.60) in any disk of the form $|z| \leqslant R$ $(r < R < 1)$. In fact, let

$$\int_{(C_R)} \Big| \zeta^m - \sum_{n=1}^{p_k} c_{nk}\varphi_n(\zeta) \Big|^2 d\zeta < \varepsilon^2.$$

By Cauchy's integral formula

$$z^m - \sum_{n=1}^{p_\varepsilon} c_{n\varepsilon}\varphi_n(z) = \frac{1}{2\pi i} \int_{(C_R)} \Big[\zeta^m - \sum_{n=1}^{p_\varepsilon} c_{n\varepsilon}\varphi_n(\zeta)\Big] \frac{d\zeta}{\zeta - z},$$

and by applying Schwarz's inequality we find that for $|z| \leqslant r$

$$\Big| z^m - \sum_{n=1}^{p_\varepsilon} c_{n\varepsilon}\varphi_n(z) \Big| < \frac{R\varepsilon}{R-r}.$$

We introduce the usual notion of scalar product for functions defined on the circumference $|z| = R$:

$$(f, \varphi) = \frac{1}{2\pi} \int_0^{2\pi} f(Re^{i\theta}) \overline{\varphi(Re^{i\theta})}\, d\theta$$

$$= \frac{1}{2\pi i} \int_{(C_R)} f(z) \bar{\varphi}\left(\frac{R^2}{z}\right) \frac{dz}{z} \qquad (\theta = \arg z). \tag{4.62}$$

We orthogonalize the system $\{\varphi_n(z)\}$ and construct the Fourier series for z^m

$$\chi_m(z) = \sum_{n=1}^{\infty} (z^m, \psi_n)\, \psi_n(z), \tag{4.63}$$

where $\{\psi_n(z)\}$ is the orthogonal system equivalent to $\{\varphi_n(z)\}$. By the Riesz-Fischer theorem this series converges in mean square on $|z| = R$ to a function of integrable square. It follows from Cauchy's integral formula that (4.63) converges uniformly in each interior disk. Therefore the function $\chi_m(z)$ is holomorphic in $|z| < R$ and belongs to L^2 on the boundary. The function

$$F(z) = z^m - \chi_m(z) = \sum_{k=0}^{\infty} a_k z^k$$

is not identically zero and is orthogonal to all the functions $\{\varphi_n(z)\}$ in the sense (4.62). Putting

$$f(z) = \bar{F}\left(\frac{R^2}{z}\right)\frac{1}{z} = \sum_{k=0}^{\infty} \frac{\bar{a}_k R^{2k}}{z^{k+1}},$$

we find that

$$\int_{(C_\rho)} f(z)\varphi_n(z)\,dz = 0 \qquad (n = 1,\ 2,\ \ldots). \tag{4.64}$$

Therefore we have constructed a function $f(z) \in \mathfrak{M}_G$ that is orthogonal to all the functions of the incomplete system[22] $\{\varphi_n(z)\}$.

We now turn to the general case of an arbitrary simply-connected region G, not containing the point at infinity and not the whole plane. Let $z = \vartheta(w)$ be the function that maps the unit disk of the w-plane on the region G in the z-plane. It is clear that the completeness of the system $\{\varphi_n(z)\}$ in G is equivalent to the completeness of $\{\varphi_n(\vartheta(w))\}$ in the unit disk.

Let the system $\{\varphi_n(z)\}$ be incomplete. Then by what has been proved there is a function $f(w)$ which is holomorphic outside some disk $|w| > R$ $(R < 1)$, is not identically zero and satisfies

$$\int_{(C_\rho)} f(w)\,\varphi_n(\vartheta(w))\,dw = 0. \tag{4.65}$$

Under the conformal mapping $z = \vartheta(w)$ the ring $R < \rho_1 < |w| < \rho_2 < 1$ is carried into a doubly-connected region Γ, which belongs to G and is bounded by two analytic contours L_1 and L_2 with equations $|w(z)| = \rho_1$ and $|w(z)| = \rho_2$. The region bounded by L_2 belongs entirely to G, and the exterior region bounded by L_1 contains the entire complement of G. The function $f(w)$ which is holomorphic in the ring $\rho_1 \leqslant |w| \leqslant \rho_2$ is carried into the function $F(z) = f[w(z)]$, which is holomorphic in Γ, and equation (4.65) can be written in the form

$$\int_{(L_\rho)} F(z)\,\varphi_n(z)\,w'(z)\,dz = 0, \tag{4.66}$$

where L_ρ is the image of the circumference $|w| = \rho$ $(\rho_1 < \rho < \rho_2)$. The function

$$\Phi(z) = F(z)\,w'(z),$$

which is holomorphic in Γ, can be represented in the form

$$\Phi(z) = -\frac{1}{2\pi i}\int_{(L_1)} \frac{\Phi(\zeta)}{\zeta - z}\,d\zeta + \frac{1}{2\pi i}\int_{(L_2)} \frac{\Phi(\zeta)}{\zeta - z}\,d\zeta.$$

The second integral, which we denote by $\theta(z)$, is holomorphic in the region bounded by L_2. Hence

$$\int_{(L_\rho)} \varphi_n(z)\,\theta(z)\,dz = 0. \tag{4.67}$$

---

[22] It is clear that instead of the unit circle we could have taken a circle of arbitrary or even of infinite radius.

The function $\Phi(z)$ cannot be holomorphic throughout the interior of L_2, since then $f(w)$ would be holomorphic in the extended plane and zero at infinity, i.e., identically zero. Therefore $\psi(z) = \Phi(z) - \theta(z)$ cannot be identically zero. If we take (4.67) into account, we can write (4.66) in the form

$$\int_{(L_\rho)} \varphi_n(z)\psi(z)\,dz = 0. \tag{4.68}$$

Thus if $\{\varphi_n(z)\}$ is not complete in G, there exists in $\mathfrak{M}_G$ a function $\psi(z)$ which is not identically zero and is orthogonal to every function of the system.

Now let $\{\varphi_n(z)\}$ be complete in G and let $\psi(z)$ be an element of $\mathfrak{M}_G$ which is orthogonal to every function of this system. We show that $\psi(z) \equiv 0$. In fact, since $\{\varphi_n(z)\}$ is complete it follows from (4.68) that

$$\int_{(L_\rho)} z^m\psi(z)\,dz = 0. \tag{4.69}$$

In (4.69) we can replace integration along L_ρ by integration along the circumference of any larger circle.

Outside some circle we have

$$\psi(z) = \sum_{k=0}^{\infty} \frac{a_k}{z^{k+1}},$$

and we obtain from (4.69) that $a_m = 0$ $(m = 0, 1, 2, \ldots)$ and consequently $\psi(z) \equiv 0$. The theorem is established.[23]

This theorem establishes the connection between completeness and uniqueness. In fact, by this theorem the completeness of the system $\{\varphi_n(z)\}$ in G means that a function $\psi(z) \in \mathfrak{M}_G$ is uniquely determined by the sequence of values of the functionals

$$\alpha_n = \int_{(L_\rho)} \psi(z)\varphi_n(z)\,dz, \qquad (n = 1, 2, \ldots).$$

Therefore every theorem on completeness generates a theorem on uniqueness and conversely.

7. Theorems on the Completeness of Some Systems of Entire Functions

In this section we use the general theorem on the connection between completeness and uniqueness to obtain some theorems on the completeness of systems of functions in the complex domain.

---

[23] For the case of a disk this theorem was proved by A. I. Markuševič [2].

We shall say that the sequence of functions $\{\varphi_n(z)\}$ *is generated by the function* $F(z, \lambda)$ *and the sequence* $\{\lambda_n\}$ if

$$\varphi_n(z) = c_n F(z, \lambda_n), \qquad \text{where } c_n \text{ is a constant.}$$

Here we suppose that $F(z, \lambda)$ is holomorphic in z in a region G_z if λ belongs to a region G_λ, and holomorphic in λ in G_λ if $z \in G_z$.[24]

In particular, G_λ may be the whole plane, and then $F(z, \lambda)$ is an entire function of λ. In what follows we shall suppose that $F(z, \lambda)$ is an entire function of λ and impose restrictions on the growth of this entire function.

We call $F(z, \lambda)$ a *closed kernel with respect to* z if when $\psi(z) \in \mathfrak{M}_G$ the equation

$$\int_{(L_\psi)} F(z, \lambda)\psi(z)\,dz = 0$$

for $\lambda \in G_\lambda$ implies $\psi(z) \equiv 0$.[25]

Consider the integral

$$\varphi(\lambda) = \int_{(L_\psi)} F(z, \lambda)\psi(z)\,dz, \tag{4.70}$$

where $\psi(z)$ is a function of $\mathfrak{M}_G$ and $F(z, \lambda)$ is a closed kernel. If the sequence $\{\lambda_n\}$ is such that for a function of the form (4.70) the equations $\varphi(\lambda_n) = 0$ $(n = 1, 2, \ldots)$ imply $\varphi(\lambda) \equiv 0$, then clearly the functions $F(z, \lambda_n)$ form a complete system[26] in G. Conversely, if a function $\varphi(\lambda)$ of the form (4.70) is not identically zero and vanishes at the points $\{\lambda_n\}$ the system $\{\varphi_n(z)\}$ is not complete in G. In particular, if the sequence $\{\lambda_n\}$ has a limit point in G_λ, the system $\varphi_n(z) = F(z, \lambda_n)$ is complete in G_z. For a kernel of the form[27]

$$F(z, \lambda) = f(z\lambda)$$

we can state the following simple criterion for closure in the disk $|z| < R$.

---

[24] This assumption does not restrict the generality of the sequence $\{\varphi_n(z)\}$. In fact, let $\{\varphi_n(z)\}$ be any sequence of functions holomorphic in G, $\bar{D}_n$ closed regions in G, $\lim n \to \infty$ $\bar{D}_n = G$, and $M_n = \max_{z \in \bar{D}_n} |\varphi_n(z)|$. Then the function

$$F(z, \lambda) = \frac{\sin \pi\lambda}{\pi} \sum_{n=1}^{\infty} \frac{(-1)^n \varphi_n(z)}{M_n n^2(\lambda - n)}$$

is a holomorphic function of z in G for each λ and an entire function of λ for $z \in G$, and the sequence $\lambda_n = n$ generates the given sequence of functions.

[25] A. I. Markuševič calls such a function a "complete function."

[26] Instead of the functionals $F(z, \lambda_n)$ we may choose any system of functionals $\varphi_n(z) = L_n[F(z, \lambda)]$. For example, we may take $\varphi_n(z) = F_\lambda^{(n)}(z, \lambda)$, and if for functions represented in the form (4.70) the equations $L_n[\varphi(\lambda)] = 0$ $(n = 1, 2, \ldots)$ imply $\varphi(\lambda) \equiv 0$, then $\{\varphi_n(z)\}$ is a complete system in G.

[27] The first problem on the completeness of systems of functions $f(\lambda_k z)$ in a disk ($f(z)$, an entire function) was set by A. O. Gel'fond [2], who obtained a theorem on the completeness of this system of functions by using the expansion of $f(\lambda z)$ in a Newton interpolation series.

If $f(z)$ is holomorphic in the disk $|z| < R$ and all the coefficients of its expansion

$$f(z) = a_0 + a_1 z + a_2 z^2 + \dots + a_n z^n + \dots \tag{4.71}$$

are different from zero, then for $|\lambda| < 1$ the function $f(\lambda z)$ is a closed kernel in this disk. In fact,

$$\frac{1}{2\pi i}\int_{(C_\rho)} f(z\lambda)\psi(z)\,dz = \sum_{k=0}^{\infty} a_k b_k \lambda^k$$

$$\left(\psi(z) = \sum_{k=0}^{\infty} \frac{b_k}{z^{k+1}},\ \rho < R\right)$$

and from the orthogonality of $\psi(z)$ with respect to the kernel it follows that $b_k = 0$ $(k = 0, 1, 2, \dots)$ and $\psi(z) \equiv 0$. Conversely, if one of the coefficients $a_m = 0$, then the kernel $f(z\lambda)$ is orthogonal to the function $\psi(z) = z^{-m-1}$. If $f(z)$ is an entire function, then clearly the nonvanishing of all its coefficients is a necessary and sufficient condition for the closure of the kernel $f(z\lambda)$ in the disk $|z| < R$ when λ belongs to any region.

In what follows we shall need an estimate for the growth of the function

$$\Phi(\lambda) = \int_{|z|=r} f(z\lambda)\psi(z)\,dz, \tag{4.72}$$

under the hypotheses that $f(z)$ is an entire function of type σ of the proximate order $\rho(r)$ and that $\psi(z)$ is bounded on the contour of integration. From the asymptotic inequality

$$\ln|f(z\lambda)| < \left(\sigma + \frac{\varepsilon}{2}\right)(r|\lambda|)^{\rho(r|\lambda|)} \qquad (\varepsilon > 0,\ |z| = r),$$

which holds for large λ, it follows that

$$\ln|f(z\lambda)| < \left(\sigma + \frac{\varepsilon}{2}\right) r^{\rho}|\lambda|^{\rho} L(r|\lambda|)$$

and by the fundamental properties of the function $L(r)$

$$\ln|f(z\lambda)| < (\sigma + \varepsilon) r^{\rho} |\lambda|^{\rho(|\lambda|)}.$$

From this and (4.72) we obtain

$$|\Phi(\lambda)| < 2\pi r M \exp\{(\sigma + \varepsilon) r^{\rho} |\lambda|^{\rho(|\lambda|)}\}$$

$$(M = \sup_{|z|=r} |\psi(z)|).$$

Finally, since $\varepsilon > 0$ it follows from this that

$$h_\Phi(\theta) \leqslant \sigma r^{\rho}. \tag{4.73}$$

This estimate makes it possible to establish the following theorem on completeness very simply.

THEOREM 18. *If an entire function $f(z)$ represented by the development* (4.71) *satisfies the condition*

$$a_n \neq 0 \qquad (n = 0, 1, 2, \ldots) \tag{4.74}$$

and has type at most σ with respect to the proximate order $\rho(r)$, and $\{\lambda_n\}$ is a sequence of complex numbers, then the system of functions

$$\varphi_n(z) = f(\lambda_n z)$$

is complete in the disk with center at zero and radius R determined by the equation

$$R^{\rho} = \frac{1}{e\rho\sigma} \overline{\lim_{n\to\infty}} \frac{n}{|\lambda_n|^{\rho(|\lambda_n|)}}. \tag{4.75}$$

PROOF. The proof is by contradiction. Suppose that the system of functions $f(\lambda_n z)$ is not complete in the disk $|z| < R$. Then by Theorem 17 there is a function $\psi(z)$, not identically zero, holomorphic in the region $|z| \geqslant r$ $(r < R)$, vanishing at infinity and orthogonal to all the functions $f(\lambda_n z)$, i.e.,

$$\int_{|z|=r} f(\lambda_n z)\,\psi(z)\,dz = 0.$$

Therefore in this case the function $\Phi(\lambda)$ defined by (4.72) vanishes at all points of the sequence $\{\lambda_n\}$.

It follows from Theorem 2 and the estimate (4.73) for the indicator of $\Phi(\lambda)$ that if $\Phi(\lambda)$ is not identically zero then

$$\overline{\lim_{n\to\infty}} \frac{n}{|\lambda_n|^{\rho(|\lambda_n|)}} \leqslant e\rho\sigma r^{\rho}.$$

For $r < R$ this is impossible. Therefore we have shown that $\Phi(\lambda) \equiv 0$ and since the kernel $f(\lambda z)$ is closed, $\psi(z) \equiv 0$ also. Thus the assumption that the system $\{f(\lambda_n z)\}$ is not complete in $|z| < R$ leads to a contradiction. The theorem is established.

If we repeat the discussion but use Theorem 3 instead of Theorem 2 to estimate the density of the set of zeros of $\Phi(z)$, we obtain the following theorem on completeness.

THEOREM 19. *Let $f(z)$ be an entire function of type σ of the proximate order $\rho(r)$; let all the coefficients of its expansion* (4.71) *be different from zero, and let $\{\lambda_n\}$ be a sequence of complex numbers. Then the system of functions $\varphi_n(z) = f(\lambda_n z)$ is complete in the disk with center at zero and radius R determined by the equation*

$$R^{\rho} = \frac{1}{\rho\sigma} \underline{\lim_{n\to\infty}} \frac{n}{|\lambda_n|^{\rho(|\lambda_n|)}}. \tag{4.76}$$

Theorem 18 was proved by A. I. Markuševič [2], and Theorem 19 is a sharpening of a theorem of A. O. Gel'fond [2]. We note that when $R = \infty$ in (4.75) or (4.76) the system of functions (4.74) is complete in the whole plane.

If the function $f(z) = \Sigma\, a_n z^n$ is such that the limit

$$\lim_{n \to \infty} \varphi(n) \sqrt[n]{|a_n|} = (e\rho\sigma)^{\frac{1}{\rho}} \tag{4.77}$$

exists, where $\varphi(n)$ is the solution of the equation

$$n = r^{\rho(r)},$$

the estimate (4.76) is, in a certain sense, exact. Indeed, in this case we can choose the sequence $\{\lambda_n\}$ in such a way that for every entire function $f(z)$ satisfying (4.77) the system $\{f(\lambda_n z)\}$ is not complete in any circle with center at zero and radius larger than the number R defined by (4.76). To show this we choose a regular set $\{\lambda_n\}$ such that the canonical function $\Phi(z)$ of this set has a constant indicator $(h_\Phi(\theta) = \sigma_\Phi)$. Then by Theorem 3 of this chapter we have

$$\lim_{n\to\infty} \frac{n}{|\lambda_n|^{\rho(|\lambda_n|)}} = \rho\sigma_\Phi. \tag{4.78}$$

Let

$$\Phi(z) = \sum_1^\infty c_n z^n.$$

Using the coefficients of this expansion, construct the function

$$\psi(z) = \sum_0^\infty \frac{c_n}{a_n z^{n+1}}.$$

This series converges outside a disk whose radius r is determined by the equation

$$r = \overline{\lim_{n\to\infty}} \frac{\sqrt[n]{|c_n|}}{\sqrt[n]{|a_n|}} = \frac{\overline{\lim\limits_{n\to\infty}}\, \varphi(n) \sqrt[n]{|c_n|}}{\lim\limits_{n\to\infty} \varphi(n) \sqrt[n]{|a_n|}} = \left(\frac{\sigma_\Phi}{\sigma_f}\right)^{\frac{1}{\rho}}.$$

If we use (4.78), we can write this equation in the form

$$r^\rho = \frac{1}{\rho\sigma_f} \lim_{n\to\infty} \frac{n}{|\lambda_n|^{\rho(|\lambda_n|)}},$$

i.e., $r = R$. The function $\Phi(z)$ can evidently be represented in the form

$$\Phi(\lambda) = \int_{|z|=R+\varepsilon} f(z\lambda)\psi(z)\,dz \qquad (\varepsilon > 0). \tag{4.79}$$

Since $\Phi(\lambda_n) = 0$ $(n = 1, 2, \ldots)$ the function $\psi(z)$, which is not identically zero, is orthogonal to all the functions $f(z\lambda_n)$ $(n = 1, 2, \ldots)$. Therefore this system of functions is not complete[28] in the disk $|z| \leqslant R + \varepsilon$ $(\varepsilon > 0)$.

The proofs of Theorems 18 and 19 depend on the uniqueness Theorems 2 and 3. Completeness theorems can also be obtained from the uniqueness theorems discussed in § 2. We shall prove the following completeness theorem which depends on Theorem 7.

THEOREM 20. *Let* $\{\lambda_n\}$ *be a sequence of points of the right hand half-plane,* $\nu(r)$ *the number of points of this sequence in the disk* $|z - r/2| < r/2$, *and*

$$d = \varliminf_{r\to\infty} \frac{\nu(r)}{r}. \tag{4.80}$$

Then the system $\{e^{\lambda_n z}\}$ *is complete in every strip I of the form*

$$y_0 < \operatorname{Im} z < y_0 + 2\pi d$$

(y_0 *is any real number*).

PROOF. Let L be any contour completely contained in I, and $\psi(z)$ a function holomorphic outside and on this contour. Then

$$\Phi(\lambda) = \int_{(L)} e^{\lambda z}\psi(z)\,dz \tag{4.81}$$

is an entire function of exponential type and the width of its indicator diagram is less than $2\pi d$. If $\psi(z)$ is orthogonal to all the functions of the system $\{e^{\lambda_n z}\}$ then $\Phi(\lambda_n) = 0$ $(n = 1, 2, \ldots)$ and, by Theorem 7 of § 2, $\Phi(\lambda) \equiv 0$. The kernel $e^{\lambda z}$ is closed, and consequently the system $\{e^{\lambda_n z}\}$ is complete inside any contour lying in I. The theorem is established.

Putting $w = e^z$, we obtain the following theorem which contains a theorem of T. Carleman [1].

The system $\{z^{\lambda_n}\}$ *is complete in the angle* $\varphi_0 < \arg z < \varphi_0 + 2\pi d$, *where* φ_0 *is any number and d is defined by* (4.80).

The formula (4.81) can be inverted (Chapter I, §20). This makes it possible to establish the following sharp theorem.

THEOREM 21. *If* $\{\lambda_n\}$ *is a sequence of complex numbers for which the limit*

$$d = \lim_{n\to\infty} \frac{n}{\lambda_n} \qquad (d > 0) \tag{4.82}$$

exists, the system $\{e^{\lambda_n z}\}$ *is complete in any region G which every straight line parallel to the imaginary axis cuts in a segment of length less than* $2\pi d$, *and the*

---

[28] One can show that Theorem 18 is sharp in the same sense. To do this one must construct a function $\Phi(\lambda)$ in the same way as the example in § 1 which established that the estimate in Theorem 2 is sharp.

system is not complete in any region which contains a segment of length $2\pi d$ parallel to the imaginary axis.[29]

PROOF. We first establish the second part of the theorem. Let G contain a line segment I of length $2\pi d$ parallel to the imaginary axis. The indicator diagram of the function

$$\Phi(\lambda) = \prod_{n=1}^{\infty}\left(1 - \frac{\lambda^2}{\lambda_n^2}\right) \tag{4.83}$$

is a segment of length $2\pi d$ of the imaginary axis. For suitable choice of the number γ the conjugate diagram of the function $\theta(\lambda) = e^{\gamma\lambda}\Phi(\lambda)$ coincides with I. All the singular points of the Borel transform $\psi(z)$ of $\theta(\lambda)$ lie on this segment.

Let L be a contour surrounding the segment I and belonging to G. Then

$$\theta(\lambda) = \frac{1}{2\pi i}\int_{(L)} e^{\lambda z}\psi(z)\,dz,$$

and the function $\psi(z)$ is not identically zero, is holomorphic outside and on L, and is orthogonal to all functions of the system $\{e^{\lambda_n z}\}$. Therefore this system is not complete inside this contour, and therefore not in G.

Now let G be intersected by every straight line parallel to the imaginary axis in a segment of length less than $2\pi d$. If the system $\{e^{\lambda_n z}\}$ is not complete in G, there is a function $\psi(z)$ which is holomorphic outside and on some contour L that belongs to G and is orthogonal to all functions of the system. On the other hand, the entire function of exponential type

$$\varphi(\lambda) = \frac{1}{2\pi i}\int_{(L)} e^{\lambda z}\psi(z)\,dz$$

vanishes at all points λ_n.

Hence it follows that the quotient

$$\chi(\lambda) = \frac{\varphi(\lambda)}{\Phi(\lambda)}$$

is a holomorphic function of exponential type in the right hand half-plane (Corollary 2 to Theorem 5 of Chapter III, §4). The function $\Phi(\lambda)$ is of completely regular growth. Hence under multiplication by it the indicators add and

$$h_\varphi(\theta) = h_\Phi(\theta) + h_\chi(\theta) = h_\chi(\theta) + \pi d\,|\sin\theta| \qquad \left(|\theta| \leqslant \frac{\pi}{2}\right).$$

Therefore the indicator diagram of $\varphi(\lambda)$ is obtained by translating the indicator diagram of $\chi(\lambda)$ by the vectors $i\pi d$ and $-i\pi d$ and taking the smallest convex region containing the two resulting regions.

[29] While this book was being prepared for publication an analogous result was obtained by A. Leont'ev [4].

Thus we find that

$$h'_{\varphi}(+0) - h'_{\varphi}(-0) \geqslant 2\pi d,$$

and consequently the boundary of the indicator diagram of $\varphi(z)$ contains a segment I perpendicular to the real axis and of length not less than $2\pi d$. The ends of the segment symmetric to I with respect to the real axis are extreme points of the conjugate diagram and hence are singular points of $\psi(z)$. This contradicts the fact that all singular points of $\psi(z)$ are in G. Thus the assumption that the system $\{e^{\lambda_n z}\}$ is not complete in G leads to a contradiction. The theorem is established.

A region in which the system $\{\varphi_n(z)\}$ is complete and which is not a proper subset of any region having the same property is naturally called a *maximal region of completeness.*

Let $y = f(x)$ be a continuous function of x. Then by Theorem 21 the curvilinear strip

$$f(x) < y < f(x) + 2\pi d$$

is a maximal region of completeness for the system $\{e^{\lambda_n z}\}$.

CHAPTER V

FUNCTIONS OF CLASS A

In applications of the theory of entire functions, an important role is played by those entire functions that satisfy the condition

$$\sum_{k=1}^{\infty}\left|\operatorname{Im}\frac{1}{a_k}\right| < \infty, \tag{5.00}$$

where the a_k are the zeros of the function. These functions are natural generalizations of a function with real zeros only.

Indeed, it follows from (5.00) that the zeros $\{a'_k\}$ lying outside the angles $|\arg z| < \delta$ and $|\arg z - \pi| < \delta$ form a convergent series

$$\sum \frac{1}{|a'_k|} < \infty$$

and hence that "almost all" zeros $\{a_k\}$ lie in the "neighborhood" of the real axis. We shall refer to these functions as *functions of class A*. In an analogous manner one can define functions of class A in a half-plane ($\operatorname{Im} z \geqslant 0$ or $\operatorname{Im} z \leqslant 0$) by assuming that the sum (5.00) is formed from the zeros of the function that lie in this half-plane.[1] The functions of greatest interest are the functions of class A of exponential type. From Lindelöf's criterion (Chapter I, § 11) $|\Sigma_{|a_n|<r} 1/a_n| < M$, it follows directly that an entire function of exponential type with zeros in one half-plane $\operatorname{Im} z \geqslant 0$ or $\operatorname{Im} z \leqslant 0$ only, is a function of class A.

In this chapter we present the basic facts concerning functions of class A.

In the study of the properties of entire functions of exponential type that depend on the behavior of the functions on the axis of reals, an important role is played by the integral

$$\int_0^R \frac{\ln|F(x)F(-x)|}{1+x^2}\,dx.$$

[1] In this case the condition (5.00) plays the same role as the condition of Blaschke $\sum_{k=1}^{\infty}(1-|a_k|)$ plays for functions defined on the unit circle. One of these conditions can be obtained from the other by mapping the unit circle on the half-plane.

In § 1 we show that the boundedness of this integral is equivalent to requiring that the entire function $F(x)$ of exponential type belong to the class A.[2]

If $F(z)$ is also of exponential type in the half-plane $\operatorname{Im} z \geqslant 0$, $h_F(0) + h_F(\pi) = 0$, and if the limit of this integral exists when $R \to \infty$, then $F(z)$ is a function of completely regular growth.[3] For functions of exponential type, these two conditions are necessary and sufficient.[4]

Besides the criterion mentioned in § 4, there exist other criteria which guarantee that a function of exponential type in a half-plane be a function of class A and be of completely regular growth. In particular, this will be the case if the integral

$$\int_{-\infty}^{\infty} \frac{\ln |F(x)|}{1+x^2}\, dx$$

or

$$\int_{-\infty}^{\infty} \frac{\ln_+ |F(x)|}{1+x^2}\, dx$$

converges. From the last criterion (which is due to Cartwright **[1]**) it follows that functions of exponential type, bounded on the axis (or belonging to $L^p(-\infty, \infty)$ $(p > 0)$), are functions of completely regular growth.

We shall show that for functions of exponential type the convergence of the second of these integrals implies that $h_F(0) + h_F(\pi) = 0$.[5]

In this book we shall also investigate the properties of the indicator diagram of functions of exponential type and of class A, and we shall consider also certain other problems related to this class of functions.

In conclusion, in § 6 we shall prove a theorem of M. G. Kreĭn that is related to these problems: *An entire function $f(z)$ of class A that admits the expansion*

$$\frac{1}{f(z)} = R(z) + \sum_{k=1}^{\infty} c_k \left(\frac{1}{z-a_k} + \frac{1}{a_k} + \frac{z}{a_k^2} + \cdots + \frac{z^p}{a_k^{p+1}} \right),$$

where $R(z)$ is a polynomial, is a function of exponential type and of completely regular growth.

1. Formula of Carleman. Criterion for an Entire Function of Exponential Type to be of Class A

The next theorem is of prime importance for what follows.[6]

---

[2] N. I. Ahiezer **[2]** and N. N. Meĭman **[3]**.

[3] B. Ja. Levin **[3]**.

[4] A. Pfluger **[5]**.

[5] While the present book was in preparation, there appeared a book by R. P. Boas **[1]** giving a somewhat different presentation of these problems based on a formula of R. Nevanlinna **[2]**.

[6] T. Carleman **[1]**.

THEOREM 1. *If $F(z)$ is a holomorphic function in the region*

$$0 < \lambda \leqslant |z| \leqslant R \quad u \quad \operatorname{Im} z \geqslant 0 \tag{5.01}$$

and if $a_k = r_k e^{i\theta_k}$ $(k = 1, 2, \ldots, n)$ are its zeros in this region, then

$$\sum_{\lambda < r_k < R} \left(\frac{1}{r_k} - \frac{r_k}{R^2}\right) \sin \theta_k = \frac{1}{\pi R} \int_0^\pi \ln |F(Re^{i\theta})| \sin \theta \, d\theta$$

$$+ \frac{1}{2\pi} \int_\lambda^R \left(\frac{1}{x^2} - \frac{1}{R^2}\right) \ln |F(x) F(-x)| \, dx + A_\lambda(F, R), \tag{5.02}$$

where

$$A_\lambda(F, R) = -\operatorname{Im} \frac{1}{2\pi} \int_0^\pi \ln F(\lambda e^{i\theta}) \left(\frac{\lambda e^{i\theta}}{R^2} - \frac{e^{-i\theta}}{\lambda}\right) d\theta. \tag{5.03}$$

This formula, together with Jensen's formula and the formulas (3.06) and (4.20), establishes a connection between the distribution of the zeros of the function $F(z)$ and the growth of the quantity $\ln |F(z)|$.

PROOF. Let us first assume that the function $F(z)$ has no zeros on the boundary of the region (5.01), and let us consider the integral

$$J = \frac{1}{2\pi i} \int_{(C)} \left(\frac{1}{R^2} - \frac{1}{z^2}\right) \ln F(z) \, dz,$$

taken along the boundary of the region (5.01) in the positive direction beginning at the point λ:

$$J = \frac{1}{2\pi i} \int_\lambda^R \left(\frac{1}{R^2} - \frac{1}{x^2}\right) \ln F(x) \, dx + \frac{1}{2\pi} \int_0^\pi \ln F(Re^{i\theta}) \left(\frac{1}{R^2} - \frac{e^{-2i\theta}}{R^2}\right) Re^{i\theta}$$

$$+ \frac{1}{2\pi i} \int_{-R}^{-\lambda} \left(\frac{1}{R^2} - \frac{1}{x^2}\right) \ln F(x) \, dx$$

$$+ \frac{1}{2\pi} \int_\pi^0 \ln F(\lambda e^{i\theta}) \left(\frac{1}{R^2} - \frac{e^{-2i\theta}}{\lambda^2}\right) \lambda e^{i\theta} \, d\theta. \tag{5.04}$$

On the other hand, integration by parts yields

$$J = \frac{1}{2\pi i} \Phi_C \left[\left(\frac{z}{R^2} + \frac{1}{z}\right) \ln F(z)\right] - \frac{1}{2\pi i} \int_{(C)} \frac{F'(z)}{F(z)} \left(\frac{z}{R^2} + \frac{1}{z}\right) dz. \tag{5.05}$$

Here, Φ_C denotes the increase of the function during one complete passage over the contour C. This increase is equal to $(\lambda/R^2 + 1/\lambda)n(R, \lambda)$, where

$n(R, \lambda)$ is the number of zeros of the function $F(z)$ within the region (5.01). The second summand in (5.05) is equal to

$$-\sum_{\lambda<r_k<R}\left(\frac{r_k e^{i\theta_k}}{R^2}+\frac{1}{r_k e^{i\theta_k}}\right).$$

Equating the imaginary parts of (5.04) and (5.05), we obtain (5.02).

If as R varies a zero occurs on the semicircle $|z| = R$, then the formula (5.02) remains valid because both parts of the formula are continuous functions of R. It is also easy to see that formula (5.02) remains true if there are zeros of $F(z)$ on the axis of reals.

REMARK. The quantity $A_\lambda(F, R)$ is bounded when $R > \lambda$, and when $R \to \infty$ we have the limit

$$A_\lambda(F, \infty) = \operatorname{Im}\frac{1}{2\pi}\int_0^\pi \ln F(\lambda e^{i\theta})\frac{e^{-i\theta}}{\lambda}\,d\theta .$$

If the function $F(z)$ is holomorphic at the origin, and if $F(0) = 1$, then

$$\lim_{\lambda\to 0} A_\lambda(F, R) = \frac{1}{2}\operatorname{Im} F'(0),$$

and we obtain the equation

$$\sum_{r_k\leqslant R}\left(\frac{1}{r_k}-\frac{r_k}{R^2}\right)\sin\theta_k = \frac{1}{\pi R}\int_0^\pi \ln|F(Re^{i\theta})|\sin\theta\,d\theta$$

$$+\frac{1}{2\pi}\int_0^R\left(\frac{1}{x^2}-\frac{1}{R^2}\right)\ln|F(x)F(-x)|\,dx+\frac{1}{2}\operatorname{Im}F'(0). \quad (5.02')$$

THEOREM 2. *In order that an entire function of exponential type may be of class A it is necessary and sufficient that for some fixed $\lambda > 0$ and for every $R > \lambda$ the following inequality be valid:*

$$\int_\lambda^R \frac{\ln|f(x)f(-x)|}{x^2}\,dx < M_{f,\lambda}, \quad (5.06)$$

where $M_{f,\lambda}$ is a constant.

It follows from this theorem, in particular, that every entire function of exponential type that is bounded on the axis of reals is of class A.

PROOF. Let us first prove the sufficiency. Suppose that condition (5.06) is satisfied. Setting

$$\psi(x) = \int_{\lambda}^{x} \frac{\ln|f(\xi)f(-\xi)|}{\xi^2}\,d\xi$$

and integrating by parts, we find that

$$\int_{\lambda}^{R} \left(1 - \frac{x^2}{R^2}\right) \frac{\ln|f(x)f(-x)|}{x^2}\,dx = \frac{2}{R^2}\int_{\lambda}^{R} \psi(x)\,x\,dx < M_{f,\lambda}, \tag{5.07}$$

and hence that the second summand on the right hand side of Carleman's formula (5.02) is bounded. Furthermore, since $f(z)$ is a function of exponential type, it follows that for $R > \lambda$ and for some constant $k_{f,\lambda}$ we have $\ln|f(Re^{i\theta})| \leqslant k_{f,\lambda}R$, and

$$\frac{1}{\pi R}\int_{0}^{\pi} \ln|f(Re^{i\theta})|\sin\theta\,d\theta < \frac{2}{\pi}k_{f,\lambda}. \tag{5.08}$$

Therefore, the function of R in the right hand part of the equation (5.02) is bounded.

On the other hand, we have

$$\begin{aligned}\sum_{\lambda < r_k < \frac{1}{2}R} \frac{\sin\theta_k}{r_k} &< \frac{4}{3}\sum_{\lambda < r_k < \frac{1}{2}R}\left(\frac{1}{r_k} - \frac{r_k}{R^2}\right)\sin\theta_k \\ &\leqslant \frac{4}{3}\sum_{\lambda < r_k < R}\left(\frac{1}{r_k} - \frac{r_k}{R^2}\right)\sin\theta_k \\ &< M_{f,\lambda} + \frac{2}{\pi}k_{f,\lambda} + A_\lambda(f, R).\end{aligned} \tag{5.09}$$

This last inequality follows from (5.02), (5.07), and (5.08). From (5.09) it follows at once that $f(z)$ is a function of class A.

For the proof of the necessity of the condition, we make use of an estimate of the modulus of an entire function from below (Chapter I, § 8, Theorem 11). From that estimate it follows that to every $\eta > 0$ there corresponds a number $H_1(\eta) > 0$ such that

$$\ln|f(Re^{i\theta})| > -H_1(\eta)R \qquad (0 \leqslant \theta < 2\pi) \tag{5.10}$$

for all values $R > R_\eta$ except possibly for values of R that lie in intervals the sum of whose lengths does not exceed ηR on any interval $(0, R)$.

Let the number R lie outside the exceptional intervals. Then

$$\frac{1}{\pi R}\int_{0}^{\pi} \ln|f(Re^{i\theta})|\sin\theta\,d\theta > -\frac{2}{\pi}H_1(\eta). \tag{5.11}$$

If $f(z)$ is of class A, then $\Sigma_{\lambda \leqslant r_k \leqslant R} \sin \theta / r_k < N$, and from (5.02) we obtain

$$\int_0^R \frac{\ln |f(x) f(-x)|}{x^2} dx < N + \frac{2}{\pi} H_1(\eta) - A_\lambda(f, R) + \frac{1}{R^2} \int_\lambda^R \ln |f(x) f(-x)| dx. \quad (5.12)$$

The last term in (5.12) is bounded because, for $x > \lambda$,

$$\ln |f(x) f(-x)| < 2k_{f,\lambda} x \quad (5.13)$$

and hence

$$\frac{1}{R^2} \int_0^R \ln |f(x) f(-x)| dx < k_{f,\lambda}. \quad (5.14)$$

Thus, if the point R lies outside the exceptional intervals, then

$$\int_\lambda^R \frac{\ln |f(x) f(-x)|}{x^2} dx < M'_{f,\lambda},$$

where the quantity $M'_{f,\lambda}$ does not depend on R. If R lies in an exceptional interval, one can find a point R' lying outside the exceptional intervals such that $R' < R$ and $R - R' < \eta R$. From the inequalities (5.13) and (5.14) we obtain

$$\int_\lambda^R \frac{\ln |f(x) f(-x)|}{x^2} dx = \int_\lambda^{R'} \frac{\ln |f(x) f(-x)|}{x^2} dx + \int_{R'}^R \frac{\ln |f(x) f(-x)|}{x^2} dx < M'_{f,\lambda} + 2k_{f,\lambda} \ln (1 + \eta). \quad (5.15)$$

REMARK. In the proof of the sufficiency of the condition (5.06) we did not make use of the fact that $f(z)$ was an entire function, but we used only the conditions that $f(z)$ was holomorphic and of exponential type in the half plane $\operatorname{Im} z \geqslant 0$. We thus established the following assertion: *If a holomorphic function $F(z)$ of exponential type satisfies condition* (5.06) *in the half-plane* $\operatorname{Im} z \geqslant 0$, *then $F(z)$ is of class A in this half-plane.*

THEOREM 3. *For an entire function $f(z)$ of exponential type, the condition*

$$\int_\lambda^R \frac{\ln |f(x) f(-x)|}{x^2} dx < M^{(0)}_{f,\lambda} \quad \text{(a)}$$

is equivalent to each of the following conditions:

$$\left|\int_\lambda^R \frac{\ln|f(x)f(-x)|}{x^2}\,dx\right| < M_{f,\lambda}^{(1)}, \tag{b}$$

$$\int_0^R \frac{\ln|f(x)f(-x)|}{1+x^2}\,dx < M_{f,\lambda}^{(2)}, \tag{c}$$

$$\left|\int_0^R \frac{\ln|f(x)f(-x)|}{1+x^2}\,dx\right| < M_{f,\lambda}^{(3)} \tag{d}$$

(*the* $M_{f,\lambda}^{(i)}$, $i = 0, 1, 2, 3$, *do not depend on* R).

PROOF. First, we shall prove that for every entire function of exponential type we have the inequality

$$\int_\lambda^R \frac{\ln|f(x)f(-x)|}{x^2}\,dx > M_{f,\lambda}^{(4)}.$$

For this we note that the formula of Carleman implies that

$$\int_\lambda^R \frac{\ln|f(x)f(-x)|}{x^2}\,dx > -\frac{1}{\pi R}\int_0^\pi \ln|f(Re^{i\theta})|\sin\theta\,d\theta - A_\lambda(f, R)$$
$$+\frac{1}{R^2}\int_\lambda^R \ln|f(x)f(-x)|\,dx.$$

If we now make use of the boundedness of $A_\lambda(f, R)$ and of the inequality (5.08), we obtain

$$\int_\lambda^R \frac{\ln|f(x)f(-x)|}{x^2}\,dx > M_{f,\lambda}^{(5)} + \frac{1}{R^2}\int_\lambda^R \ln|f(x)f(-x)|\,dx. \tag{5.16}$$

Next, setting

$$J_\lambda^R(\theta) = \int_\lambda^R \frac{\ln|f(xe^{i\theta})|}{x}\,dx$$

and integrating by parts, we find that

$$\frac{1}{R^2}\int_\lambda^R \ln|f(x)f(-x)|\,dx$$
$$= \frac{x}{R^2}[J_\lambda^x(0)+J_\lambda^x(\pi)] - \frac{1}{R^2}\int_\lambda^R [J_\lambda^x(0)+J_\lambda^x(\pi)]\,dx. \tag{5.17}$$

In Chapter III (Lemma 4) we established an inequality which for entire functions of exponential type has the form

$$|J_\lambda^x(\theta)| < c_{f,\lambda}|x| \qquad (c_{f,\lambda} \text{ is a constant}). \tag{5.18}$$

From this inequality and (5.17) we obtain

$$\left|\frac{1}{R^2}\int_\lambda^R \ln|f(x)f(-x)|\,dx\right| < 3c_{f,\lambda},$$

and after substitution into (5.16) we have

$$\int_\lambda^R \frac{\ln|f(x)f(-x)|}{x^2}\,dx > M_\lambda + 3c_{f,\lambda} = M_{f,\lambda}^{(1)}.$$

Combining this with (a) we obtain (b). Thus, the conditions (a) and (b) are seen to be equivalent.

For the establishment of the equivalence of the conditions (b) and (a) with those of (c) and (d), we consider the integral

$$\int_\lambda^R \left(\frac{1}{x^2}-\frac{1}{1+x^2}\right)\ln|f(x)f(-x)|\,dx$$

$$= \int_\lambda^R \frac{1}{x(1+x^2)}\,\frac{\ln|f(x)f(-x)|}{x}\,dx.$$

Integration by parts yields

$$\int_\lambda^R \left(\frac{1}{x^2}-\frac{1}{1+x^2}\right)\ln|f(x)f(-x)|\,dx$$

$$=\frac{1}{R(1+R^2)}[J_\lambda^R(0)+J_\lambda^R(\pi)]+\int_\lambda^R \frac{3x^2+1}{x^2(1+x^2)^2}[J_\lambda^x(0)+J_\lambda^x(\pi)]\,dx.$$

Next, applying the inequality (5.18) we obtain, for $R > 1$,

$$\left|\int_\lambda^R \left(\frac{1}{x^2}-\frac{1}{1+x^2}\right)\ln|f(x)f(-x)|\,dx\right| < 4c_{f,\lambda}.$$

Rewriting this inequality in the form

$$\int_\lambda^R \frac{\ln|f(x)f(-x)|}{x^2}\,dx - 4c_{f,\lambda} < \int_\lambda^R \frac{\ln|f(x)f(-x)|}{1+x^2}\,dx$$

$$< \int_\lambda^R \frac{\ln|f(x)f(-x)|}{x^2}\,dx + 4c_{f,\lambda},$$

we easily deduce the equivalence of condition (a) to (c), and of (b) to (d). This establishes the truth of the theorem. Thus each of the conditions (a), (b), (c) and (d) is necessary and sufficient that an entire function of exponential type belong to class A.

The integrals occurring in these conditions play an important role in the criteria that an entire function of exponential type not only belong to class A but also be of completely regular growth. For the establishment of these criteria we shall derive a special representation of an entire function of class A. This representation is analogous to a known expression for a holomorphic function bounded in a circle. The next two sections are devoted to this representation.

2. Representation of a Function Harmonic in a Half-Plane

In this section we establish some sufficient conditions for the representability of a function harmonic in a half-plane as a Poisson integral

$$\frac{1}{\pi}\int_{-\infty}^{\infty}\frac{yu(t)}{(t-x)^2+y^2}\,dt.$$

LEMMA 1 (N. G. ČEBOTAREV [6]). *An entire real*[7] *function* $f(z)$, *satisfying the condition* $\operatorname{Im} f(z) > 0$ *when* $\operatorname{Im} z > 0$, *has the form* $f(z) = \alpha z + \beta$ $(\alpha > 0$, β *is a real number*).

PROOF. From the principle of symmetry, and from the hypotheses of the lemma, it follows that $\operatorname{Im} f(z) \leqslant 0$ when $\operatorname{Im} z \leqslant 0$. The increase of $\arg f(z)$ on any circumference does not exceed 2π. Hence $f(z)$ has at most one zero. This single zero x_0 is real. The function $\varphi(z) = (z - x_0)^{-1}f(z)$ has no zero, and $|\arg \varphi(z)| \leqslant \pi$ in the whole plane. Hence the entire function $w = \ln \varphi(z)$ maps the entire plane z on the strip $|\operatorname{Im} w| \leqslant \pi$, and therefore $\varphi(z) = \text{const.}$[8]

LEMMA 2. *If a continuous real function* $u(t)$, *defined on the real axis, satisfies the condition*

$$\overline{\lim_{t\to\pm\infty}}\frac{u(\pm t)}{t} = \sigma < \infty \tag{5.19}$$

and if the integral

$$J = \frac{1}{\pi}\int_{-\infty}^{\infty}\frac{u(t)}{1+t^2}\,dt \tag{5.20}$$

[7] An entire function is said to be *real* if it takes on only real values on the axis of reals.

[8] It is sufficient to note that the function $u = \dfrac{1}{w - 2\pi i}$ is bounded on the entire plane z $(|u| < \pi)$ and to make use of a familiar theorem of Liouville.

exists, then the integral

$$v(z) = \frac{y}{\pi} \int_{-\infty}^{\infty} u(t) \frac{dt}{(t-x)^2 + y^2} \tag{5.21}$$

will exist for all real values of the variables x, y ($y \neq 0$).

PROOF. We shall first prove that conditions (5.19) and (5.20) imply the existence of the integral

$$\int_1^{\infty} |u(\pm t)| \frac{dt}{t^3}. \tag{5.22}$$

Setting

$$\psi(x) = \int_0^x u(t) \frac{dt}{1+t^2},$$

we obtain from the existence of the integral (5.20) the inequality $|\psi(x)| \leqslant M$ $(-\infty < x < \infty)$, and for every $a > 0$ and $h > 0$ we have

$$\int_a^{a+h} u(t) \frac{dt}{t^3} = \left[\psi(t)\left(\frac{1}{t} + \frac{1}{t^3}\right)\right]_a^{a+h} + \int_a^{a+h} \psi(t)\left(\frac{1}{t^2} + \frac{3}{t^4}\right) dt = O\left(\frac{1}{a}\right).$$

We have thus established the existence of the integral

$$\int_1^{\infty} u(\pm t) t^{-3} \, dt.$$

On the other hand, it follows from (5.19) that[9]

$$\int_1^{\infty} u_+(\pm t) t^{-3} \, dt < \infty.$$

This implies the existence of the integrals

$$\int_1^{\infty} \frac{u_-(\pm t)}{t^3} dt = -\int_1^{\infty} \frac{u(\pm t) - u_+(\pm t)}{t^3} dt,$$

and, since $|u(t)| = u_+(t) + u_-(t)$, also the existence of the integral (5.22).

[9] The functions $u_+(t)$ and $u_-(t)$ are defined as usual by

$$u_+(t) = \begin{cases} u(t) \text{ when } u(t) > 0, \\ 0 \text{ when } u(t) \leqslant 0, \end{cases}$$

$$u_-(t) = \begin{cases} -u(t) \text{ when } u(t) < 0, \\ 0 \text{ when } u(t) \geqslant 0. \end{cases}$$

Furthermore, the equation

$$\frac{y}{(t-x)^2+y^2}-\frac{y}{t^2+1}=O\,(t^{-3})$$

and the existence of the integral (5.22) imply that the existence of the integral (5.20) is equivalent to the existence of (5.21) for all x and y. This completes the proof of the lemma.

In the next lemma we give an estimate of a function represented by Poisson's integral.

LEMMA 3. *The integral*

$$v\,(z)=\frac{y}{\pi}\int_{-\infty}^{\infty}\frac{u\,(t)\,dt}{(t-x)^2+y^2} \tag{5.21}$$

defines a holomorphic function $v(z)$ in the half-plane $y>0$ whose limiting values on the real axis coincide with the values of the function $u(x)$, and which satisfies the inequality

$$|v\,(z)|<c_\varepsilon+\varepsilon\frac{r^2}{|y|}\qquad (r=|z|), \tag{5.23}$$

where ε is an arbitrarily small positive number, and c_ε is a constant depending on ε.

PROOF. The existence of the integral (5.22) implies that the integral

$$v\,(z)-Jy=\frac{1}{\pi}\int_{-\infty}^{\infty}u\,(t)\left[\frac{y}{(t-x)^2+y^2}-\frac{y}{1+t^2}\right]dt \tag{5.24}$$

converges uniformly on any closed subset of the half-plane $y>0$, and therefore represents a harmonic function for $y>0$. The equation $v(x)=u(x)$ is established in the usual way. Let us now proceed with the proof of the validity of the inequality (5.23). Let us set

$$\chi_N\,(t)=\begin{cases}u\,(t) & \text{when } |t|>N,\\ 0 & \text{when } |t|\leqslant N,\end{cases}\qquad (N>0)$$

and

$$\psi_N\,(x)=\int_0^x\chi_N\,(t)\,t^{-2}\,dt. \tag{5.25}$$

Since $1/t^2-1/t^2+1=O(t^{-4})$, it is true for every $\eta>0$ and $N>N_\eta$ that $|\psi_N(t)|<\eta$.

Integrating (5.21) by parts, we obtain

$$v(z) = O(1) + \frac{y}{\pi}\left\{\left[\psi_N(t)\frac{t^2}{(t-x)^2+y^2}\right]_{-\infty}^{-N} + \left[\psi_N(t)\frac{t^2}{(t-x)^2+y^2}\right]_N^{\infty}\right\}$$

$$+ \frac{y}{\pi}\int_{-\infty}^{\infty}\psi_N(t)\frac{d}{dt}\left[\frac{t^2}{(t-x)^2+y^2}\right]dt. \qquad (5.26)$$

When $r > 2N$, the following inequalities hold:

$$\frac{N^2}{(N-x)^2+y^2} < 1, \qquad \frac{N^2}{(N+x)^2+y^2} < 1.$$

Therefore the absolute value of the expression in braces of (5.26) is less than 4η.

For the evaluation of the third term of (5.26) we note that $\frac{d}{dt}\left[\frac{t^2}{(t-x)^2+y^2}\right]$ changes its sign when $t = 0$ and $t = x + y^2/x$, and hence we have the result

$$\left|\int_{-\infty}^{\infty}\psi_N(t)\frac{d}{dt}\left[\frac{t^2}{(t-x)^2+y^2}\right]dt\right| \leqslant \eta\int_{-\infty}^{\infty}\left|\frac{d}{dt}\left[\frac{t^2}{(t-x)^2+y^2}\right]\right|dt$$

$$= \eta\left\{\left[\frac{t^2}{(t-x)^2+y^2}\right]_0^{-\infty} + \left[\frac{t^2}{(t-x)^2+y^2}\right]_0^{x+\frac{y^2}{x}}\right.$$

$$\left.+\left[\frac{t^2}{(t-x)^2+y^2}\right]_{\infty}^{x+\frac{y^2}{x}}\right\} = 2\eta\frac{x^2+y^2}{y^2}.$$

With the aid of these inequalities we derive from (5.26) the inequality

$$|v(z)| < O(1) + \frac{6\eta}{\pi}\frac{r^2}{y}.$$

This completes the proof of the lemma.

THEOREM 4. *In order that a function $u(z)$ which is holomorphic in the half-plane $y > 0$ and is continuous in the closed half-plane $y \geqslant 0$ and also satisfies the condition*

$$\text{(a)} \qquad \overline{\lim_{|z|\to\infty}}\frac{u(z)}{|z|} < \infty$$

be representable in the form

$$\text{(b)} \qquad u(z) = \frac{y}{\pi}\int_{-\infty}^{\infty}u(t)\frac{dt}{(t-x)^2+y^2} + ky \qquad (z = x + iy),$$

where

(c) $$k = \overline{\lim_{y \to +\infty}} \frac{u(iy)}{y} > -\infty,$$

it is necessary and sufficient that the integral

$$\int_{-\infty}^{\infty} \frac{u(t)}{1+t^2} dt$$

be convergent.[9a]

PROOF. The necessity is obtained directly from the equation (c) if one sets $z = i$. For the proof of the sufficiency we note that from Lemmas 2 and 3, and from the representation of the function (5.21), it follows easily that the function

$$w(z) = u(z) - v(z) - k_+ y \qquad (k_+ = \max(k, 0)) \tag{5.27}$$

is harmonic when $y > 0$ and satisfies the conditions

(1) $$w(x) = 0,$$

(2) $$w(iy) \leqslant o(y),$$

(3) $$w(z) \leqslant O(r) + \frac{o(r^2)}{y}.$$

From condition (1) it follows on the basis of the principle of symmetry that the function $w(z)$ can be extended harmonically into the lower half-plane as an odd function of y. The harmonic function

$$w_h(z) = \frac{1}{4h^2} \int_{y-h}^{y+h} \int_{\tau-h}^{\tau+h} w(x + it)\, dt\, d\tau$$

satisfies the conditions (1) and (2), while condition (3) is replaced by the condition

(3′) $$w_h(z) \leqslant o(r^2).$$

By the application of a theorem of Phragmén and Lindelöf[10] inside the angles $0 \leqslant \arg z \leqslant \pi/2$ and $\pi/2 \leqslant \arg z \leqslant \pi$, we find that $w_h(z) - \varepsilon y$ is bounded in the half-plane $y \geqslant 0$ and that $w_h(x) = 0$. Applying again the theorem of Phragmén and Lindelöf to the function $w_h(z) - \varepsilon y$ in the half-plane $y \geqslant 0$, we obtain

$$w_h(z) \leqslant \varepsilon y \qquad (y \geqslant 0).$$

[9a] This representation of a harmonic function in a half-plane was established by R. Nevanlinna [2] under the hypothesis that

$$\int_{-\infty}^{\infty} \frac{u_+(t)}{1+t^2} dt < \infty.$$

[10] The theorem of Phragmén and Lindelöf for harmonic functions can be obtained from Theorem 22, § 14, Chapter I, by noting that a harmonic function is the logarithm of the modulus of a holomorphic function.

Since $\varepsilon > 0$ is arbitrary,

$$w_h(z) \leqslant 0 \quad \text{when} \quad \operatorname{Im} z \geqslant 0.$$

Next, taking the limit as $h \to 0$, we obtain the inequality

$$w(z) \leqslant 0 \quad \text{when} \quad \operatorname{Im} z > 0.$$

Denoting by $X(z)$ the conjugate harmonic function, we construct the entire real function

$$\varphi(z) = X(z) - iw(z),$$

which maps the upper half-plane into the upper half-plane.

By a lemma of N. G. Čebotarev this function has the form $\varphi(z) = \alpha z + \beta$ and hence

$$w(z) = -\alpha y.$$

Substituting this into (5.27), we obtain

$$u(z) = v(z) + (k_+ - \alpha)y.$$

It follows from Lemma 3 that $k = k_+ - \alpha > -\infty$. This proves the theorem.

$$\overline{\lim_{y \to +\infty}} \frac{w(iy)}{y} = 0.$$

Therefore $\alpha = 0$, and $w(z) \equiv 0$. This proves the theorem.

COROLLARY 1. *If the function $u(z)$ is harmonic in the half-plane* $\operatorname{Im} z > 0$, *is continuous when* $\operatorname{Im} z \geqslant 0$, *and satisfies the condition*

(a) $$\overline{\lim_{|z| \to \infty}} \frac{u(z)}{|z|} < \infty$$

and if the integral

$$\int_{-\infty}^{\infty} \frac{u(t)}{1+t^2}\, dt$$

converges, then

$$\lim_{r \to \infty} \frac{u(re^{i\theta})}{r} = k \sin \theta \qquad (0 < \theta < \pi). \tag{5.28}$$

This corollary is obtained directly from the representation (b) and from Lemma 3.

COROLLARY 2. *If the function $u(z)$ is harmonic when* $\operatorname{Im} z > 0$, *is continuous when* $\operatorname{Im} z \geqslant 0$, *satisfies the condition* (a) *and the conditions*

(c′) $$\int_{0}^{\infty} \frac{u(t) + u(-t)}{1+t^2}\, dt \quad \textit{exists}$$

and

(d) $$\lim_{|x| \to \infty} \frac{u(x)}{|x|} \leqslant 0,$$

then

$$\lim_{r\to\infty} \frac{u(re^{i\theta})}{r} = k\sin\theta \qquad (0<\theta<\pi). \tag{5.28'}$$

PROOF. By Theorem 4, the function $u(z)+u(-\bar{z})$ is representable in the form (b), and therefore

$$\lim_{r\to\infty} \frac{u(re^{i\theta})+u(re^{i(\pi-\theta)})}{r} = 2k\sin\theta \qquad (0<\theta<\pi). \tag{5.29}$$

Furthermore, condition (d) implies that the indicator diagram corresponding to the function $u(z)$ is an interval (segment) on the axis Oy, namely,

$$\overline{\lim_{r\to\infty}} \frac{u(re^{i\theta})}{r} = k\sin\theta. \tag{5.30}$$

On the other hand, the relations (5.29) and (5.30) imply that

$$\underline{\lim_{r\to\infty}} \frac{u(re^{i\theta})}{r} = \lim_{r\to\infty} \frac{u(re^{i\theta})+u(re^{i(\pi-\theta)})}{r} - \overline{\lim_{r\to\infty}} \frac{u(re^{i(\pi-\theta)})}{r}$$
$$= k\sin\theta.$$

This result and (5.30) imply equation (5.28').

REMARK. The results of this section clearly remain true if we replace the hypothesis that $u(z)$ is continuous in the upper half-plane by the hypothesis that $u(z)$ is harmonic in $y \geqslant 0$ except perhaps for isolated logarithmic singular points on the real axis.

3. Representation of a Function of Exponential Type and of Class *A* in the Upper Half-plane

If a function $F(z)$ of exponential type in the half-plane $y \geqslant 0$ has no zeros in this half-plane, and if the integral

$$\int_{-\infty}^{\infty} \frac{\ln|F(t)|}{t^2+1}\,dt$$

converges, then Theorem 4 of the preceding section implies that this function has the representation

$$\ln|F(z)| = \frac{y}{\pi}\int_{-\infty}^{\infty} \frac{\ln|F(t)|}{(t-x)^2+y^2}\,dt + ky,$$

where $k = h_F(\pi/2)$. In order to derive criteria for the complete regularity of the growth of functions of class A and of exponential type, we shall need a representation of such functions in the more general case, when the function has zeros in the upper half-plane. In this section we shall obtain such a representation. First we shall prove two lemmas.

LEMMA 4. *If the series*

$$\sum_{k=1}^{\infty} \left|\operatorname{Im}\frac{1}{a_k}\right| \tag{5.31}$$

converges, then the infinite product

$$\chi(z)=\prod_{k=1}^{\infty}\left(1-\frac{z}{a_k}\right)\left(1-\frac{z}{\bar{a}_k}\right)^{-1} \tag{5.32}$$

converges uniformly and absolutely in any finite closed part of the plane that does not contain points $\bar{a}_k$.

PROOF. Let us write $\chi(z)$ in the form

$$\chi(z)=\prod_{k=1}^{\infty}\left[1+z\left(1-\frac{z}{\bar{a}_k}\right)^{-1}\left(\frac{1}{\bar{a}_k}-\frac{1}{a_k}\right)\right]. \tag{5.33}$$

By a well-known criterion, uniform and absolute convergence of the infinite product (5.32) is equivalent to uniform and absolute convergence of the series

$$\sum_{k=1}^{\infty}\left|z\left(1-\frac{z}{\bar{a}_k}\right)^{-1}\right|\,\left|\operatorname{Im}\frac{1}{a_k}\right|,$$

while the convergence of this series is equivalent to the convergence of the series $\Sigma_{k=1}^{\infty}|\operatorname{Im} 1/a_k|$. This establishes the lemma.

LEMMA 5. *If* $\{a_k\}$ *is a sequence of complex numbers such that* $\operatorname{Im} a_k>0$ *and* $\Sigma_1^{\infty}|\operatorname{Im} 1/a_k|<\infty$, *then*

$$\lim_{r\to\infty}{}^{*}\frac{\ln|\chi(re^{i\theta})|}{r}=0,$$

where $\chi(z)$ *is a meromorphic function defined by the equation* (5.32).[11]

PROOF. We note first of all that the modulus of each factor in the product (5.32) is less than unity when z lies in the upper half-plane. Therefore,

$$|\chi(z)|<1 \qquad (\operatorname{Im} z>0),$$

and hence

$$\overline{\lim_{r\to\infty}}\frac{\ln|\chi(re^{i\theta})|}{r}\leqslant 0 \qquad (0<\theta<\pi). \tag{5.34}$$

[11] B. Ja. Levin [5], N. N. Ahiezer [2]. In the article [1] of Ahlfors and Heins the following more precise fact is proved: the $\lim_{r\to\infty}\dfrac{\ln|\chi(re^{i\theta})|}{r}=0$ under the condition that r goes to infinity without taking on values in some set E of finite logarithmic length, i.e., a set E such that $\int_E\dfrac{dx}{x}<\infty$. We call attention to the fact that a set of finite logarithmic length has zero relative measure, but not every set of zero relative measure has finite logarithmic length. Hayman [1] established the stronger result that $(\ln|\chi(re^{i\theta})|)/r \rightrightarrows 0$ as $r\to\infty$, $r\,\overline{\in}\,E$, $0<\theta<\pi$. Azarin [3] noted that the exceptional set E can be replaced by an exceptional set of disks c_j such that $\Sigma_j\alpha_j/r_j<\infty$, where α_j are the diameters of the disks and r_j are their distances from the origin. We shall use Lemma 5 in what follows, but replacing it by the theorems of Hayman and Azarin leads to somewhat stronger results. In the cited paper Azarin extended the result to functions of n variables that are subharmonic in a cone. [*Translator's remark*: See also J. L. Doob, [1].]

Next, we need to find a lower bound of $|\chi(z)|$. For this purpose we express $\chi(z)$ in the form

$$\chi(z)=\chi_1(z)\chi_2(z), \tag{5.35}$$

where

$$\chi_1(z)=\prod_{\frac{\pi}{2}-\delta<\left|\arg a_k-\frac{\pi}{2}\right|<\frac{\pi}{2}}\left(1-\frac{z}{a_k}\right)\left(1-\frac{z}{\bar a_k}\right)^{-1},$$

$$\chi_2(z)=\prod_{\left|\arg a_k-\frac{\pi}{2}\right|\leqslant\frac{\pi}{2}-\delta}\left(1-\frac{z}{a_k}\right)\left(1-\frac{z}{\bar a_k}\right)^{-1}.$$

Let us first examine the function $\chi_1(z)$. For this purpose we represent $\chi_1^{-1}(z)$ in the form

$$\chi_1^{-1}(z)=\prod_{\frac{\pi}{2}-\delta<\left|\arg a_k-\frac{\pi}{2}\right|<\frac{\pi}{2}}\left(1+\frac{2iz\operatorname{Im}\frac{1}{a_k}}{1-\frac{z}{a_k}}\right).$$

From the inequality $\ln|1+x|\leqslant|x|$ it follows that

$$-\ln|\chi_1(z)|\leqslant 2\sum_{\frac{\pi}{2}-\delta<\left|\arg a_k-\frac{\pi}{2}\right|<\frac{\pi}{2}}\frac{|z|}{\left|1-\frac{z}{a_k}\right|}\left|\operatorname{Im}\frac{1}{a_k}\right|.$$

When $|\arg z-\pi/2|\leqslant\pi/2-2\delta$, the inequality $\dfrac{|a_k|}{|a_k-z|}\leqslant(\sin\delta)^{-1}$ holds, and therefore within this angle and for some $N>0$ we have

$$\sum_{\frac{\pi}{2}-\delta<\left|\arg a_k-\frac{\pi}{2}\right|<\frac{\pi}{2},\ |a_k|>N}\frac{|z|}{\left|1-\frac{z}{a_k}\right|}\left|\operatorname{Im}\frac{1}{a_k}\right| \leqslant(\sin\delta)^{-1}\sum_{|a_k|>N}\left|\operatorname{Im}\frac{1}{a_k}\right||z|.$$

For a given $\varepsilon>0$, and for a sufficiently large N, the right hand side will not exceed $\varepsilon|z|/4$. After N has been chosen, one can select the number R_ε so that if $|z|>R_\varepsilon$ then the following inequality holds:

$$\sum_{\frac{\pi}{2}-\delta<\left|\arg a_k-\frac{\pi}{2}\right|<\frac{\pi}{2},\ |a_k|\leqslant N}\frac{|z|}{\left|1-\frac{z}{a_k}\right|}\left|\operatorname{Im}\frac{1}{a_k}\right|<\frac{\varepsilon}{4}|z|.$$

We have thus proved that if $|z|>R_\varepsilon$, then within the angle $|\arg z-\pi/2|\leqslant\pi/2-2\delta$

$$-\ln|\chi_1(z)|<\varepsilon|z|,$$

and therefore

$$\varlimsup_{r\to\infty} \frac{\ln|\chi_1(re^{i\theta})|}{r} \geqslant 0. \tag{5.36}$$

Let us now consider the second factor in (5.35). All the zeros of the function $\chi_2(z)$ lie within the angle $|\arg z - \pi/2| \leqslant \pi/2 - \delta$. Therefore they satisfy the inequality $\operatorname{Im} a_k \geqslant \sin\delta\,|a_k|$, or

$$\left|\operatorname{Im}\frac{1}{a_k}\right| \geqslant \frac{\sin\delta}{|a_k|}.$$

From this, and from (5.31) we see that the series

$$\sum_{\left|\arg a_k - \frac{\pi}{2}\right| \leqslant \frac{\pi}{2} - \delta} \frac{1}{|a_k|} \tag{5.37}$$

converges and hence that

$$\varphi(z) = \prod_{\left|\arg a_k - \frac{\pi}{2}\right| \leqslant \frac{\pi}{2} - \delta} \left(1 - \frac{z}{a_k}\right)$$

is an entire function of genus zero. Therefore $\Delta_\varphi(\theta) = 0$, $\delta_\varphi = 0$, and by Theorem 2, Chapter II,

$$\lim_{r\to\infty} \frac{\ln|\varphi(re^{i\theta})|}{r} = 0,$$

where r tends to infinity without taking on any values of a set E of zero relative measure. The function

$$\bar\varphi(z) = \prod_{\left|\arg a_k - \frac{\pi}{2}\right| \leqslant \frac{\pi}{2} - \delta} \left(1 - \frac{z}{\bar a_k}\right)$$

is also of genus zero and has no zeros in the upper half-plane, and hence

$$\lim_{r\to\infty} \frac{\ln|\bar\varphi(re^{i\theta})|}{r} = 0$$

when $0 \leqslant \theta < \pi$. Since

$$\chi_2(z) = \frac{\varphi(z)}{\bar\varphi(z)},$$

the last two equations imply that

$$\lim_{\substack{r\to\infty\\ r\bar{\in} E}} \frac{\ln|\chi_2(re^{i\theta})|}{r} = 0.$$

From this and from (5.35) and (5.36) it follows that

$$\varliminf_{\substack{r\to\infty \\ r\notin E}} \frac{\ln|\chi(re^{i\theta})|}{r} = 0.$$

Comparing this with (5.34), we establish the lemma.

REMARK. *If an angle* $0 < \theta_1 \leqslant \theta \leqslant \theta_2 < \pi$ *does not contain points of the set* $\{a_k\}$ *then within this angle the following limit exists in the ordinary sense*:

$$\lim_{r\to\infty} \frac{\ln|\chi(re^{i\theta})|}{r} = 0 \qquad (\theta_1 < \theta < \theta_2).$$

Indeed, from (5.34) (where one can obviously replace $\chi(z)$ by $\chi_1(z)$) and from (5.36) it follows that for $\delta > 0$ sufficiently small

$$\lim_{r\to\infty} \frac{\ln|\chi_1(re^{i\theta})|}{r} = 0.$$

The function $\chi_2(z)$ is of completely regular growth with exponent $\rho = 1$ and has no zeros. For $\theta_1 \leqslant \theta \leqslant \theta_2$ the limit

$$\lim_{r\to\infty} \frac{\ln|\chi_2(re^{i\theta})|}{r} = 0 \qquad (\theta_1 < \theta < \theta_2)$$

exists, and hence

$$\lim_{r\to\infty} \frac{\ln|\chi(re^{i\theta})|}{r} = 0.$$

The next theorem yields a representation of special functions which are holomorphic in a half-plane.

THEOREM 5. *If the function* $F(z)$ *is holomorphic and of exponential type in the half-plane* $\operatorname{Im} z \geqslant 0$, *and if the integral*

$$\int_{-\infty}^{\infty} \frac{\ln|F(t)|}{1+t^2}\,dt \tag{5.38}$$

exists, then

$$\ln|F(z)| = \frac{y}{\pi}\int_{-\infty}^{\infty} \ln|F(t)|\frac{dt}{(t-x)^2+y^2} + ky + \ln|\chi(z)|, \tag{5.39}$$

where $k = h_F(\pi/2)$ *and*

$$\chi(z) = \prod_{k=1}^{\infty}\left(1-\frac{z}{a_k}\right)\left(1-\frac{z}{\bar{a}_k}\right)^{-1}, \tag{5.32}$$

the a_k *being the zeros of the function* $F(z)$ *in the half-plane* $y > 0$.[12]

[12] R. Nevanlinna [2] established this proposition under somewhat different hypotheses.

PROOF. First of all we note that the product $\chi(z)$ on the right hand side of (5.39) exists. Indeed, the convergence of the integral (5.38) implies by Theorem 1 that $F(z)$ is of class A and by Theorem 4 that the product $\chi(z)$ converges in every finite closed region of the upper half-plane free of points a_k.

Let us now construct the function

$$u(z)=\frac{y}{\pi}\int_{-\infty}^{\infty}\ln|F(t)|\frac{dt}{(t-x)^2+y^2}+ky. \tag{5.40}$$

By Theorem 4 and the remark at the end of §2, this function is harmonic in the upper half-plane and takes on the value $u(x) = \ln|F(x)|$ on the real axis. Let $v(z)$ be the harmonic function (when $y > 0$) which is the conjugate of $u(z)$.

The function

$$\psi(z)=F(z)\,e^{-u(z)-iv(z)} \tag{5.41}$$

is obviously holomorphic in the upper half-plane, and its modulus has the limiting value one on the real axis.

By the principle of symmetry, this function can be extended over the lower half-plane, and it has poles at the points $\bar{a}_k$ of the same multiplicity as the zeros a_k of the function $F(z)$.

In order to obtain an estimate of the function $|\psi(z)|$ in the upper half-plane, we note that by Lemma 5

$$|u(z)-ky|<c_\varepsilon+\varepsilon\frac{r^2}{y}.$$

Since $F(z)$ is by hypothesis of exponential type,

$$\ln|\psi(z)|<c_\varepsilon+\varepsilon\frac{r^2}{y}+(\sigma+\varepsilon)r. \tag{5.42}$$

Finally, it follows from (5.41) and from the equation $k = h_F(\pi/2)$ that $h_\psi(\pi/2) \leqslant 0$. In order to sharpen the inequality (5.42) we shall apply the procedure used in the proof of Theorem 4.

We introduce the averaged (mean) function

$$w_h(z)=\frac{1}{4h^2}\int_{y-h}^{y+h}\int_{\tau-h}^{\tau+h}\ln|\psi(x+it)|\,dt\,d\tau,$$

This is a subharmonic function when $y > 0$ and satisfies the conditions

1) $w_h(x) = 0$,
2) $w_h(iy) = o(y)$,
3) $w_h(z) = o(r^2)$.

Applying the Phragmén-Lindelöf theorem to the function $w_h(z) - \varepsilon y$ in each of the angles $0 \leqslant \arg z \leqslant \pi/2$, $\pi/2 \leqslant \arg z \leqslant \pi$, we find that for every

$\varepsilon > 0$ the function $w_h(z) - \varepsilon y$ is bounded in the half-plane $y \geqslant 0$. Again, applying the Phragmén-Lindelöf theorem to this function in the half-plane $y \geqslant 0$, we see from (1) that $w_h(z) - \varepsilon y \leqslant 0$ for every $\varepsilon > 0$, and $y > 0$. Since $\varepsilon > 0$ is arbitrary, it follows that

$$w_h(z) \leqslant 0 \qquad (y \geqslant 0).$$

Taking the limit as $h \to 0$, we obtain

$$\ln |\psi(z)| \leqslant 0 \qquad (y \geqslant 0),$$

namely,

$$|\psi(z)| \leqslant 1$$

in the half-plane $y \geqslant 0$.

We shall next show that the function $\psi(z)$ is equal to the product $\chi(z)$ multiplied by a constant. For this purpose we set

$$\chi_n(z) = \prod_{k=1}^{n} \left(1 - \frac{z}{a_k}\right)\left(1 - \frac{z}{\bar{a}_k}\right)^{-1}.$$

The function

$$\psi_n(z) = \psi(z)\chi_n^{-1}(z)$$

is bounded in the upper half-plane, and since $|\psi_n(x)| = 1$ on the real axis, we conclude from the Phragmén-Lindelöf theorem that $|\psi_n(z)| \leqslant 1$ when $\operatorname{Im} z \geqslant 0$. Taking the limit as $n \to \infty$ and setting $\psi_\infty(z) = \psi(z)\chi^{-1}(z)$ we obtain

$$|\psi_\infty(z)| \leqslant 1 \qquad (\operatorname{Im} z \geqslant 0).$$

This function has no zeros or poles and hence

$$\varphi(z) = -i \ln \psi_\infty(z)$$

is also a real entire function for $\operatorname{Im} z > 0$, $\operatorname{Im} \varphi(z) > 0$. By the lemma of N. G. Čebotarev, $\varphi(z) = \alpha z + \beta$ ($\alpha \geqslant 0$, and $\operatorname{Im} \beta = 0$).

To determine α we select a sequence $y_j \to +\infty$ such that

$$\lim_{y_j \to +\infty} \frac{\ln |F(iy_j)|}{y_j} = k.$$

From the inequality $\operatorname{Im} \varphi(iy) \leqslant u(iy) - \ln |F(iy)|$ it follows that

$$\lim_{y_j \to +\infty} \frac{\operatorname{Im} \varphi(iy_j)}{y_j} \leqslant 0.$$

This implies that $\alpha = 0$ and $\psi_\infty(z) = e^{i\beta}$ or $\psi(z) = e^{i\beta}\chi(z)$. To complete the proof of the theorem, one needs only to substitute this expression into (5.41).

4. Functions of Class A and of Completely Regular Growth

On the basis of the representation (5.39), we shall obtain in this section a number of theorems which yield useful criteria to determine whether a function of exponential type is a function of class A and of completely regular growth.

It is useful to note that the density (when $\rho(r) = 1$) of the set of zeros of an entire function of class A is equal to zero within any angle of the form $0 < \delta \leqslant |\arg z| \leqslant \pi - \delta$ (see (5.37)). Therefore, for every entire function of exponential type, of class A, and of completely regular growth, the density $\Delta(\psi)$ of the set of zeros is a piece-wise constant function which has only two jumps of equal height at the points 0 and π. From this it follows, that *the indicator diagram of such a function is an interval (straight line segment) parallel to the imaginary axis.*

THEOREM 6. *If $F(z)$ is a function of exponential type in the half-plane* $\operatorname{Im} z \geqslant 0$, *and if the integral*

$$\int_{-\infty}^{\infty} \frac{\ln |F(t)|}{1+t^2}\, dt \tag{5.43}$$

converges, then $F(z)$ is a function of class A, and of completely regular growth ($0 < \arg z < \pi$), *whereby* $h_F(\theta) = k \sin \theta$.

PROOF. Under the hypotheses of the theorem the function $\ln |F(z)|$ admits the representation (5.39) in the half-plane $\operatorname{Im} z \geqslant 0$. By Lemma 3, the first summand on the right hand side of (5.39) (after division by r) tends to zero when $r \to \infty$ and $|\theta - \pi/2| \leqslant \pi/2 - \delta$ ($\delta > 0$). From Lemma 5 we obtain

$$\lim_{\substack{r \to \infty \\ r \overline{\in} E}} \frac{\ln |\chi(re^{i\theta})|}{r} = 0,$$

where E is some E^0-set, and therefore

$$\lim_{\substack{r \to \infty \\ r \overline{\in} E}} \frac{\ln |F(re^{i\theta})|}{r} = k \sin \theta \tag{5.44}$$

when $\delta \leqslant \theta \leqslant \pi - \delta$. Since $\delta > 0$ is arbitrary, it follows that for some set E of zero relative measure equation (5.44) is valid within the entire angle $0 < \theta < \pi$ (Chapter III, Theorem 1).

THEOREM 7 (M. CARTWRIGHT [1]). *If $F(z)$ is a function of exponential type in the half-plane* $\operatorname{Im} z \geqslant 0$, *and if the integral*

$$\int_{-\infty}^{\infty} \frac{\ln_+ |F(t)|}{1+t^2}\, dt \tag{5.45}$$

converges, then $F(z)$ is of class A and of completely regular growth when $0 <$ $\arg z < \pi$, and $h_F(\theta) = k \sin\theta$ $(0 < \theta < \pi)$.

PROOF. We shall prove that for a function of exponential type, the convergence of the integral (5.45) implies the convergence of the integral (5.43). Then the assertion of Theorem 7 will follow directly from Theorem 6. For this purpose we make use of the formula of Carleman, which can be written in the form

$$\frac{1}{2\pi}\int_{\lambda}^{R} [\ln_-|F(t)| + \ln_-|F(-t)|]\left(\frac{1}{t^2}-\frac{1}{R^2}\right)dt$$

$$+\sum_{\lambda<r_k\leqslant R}\left(\frac{1}{r_k}-\frac{r_k}{R^2}\right)\sin\theta_k = \frac{1}{2\pi}\int_{\lambda}^{R}[\ln_+|F(t)|+\ln_+|F(-t)|]$$

$$\times\left(\frac{1}{t^2}-\frac{1}{R^2}\right)dt + \frac{1}{\pi R}\int_0^{\pi}\ln|F(Re^{i\theta})|\sin\theta\, d\theta + A_\lambda(F, R).$$

For a function of exponential type, the sum of the last two terms is bounded from above. We note also that

$$\sum_{\lambda<r_k\leqslant R}\left(\frac{1}{r_k}-\frac{r_k}{R^2}\right)\sin\theta_k \geqslant 0,$$

and obtain

$$\int_{\lambda}^{R}[\ln_-|F(t)|+\ln_-|F(-t)|]\left(\frac{1}{t^2}-\frac{1}{R^2}\right)dt$$

$$< \int_{\lambda}^{R}[\ln_+|F(t)|+\ln_+|F(t)|]\left(\frac{1}{t^2}-\frac{1}{R^2}\right)dt + O(1).$$

Next, from this and from the obvious inequality

$$\int_{\lambda}^{\frac{R}{2}}[\ln_-|F(t)|+\ln_-|F(-t)|]\frac{dt}{t^2}$$

$$< \frac{4}{3}\int_{\lambda}^{R}[\ln_-|F(t)|+\ln_-|F(-t)|]\left(\frac{1}{t^2}-\frac{1}{R^2}\right)dt$$

it follows that

$$\int_{\lambda}^{\frac{R}{2}}[\ln_-|F(t)|+\ln_-|F(-t)|]\frac{dt}{t^2}$$

$$< \frac{4}{3}\int_{\lambda}^{R}[\ln_+|F(t)|+\ln_+|F(-t)|]\frac{dt}{t^2} + O(1)$$

and finally

$$\int_{\lambda}^{\frac{R}{2}} \frac{\ln_-|F(t)|+\ln_-|F(-t)|}{1+t^2}\,dt$$
$$< c_\lambda \int_{\lambda}^{R} \frac{\ln_+|F(t)|+\ln_+|F(-t)|}{1+t^2}\,dt + O(1),$$

where c_λ is a constant. Under the hypotheses of the theorem, the last inequality implies

$$\int_{-\infty}^{\infty} \frac{|\ln|F(t)||}{1+t^2}\,dt < \infty,$$

i.e., the absolute convergence of the integral (5.43).

COROLLARY. *The functions of exponential type in the half-plane* $\operatorname{Im} z \geqslant 0$ *that are bounded on the real axis (or belong to* $L^p(-\infty, \infty)$*) are functions of class* A*, and are of completely regular growth in this half-plane.*

Indeed, such functions obviously satisfy the conditions of the theorem of Cartwright.[13]

THEOREM 8. *If* $F(z)$ *is holomorphic and of exponential type in the half-plane* $\operatorname{Im} z \geqslant 0$*, and if it satisfies the conditions*

$$\int_{0}^{\infty} \frac{\ln|F(x)F(-x)|}{1+x^2}\,dx \quad exists \tag{α}$$

$$h_F(0)+h_F(\pi)=0, \tag{β}$$

then the function $F(z)$ *is of class* A *and of completely regular growth in the upper half-plane*[14] $(0 < \theta < \pi)$.

PROOF. Let us set $\bar{F}(z) = \overline{F(\bar{z})}$ $(\operatorname{Im} z \leqslant 0)$. The function $\varphi(z) = F(z)\bar{F}(-z)$ satisfies the hypotheses of Theorem 6, and hence

$$h_\varphi(\theta) = \lim_{r\to\infty}{}^{*} \frac{\ln|\varphi(re^{i\theta})|}{r} = 2k\sin\theta, \tag{5.46}$$

where $k = h_F(\pi/2)$. On the other hand, from condition (β) it follows that the indicator diagram of the function $F(z)$ is an interval parallel to the y-axis. Without loss of generality, one may assume that this interval lies on the y-axis

---

[13] For a function of $L^p(-\infty, \infty)$, this follows from the inequality $p\ln_+|f(x)| < |f(x)|^p$.

[14] A theorem similar to Theorem 8 was proved by N. Levinson for entire functions of exponential type (see [1], 25–42).

(it is sufficient to multiply the function by $e^{\gamma z}$ where γ is some real number). Thus

$$h_F(\theta) = \overline{\lim_{r\to\infty}} \frac{\ln|F(re^{i\theta})|}{r} = k\sin\theta. \tag{5.47}$$

From (5.46) and (5.47) we have

$$\underline{\lim_{\substack{r\to\infty\\ r\,\overline{\in}\,E^0}}} \frac{\ln|F(re^{i\theta})|}{r} = \lim_{\substack{r\to\infty\\ r\,\overline{\in}\,E^0}} \frac{\ln|\varphi(re^{i\theta})|}{r} - \overline{\lim_{\substack{r\to\infty\\ r\,\overline{\in}\,E^0}}} \frac{\ln|F(re^{i(\pi-\theta)})|}{r} = k\sin\theta$$

$$(0 < \theta < \pi).$$

From this and from (5.47) it follows that

$$\lim_{r\to\infty}{}^{*} \frac{\ln|F(re^{i\theta})|}{r} = k\sin\theta.$$

This completes the proof of the theorem.

THEOREM 9.[15] *In order that an entire function $f(z)$ of exponential type be of class A and that at the same time the density*

$$\Delta_f = \lim_{r\to\infty} \frac{n_f(r)}{r} \tag{5.48}$$

exist, it is necessary and sufficient that the integral

$$\int_0^\infty \frac{\ln|f(x)f(-x)|}{1+x^2}\,dx \tag{5.49}$$

exist.

PROOF. If the integral (5.49) exists then, by Theorem 6, the function

$$\varphi(z) = f(z)f(-z)$$

is of completely regular growth in the half-planes $0 < \arg z < \pi$, and $-\pi < \arg z < 0$, and by the corollary of Theorem 1, Chapter III, this function is of completely regular growth in the entire plane. The set of zeros of a function of completely regular growth has an angular density, and therefore the limit

$$2\Delta_f = \lim_{r\to\infty} \frac{n_\varphi(r)}{r}$$

exists.

For the establishment of (5.48) it is sufficient to note that $n_\varphi(r) = 2n_f(r)$. Furthermore, the existence of the integral (5.49) implies that the function $f(z)$ is of class A. This is a consequence of the criterion for membership in class A formulated in Theorems 2 and 3.

[15] A theorem of A. Pfluger [5].

Let us next prove the necessity. From the fact that the entire function $f(z)$ of exponential type belongs to class A, and from the condition (5.48), it follows that $\varphi(z) = f(z)f(-z)$ is an entire, even function of exponential type and of class A. The set of zeros of this function has a density. We recall (see (5.37)) that all zeros of a function of class A, except possibly for a set of zero density, lie within arbitrarily small angles containing the rays $\arg z = 0$, $\arg z = \pi$. Therefore the set of zeros of the even function $\varphi(z)$ has an angular density. Thus the zeros of $\varphi(z)$ are regularly distributed, and hence $\varphi(z)$ is of completely regular growth.

For the proof of the convergence of the integral (5.49) we make use of Carleman's formula, applying it to $\varphi(z)$ in the upper half-plane. First of all, we examine the quantity

$$\sum_{\lambda < r_k < R} \left(\frac{1}{r_k} - \frac{r_k}{R^2}\right) \sin \theta_k \qquad (0 < \theta_k < \pi),$$

where $b_k = r_k e^{i\theta_k}$ are the zeros of $\varphi(z)$ in the upper half-plane. From the inequality

$$\frac{1}{R^2} \sum_{r_\varepsilon < r_k < R} r_k \sin \theta_k < \sum_{r_k > r_\varepsilon} \frac{\sin \theta_k}{r_k}$$

it follows that for a sufficiently large r_ε

$$\frac{1}{R^2} \sum_{r_\varepsilon < r_k < R} r_k \sin \theta_k < \varepsilon.$$

This implies that

$$\lim_{R \to \infty} \frac{1}{R^2} \sum_{r_k < R} r_k \sin \theta_k = 0.$$

In this manner we obtain

$$\lim_{R \to \infty} \sum_{\lambda < r_k < R} \left(\frac{1}{r_k} - \frac{r_k}{R^2}\right) \sin \theta_k = - \sum_{|b_k| > \lambda} \operatorname{Im} \frac{1}{b_k}. \tag{5.50}$$

To investigate the quantity

$$\frac{1}{\pi R} \int_0^\pi \ln |\varphi(Re^{i\theta})| \sin \theta \, d\theta$$

we make use of the fact that $\varphi(z)$ is of completely regular growth. From Theorem 3 of Chapter IV it follows that

$$\lim_{R \to \infty} \frac{n_\varphi(R)}{R} = \frac{1}{2\pi} \int_0^{2\pi} h_\varphi(\theta) \, d\theta.$$

Hence

$$\lim_{R \to \infty} \frac{N_\varphi(R)}{R} = \frac{1}{2\pi} \int_0^{2\pi} h_\varphi(\theta) \, d\theta,$$

and by Jensen's theorem

$$\lim_{r\to\infty}\int_0^{2\pi}\left[\frac{\ln|\varphi(Re^{i\theta})|}{R}-h_\varphi(\theta)\right]d\theta=0. \tag{5.51}$$

If one makes use of the asymptotic relation (Chapter I, Theorem 28)

$$\frac{\ln|\varphi(Re^{i\theta})|}{R}<h_\varphi(\theta)+\varepsilon \qquad (\varepsilon>0),$$

then it follows from (5.51) that

$$\lim_{r\to\infty}\int_0^{2\pi}\left|\frac{\ln|\varphi(Re^{i\theta})|}{R}-h_\varphi(\theta)\right|d\theta=0.$$

From this we obtain the equation

$$\lim_{R\to\infty}\frac{1}{\pi R}\int_0^{\pi}\ln|\varphi(Re^{i\theta})|\sin\theta\,d\theta=\frac{1}{\pi}\int_0^{\pi}h_\varphi(\theta)\sin\theta\,d\theta. \tag{5.52}$$

Since the quantity $A_\lambda(\varphi, R)$ also has a limit when $R\to\infty$, it follows from (5.50) and (5.52) that the limit

$$\lim_{R\to\infty}\int_\lambda^R\left(\frac{1}{x^2}-\frac{1}{R^2}\right)\ln|\varphi(x)|\,dx \tag{5.53}$$

exists. It remains to prove that the limit

$$\lim_{R\to\infty}\frac{1}{R^2}\int_\lambda^R\ln|f(x)f(-x)|\,dx$$

exists. We have

$$\frac{1}{R^2}\int_\lambda^R x\,\frac{\ln|\varphi(x)|}{x}\,dx=\left[\frac{x}{R^2}\int_\lambda^x\frac{\ln|\varphi(t)|}{t}\,dt\right]_\lambda^R-\frac{1}{R^2}\int_\lambda^R J_\varphi^x(0)\,dx.$$

By Lemma 2 of Chapter III the limit

$$\lim_{R\to\infty}\frac{1}{R}J_\varphi^R(0)=h_\varphi(0)$$

exists, and therefore

$$\lim_{R\to\infty}\frac{1}{R^2}\int_\lambda^R\ln|\varphi(x)|\,dx=\frac{1}{2}h_\varphi(0). \tag{5.54}$$

From this, and from (5.53) it follows that the integral

$$\int_{\lambda}^{\infty} \frac{\ln|f(x)f(-x)|}{x^2}\,dx$$

converges. In order to deduce from this the convergence of the integral (5.49), we set

$$\psi(x) = \int_{\lambda}^{x} \frac{\ln|f(t)f(-t)|}{t^2}\,dt.$$

Integrating by parts, we obtain

$$\int_{0}^{R} \frac{\ln|f(x)f(-x)|}{1+x^2}\,dx = \psi(R)\frac{R^2}{R^2+1} - \int_{\lambda}^{R} \psi(x)\left(\frac{x^2}{1+x^2}\right)' dx,$$

and the existence of the limit when $R \to \infty$ is obvious. This proves the theorem.

If $f(z)$ is an entire function of exponential type and of class A, if its set of zeros has density, and if[16] $f(0) = 1$ then

$$\frac{1}{\pi}\int_{0}^{\infty} \frac{\ln|f(x)f(-x)|}{x^2}\,dx = \sum_{1}^{\infty}\left|\operatorname{Im}\frac{1}{a_k}\right| - \frac{1}{2}\operatorname*{lim^*}_{y\to\infty} \frac{\ln|f(iy)| + \ln|f(-iy)|}{y}. \tag{5.55}$$

For the proof it is natural to start from formula (5.02′). From the equations $\varphi'(0) = 0$, $h_\varphi(\theta) = h_\varphi(\pi/2)\sin\theta$ $(0 < \theta < \pi)$, (5.52) and (5.54) we obtain the equation

$$\frac{1}{\pi}\int_{0}^{\infty} \frac{\ln|\varphi(x)|}{x^2}\,dx = -\sum_{\operatorname{Im} b_k > 0} \operatorname{Im}\frac{1}{b_k} - \frac{1}{2}h_\varphi\left(\frac{\pi}{2}\right), \tag{5.56}$$

which yields directly (5.55).

We note that in view of (5.55) the equation

$$\int_{0}^{\infty} \frac{\ln|f(x)f(-x)|}{x^2}\,dx = -\frac{\pi}{2}\operatorname*{lim^*}_{y\to\infty} \frac{\ln|f(iy)| + \ln|f(-iy)|}{y}$$

is equivalent to stating that all the zeros of $f(z)$ are real.[17] We note also that the

[16] This condition obviously does not reduce the generality of the assertion.

[17] This is a theorem of Paley and Wiener [1].

existence of the integral on the left side of (5.55) for an entire function $f(z)$ of exponential type implies the convergence of the series and the existence of the limit on the right side of the equation.[18]

The inequality (5.55) can be transformed by noting that by Theorem 3 of Chapter III

$$\lim_{r\to\infty} \frac{n_\varphi(r)}{r} = \frac{1}{2\pi}\int_0^{2\pi} h_\varphi(\theta)\,d\theta = \frac{2}{\pi} h_\varphi\left(\frac{\pi}{2}\right),$$

and hence

$$\lim_{r\to\infty} \frac{n_f(r)}{r} = \frac{1}{\pi} h_\varphi\left(\frac{\pi}{2}\right).$$

Substituting this into equation (5.56), we obtain

$$\frac{1}{\pi}\int_0^\infty \frac{\ln|f(x)f(-x)|}{x^2}\,dx = \sum_1^\infty \left|\operatorname{Im}\frac{1}{a_k}\right| - \frac{\pi}{2}\lim_{r\to\infty}\frac{n_f(r)}{r}.$$

Here, by Theorem 9, the convergence of the integral on the left side of the equation is a necessary and sufficient condition for the convergence of the series, and for the existence of the limit on the right hand side.

THEOREM 10.[19] *In order that an entire function $f(z)$ of exponential type be of class A and of completely regular growth, it is necessary and sufficient that it satisfy the following conditions*:

$$\int_0^\infty \frac{\ln|f(x)f(-x)|}{1+x^2}\,dx \quad \textit{exists} \tag{α}$$

$$h_f(0)+h_f(\pi)=0. \tag{β}$$

PROOF. The sufficiency of the conditions follows from Theorem 8 since complete regularity of growth when $0<\theta<\pi$ and $0>\theta>-\pi$ implies regularity of growth in the whole plane (Chapter III, Theorem 1). The necessity of the condition (α) follows from Theorem 9 because the set of the zeros of an entire function of completely regular growth has a density. The necessity of condition (β) is obvious since the indicator diagram of an entire function of exponential type, of class A, and of completely regular growth is an interval parallel to the imaginary axis.

Let us now combine some of these results into one theorem.

---

[18] The converse can also be proved.

[19] A theorem of A. Pfluger. Govorov [**1**; **2**] studied functions $f(z)$ that are holomorphic and of exponential type in the half-plane $\operatorname{Im} z>0$ and showed that Pfluger's conditions (α) and (β) are necessary and sufficient for $f(z)$ to be of completely regular growth in the closed half-plane. For $f(z)$ to be of completely regular growth in the open half-plane, it is necessary and sufficient that (α) is satisfied and that the two limits

$$\lim_{r\to\infty}\frac{1}{r}\int_1^r \frac{\ln|f(x)|}{x}\,dx, \qquad \lim_{r\to\infty}\frac{1}{r}\int_{-1}^{-r}\frac{\ln|f(x)|}{x}\,dx \qquad \text{exist.}$$

THEOREM 11. *If an entire function of exponential type satisfies one of the following conditions:*[20]

(a) *the integral* $\int_0^\infty \frac{\ln |f(x)f(-x)|}{1+x^2}\,dx$ *exists, and* $h_f(0) = h_f(\pi) = 0$,

(b) *the integral* $\int_{-\infty}^{\infty} \frac{\ln |f(x)|}{1+x^2}\,dx$ *exists,*

(c) $$\int_{-\infty}^{\infty} \frac{\ln_+ |f(x)|}{1+x^2}\,dx < \infty,$$

(d) $|f(x)|$ *is bounded on the real axis,*

(e) $$|f(x)| \in L^p(-\infty, \infty),$$

then

(1) $f(z)$ *is of class A and is of completely regular growth, and its indicator diagram is an interval on the imaginary axis;*

(2) *all the zeros of the function* $f(z)$, *except possibly those of a set of zero density, lie inside arbitrarily small angles* $|\arg z| < \varepsilon$ *and* $|\arg z - \pi| < \varepsilon$, *where the density*

$$\Delta_i = \lim_{t\to\infty} \frac{n_i(t)}{t} \qquad (i = 1,\ 2)$$

of the set of zeros within each of these angles is equal to $d/2\pi$, *where d is the length of the indicator diagram. Furthermore, the limit* $\delta = \lim_{r\to\infty} \delta(r)$ *exists, where*

$$\delta(r) = \sum_{|a_n|<r} a_n^{-1};$$

(3) *the function* $f(z)$ *can be represented in the form*

$$f(z) = cz^m e^{ikz} \lim_{R\to\infty} \prod_{|a_n|<R} \left(1 - \frac{z}{a_n}\right)$$

(*k is a real number*).

PROOF. Assertion (1) expresses essentially the assertions of Theorems 6, 7, and 8. Assertion (2) follows from Theorems 3 and 4, § 3, Chapter III, on the

---

[20] These conditions are obviously not independent. Each one (after the first) implies the preceding ones. Indeed, by Theorem 6, it follows from (b) that $h_f(\theta) = k \sin\theta$ $(0 < \theta < \pi)$, namely, $h(0) = h(\pi) = 0$. In the proof of Theorem 7 we showed that (e) implies (b). It is obvious that (d) implies (c). It is not difficult to show also that (e) implies (d) (this follows directly from a theorem of S. Bernšteĭn, which we shall prove in Chapter IX).

In an article by Boas [2] a detailed investigation is given of the relationships among various conditions of this type; see also Boas [1].

distribution of the zeros of a function of completely regular growth. Assertion (3) is new. For its proof we use the representation of an entire function of exponential type and of completely regular growth

$$f(z) = z^m e^{az+b} \lim_{R\to\infty} \prod_{|a_k|<R} \left(1 - \frac{z}{a_k}\right),$$

which was given in the remark after Theorem 4 of Chapter III. We must show that the a occurring in this representation is purely imaginary. For this we note that the function

$$\varphi(z) = \lim_{R\to\infty} \prod_{|a_k|<R} \left(1 - \frac{z}{a_k}\right)$$

can be represented in the form

$$\varphi(z) = e^{[\delta(r)-\delta]z} \prod_{|a_k|<r} \left(1 - \frac{z}{a_k}\right) \prod_{|a_k|\geqslant r} \left(1 - \frac{z}{a_k}\right) e^{\frac{z}{a_k}} \quad (|z| = r).$$

Since $\lim_{r\to\infty} |\delta(r) - \delta| = 0$, it follows from Lemma 5 of Chapter II that the indicator diagram of the function $f(z)$ can be obtained from the indicator diagram of $\varphi(z)$ by a translation given by the vector $\bar{a}$. Since this displacement does not move the diagram from the imaginary axis, Re $\bar{a} = 0$.

5. Indicator Diagram of an Entire Function of Exponential Type, and of Class A

Let us examine in more detail the indicator diagram of a function of exponential type and of class A. It follows readily from Lemma 2 of this chapter that *the indicator diagram of an entire function $f(z)$ of exponential type and of class A has an axis of symmetry parallel to the real axis.*[21] Indeed, if

$$f(z) = z^p e^{az+b} \prod_{k=1}^{\infty} \left(1 - \frac{z}{a_k}\right) e^{\frac{z}{a_k}}, \tag{5.57}$$

then

$$\frac{f(z)}{\bar{f}(z)} = e^{i\gamma z + i\beta} \prod_{k=1}^{\infty} \left(1 - \frac{z}{a_k}\right)\left(1 - \frac{z}{\bar{a}_k}\right)^{-1}, \tag{5.58}$$

where $\gamma = 2 \operatorname{Im} a + 2 \sum_{k=1}^{\infty} \operatorname{Im} 1/a_k$, $\beta = 2 \operatorname{Im} b$, and from Lemma 5 it is easy to deduce that

$$\operatorname{lim}^*_{r\to\infty} \frac{\ln|f(re^{i\theta})| - \ln|\bar{f}(re^{i\theta})|}{r} = -\gamma \sin\theta. \tag{5.59}$$

Therefore,

$$h_f(\theta) = h_f(-\theta) - \gamma \sin\theta, \tag{5.60}$$

namely, the indicator diagram has the axis of symmetry $y = -\gamma/2$.

[21] This fact was first pointed out by M. G. Krein.

Suppose that $\alpha_k = \operatorname{Re} a_k$. We shall show that the indicator diagram of the function $f(z)$ and the function

$$f_1(z) = \prod_{k=1}^{\infty} \left(1 - \frac{z}{\alpha_k}\right) e^{\frac{z}{\alpha_k}} \tag{5.61}$$

can be obtained from each other by means of a parallel shift along some vector. For this we note that if we set

$$\alpha = \sum_{k=1}^{\infty} \left(\frac{1}{a_k} - \frac{1}{\alpha_k}\right) \quad \text{and} \quad \chi_1(z) = \prod_{k=1}^{\infty} \left(1 - \frac{z}{a_k}\right)\left(1 - \frac{z}{\alpha_k}\right)^{-1},$$

then

$$\frac{f(z)}{f_1(z)} = e^{\alpha z + \beta} \chi_1(z).$$

By an unessential variation of the argument used in the proof of Lemma 5 of this chapter one can show that[22]

$$\lim_{r \to \infty}{}^{*} \frac{\ln |\chi_1(re^{i\theta})|}{r} = 0, \tag{5.62}$$

and hence that

$$h_f(\theta) = h_{f_1}(\theta) + |\alpha| \cos(\theta + \theta_0) \quad (\theta_0 = \arg \alpha). \tag{5.63}$$

If the entire function of exponential type and of class A has completely regular growth then, as was shown, the indicator diagram is an interval parallel to the imaginary axis. One might conjecture that this is a property of all entire functions of exponential type and of class A. We shall show that this is not so by constructing the following example.

Example. Let us select a sequence of positive numbers $\{b_n\}$ $(b_1 < b_2 < b_3 < \ldots)$ of density Δ, that is,

$$\lim_{n \to \infty} \frac{n}{b_n} = \Delta,$$

and a sequence of real numbers $\{c_n\}$ $(|c_1| \leqslant |c_2| < |c_3| < \ldots)$ of zero density and such that the series

$$\sum_{1}^{\infty} \frac{1}{|c_n|}$$

diverges. (For example, one can take $b_n = n$ and $c_n = \pm n \ln n$.) We then select the signs of the c_n so that the limits of indeterminacy of the quantity

$$\delta_c(r) = \sum_{|c_n| < r} \frac{1}{c_n}$$

are equal to the numbers a and b $(a < b)$.

---

[22] Noble [1].

Let us rearrange the sets of numbers $\{\pm b_n\}$ and $\{c_n\}$ in the order of ascending moduli and denote them by $\{a_n\}$. The set of points $\{a_n\}$ has density Δ on the positive and on the negative half of the real axis. Furthermore, for the quantity

$$\delta(r) = \sum_{|a_n|<r} \frac{1}{a_n}$$

the following equations are valid:

$$\underline{\lim}_{r\to\infty} \delta(r) = a, \qquad \overline{\lim}_{r\to\infty} \delta(r) = b. \tag{5.64}$$

The canonical product

$$f(z) = \prod_{n=1}^{\infty} \left(1 - \frac{z}{a_n}\right) e^{\frac{z}{a_n}}$$

is an entire function of class A. By the theorem of Lindelöf (Chapter I, § 14) it is of exponential type. Let us express $f(z)$ in the form

$$f(z) = e^{\delta(r)z} f_r(z),$$

where

$$f_r(z) = \prod_{|a_n|<r} \left(1 - \frac{z}{a_n}\right) \prod_{|a_n|\geqslant r} \left(1 - \frac{z}{a_n}\right) e^{\frac{z}{a_n}}.$$

We find that

$$\frac{\ln|f(re^{i\theta})|}{r} = \delta(r)\cos\theta + \frac{\ln|f_r(re^{i\theta})|}{r}.$$

From Lemma 5, Chapter II, with $\theta \neq 0, \pi$, we deduce

$$\lim_{r\to\infty} \frac{\ln|f_r(re^{i\theta})|}{r} = \pi\Delta|\sin\theta|.$$

Bearing in mind (5.64), we obtain

$$h_f(\theta) = \pi\Delta|\sin\theta| + b\cos\theta \qquad \left(|\theta| \leqslant \frac{\pi}{2}\right),$$

$$h_f(\theta) = \pi\Delta|\sin\theta| + a\cos\theta \qquad \left(|\theta - \pi| \leqslant \frac{\pi}{2}\right).$$

Thus, the indicator diagram of this function is the rectangle $a \leqslant x \leqslant b$, $-\Delta \leqslant y \leqslant \Delta$. If we change the signs at all the points $\{c_n\}$, we obtain a function $f_1(z)$ which has for its indicator diagram the rectangle $-b \leqslant x \leqslant -a$, $-\Delta \leqslant y \leqslant \Delta$. The product $f_1(z)f(z)$ is a function of completely regular growth with the indicator $h(\theta) = 2\pi\Delta|\sin\theta|$.

We have thus constructed an example of functions whose indicators are not added when the functions are multiplied.

In this example the indicator diagram of the function $f(z)$ was broadened in the direction of the real axis because of the oscillation of the quantity $\delta(r)$.

One can, however, construct an even function of exponential type (for which $\delta(r) = 0$) with real zeros which has an indicator diagram that is not an interval. The set of zeros of such a function obviously can not have a density, since otherwise it would be of regular growth and its indicator diagram would be an interval. Hence, the indicator diagram is broadened when the distribution of the zeros ceases to be regular. However, in this case the indicator diagram can not be arbitrarily wide in the direction of the real axis. The exact result is given in the following theorem.

THEOREM 12. *The indicator diagram of an even entire function $f(z)$ of exponential type and of class A is situated, except for a parallel shift, within a convex region K with the supporting function*

$$k(\theta) = \begin{cases} \Delta \cos\theta \cdot \ln(1 + 2\cos\theta\sqrt{2\cos 2\theta} + 2\cos 2\theta) \\ +\Delta \sin\theta\left[\operatorname{arctg}\dfrac{1-\cos\theta\sqrt{2\cos 2\theta}}{\sin\theta\sqrt{2\cos 2\theta}} + \operatorname{arctg}\dfrac{1+\cos\theta\sqrt{2\cos 2\theta}}{\sin\theta\sqrt{2\cos 2\theta}}\right] \\ \qquad \text{when } |\theta| < \dfrac{\pi}{4} \text{ and } |\theta - \pi| < \dfrac{\pi}{4}, \\ \pi\Delta\,|\sin\theta| \quad \text{when } \left|\theta \pm \dfrac{\pi}{2}\right| \leqslant \dfrac{\pi}{4}. \end{cases} \tag{5.65}$$

Here

$$\Delta = \overline{\lim_{r\to\infty}}\, \frac{n_f(r)}{r}. \tag{5.66}$$

The region K is symmetric with respect to the y-axis (Fig. 8).

The inequality $h_f(\theta) \leqslant k(\theta)$ is sharp since there exists an even entire function $f(z)$ of finite degree and of class A whose indicator diagram coincides with K.

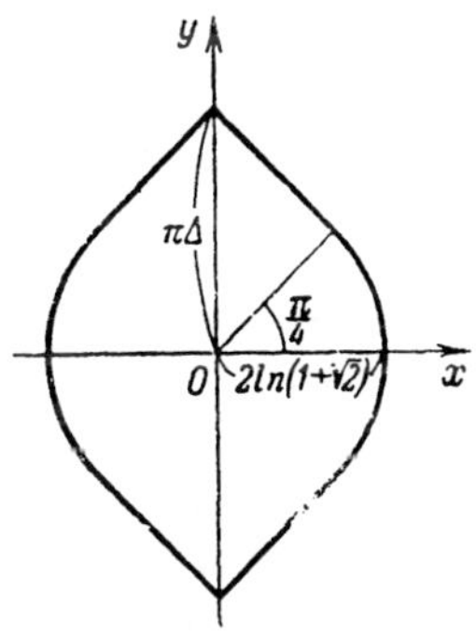

Fig. 8

PROOF. From (5.63) it follows that it is sufficient to consider only even entire functions with real zeros, i.e.,

$$f(z) = \prod_{n=1}^{\infty}\left(1 - \frac{z^2}{a_n^2}\right) \qquad (\operatorname{Im} a_n = 0). \tag{5.67}$$

We have

$$\ln|f(re^{i\theta})| = \int_0^\infty \ln\left|1 - \frac{r^2e^{2i\theta}}{t^2}\right| dn(t).$$

The integrand $\ln|1 - r^2e^{2i\theta}/t^2|$ takes on negative values when $t > r(2\cos 2\theta)^{-1/2}$ ($|\theta| < \pi/4$). Therefore

$$\ln|f(re^{i\theta})| < \int_0^{r(2\cos 2\theta)^{-\frac{1}{2}}} \ln\left|1 - \frac{r^2}{t^2}e^{2i\theta}\right| dn(t) \qquad \left(|\theta| \leqslant \frac{\pi}{4}\right).$$

Integration by parts yields

$$\ln|f(re^{i\theta})| < -\int_0^{r(2\cos\theta)^{-\frac{1}{2}}} \frac{d}{dt}\left\{\ln\left|1 - \frac{r^2}{t^2}e^{2i\theta}\right|\right\} n(t)\,dt$$

$$\left(|\theta| < \frac{\pi}{4}\right).$$

The integrand is negative on the entire interval of integration. Furthermore, $n(t) < (\Delta + \varepsilon)t$ asymptotically. Thus we have the asymptotic inequality

$$\ln|f(re^{i\theta})| < (\Delta + \varepsilon)\int_0^{r(2\cos 2\theta)^{-\frac{1}{2}}} \frac{d}{dt}\left\{-\ln\left|1 - \frac{r^2}{t^2}e^{2i\theta}\right|\right\} t\,dt$$

$$\left(|\theta| < \frac{\pi}{4}\right)$$

and hence

$$\ln|f(re^{i\theta})| < (\Delta + \varepsilon)\,\mathrm{Re}\int_0^{r(2\cos 2\theta)^{-\frac{1}{2}}} \ln\left(1 - \frac{r^2}{t^2}e^{2i\theta}\right) dt$$

$$\left(|\theta| < \frac{\pi}{4}\right).$$

Finally, setting $t = \tau r$, we obtain

$$\frac{\ln|f(re^{i\theta})|}{r} < (\Delta + \varepsilon)\,\mathrm{Re}\int_0^{(2\cos 2\theta)^{-\frac{1}{2}}} \ln\left(1 - \frac{e^{2i\theta}}{\tau^2}\right) d\tau$$

$$\left(|\theta| \leqslant \frac{\pi}{4}\right). \tag{5.68}$$

When $|\theta \pm \pi/2| \leqslant \pi/4$, the function

$$\ln\left|1 - \frac{r^2e^{2i\theta}}{t^2}\right|$$

is positive, and it is monotone decreasing when t is positive. Therefore, if in the preceding integrals we replace the upper limit by infinity, we obtain the asymptotic inequality

$$\frac{\ln|f(re^{i\theta})|}{r} < (\Delta+\varepsilon)\,\mathrm{Re}\int_0^\infty \ln\left(1-\frac{e^{2i\theta}}{\tau^2}\right)d\tau$$

$$\left(\left|\theta\pm\frac{\pi}{2}\right|\leqslant\frac{\pi}{4}\right). \tag{5.69}$$

Evaluating the integrals (5.68) and (5.69), and making use of the arbitrariness of ε, we deduce that $h_f(\theta)\leqslant k(\theta)$. This proves the first part of the theorem.

Let us now construct an example of an even entire function of exponential type with real zeros and with an indicator equal to $k(\theta)$. For this we select a sequence of positive numbers $p_1 < p_2 < p_3 < \ldots$ such that

$$\lim_{n\to\infty}\frac{p_{n+1}}{p_n}=\infty,$$

and construct the infinite product

$$\varphi(z)=\prod_{k=1}^{\infty}\left(1-\frac{z^2}{a_k^2}\right),$$

where the a_k are integers which lie in the intervals (p_{2q-1}, p_{2q}). It is obvious that

$$\overline{\lim_{t\to\infty}}\frac{n(t)}{t}=1.$$

For the determination of the indicator of the function $f(z)$ when $|\theta| < \pi/4$, we select a sequence of values r_q,

$$r_q=p_{2q-1}(2\cos 2\theta)^{-\frac{1}{2}}.$$

It is easy to see that[23] as $q\to\infty$

$$\frac{1}{r_q}\int_{p_{2q}}^{\infty}\ln\left|1-\frac{r_q^2e^{2i\theta}}{t^2}\right|dn(t)\to 0$$

and

$$\frac{1}{r_q}\int_0^{p_{2q-1}}\ln\left|1-\frac{r_q^2e^{2i\theta}}{t^2}\right|dn(t)\to k(\theta)\qquad\left(|\theta|<\frac{\pi}{4}\right).$$

[23] This is easily established if one integrates by parts and replaces t by τp_{2q-1}.

When $\pi/4 \leqslant |\theta| \leqslant \pi/2$, one must take $r_q = (p_{2q} + p_{2q-1})/2$. Then it is easily seen that

$$\frac{\ln|\varphi(r_q e^{i\theta})|}{r_q} \geqslant \frac{1}{r_q} \int_{p_{2q-1}+1}^{p_{2q}} \ln\left|1 - \frac{r_q^2 e^{2i\theta}}{t^2}\right| dn(t) \approx \int_{\frac{p_{2q-1}+1}{r_q}}^{\frac{p_{2q}}{r_q}} \ln\left|1 - \frac{e^{2i\theta}}{u^2}\right| du,$$

or

$$\lim_{q\to\infty} \frac{\ln|\varphi(r_q e^{i\theta})|}{r_q} \geqslant \int_0^\infty \ln\left|1 - \frac{e^{2i\theta}}{u^2}\right| du = \pi|\sin\theta|.$$

Combining this with the inequality $h_\varphi(\theta) \leqslant k(\theta)$, we obtain

$$h_\varphi(\theta) = k(\theta) \qquad (0 \leqslant \theta < 2\pi).$$

REMARK. The function $\varphi_1(z)$ constructed in the same manner as $\varphi(z)$ with zeros in the complementary intervals (p_{2q}, p_{2q+1}) will obviously have the same indicator diagram as $\varphi(z)$. The product of these functions is $\sin \pi z$ and its indicator diagram is an interval (a straight line segment). Thus we have an example of even functions of exponential type whose indicators are not added when the functions are multiplied.

6. Theorem of M. G. Kreĭn on the Expansion of the Reciprocal of an Entire Function

An interesting subclass of the set of entire functions was introduced by M. G. Kreĭn [2; 3] in connection with his development of the theory of entire operators in a Hilbert space. We are speaking here of entire functions $f(z)$ which admit an expansion of the type

$$\frac{1}{f(z)} = R(z) + \sum_{k=1}^{\infty} \frac{1}{f'(a_k)}\left[\frac{1}{z-a_k} + \frac{1}{a_k} + \frac{z}{a_k^2} + \dots + \frac{z^{p-1}}{a_k^p}\right], \tag{5.70}$$

where $R(z)$ is a polynomial and

$$\sum_{k=0}^{\infty} \frac{1}{|a_k|^{p+1}|f'(a_k)|} < \infty. \tag{5.71}$$

The fundamental theorem of M. G. Kreĭn is as follows.

THEOREM 13. *If an entire function $f(z)$ of class A admits an expansion of the type* (5.70), *then*

(1) $f(z)$ *is a function of exponential type and*

(2) $\int_{-\infty}^{\infty} \frac{\ln_+ |f(t)|}{1+t^2}\, dt < \infty$.

Before proving the theorem, we note that (1) and (2) imply that $f(z)$ is a function of completely regular growth and that its indicator diagram is an interval.[24]

PROOF. We note that without loss of generality we can assume that $R(z) \equiv 0$ and $p = 0$. Indeed, it is easily seen that if the function $f^{-1}(z)$ is expressed in the form (5.70) then, multiplying it by a polynomial $P(z)$ of degree $q > p$, and q also greater than the degree of $R(z)$, and such that its zeros are distinct from those of $f(z)$, we obtain $f_1(z) = P_q(z)f(z)$, which admits the expansion

$$f_1^{-1}(z) = \sum_{k=1}^{\infty} \frac{c_k}{z - a_k}, \tag{5.72}$$

where the a_k are all the zeros of $f_1(z)$, and where $\Sigma_{k=1}^{\infty} |c_k| < \infty$.

Let us next find an estimate of the function $f(z)$ which admits the representation (5.72) in the half-plane $\operatorname{Im} z \geqslant \eta > 0$. For this purpose we circumscribe around each a_k as center a circle of radius $d_k = (1/16A)\, |\operatorname{Im} 1/a_k|\, |\operatorname{Im} a_k|$, where $A = \Sigma_{k=1}^{\infty} |\operatorname{Im} 1/a_k|$. Let us denote by $C(x, h/2)$ the circle $|z - x - ih| < h/2$ when x and h are fixed, and by C the set of all exceptional circles.

The sum of the radii of the circles corresponding to the zeros that lie in the circle $C(x, h/2)$ satisfies the inequality

$$\sum_{|a_k - x - ih| < \frac{h}{2}} d_k < \frac{1}{16A}\, \frac{3}{2}\, h \sum_{k=1}^{\infty} \left| \operatorname{Im} \frac{1}{a_k} \right| < \frac{1}{4} h. \tag{5.73}$$

From this it follows that if the point $x + ih$ lies in a region covered by the exceptional circles, then the entire region lies in the circle $C(x, h/2)$. For every point z of the circle $C(x, h/4)$ that does not lie in the exceptional circle we have the inequality

$$|f^{-1}(z)| \leqslant \left| \sum_{|a_k - x - ih| > \frac{h}{2}} \frac{c_k}{z - a_k} \right| + \left| \sum_{|a_k - x - ih| \leqslant \frac{h}{2}} \frac{c_k}{z - a_k} \right|$$

$$\leqslant \frac{4}{h} \sum_{k=1}^{\infty} |c_k| + \sum_{|a_k - x - ih| < \frac{h}{2}} |c_k|\, d_k^{-1}$$

[24] This theorem was used by M. G. Kreĭn [4] in his study of the spectrum of the Sturm-Liouville boundary value problem. Further generalizations of this theorem are given in Ostrovskiĭ [1–3], Macaev [1] and Sergienko [1; 2]. See also Gol'dberg and Ostrovskiĭ [1], Chapter VI.

If one now takes into account the inequality $d_k^{-1} \leqslant 8A\,|a_k|^2\,4/h^2$, then one sees that for every $z \bar{\in} C$ the following inequality holds:

$$|f^{-1}(z)| \leqslant \frac{g}{h^2}(|z|^2+1) \qquad \left(\frac{h}{2} < \operatorname{Im} z < \frac{3h}{2}\right), \tag{5.74}$$

where g is a constant that depends on $f(z)$.

Let us define the function $\chi(z)$ by the equation

$$\chi(z) = \prod_{k=1}^{\infty}\left(1-\frac{z}{a_k}\right)\left(1-\frac{z}{\bar{a}_k}\right)^{-1} \qquad (\operatorname{Im} a_k > 0).$$

In the upper half-plane $|\chi(z)| \leqslant 1$, and from (5.74) one can obtain, by the principle of the maximum, the inequality

$$|f^{-1}(z)\chi(z)| < \frac{g}{|\operatorname{Im} z|^n}(|z|^2+1) \qquad (\operatorname{Im} z > 0). \tag{5.75}$$

From this it follows that if $\varphi(z) = f^{-1}(z)\chi(z)$, then the function $\ln|\varphi(z)|$ can be represented by a Poisson integral (see Theorem 4) in the half-plane $\operatorname{Im} z \geqslant \eta > 0$; namely,

$$\ln|\varphi(z+i\eta)| = \frac{1}{\pi}\int_{-\infty}^{\infty}\frac{(y-\eta)\ln|\varphi(t+i\eta)|}{(t-x)^2+(y-\eta)^2}\,dt + k\,(y-\eta), \tag{5.76}$$

where

$$k = \varlimsup_{y\to+\infty}\frac{\ln|\varphi(iy)|}{y}.$$

By (5.75) we find that

$$\int_{-\infty}^{\infty}\frac{\ln_+|\varphi(t+i\eta)|}{1+t^2}\,dt < 2\pi g\ln\frac{1}{\eta} + g_1\ (g_1 \text{ is a constant}). \tag{5.77}$$

Substituting $z = x + i\eta$ into (5.76), we obtain

$$-\int_{-\infty}^{\infty}\frac{\ln|\varphi(t+i\eta)|}{1+t^2}\,dt = \pi k - \pi\ln|\varphi[(2\eta+1)i]|$$

or

$$-\int_{-\infty}^{\infty}\frac{\ln_+|\varphi(t+i\eta)|\,dt}{1+t^2} + \int_{-\infty}^{\infty}\frac{\ln_+|\varphi^{-1}(t+i\eta)|}{1+t^2}\,dt$$

$$= \pi k - \pi\ln|[\varphi(2\eta+1)i]|.$$

Combining this with (5.77), we find that for $0 < \eta \leqslant 2$

$$\int_{-\infty}^{\infty}\frac{\ln_+|f(t+i\eta)\chi^{-1}(t+i\eta)|\,dt}{1+t^2} < 2g\pi\ln\frac{1}{\eta} + g_2$$

(g_2 is a constant),

and bearing in mind that $|\chi^{-1}(t + i\eta)| \geqslant 1$ we see that

$$\int_{-\infty}^{\infty} \frac{\ln_+ |f(t+i\eta)|}{1+t^2}\, dt < 2g\pi \ln \frac{1}{\eta} + g_2. \tag{5.78}$$

An analogous inequality holds also when $-2 \leqslant \eta < 0$. Integrating the terms of the inequality (5.78) with respect to η from -2 to 2, we obtain

$$\int_{-2}^{2} \int_{-\infty}^{\infty} \frac{\ln_+ |f(t+i\eta)|}{1+t^2}\, dt\, d\eta < g_3 \quad (g_3 \text{ is a constant}). \tag{5.79}$$

Making use of the theorem of Jensen for the entire function $f(z)$, we obtain

$$\ln |f(z_0)| \leqslant \frac{1}{2\pi} \int_0^{2\pi} \ln |f(z_0 + re^{i\theta})|\, d\theta.$$

If we multiply each term of this inequality by r and integrate the result with respect to r from zero to one, we obtain

$$\ln |f(z_0)| \leqslant \frac{1}{\pi} \int_0^{2\pi} \int_0^1 \ln |f(z_0 + re^{i\theta})|\, r\, dr\, d\theta.$$

Finally

$$\ln_+ |f(z_0)| \leqslant \frac{1}{\pi} \int_0^{2\pi} \int_0^1 \ln_+ |f(z_0 + re^{i\theta})|\, r\, dr\, d\theta$$

$$< \frac{1 + (|x_0| + 1)^2}{\pi} \int_{y_0-1}^{y_0+1} \int_{x_0-1}^{x_0+1} \frac{\ln_+ |f(x+iy)|}{1+x^2}\, dx\, dy,$$

and from (5.79), with $|y_0| \leqslant 1$, we obtain the estimate

$$\ln_+ |f(z_0)| < \frac{g_3}{\pi} [1 + (|x_0| + 1)^2]. \tag{5.80}$$

We shall prove that $f(z)$ is of exponential type.

The inequality (5.80) gives an estimate for the function $f(z)$ in the strip $|\text{Im } z| \leqslant 1$. In order to obtain an estimate of this function outside this strip, one can use the representation (5.76) by setting, for example, $\eta = \frac{1}{2}$. Noting that $|f(z)| \leqslant |\varphi(z)|^{-1}$ when $\text{Im } z > 0$, we can conclude, on the basis of the representation (5.76) and by Lemma 3, that when $r > r_\varepsilon$ and $|y| > \frac{1}{2}$ then

$$\ln |f(z)| < k|y| + \varepsilon \frac{r^2}{\left| |y| - \frac{1}{2} \right|}. \tag{5.81}$$

It now follows that for $y \geqslant 1$

$$\ln |f(z)| = o(r^2). \tag{5.82}$$

Combining this with the inequality (5.80) we see that (5.82) is valid in the whole plane. By applying the Phragmén-Lindelöf theorem in each of the angles $|\arg z| \leqslant \delta$ and $|\arg z - \pi| < \delta$, we establish the inequality

$$|f_1(z)| < M_1 e^{(k+\varepsilon)y},$$

where M_1 is a constant. Therefore $f(z)$ is of exponential type in the whole plane.

In order to obtain the second assertion of the theorem, we note that $\ln_+ |f(t+z)|$ is a subharmonic function[25] of z. From this it follows that the function

$$\Psi_R(z) = \int_{-R}^{R} \frac{\ln_+ |f(t+z)|}{t^2+1}\, dt$$

is also subharmonic. As a direct consequence of the definition of a subharmonic function we obtain the inequality

$$\Psi_R(0) \leqslant \frac{1}{2\pi} \int_0^{2\pi} \Psi_R(e^{i\theta})\, d\theta .$$

The estimate (5.78) now yields

$$\int_{-R}^{R} \frac{\ln_+ |f(t)|}{1+t^2}\, dt < 4g\pi \ln 2 + g_2,$$

from which it follows that

$$\int_{-\infty}^{\infty} \frac{\ln_+ |f(t)|}{1+t^2}\, dt < \infty .$$

This proves the theorem.

As was indicated earlier, conditions (1) and (2) imply that $f(z)$ is of completely regular growth and that its indicator diagram is an interval on the imaginary axis. But one can say more than this about the growth of functions of class A which admit the representation (5.72).

We shall confine ourselves to the case when all the a_k are real.

Indeed, the indicator of an entire function of exponential type

$$f(z)\, e^{ikz} \qquad \left(k = h_f\left(\frac{\pi}{2}\right)\right)$$

---

[25] A subharmonic function is a real function $u(z)$ which satisfies the inequality

$$u(z) \leqslant \frac{1}{2\pi} \int_0^{2\pi} u(z + Re^{i\theta})\, d\theta$$

for all z and for $R > 0$.

is zero when $0 \leqslant \theta \leqslant \pi$. From the complete regularity of the growth of the function $f(z)$ it follows that the indicator of a function

$$\varphi(z) = f^{-1}(z)\, e^{-ikz},$$

which is holomorphic and of exponential type in the half-plane $\operatorname{Im} z > 0$, is also zero. From the representation (5.72) we can deduce directly that

$$|\varphi(x+ih)| < \frac{e^{kh}}{h} \sum_{1}^{\infty} |c_n| = M \qquad (h > 0).$$

This implies, by the Phragmén-Lindelöf theorem for the half-plane, that if $\operatorname{Im} z \geqslant h > 0$, then

$$|\varphi(z)| < M$$

and hence[26]

$$|f(z)| > M e^{h_f\left(\frac{\pi}{2}\right) y}). \tag{5.83}$$

An analogous estimate applies in the lower half-plane.

It is interesting to note that if an entire function which is of exponential type and has its indicator diagram $[-i\mu, i\mu]$ on the imaginary axis satisfies the inequality

$$|f(z)| > c\,|z|\, e^{\mu |y|} \qquad (c > 0) \tag{5.84}$$

in a sequence of circumferences, then it admits an expansion of the type (5.72). This follows easily from Cauchy's method for the expansion of meromorphic functions.

Thus, for entire functions $f(z)$ with real zeros, the necessary condition (5.83) differs little from the sufficient condition (5.84) for the expansion of the function $1/f(z)$ into a series of the type (5.72).

[26] This inequality was established by M. G. Kreĭn [2].

CHAPTER VI

ROOTS OF EXPONENTIAL SUMS

This chapter consists of two parts. In the first four sections we consider the distribution of roots for holomorphic almost-periodic functions. In particular, in §§ 2 and 4 we study the distribution of roots for almost-periodic functions with semibounded and with bounded spectrum.

In the second part, consisting of §§ 5, 6 and 7, we investigate the distribution of roots for entire functions of exponential type, which are approximated in a certain sense by finite sums of the form

$$P(z) = \sum_{k=1}^{n} a_k e^{\lambda_k z} \tag{6.00}$$

with complex exponents.

The almost-periodic functions with bounded spectrum constitute a very special form of such entire functions. The investigation of the distribution of the roots for these functions is based on the results of the first four sections of this chapter and on the theorems and methods presented in the preceding chapters.

The fundamental result is the theorem according to which an entire function $f(z)$, represented by the absolutely convergent series

$$f(z) = \sum_{k=1}^{\infty} a_k e^{\lambda_k z} \qquad \left(\sum_{k=1}^{\infty} |a_k| < \infty\right), \tag{6.01}$$

in which the λ_k are points of the boundary of some bounded convex region $\bar{I}$, is a function of completely regular growth.

Thus, by Theorem 3 of Chapter III there is an asymptotic formula for the distribution of the roots of these functions.

Analogous theorems are proved for functions of more general form. Finite sums of the type (6.00) will be called *exponential polynomials*.

1. Information from the Theory of Almost-Periodic Functions

We shall need certain basic definitions and theorems from the theory of almost-periodic functions. We recall them for the reader's convenience.[1]

[1] A detailed exposition of the theory of almost-periodic functions is given in the book of B. M. Levitan [1]. See also H. Bohr [1].

A set $\{\tau\}$ of real numbers is said to be *relatively dense* if any interval whose length is not less than some positive number l contains at least one number of the set. An interval of length l is called an *enclosing interval* of the set.

The number τ is called an *ε-shift of the function* $f(x)$ if for all values of the variable x the inequality

$$|f(x+\tau)-f(x)|<\varepsilon. \tag{6.02}$$

is satisfied.

The function $f(x)$ is taken to be complex-valued but defined for real values of x.

The function $f(x)$ is said to be *almost-periodic* if it is continuous on the entire real axis, and if to every real number ε there corresponds a relatively dense set of ε-shifts. We shall enumerate a number of properties of almost-periodic functions.

(a) A necessary and sufficient condition for a function $f(x)$ to be almost-periodic is the compactness on the whole axis of the functions $f_h(x) = f(x + h)$, depending on the arbitrary real parameter h (S. Bochner).

A set is called compact if every infinite subset contains a uniformly convergent sequence.

(b) An almost-periodic function is bounded and uniformly continuous on the entire numerical axis. The sum and product and also the limit of a uniformly convergent sequence of almost-periodic functions is an almost-periodic function.

(c) A generalized trigonometric polynomial, i.e., a function of the form

$$P(x)=\sum_{k=0}^{n} a_k e^{i\lambda_k x} \tag{6.03}$$

(λ_k a real number),

is an almost-periodic function, since it is the sum of periodic functions. The uniform limit of such polynomials is also an almost-periodic function.

(d) For each almost-periodic function there exists a "mean," i.e.,

$$M\{f\}=\lim_{T\to\infty}\frac{1}{2T}\int_{-T+\alpha}^{T+\alpha} f(t)\,dt, \tag{6.04}$$

where the convergence is uniform relative to α.

The "Fourier coefficients"

$$a(\lambda)=M\{f(x)e^{-i\lambda x}\} \tag{6.05}$$

differ from zero only for a countable set of values λ. This countable set $\lambda_1, \lambda_2, \ldots, \lambda_n, \ldots$ is called the "spectrum of the function $f(x)$" or the set of its Fourier exponents, and the series

$$f(x)\sim\sum_{k=1}^{\infty} a_k e^{i\lambda_k x} \qquad (a_k=a(\lambda_k)) \tag{6.06}$$

is the Fourier series of the function $f(x)$.

Parseval's equation holds:

$$M\{|f|^2\} = \sum_{k=1}^{\infty} |a_k|^2 \qquad (6.07)$$

This is equivalent to the uniqueness theorem, i.e., the theorem according to which different almost-periodic functions have different Fourier series.

(e) There exists a sequence of multipliers $b_k^{(n)}$ $(k = 1, 2, \ldots, n, 0 \leqslant b_k^{(n)} < 1)$ and $b_k^{(n)} \to 1$ as $n \to \infty$ such that for any $\varepsilon > 0$ and $n > N(\varepsilon)$

$$\left| f(x) - \sum_{k=1}^{n} b_k^{(n)} a_k e^{i\lambda_k x} \right| < \varepsilon \qquad (-\infty < x < \infty). \qquad (6.08)$$

It follows from this fundamental theorem that the class of almost-periodic functions is the closure of the class of generalized trigonometric polynomials.

(f) If the sequence of generalized trigonometric polynomials

$$P_n(x) = \sum_{k=1}^{n} a_k^{(n)} e^{i\lambda_k x}$$

converges uniformly to $f(x)$, then $a_k^{(n)} \to a_k$, the corresponding Fourier coefficient of the function $f(x)$.

A function $f(z)$, holomorphic in a closed strip $\alpha \leqslant y \leqslant \beta$ $(y = \operatorname{Im} z)$ is said to be almost-periodic in that strip if to each positive number ε there corresponds a relatively dense set of ε-shifts of the function $f(z)$, i.e.,

$$|f(z+\tau) - f(z)| < \varepsilon \quad \text{for} \quad \alpha \leqslant y \leqslant \beta. \qquad (6.09)$$

The function $f(z)$ is said to be almost-periodic in the open strip $-\infty \leqslant \alpha < y < \beta \leqslant \infty$ if it is almost-periodic in every closed strip lying in the strip $\alpha < y < \beta$. For almost-periodic functions in a strip we have the theorem of S. Bochner, i.e.,

(a′) the almost periodicity of the holomorphic function $f(z)$ in the strip $\alpha \leqslant y \leqslant \beta$ means that the set of functions $f(z + h)$ for real h is compact in the sense of uniform convergence in that strip.

Moreover, $f(z)$ may be uniformly approximated in the strip by exponential polynomials.

THEOREM 1. *Suppose that $f(x)$ is an almost-periodic function defined on the real axis. For there to exist a holomorphic function of exponential type Δ in the half-plane* $\operatorname{Im} z > 0$ *which on the real axis has limit values coinciding with the values of $f(x)$, it is necessary and sufficient that the greatest lower bound of the spectrum of $f(x)$ be equal to $-\Delta$. In addition $f(z)$ will be almost-periodic in the half-plane* $\operatorname{Im} z \geqslant 0$.

PROOF. Suppose that $f(x)$ is an almost-periodic function and $-\mu = \inf \lambda_k$. Because of Theorem (e) there exists a sequence of polynomials

$$P_n(x) = \sum_{k=1}^{n} a_k^{(n)} e^{i\lambda_k x}$$

which converge uniformly to the function $f(x)$ on the real axis, while the numbers λ_k are the Fourier exponents of the function $f(x)$. Since $|e^{i\lambda_k z}| = e^{-\lambda_k y} \leqslant e^{\mu y}$ ($y = \operatorname{Im} z$), the polynomials $e^{i\mu z}P_n(z)$ are bounded in the half-plane $\operatorname{Im} z > 0$.

By the Phragmén-Lindelöf principle, the moduli of these polynomials are bounded in the half-plane $\operatorname{Im} z \geqslant 0$ by the same number as on the real axis. Thus from the inequality

$$|P_n(x) - P_m(x)| < \varepsilon \qquad (-\infty < x < \infty)$$

there follows the inequality

$$|e^{i\mu z}P_n(z) - e^{i\mu z}P_m(z)| < \varepsilon \quad \text{for } \operatorname{Im} z \geqslant 0,$$

and the sequence of polynomials $e^{i\mu z}P_n(z)$ converges uniformly in the half-plane $\operatorname{Im} z \geqslant 0$ to a bounded function $e^{i\mu z}f(z)$. The function $f(z)$ is holomorphic for $\operatorname{Im} z > 0$ and

$$|f(z)| \leqslant Me^{\mu y}$$

$$\left(M = \sup_{-\infty < x < \infty} |f(x)|, \quad y = \operatorname{Im} z \geqslant 0\right), \tag{6.10}$$

i.e., $f(z)$ is of exponential type $\Delta \leqslant \mu$.

Now suppose we are given that the almost-periodic function $f(x)$ is a limiting function for the holomorphic function $f(z)$ of exponential type Δ in the half-plane $\operatorname{Im} z > 0$. This function is bounded on the real axis and accordingly, by the Phragmén-Lindelöf principle,

$$|f(z)| \leqslant Me^{\Delta y}. \tag{6.11}$$

By Cauchy's theorem, the integral around the rectangular contour $(-T, T, T + iy, -T + iy, -T)$ of the holomorphic function $f(z)e^{i\lambda z}$ is equal to zero, i.e.,

$$\frac{1}{2T}\int_{-T}^{T} f(x)e^{-i\lambda x}dx - \frac{1}{2T}\int_{-T}^{T} f(x+iy)\, e^{-i\lambda(x+iy)}\, dx$$

$$+\frac{1}{2T}\int_{0}^{y} f(T+it)\, e^{-i\lambda(T+it)}\, dt$$

$$-\frac{1}{2T}\int_{0}^{y} f(-T+it)e^{-i\lambda(-T+it)}\, dt = 0. \tag{6.12}$$

From inequality (6.11) it follows that as $T \to \infty$ the last two terms converge to zero and accordingly

$$a(\lambda) = M\{f(x)\, e^{-i\lambda x}\} = M\{f(x+iy)\, e^{-i\lambda(x+iy)}\}.$$

Hence, from inequality (6.11), it follows that

$$|a(\lambda)| \leqslant Me^{(\Delta+\lambda)y}.$$

If $\lambda < -\Delta$, then as $y \to +\infty$ we obtain $a(\lambda) = 0$. Thus the spectrum of $f(x)$ is bounded below and $-\mu = \inf \lambda_k \geqslant -\Delta$. The theorem is proved.

COROLLARY. *For the almost-periodic function $f(z)$ to be an entire function of exponential type it is necessary and sufficient that its spectrum be bounded.*

We observe further that, in addition, the smallest segment of the imaginary axis containing the whole spectrum of the function $f(x)$ coincides with its conjugate diagram.

2. Roots of an Almost-Periodic Function with a Bounded Spectrum

If in the sum (6.01) all the exponents λ_k lie on the imaginary axis, so that the convex region I is a segment of the imaginary axis, then it follows from the corollary to Theorem 1 that $f(z)$ is an entire almost-periodic function of exponential type. Thus in this case our problem comes down to that of the distribution of the roots of an entire almost-periodic function. The corresponding theorems will be presented in this and the following two sections. These theorems are needed not only when the convex region I is a segment of the imaginary axis, but also in the more general case when the boundary of I contains a rectilinear segment. We begin by proving some lemmas.

LEMMA 1. *Suppose that $f(z)$ is a holomorphic almost-periodic function in the strip $|y| < h$ not identically equal to zero. Then to any number $\delta > 0$ there corresponds a number $m(\delta) > 0$ such that*

$$|f(z)| \geqslant m(\delta)$$

for z lying in the strip $|y| < h - \delta$ and outside circles of radius δ with centers at the roots of the function $f(z)$.

PROOF. Suppose that the greatest lower bound of the values of the function $|f(z)|$ in the given region is equal to zero, i.e., that for some sequence $\{z_j\}$ $(j = 1, 2, \ldots)$, taken from this region, $\lim_{j\to\infty} f(z_j) = 0$.

By Bochner's theorem (a′) one may select from the sequence of functions $\varphi_j(z) = f(z + x_j)$ $(z_j = x_j + iy_j)$ a sequence $\varphi_{n_j}(z) = f(z + x_{n_j})$, which converges uniformly in the strip $|y| \leqslant h - \delta$ to some function $f_0(z)$. Suppose that $z = iy_0$ is a condensation point for the points iy_{n_j}. From the condition

$$\lim_{j\to\infty} \varphi_{n_j}(iy_{n_j}) = \lim_{j\to\infty} f(x_{n_j} + iy_{n_j}) = 0$$

it follows that $f_0(iy_0) = 0$. From the well-known Hurwitz theorem on the roots of uniformly convergent sequences of holomorphic functions, all the terms of the sequence $\varphi_{n_j}(z)$ from a certain point onward must have roots in a δ-neighborhood of the point iy_0. In other words, the function $f(z)$ has an infinite set of roots in the circles $|z - z_{n_j}| < \delta$. But that is impossible, since the points

z_{n_j} lie at a distance larger than δ from the set of roots of the function $f(z)$. The lemma is proved.

LEMMA 2. *The number of roots of a holomorphic and almost-periodic function in the strip $|y| < h$ occurring in the rectangle $t < x \leqslant t + 1$, $|y| \leqslant h - \delta$, is bounded by some number $k(\delta)$ not depending on t.*

PROOF. Suppose that there exists an infinite sequence of numbers t_j such that the number of roots of the function $f(z)$ in the corresponding rectangles increases without limit. We select from the sequence of functions $f(z + t_j)$ a uniformly convergent subsequence. The limit function $f_0(z)$ must have an infinite set of roots in the rectangle $0 \leqslant x \leqslant 1$, $|y| \leqslant h - \delta$, and thus it is identically equal to zero. But from the uniform convergence of the sequence of functions $f(z + t_{n_j})$ to $f_0(z)$ it follows that

$$\sup_{-\infty<x<\infty} |f_0(z)| = \sup_{-\infty<x<\infty} |f(z)| \qquad (|y| \leqslant h - \delta),$$

i.e., $f(z) \equiv 0$. The lemma is proved.

LEMMA 3. *If the spectrum of an almost-periodic function $f(z)$ is bounded below and the greatest lower bound $-\Delta$ enters into the spectrum, then*

$$\lim_{y\to+\infty} e^{i\Delta(x+iy)} f(x+iy) = a(-\Delta),$$

while the convergence to the limit is uniform relative to x.

PROOF. Suppose that

$$P_n(x) = \sum_1^n a^{(n)}(\lambda_k) e^{i\lambda_k x}$$

is a sequence of almost-periodic polynomials, uniformly converging on the whole real axis to the function $f(x)$. In the proof of Theorem 1 it was shown that the sequence of almost-periodic polynomials

$$Q_n(z) = e^{i\Delta z} P_n(z)$$

converges uniformly in the half-plane $\operatorname{Im} z \geqslant 0$ to the function

$$e^{i\Delta z} f(z).$$

As $y \to +\infty$ the almost-periodic polynomial $Q_n(z)$ converges uniformly to its Fourier coefficient $a^{(n)}(-\Delta)$. On the other hand,

$$\lim_{n\to\infty} a^{(n)}(-\Delta) = a(-\Delta),$$

where $a(-\Delta)$ is the Fourier coefficient of the function $f(x)$.

Choosing the number n so that

$$|a^{(n)}(-\Delta) - a(-\Delta)| < \frac{\varepsilon}{3}$$

and

$$|f(x) - P_n(x)| < \frac{\varepsilon}{3},$$

we find that for all sufficiently large values of y

$$|e^{i\Delta z} f(z) - a(-\Delta)| \leqslant |e^{i\Delta z}[f(z) - P_n(z)]| + |e^{i\Delta z} P_n(z) - a^{(n)}(-\Delta)| + |a^{(n)}(-\Delta) - a(-\Delta)| < \varepsilon.$$

Thus, as $y \to +\infty$ the function

$$f(z)\, e^{i\Delta z}$$

uniformly converges to $a(-\Delta)$.

THEOREM 2 (H. BOHR). *In order that all the roots $\{a_k\}$ of an almost-periodic function of exponential type Δ in the upper half-plane* ($\operatorname{Im} z > 0$) *be situated below some line, parallel to the real axis* ($\operatorname{Im} a_k < h$), *it is necessary and sufficient that the greatest lower bound $-\Delta$ of the spectrum enter into the spectrum.*

PROOF. If $-\Delta$ enters into the spectrum, then $a(-\Delta) \neq 0$, and by Lemma 3 the function $f(z)e^{i\Delta z}$ does not vanish for sufficiently large y ($y \geqslant h$).

On the other hand, if all the roots of the function $f(z)$ lie below the line $y = h$, then the function $\ln |f(z)e^{i\Delta z}|$ is harmonic and is bounded above in the half-plane $y \geqslant h_1$ ($h_1 = h + \delta$, $\delta > 0$) and accordingly, from Theorem 4 of Chapter V, it may be represented in the form

$$\ln |f(z)\, e^{i\Delta z}| = \frac{y - h_1}{\pi} \int_{-\infty}^{\infty} \ln [\,|f(t + ih_1)|\, e^{-\Delta h_1}]\, \frac{dt}{(t - x)^2 + (y - h_1)^2}. \qquad (6.13)$$

In this connection we make use of the fact that, from Theorem 1 of this chapter,

$$\overline{\lim_{y \to +\infty}} \frac{\ln |f(iy)|}{y} = \Delta.$$

By Lemma 1 there exists a number $m > 0$ such that $\ln |f(t + ih_1)| \geqslant \ln m$ for $-\infty < t < \infty$, and from the equation (6.13) we find that

$$|f(z)\, e^{i\Delta z}| > m e^{-\Delta h_1} \quad \text{for} \quad y > h_1.$$

Thus, $\lim_{y \to \infty} f(z)e^{i\Delta z} \neq 0$, so that the number $-\Delta$ enters into the spectrum.

COROLLARY 1. *If the lower bound of the spectrum of an almost-periodic function $f(z)$ is negative and if it does not enter into the spectrum of the function, then $f(z)$ takes on every value in the half-plane $y > h > 0$.*

In fact, for any complex number A the greatest lower bound of the spectrum of the function $f(z) - A$ (the same as for $f(z)$), does not enter into the spectrum, and accordingly this function has a zero with arbitrarily large imaginary part.

Corollary 1 is a well-known theorem of Bohr.

COROLLARY 2. *In order that all the roots of an entire almost-periodic function of exponential type be situated in some strip parallel to the real axis, it is necessary and sufficient that the upper and lower bounds of the spectrum (the "ends") enter into the spectrum.*

We shall call such functions *functions of class* $[\Delta]$ if the width of the indicator diagram is equal to 2Δ. A function of class $[\Delta]$ is, up to a multiplier of the form $e^{i\lambda z}$, a function of degree Δ.

LEMMA 4. *If the lower bound* $-\Delta$ *of the spectrum of the positive function* $f(z)$ *enters into that spectrum, then there exists a positive real number* H *such that if* $y > H$ *the function* $\arg f(x + iy)$, *considered as a function of* x, *may be represented in the form*

$$\arg f(x+iy) = -\Delta x + \varphi_y(x), \tag{6.14}$$

where $\varphi_y(x)$ *is an almost-periodic function.*

Here we have, of course, selected out of the infinite set of values of $\arg f(x + iy)$ some continuous branch. Such a selection is possible since, for $y > H, f(x + iy) \neq 0$.

PROOF. Represent the function $f(z)$ in the form

$$f(z) = a(-\Delta)e^{-i\Delta z}[1 + \zeta(z)].$$

This is possible, since by definition $a(-\Delta) \neq 0$. The Fourier exponents of the function $\zeta(z)$ are all positive and accordingly, for sufficiently large $H > 0$ and $y > H$, will satisfy the inequality $|\zeta(x + iy)| < 1$.

Taking a univalent branch of the function $\ln [1 + \zeta(z)]$, for example putting $|\arg [1 + \zeta(z)]| < \pi$ for $|\zeta(z)| < 1$, we find that

$$\theta_y(x) = \ln [1 + \zeta(x+iy)]$$

is a continuous function of the almost-periodic function $1 + \zeta(x + iy)$ and, as follows directly from the definition, is itself almost-periodic. Thus we have

$$\arg f(x+iy) = -\Delta x + \operatorname{Im} [\theta_y(x) + \ln a(-\Delta)],$$

while $\varphi_y(x) = \operatorname{Im} [\theta_y(x) + \ln a(-\Delta)]$ is an almost-periodic function.

THEOREM 3. *The roots of a function* $f(z)$ *of class* $[\Delta]$ *lie in the strip* $|y| \leqslant h$, *and if* $n(t)$ *is the number of roots in the rectangle* $|y| \leqslant h$, $0 < x \leqslant t$, *then*

$$n(t) = \frac{\Delta}{\pi} t + \omega(t), \tag{6.15}$$

while the function $\omega(t)$ *is bounded.*[2]

PROOF. From Corollary 2 to Theorem 2, all of the roots of $f(z)$ lie in the strip. Take H sufficiently large, and apply to the rectangle $|y| \leqslant H, 0 < x \leqslant t$,

[2] In a supplement we present the work of M. G. Kreĭn and the author, in which we gave a complete characterization of the set of roots of a function of class $[\Delta]$.

the well-known formula for the number of roots of a function inside a contour. We obtain

$$n(t)=\frac{1}{2\pi i}\int_0^t \frac{f'(u-iH)}{f(u-iH)}\,du+\frac{1}{2\pi}\int_{-H}^{H}\frac{f'(t+is)}{f(t+is)}\,ds$$

$$-\frac{1}{2\pi i}\int_0^t \frac{f'(u+iH)}{f(u+iH)}\,du-\frac{1}{2\pi}\int_{-H}^{H}\frac{f'(is)}{f(is)}\,ds.$$

Here we are assuming that on the lines $x = 0$ and $x = t$ there are no roots of the function $f(z)$.

The first of these assumptions may be made without destroying the generality of the considerations, since one may consider the function $f(z + \varepsilon)$ instead of the function $f(z)$. Taking the real parts of both sides of the equation and making use of Lemma 4, we obtain

$$2\pi n(t)=2\Delta t-\varphi_H(t)+\varphi_{-H}(t)+\operatorname{Re}\int_{-H}^{H}\frac{f'(t+is)}{f(t+is)}\,ds+c, \tag{6.16}$$

where $\varphi_H(t)$ and $\varphi_{-H}(t)$ are almost-periodic functions and

$$c=-\operatorname{Re}\int_{-H}^{H}\frac{f'(is)}{f(is)}\,ds+\varphi_H(0)-\varphi_{-H}(0).$$

We shall prove the boundedness of the function

$$\omega_1(t)=2\pi n(t)-2\Delta t+\varphi_H(t)-\varphi_{-H}(t)-c.$$

We shall call *exceptional* those values of t for which the line $\operatorname{Re} z = t$ intersects circles of radius $\delta > 0$ with centers at the roots a_k of the function $f(z)$ These values evidently form intervals, which we shall call *exceptional intervals.* The boundedness of the function $\omega(t)$ outside the excluded intervals follows easily from its integral representation

$$\omega_1(t)=\operatorname{Re}\int_{-H}^{H}\frac{f'(t+is)}{f(t+is)}\,ds.$$

In fact, from the formula

$$f'(z)=\frac{1}{2\pi i}\int_{|\zeta-z|=1}\frac{f(\zeta)}{(\zeta-z)^2}\,d\zeta$$

there follows for $|y| \leqslant H$ the inequality

$$|f'(z)|\leqslant M,$$

in which

$$M=\sup_{|y|<H+1}|f(z)|.$$

On the other hand, for t lying outside the exceptional intervals, from Lemma 1 $|f(t + is)| \geqslant m(\delta) > 0$, and accordingly

$$|\omega_1(t)| \leqslant 2HMm^{-1}(\delta).$$

In order to estimate the function $\omega(t)$ in the exceptional intervals we observe that by Lemma 2 their total length in any interval $(x, x + 1)$ does not exceed $2k\delta$, where k is the largest number of roots of the function $f(z)$ in the rectangle $t < x \leqslant t + 1, |y| \leqslant H$. We shall suppose that $\delta < 1/4k$. Denote by $\chi(s)$ the least upper bound of the oscillation of the function $2\Delta t - \varphi_H(t) + \varphi_{-H}(t)$ on the interval $(x, x + s)$ for $-\infty < x < +\infty$. Because of the uniform continuity of the functions $\varphi_H(t)$ and $\varphi_{-H}(t)$, the function $\chi(s)$ is bounded. For a given value of t we choose the number $s(|s| < \frac{1}{2})$ such that the point $t + s$ lies outside the exceptional intervals. Then

$$|\omega_1(t+s)-\omega_1(t)| \leqslant 2\pi k+\chi\left(\frac{1}{2}\right)$$

and accordingly

$$|\omega_1(t)| < 2HMm^{-1}(\delta)+2\pi k+\chi\left(\frac{1}{2}\right)$$

for all values of t. The theorem is proved.

3. A General Theorem on Roots and on Mean Motion for Holomorphic Almost-Periodic Functions

The mean motion for a complex-valued function $f(x)$, defined on the real axis and different from zero, is the limit

$$c_f = \lim_{x-y\to\infty} \frac{\arg f(x)-\arg f(y)}{x-y}.$$

Here we have chosen some continuous branch of the function $\arg f(x)$. If the function $f(x)$, holomorphic on the real axis, vanishes at certain points, one may distinguish "right" and "left" arguments of the function $f(x)$ at these points. The right argument is defined as follows: at any point x_0 one chooses arbitrarily a value of $\arg f(x)$ and continuously extends it to the right up to the nearest root of $f(x)$. One goes around this root in a small semicircle below it and one again extends it continuously to the right along the real axis. In the same way one extends $\arg f(x)$ to the left, going around and below each root. Evidently on passing from left to right around a root of the function of multiplicity p the right argument is increased by $p\pi$. The left argument is obtained in the same way by going around the roots of the function in a semicircle above them. Evidently on going around a root of multiplicity p, the left argument is increased by $-\pi p$. Correspondingly one defines right and left mean motions

c_f^+ and c_f^-. If $c_f^+ = c_f^-$, then this quantity is called the mean motion and is also denoted by c_f.[3]

In this section we shall present some of the theorems of Jessen on the mean motion and the distribution of the roots of holomorphic almost-periodic functions.

LEMMA 5. *For every almost-periodic function $f(z)$ holomorphic in the strip $\alpha < y < \beta$ and not identically equal to zero, there exists a mean*

$$\varphi(y) = \underset{x}{M}\{\ln|f(x+iy)|\} \tag{6.17}$$

for $\alpha < y < \beta$.

PROOF. Put $|f(z)|_m = \max\{|f(z)|; m\}$. The function $\ln|f(x+iy)|_m$ evidently is an almost-periodic function of x and therefore has a mean

$$\varphi_m(y) = \underset{x}{M}\{\ln|f(x+iy)|_m\}. \tag{6.18}$$

Construct the square (I): $|x - x_0| < 1$ and $|y - y_0| < 1$, and denote by k the largest possible number of roots in the intersection of this square with the

---

[3] The first problem related to mean motion was posed by Lagrange in 1782, in connection with an astronomical problem. It may be formulated as follows. Prove that every generalized trigonometrical polynomial has a mean motion.

Many subtle investigations have been devoted to this problem. Here we shall mention a few of them. Lagrange himself considered only the trivial case when the modulus of one of the coefficients exceeded the sum of the moduli of all the others. The Riga mathematician Bohl showed that a generalized trigonometric polynomial consisting of three terms always has a mean motion.

In 1961 H. Weyl showed that if in the generalized trigonometrical polynomial

$$P(x) = \sum_{k=1}^{n} a_k e^{i\lambda_k x}$$

the numbers λ_k were linearly independent (i.e., there are no integers m_k such that $\sum_{k=1}^{n} \lambda_k m_k = 0$ and $\sum_{k=1}^{n} m_k^2 \neq 0$), then the polynomial has a mean motion. Subsequently, developing his method, H. Weyl [1] proved that the Lagrange problem has a positive solution. However, the proof of Weyl contained a mistake. Making the necessary corrections, Jessen and Tornhave [1] gave in 1945 a complete solution of the question. In this paper Jessen and Tornhave also presented a detailed history of the question, with the fundamental results and with a complete bibliography.

The problem of mean motion may be posed not only for generalized trigonometrical polynomials, but also for arbitrary almost-periodic functions. The following theorem of Bohr [2] is fundamental here.

The argument of an almost-periodic function $f(t)$, satisfying the inequality $|f(t)| \geqslant m > 0$ on the entire numerical axis, may be represented in the form

$$\arg f(t) = ct + \varphi(t),$$

where c is a constant and $\varphi(t)$ an almost-periodic function.

strip $\alpha + \delta \leqslant y \leqslant \beta - \delta (\delta > 0)$ for a fixed y_0 and with x_0 varying from $-\infty$ to ∞ (see Lemma 2).

Choose a number $\delta_1 > 0$ so small that the inequality $2k\delta_1 < \frac{1}{2}$ is satisfied. The set covered by circles of radius δ_1 with centers at the roots of the function $f(z)$ consists of regions. The diameter of each such region ω_j $(j = 1, 2, \ldots, p)$ intersecting the square $|y - y_0| \leqslant \frac{1}{2}$, $|x - x_0| \leqslant \frac{1}{2}$ evidently does not exceed the quantity $2k\delta_1 < \frac{1}{2}$ and therefore each such region will lie in the square (I). Outside of these exceptional regions we have, from Lemma 1,

$$|f(z)| \geqslant m(\delta_1) > 0 \qquad (\alpha + \delta \leqslant y \leqslant \beta - \delta).$$

Consider the integral

$$J = \int_{x_0 - \frac{1}{2}}^{x_0 + \frac{1}{2}} [\ln |f(x + iy)|_m - \ln |f(x + iy)|]\, dx$$

in which $m = m(\delta)$ and $|y - y_0| \leqslant \frac{1}{2}$. Outside the exceptional regions ω_j $(j = 1, 2, \ldots, p)$ the integrand is equal to zero. Construct the polynomial

$$P(z) = \prod_{s=1}^{q} (z - z_s).$$

in which the z_s are the roots of the function $f(z)$ in the square (I) $(q \leqslant k)$. Evidently the inequality $|P(z)| \leqslant (2\sqrt{2})^k$ holds throughout (I). On the boundary of the exceptional region ω_j we will have

$$|f(z) P^{-1}(z)| \geqslant m (2\sqrt{2})^{-k}.$$

By the maximum principle, this inequality will hold in the whole square $|x - x_0| \leqslant \frac{1}{2}$, $|y - y_0| \leqslant \frac{1}{2}$, and accordingly

$$\ln |f(x + iy)| \geqslant \ln m - \frac{3}{2} k \ln 2 + \ln |P(x + iy)|$$
$$\left(|x - x_0| \leqslant \frac{1}{2},\ \alpha + \delta \leqslant y \leqslant \beta - \delta\right),$$

and in the regions ω_j we will have the inequality

$$\ln |f(x + iy)|_m - \ln |f(x + iy)| < \frac{3}{2} k \ln 2 - \sum_{s=1}^{q} \ln |z - z_s|.$$

Since the diameter ω_j does not exceed $2k\delta_1$, we therefore obtain

$$|J| < 3k^2\delta_1 \ln 2 - k \int_{-2k\delta_1}^{2k\delta_1} \ln |t|\, dt \quad (\alpha + \delta \leqslant y \leqslant \beta - \delta).$$

This estimate holds for any x_0.

Letting δ tend to zero, and therefore also $m(\delta) \to 0$, we find that $J \to 0$ uniformly relative to $\alpha + \delta \leqslant y \leqslant \beta - \delta$. Writing

$$\varphi(y, x_1, x_2) = \frac{1}{x_2 - x_1} \int_{x_1}^{x_2} \ln |f(x+iy)|\, dx$$

and constructing $\varphi_m(y, x_1, x_2)$ analogously from $\ln |f(x + iy)|_m$, we obtain

$$|\varphi(y, x_1, x_2) - \varphi_m(y, x_1, x_2)| < \varepsilon$$

for $m < m_\varepsilon$ and arbitrary x_1 and x_2 $(x_2 > x_1 + 1)$.

Thus we obtain for sufficiently small m

$$\varphi_m(y) - \varepsilon < \varliminf_{x_2 - x_1 \to \infty} \varphi(y, x_1, x_2) \leqslant \varlimsup_{x_2 - x_1 \to \infty} \varphi(y, x_1, x_2) < \varphi_m(y) + \varepsilon.$$

Therefore the existence of $\varphi(y)$ follows. The lemma is proved.

Let us study the function $\varphi(y)$. To this end it is convenient to use the argument principle. Denoting by $n(x_1, x_2, s, t)$ the number of roots of the function $f(z)$ in the rectangle $x_1 < \operatorname{Re} z \leqslant x_2$, $s \leqslant \operatorname{Im} z \leqslant t$, situated in the strip of almost-periodicity of $f(z)$, and applying the argument principle, we may write

$$2\pi n(x_1, x_2, s, t) = \int_{x_1}^{x_2} \frac{d \arg f(x+is)}{dx}\, dx - \int_{x_1}^{x_2} \frac{d \arg f(x+it)}{dx}\, dx$$
$$+ \operatorname{Im} \int_s^t \left[\frac{f'(x_2+iy)}{f(x_2+iy)} - \frac{f'(x_1+iy)}{f(x_1+iy)} \right] dy. \tag{6.19}$$

The last term in this equation, which we will call $C(x_1, x_2, s, t)$, is a function bounded in the relatively dense set E of values of the variables x_1 and x_2 (cf. the discussion in Theorem 3).

From the Cauchy-Riemann conditions it follows that

$$\frac{d[\arg f(x + is)]}{dx} = -\frac{d \ln |f(x+is)|}{ds}$$

and after substitution we obtain

$$2\pi n(x_1, x_2, s, t)$$
$$= -\int_{x_1}^{x_2} \frac{d \ln |f(x+is)|}{ds}\, dx + \int_{x_1}^{x_2} \frac{d \ln |f(x+it)|}{dt}\, dx + C(x_1, x_2, s, t). \tag{6.20}$$

The integrals on the right side of equation (6.20) exist for all values of the quantities s and t with the possible exception of a countable set.

Let $\alpha < y_1 < y_2 < y_3 < \beta$. Integrating both sides of equation (6.20) in s between the limits y_1 and y_2, and in t between the limits y_2 and y_3, and dividing by $x_2 - x_1$, we obtain

$$\frac{2\pi}{x_2 - x_1} \int_{y_2}^{y_3} \int_{y_1}^{y_2} n(x_1, x_2, s, t)\, ds\, dt$$

$$= -\left\{\frac{1}{x_2 - x_1} \int_{x_1}^{x_2} [\ln|f(x+iy_2)| - \ln|f(x+iy_1)|]\, dx\right\}(y_3 - y_2)$$

$$+ \left\{\frac{1}{x_2 - x_1} \int_{x_1}^{x_2} [\ln|f(x+iy_3)| - \ln|f(x+iy_2)|]\, dx\right\}(y_2 - y_1)$$

$$+ \frac{1}{x_2 - x_1} \int_{y_2}^{y_3} \int_{y_1}^{y_2} C(x_1, x_2, s, t)\, ds\, dt.$$

Passing thereupon to the limit as $x_2 - x_1 \to +\infty$ and recalling that x_1 and x_2 belong to the relatively dense set E on which the function $C(x_1, x_2, s, t)$ is bounded, we obtain

$$-[\varphi(y_2) - \varphi(y_1)](y_3 - y_2) + [\varphi(y_3) - \varphi(y_2)](y_2 - y_1) \geqslant 0 \tag{6.21}$$

or

$$\varphi(y_1)(y_2 - y_3) + \varphi(y_2)(y_3 - y_1) + \varphi(y_3)(y_1 - y_2) \leqslant 0. \tag{6.22}$$

Thus the function $\varphi(y)$ is convex. By the known properties of convex functions, $\varphi(y)$ has at each point a right and left derivative, while the right derivative at a given point is not less than the left. It is also known that the right derivative of a convex function is equal to the left at every point with the possible exception of a countable set, and $\varphi'(y)$ is a nondecreasing function.[4]

THEOREM 4. *If the derivative $\varphi'(y)$ exists for $y = y_1$ and $y = y_2$ $(y_2 > y_1)$ then the limit*

$$2\pi \lim_{x_2 - x_1 \to \infty} \frac{n(x_1, x_2, y_1, y_2)}{x_2 - x_1} = \varphi'(y_2) - \varphi'(y_1) \tag{6.23}$$

exists.

PROOF. Choose the number h sufficiently small so that

$$\left|\frac{\varphi(y_2 + h) - \varphi(y_2)}{h} - \varphi'(y_2)\right| < \frac{\varepsilon}{2}$$

and

$$\left|\frac{\varphi(y_1) - \varphi(y_1 - h)}{h} - \varphi'(y_1)\right| < \frac{\varepsilon}{2}.$$

[4] In Chapter I we obtained analogous theorems for trigonometrically convex functions.

Integrate the equation (6.20) in s between the limits $y_1 - h$, y_1 and in t between the limits y_2, $y_2 + h$, divide by $x_2 - x_1$, and pass to the limit as $x_2 - x_1 \to \infty$, recalling that x_1 and x_2 take on values in the relatively dense set E on which the function $C(x_1, x_2, s, t)$ is bounded. We obtain

$$\lim_{x_2-x_1\to\infty} \frac{2\pi}{h^2} \int_{y_2}^{y_2+h} \int_{y_1-h}^{y_1} \frac{n(x_1, x_2, s, t)}{x_2 - x_1}\, ds\, dt$$
$$= \frac{\varphi(y_2+h) - \varphi(y_2)}{h} - \frac{\varphi(y_1) - \varphi(y_1 - h)}{h}. \tag{6.24}$$

For $h > 0$ the integrand is not less than $n(x_1, x_2, y_1, y_2)(x_2 - x_1)^{-1}$, and we obtain

$$2\pi \varlimsup_{x_2-x_1\to\infty} \frac{n(x_1, x_2, y_1, y_2)}{x_2 - x_1} \leqslant \varphi'(y_2) - \varphi'(y_1) + \varepsilon.$$

Analogously for $h < 0$ we obtain

$$2\pi \varliminf_{x_2-x_1\to\infty} \frac{n(x_1, x_2, y_1, y_2)}{x_2 - x_1} \geqslant \varphi'(y_2) - \varphi'(y_1) - \varepsilon.$$

The desired (6.23) follows from these two inequalities. Moreover, we may free ourselves from the requirement that the quantities x_1 and x_2 take on values only in E. Observe that for any x_1 and x_2 we may pick out four numbers x_1', x_1'', x_2', and x_2'' belonging to E and such that $|x_1'' - x_1'| < 1$, $|x_2'' - x_2'| < 1$ and $x_1' < x_1 < x_1''$, $x_2' < x_2 < x_2''$.

It follows immediately from the definition of the function $n(x_1, x_2, y_1, y_2)$ that

$$\frac{x_2' - x_1''}{x_2 - x_1} \cdot \frac{n(x_1'', x_2', y_1, y_2)}{x_2' - x_1''} \leqslant \frac{n(x_1, x_2, y_1, y_2)}{x_2 - x_1}$$
$$\leqslant \frac{n(x_1', x_2'', y_1, y_2)}{x_2'' - x_1'} \cdot \frac{x_2'' - x_1'}{x_2 - x_1}, \tag{6.25}$$

and evidently all three quantities have a common limit as $x_2 - x_1 \to \infty$.

THEOREM 5. *If for $y = y_0$ there exists a derivative $\varphi'(y_0)$, then the function $f(x + iy_0)$ has a mean motion $c_f(y_0) = -\varphi'(y_0)$.*

PROOF. By Theorem 4 the function $n(x_1, x_2, y_0, t)(x_2 - x_1)^{-1}$, monotonic in t, has a limit as $x_2 - x_1 \to \infty$ for all values of the variable t with the possible exclusion of a countable set. Moreover, since that function is bounded for $y_1 \leqslant t \leqslant y_1 + h$, there exists a limit

$$\lim_{x_2-x_1\to\infty} (x_2 - x_1)^{-1} \int_{y_1}^{y_1+h} n(x_1, x_2, y_0, t)\, dt\,.$$

Putting $s = y_0$ in formula (6.20) and integrating both sides in t from y_1 to $y_1 + h$, we obtain

$$h\,\frac{\arg f(x_2+iy_0)-\arg f(x_1+iy_0)}{x_2-x_1}$$

$$=-(x_2-x_1)^{-1}\int_{x_1}^{x_2}\ln|f[x+i(y_1+h)]|\,dx$$

$$+(x_2-x_1)^{-1}\int_{x_1}^{x_2}\ln|f(x+iy_1)|\,dx$$

$$-(x_2-x_1)^{-1}\int_{y_1}^{y_1+h}C(x_1,\ x_2,\ y_0,\ t)\,dt$$

$$+2\pi(x_2-x_1)^{-1}\int_{y_1}^{y_1+h}n(x_1,\ x_2,\ y_0,\ t)\,dt. \tag{6.26}$$

All the quantities in the right side of the equation have limits as $x_2 - x_1 \to \infty$. Passing to the limit we obtain from Theorem 4

$$c_f(y) = -\varphi'(y). \tag{6.27}$$

4. A Theorem on Mean Motion for an Almost-Periodic Function with Semibounded Spectrum

It follows from (6.27) that $c_f(y)$ is a monotonically decreasing function. If the greatest lower bound $-\Delta$ of the spectrum of the function is finite, then the function $e^{i\Delta z}f(z)$ is bounded in the upper half-plane, and thus the convex function $\varphi(y) - \Delta y$ is also bounded for $y > 0$. This is possible only under the condition $\varphi'(y) - \Delta \leqslant 0$. Thus, in this case the monotonically decreasing function $c_f(y)$ is bounded below by the number $-\Delta$ and accordingly has a limit

$$\lim_{y\to+\infty} c_f(y) \geqslant -\Delta. \tag{6.28}$$

It follows from Lemma 3 that if the number $-\Delta$ enters into the spectrum of the function $f(x)$, then equality holds in (6.28). *It turns out that this limit is equal to $-\Delta$ also in the case when $-\Delta$ does not enter into the spectrum of the function $f(x)$.*

This theorem seems very natural, but its proof is rather complicated.[5] In what follows we shall use the theorem for the study of general exponential sums. For its proof we shall need a series of lemmas. First of all we prove the following general lemma.

---

[5] B. Ja. Levin [4].

LEMMA 6. *If the function $f(z)$, holomorphic in the circle $|z - z_0| \leqslant r$, satisfies the inequality $|f(z)| \leqslant 1$ and if moreover*

$$|f(z_0)| \geqslant e^{-\gamma}, \tag{6.29}$$

then the variation of the function $\arg f(z)$ *on any rectilinear segment lying in the circle $|z - z_0| \leqslant p = r(2e)^{-1}$ does not exceed the quantity* $(4 + 2\pi)\gamma$.

PROOF. Putting $\sigma = re^{-1}$ and denoting by $n(f, \sigma, z_0)$ the number of roots of the function $f(z)$ in the circle $|z - z_0| < \sigma$, and using Lemma 4 of § 5 of Chapter I, we obtain

$$n(f, \sigma, z_0) < \gamma. \tag{6.30}$$

If we set up for these roots the function

$$\chi(z, z_0) = \prod_{|u_k| < \sigma} \frac{\sigma(u - u_k)}{\sigma^2 - \bar{u}_k u},$$

where $u = z - z_0$ and $u_k = a_k - z_0$ (a_k a root of $f(z)$), the function

$$\varphi(z, z_0) = f(z)\chi^{-1}(z, z_0), \tag{6.31}$$

will be holomorphic and will not have roots in the circle $|z - z_0| \leqslant \sigma$, and will satisfy on the boundary and therefore inside this circle the inequality

$$|\varphi(z, z_0)| \leqslant 1$$

Since inside this circle $|\chi(z, z_0)| < 1$, therefore $|\varphi(z_0, z_0)| > e^{-\gamma}$.

Applying to the function $\ln \varphi(z, z_0)/\varphi(z_0, z_0)$, holomorphic for $|z - z_0| < \sigma$, the Carathéodory inequality (Chapter I, § 6) with $R = 2r$, we obtain

$$|\ln \varphi(z, z_0) - \ln \varphi(z_0, z_0)| < -2 \ln |\varphi(z_0, z_0)|$$

$$\text{for } |z - z_0| < \frac{\sigma}{2} = p.$$

Thus, if z_1 and z_2 are two points of the circle $|z - z_0| < p$, then

$$|\ln \varphi(z_2, z_0) - \ln \varphi(z_1, z_0)| < 4\gamma$$

and moreover

$$|\arg \varphi(z_2, z_0) - \arg \varphi(z_1, z_0)| < 4\gamma. \tag{6.32}$$

The variation of the argument of the function $\sigma(u - u_k)/\sigma^2 - \bar{u}_k u$ on any rectilinear segment is smaller than 2π, so that

$$|\arg \chi(z_2, z_0) - \arg \chi(z_1, z_0)| < 2\pi n(f, \sigma, z_0). \tag{6.33}$$

Comparing (6.31), (6.32), (6.33) and (6.30), we find that

$$|\arg f(z_2) - \arg f(z_1)| < (4 + 2\pi)\gamma.$$

Here it evidently makes no difference whether one takes as $\arg f(z)$ the right or left argument as the point z moves over the segment $[z_1, z_2]$. If the spectrum is bounded below, then one may assume without loss of generality that its lower bound is equal to zero. If in addition zero does not enter into the spectrum, then from Lemma 3 the function $f(x+iy)$ converges uniformly to zero as $y \to +\infty$. Therefore without loss of generality one may assume that $|f(x)| \leqslant 1$ and thus that in the entire half-plane $y \geqslant 0$ we have $|f(z)| \leqslant 1$.

To estimate the variation of the argument of the function $f(z)$ on the line $y = R$ we use Lemma 6. In order to do this we must prove that in the neighborhood of each such line we can find a sufficiently dense set of points (supporting points) in which an inequality of type (6.29) is satisfied.

The following material is devoted to the proof of the existence of such supporting points.

LEMMA 7. *Suppose that* $\{a_k\}$ *is a sequence of numbers*

$$\operatorname{Im} a_k > 0, \quad \sum_{k=1}^{\infty} \left| \operatorname{Im} \frac{1}{a_k} \right| < \infty$$

and

$$\chi(z) = \prod_{k=1}^{\infty} \left(1 - \frac{z}{a_k}\right)\left(1 - \frac{z}{\bar{a}_k}\right)^{-1} \tag{6.34}$$

Then the quantity

$$J(\chi, R) = \frac{2}{\pi R} \int_0^{\pi} \ln |\chi(Re^{i\theta})| \sin\theta \, d\theta$$

converges to zero as R increases.

PROOF. Noting that $|\chi(x)| = 1$ and using the Carleman formula (5.02′), we may write

$$\frac{1}{2} J(\chi, R) + \frac{1}{2} \operatorname{Im} \chi'(0) = \sum_{r_k < R} \left(\frac{1}{r_k} - \frac{r_k}{R^2}\right) \sin\theta_k,$$

where $a_k = r_k e^{i\theta_k}$. But $\chi'(0) = -2i \sum_{k=1}^{\infty} \operatorname{Im} 1/a_k$ and accordingly

$$\frac{1}{2} J(\chi, R) = \sum_{r_k > R} \frac{\sin\theta_k}{r_k} - \frac{1}{R^2} \sum_{r_k < R} r_k \sin\theta_k > -\frac{1}{R^2} \sum_{r_k < R} r_k \sin\theta_k .$$

For some $r_\varepsilon > 0$ we will have

$$\frac{1}{R^2} \sum_{r_\varepsilon < r_k < R} r_k \sin\theta_k < \sum_{r_k > r_\varepsilon} \frac{\sin\theta_k}{r_k} < \frac{\varepsilon}{2}.$$

Moreover, for some $R_\varepsilon > 0$ and $R > R_\varepsilon$,

$$\frac{1}{R^2} \sum_{r_k < r_\varepsilon} r_k \sin\theta_k < \frac{\varepsilon}{2},$$

so that asymptotically

$$-2\varepsilon < J(\chi, R) < 0.$$

The right inequality results from the inequality $|\chi(z)| < 1$ for $\operatorname{Im} z > 0$.

LEMMA 8. *If the function $\psi(z)$ is holomorphic for* Im $z > 0$, *is continuous in the closed half-plane, is of exponential type, and satisfies the inequality*

$$\ln|\psi(z)| = \frac{y}{\pi}\int_{-\infty}^{\infty}\frac{\ln|\psi(t)|}{(t-x)^2+y^2}\,dt + ky, \tag{6.35}$$

then

$$\lim_{R\to\infty} J(\psi, R) = k.$$

PROOF. Setting up the expression $J(\psi, R)$ and changing the order of integration, we obtain

$$J(\psi, R) = \frac{2}{\pi}\int_{-\infty}^{\infty}\ln|\psi(t)|\,k(t, R)\,dt + k,$$

where

$$k(t, R) = \frac{1}{\pi}\int_0^{\pi}\frac{\sin^2\theta\,d\theta}{R^2+t^2-2Rt\cos\theta},$$

which on calculation yields

$$k(t, R) = \begin{cases} \dfrac{1}{2R^2} & \text{for } |t| < R, \\[2ex] \dfrac{1}{2t^2} & \text{for } |t| > R. \end{cases}$$

Thus,

$$J(\psi, R) = \frac{1}{\pi R^2}\int_0^R \ln|\psi(t)\psi(-t)|\,dt + \frac{1}{\pi}\int_R^{\infty}\frac{\ln|\psi(t)\psi(-t)|}{t^2}\,dt + k. \tag{6.36}$$

The existence of the second integral may be obtained immediately from the existence of the integral (6.35) by integrating by parts, just as in § 3 of Chapter V.

Introducing the function

$$\varphi_\lambda(x) = \int_\lambda^{\infty}\frac{\ln|\psi(t)\psi(-t)|}{t^2}\,dt \qquad (x > \lambda),$$

we easily find that

$$\frac{1}{R^2}\int_\lambda^R \ln|\psi(t)\psi(-t)|\,dt = \varphi_\lambda(t)\frac{t^2}{R^2}\Bigg|_\lambda^R - \frac{2}{R^2}\int_\lambda^R \varphi_\lambda(t)\,t\,dt,$$

and recalling that $\varphi_\lambda(t)$ tends to zero for $t > \lambda$ and $\lambda \to \infty$, we find that for some $\lambda_\varepsilon > 0$ and $R > \lambda \geqslant \lambda_\varepsilon$

$$\left|\frac{1}{\pi R^2}\int_\lambda^R \ln|\psi(t)\psi(-t)|\,dt\right| < \frac{\varepsilon}{3}. \tag{6.37}$$

Choosing the number R_ε sufficiently large, we will also have for $R > R_\varepsilon$

$$\left.\begin{aligned} &\left|\frac{1}{\pi R^2}\int_0^\lambda \ln|\psi(t)\psi(-t)|\,dt\right| < \frac{\varepsilon}{3} \\ \text{and}\quad &\left|\frac{1}{\pi}\int_R^\infty \frac{\ln|\psi(t)\psi(-t)|}{t^2}\,dt\right| < \frac{\varepsilon}{3}. \end{aligned}\right\} \tag{6.38}$$

The assertion of the lemma follows from (6.36), (6.37) and (6.38). Combining Lemmas 7 and 8 we get

LEMMA 9. *If the function $f(z)$ is holomorphic and of exponential type in the half-plane* Im $z \geqslant 0$, *while*

$$\int_{-\infty}^{\infty} \frac{\ln|f(t)|}{1+t^2}\,dt \tag{6.39}$$

exists and

$$\overline{\lim_{y\to+\infty}} \frac{\ln|f(iy)|}{y} = k, \tag{6.40}$$

then

$$\lim_{R\to\infty} J(f, R) = k. \tag{6.41}$$

PROOF. From Theorem 5 of § 3, Chapter V, it follows from the hypotheses of this lemma that the function $f(z)$ may be represented in the form

$$f(z) = \psi(z)\chi(z),$$

where $\chi(z)$ is defined by formula (6.34) (the a_k are the roots of the function $f(z)$) and $\psi(z)$ admits a representation (6.35). Therefore we obtain

$$J(f, R) = J(\psi, R) + J(\chi, R), \tag{6.42}$$

and the result follows from Lemmas 6 and 7.

LEMMA 10. *Suppose that $f(x)$ is an almost-periodic function with the greatest lower bound of its spectrum equal to zero, and let*

$$J(f, \tau, R) = \frac{2}{\pi R}\int_0^\pi \ln|f(\tau + Re^{i\theta})|\sin\theta\,d\theta. \tag{6.43}$$

Then to each number $\varepsilon > 0$ there correspond a relatively dense set of real numbers τ and a number R_ε such that for $R > R_\varepsilon$

$$-\varepsilon < J(f, \tau, R) < 0 \tag{6.44}$$

for the entire set of numbers $\{\tau\}$.

PROOF. Without loss of generality one may suppose that $|f(x)| < 1$ (it is sufficient to multiply the function $f(x)$ by some constant). From Carleman's formula we have

$$2 \sum_{\lambda < r_k(\tau) < R} \left(\frac{1}{r_k(\tau)} - \frac{r_k(\tau)}{R^2}\right) \sin \theta_k(\tau)$$

$$- \frac{1}{\pi} \int_\lambda^R \ln |f(t+\tau) f(\tau - t)| \left(\frac{1}{t^2} - \frac{1}{R^2}\right) dt$$

$$= J(f, \tau, R) + A_\lambda(f, \tau, R), \tag{6.45}$$

where $r_k(\tau)e^{i\theta_k(\tau)} = a_k - \tau$ (a_k a root of $f(z)$), and

$$A_\lambda(f, \tau, R) = -\operatorname{Im} \frac{1}{\pi} \int_0^\pi \ln f(\tau + \lambda e^{i\theta}) J\left(\frac{\lambda e^{i\theta}}{R^2} - \frac{e^{-i\theta}}{\lambda}\right) d\theta.$$

The left side of equation (6.45) is a function monotonically increasing as R increases for any fixed τ. We will denote it by $\psi(R, \tau)$. From Lemma 9, we have for some $R'_\varepsilon > 0$ and $R > R'_\varepsilon$

$$-\frac{\varepsilon}{3} < J(f, 0, R) < 0. \tag{6.46}$$

Suppose moreover that R'_ε is so large that for $R \geqslant R'_\varepsilon$ the inequality

$$|B_\lambda(f, 0, R)| < \frac{\varepsilon}{6} \tag{6.47}$$

is satisfied, where

$$B_\lambda(f, \tau, R) = A_\lambda(f, \tau, R) - A_\lambda(f, \tau, \infty)$$

$$= -\operatorname{Im} \frac{1}{\pi R^2} \int_0^\pi [\ln f(\tau + \lambda e^{i\theta})] \lambda e^{i\theta} d\theta.$$

Choose the numbers $0 < \lambda < 1$ and $R_\varepsilon \geqslant R'_\varepsilon > \lambda$ so that on the semicircles $z = \lambda e^{i\theta}$ and $z = R_\varepsilon e^{i\theta}$ $(0 \leqslant \theta \leqslant \pi)$ there are no roots of the function $f(z)$. Denote the greatest lower bound of $|f(z)|$ on these semicircles by $m > 0$ and choose the number δ so that $\delta < m_\varepsilon/12$.

If τ is a δ-shift of the function $f(z)$ in the strip $0 \leqslant y \leqslant R$ and the point z belongs to the circumference of one of the semicircles, then

$$|\ln f(z+\tau) - \ln f(z)| = \left|\ln\left[1 - \frac{f(z) - f(z+\tau)}{f(z)}\right]\right| < \frac{2\delta}{m} < \frac{\varepsilon}{6}. \tag{6.48}$$

Thus for $R \geqslant R_\varepsilon$ and the indicated choice of the numbers τ

$$\left.\begin{aligned} &|J(f, \tau, R_\varepsilon) - J(f, 0, R_\varepsilon)| < \frac{\varepsilon}{3} \\ \text{and} \\ &|B_\lambda(f, \tau, R_\varepsilon) - B_\lambda(f, 0, R_\varepsilon)| < \frac{\varepsilon}{6}. \end{aligned}\right\} \tag{6.49}$$

From the first of these inequalities and (6.46) it follows that

$$-\frac{2\varepsilon}{3} < J(f, \tau, R_\varepsilon) < 0. \tag{6.50}$$

From Carleman's formula the function $J(f, \tau, R)$ is the sum of a monotonic increasing function $\psi(R, \tau)$ and a function $A_\lambda(f, \tau, R)$ whose oscillation as R varies from R_ε to ∞ does not, in view of the inequalities (6.47) and (6.49), exceed the magnitude $\varepsilon/3$. Thus for $R \geqslant R_\varepsilon$

$$-\varepsilon < J(f, \tau, R) < 0, \tag{6.51}$$

where $\{\tau\}$ is a relatively dense set of δ-shifts of the function $f(z)$ in the strip $0 \leqslant y \leqslant R$. The lemma is proved.

From Lemma 10 it easily follows that on each arc $z = Re^{i\theta} + \tau$, $|\theta - (\pi/2)| < \alpha$ $(\alpha > 0)$ and $R > R_\varepsilon$ there exists a point z_τ at which

$$\ln|f(z_\tau)| \geqslant -\eta R \qquad \left(\eta = \frac{\pi\varepsilon}{4 \sin\alpha}\right). \tag{6.52}$$

Indeed, in the contrary case we would have

$$J(f, \tau, R) < \frac{2}{\pi R} \int_{\frac{\pi}{2}-\alpha}^{\frac{\pi}{2}+\alpha} \ln|f(Re^{i\theta})| \sin\theta \, d\theta$$

$$< -\frac{4}{\pi}\eta \sin\alpha = -\varepsilon, \tag{6.53}$$

which contradicts (6.44). Thus from the last four lemmas we obtain a corollary which we formulate as a separate lemma.

LEMMA 11. *If the greatest lower bound of the spectrum of the almost-periodic function $f(x)$ is equal to zero, then to each pair of numbers $\varepsilon > 0$ and $\zeta > 0$ there correspond a relatively dense set of numbers $\{\tau\}$ and a number $R(\varepsilon, \zeta) > 0$ such that if $R > R(\varepsilon, \zeta)$ each circle with center at the point $\tau + iR$ and of radius ζR contains some point z_τ at which*

$$\ln|f(z_\tau)| > -\varepsilon R. \tag{6.54}$$

The preceding four lemmas were necessary only in the proof of Lemma 11, of which we now take advantage.

Using Lemmas 6 and 11 we may prove the following theorem.

THEOREM 6. *If the greatest lower bound of the spectrum of the almost-periodic function $f(x)$ is equal to $-\Delta$, then*[6]

$$\lim_{y \to +\infty} c_f(y) = \Delta. \tag{6.55}$$

PROOF. One evidently has to consider only the case when the number $-\Delta$ does not enter into the spectrum. Without loss of generality we may take $\Delta = 0$ and $|f(x)| \leqslant 1$ $(-\infty < x < \infty)$, since if necessary the function $f(x)$ may be multiplied by $ke^{i\Delta x}$, with k a constant.

Suppose that $\varepsilon > 0$ is given. Choose a number $\zeta > 0$ such that $\zeta < (1-\zeta)/4e$, define a relatively dense set $\{\tau\}$ and a number $R(\varepsilon, \zeta)$ as in Lemma 11, and choose a number $R'_\varepsilon \geqslant \max\{R(\varepsilon, \zeta), 4el[\sqrt{3}\,(1-\zeta)]\}$ where l is the enclosing interval for the set τ. We describe around each of the points z_τ, whose existence was asserted in Lemma 11, a circle $K_{\tau,R}$ of radius $d = R(2e)^{-1}(1-\zeta)$ with center at that point. With the indicated choice of the number ζ and $R > R'_\varepsilon$ these circles completely cover the line $y = R$.

From Lemma 11 we have $|f(z_\tau)| > e^{-\varepsilon R}$, and moreover $|f(z)| \leqslant 1$ in the entire half-plane $\operatorname{Im} z \geqslant 0$ and accordingly in any circle $|z - z_\tau| \leqslant (1-\zeta)R$.

By Lemma 6 the variation of $\arg f(z)$ on any rectilinear segment in the circle $|z - z_\tau| < d$ does not exceed the quantity $(2\pi + 4)\varepsilon R$. Each segment of the line $y = R$ whose length does not exceed $\sqrt{3}d$ is completely covered by two circles $K_{\tau,R}$ and accordingly the increase of the function $\arg f(x + iR)$ on this segment does not exceed the quantity $4(\pi + 2)\varepsilon R$. Thus for any $\varepsilon > 0$ and $y > R'_\varepsilon$

$$|c_f(y)| < \frac{8(\pi+2)e}{\sqrt{3}}\varepsilon \tag{6.56}$$

and accordingly

$$\lim_{y \to +\infty} c_f(y) = 0. \tag{6.57}$$

From the theorem just proved and from Theorems 4 and 5 we obtain the

COROLLARY. *If the spectrum of an almost-periodic function is bounded and*

$$\sigma(y_1, y_2) = \lim_{x_2 - x_1 \to +\infty} \frac{n(x_1, x_2, y_1, y_2)}{x_2 - x_1},$$

then

$$\lim_{\substack{y_1 \to -\infty \\ y_2 \to +\infty}} \sigma(y_1, y_2) = \frac{d}{2\pi}, \tag{6.58}$$

where d is the length of the smallest segment containing the entire spectrum of the function $f(x)$.

---

[6] Here, as usual, $c_f(y)$ denotes the mean motion for the function $f(x + iy)$, i.e.,

$$c_f(y) = \lim_{x_2 - x_1 \to +\infty} \frac{\arg f(x_2 + iy) - \arg f(x_1 + iy)}{x_2 - x_1}.$$

In fact, by Theorem 4 $\sigma(y_1, y_2)$ exists for all values y_1 and y_2 except possibly a countable set, while $\sigma(y_1, y_2) = \varphi'(y_2) - \varphi'(y_1)$, or what is the same thing,

$$\sigma(y_1, y_2) = \frac{1}{2\pi}[c_f(y_1) - c_f(y_2)]. \tag{6.59}$$

As $y_1 \to -\infty$ or $y_2 \to +\infty$ the quantities $c_f(y_1)$ and $c_f(y_2)$ tend, respectively, to the lower and upper bounds of the spectrum of the function.[7]

5. Functions Approximated by Exponential Polynomials

The theory presented in the preceding chapters makes it possible to carry out a very thorough investigation of the question of the distribution of the roots for entire functions of exponential type of the form

$$f(z) = \sum_1^\infty a_k e^{\lambda_k z} \qquad \left(\sum_1^\infty |a_k| < \infty\right), \tag{6.01}$$

for which the exponents λ_k belong to the boundary of some bounded convex region $\bar{I}$.[8] In this investigation we make use of the results presented in the first four sections of this chapter.

We shall carry out an investigation for functions of a wider class than those represented by the series (6.01). In order to define this class we introduce a norm for entire functions of exceptional type which is connected with a convex region I. Let $\bar{I}$ be any bounded convex region and $h(-\theta)$ its supporting function. By means of $h(\theta)$ we define a norm $\|F\|_I$ of the entire function $F(z)$ of exponential type in the following way:

$$\|F\|_I = \sup_{\substack{r > 0, \\ 0 < \theta < 2\pi}} (|F(re^{i\theta})|\, e^{-h(\theta) r}). \tag{6.60}$$

The set of entire functions of exponential type which are bounded in this norm forms *a linear space* B_I.

If a sequence of entire functions bounded in this norm converges in the norm, then it evidently converges uniformly in any finite portion of the plane, and accordingly there exists a limit entire function with bounded norm.

Thus to each bounded convex region $\bar{I}$ there corresponds a *complete* space B_I of functions of exponential type. In it are included only those entire functions of exponential type with indicator diagrams which belong to the convex region I. (I is symmetric to the region $\bar{I}$ relative to the real axis.) It is easy to see that all exponential polynomials (6.00), whose exponents belong to the boundary of

[7] One may show that if one of the ends does not enter the spectrum, then $\sigma(y_1, y_2) < \dfrac{d}{2\pi}$ for any y_1, y_2 $(y_2 > y_1)$.

[8] This question has been investigated in many papers. See, for example, M. Regensburger [1] and G. Mattison [1].

the region $\bar{I}$, enter into B_I and $\|P\|_I \leqslant \Sigma_1^n |a_k|$. This follows immediately from the relations

$$|P(re^{i\theta})| \leqslant \left(\sum_1^n |a_k|\right) \max_k e^{|\lambda_k| r \cos(\theta+\alpha_k)} = \left(\sum_1^n |a_k|\right) e^{h(\theta) r}$$
$$(\alpha_k = \arg \lambda_k).$$

In exactly the same way one shows that a function $f(x)$ of type (6.01) belongs to B_I if as the norming region $\bar{I}$ one takes the smallest convex region containing all the exponents λ_k. It is also evident that the series (6.01) converges in the norm of B_I.

We define *the class E_I of entire functions* approximated in the norm of B_I by exponential polynomials whose exponents belong to the frontier of $\bar{I}$. This class of functions is wider than the class of functions represented by the series (6.01).

We shall now find a representation of functions of the class E_I which is important for what follows. Let λ be any point of the boundary of $\bar{I}$, and θ a direction such that the point $\bar{\lambda}$ is a supporting point for the line l_θ: $x \cos\theta + y \sin\theta - h(\theta) = 0$. The direction $\arg z = \theta$ perpendicular to the line l_θ will be called the direction corresponding to the point λ. If λ is not an angular point of the boundary of the region $\bar{I}$, then the direction corresponding to it is unique. In the opposite case directions form an angle corresponding to the point λ.

Construct the function

$$F_\theta^\lambda(z) = F(ze^{i\theta}) \exp(-\lambda e^{i\theta} z), \tag{6.61}$$

where λ is a point of the boundary of $\bar{I}$, and θ is a direction corresponding to this point. The conjugate diagram $\bar{I}_\theta^\lambda$ of this function is obtained from the conjugate diagram $\bar{I}$ of the function $F(z)$ by shifting by the vector $-\lambda$ and rotating through the angle θ. Here the point λ goes into the point zero, the entire diagram $\bar{I}_\theta^\lambda$ is situated in the left half-plane, and the imaginary axis is supported by $\bar{I}_\theta^\lambda$ at the point zero. The supporting function of the region $\bar{I}_\theta^\lambda$ will be denoted by $h_\theta^\lambda(\varphi)$. In the general case we will have a supporting segment belonging to the imaginary axis and containing the point zero.

Each exponential polynomial $P_{n,\theta}^\lambda(z)$ may be represented in the form

$$P_{n,\theta}^\lambda(z) = Q_{n,\theta}^\lambda(z) + R_{n,\theta}^\lambda(z) \tag{6.62}$$

in such a way that the exponents of the exponential polynomial $Q_{n,\theta}^\lambda(z)$ lie on a segment of the imaginary axis, and the real parts of the exponents of the exponential polynomial $R_{n,\theta}^\lambda(z)$ are negative. For the given situation of the diagram $\bar{I}_\theta^\lambda$ $h_\theta^\lambda(0) = 0$, so that since the sequence of polynomials $P_n(z)$ converges in norm, we have for $x > 0$ and $n, m > N_\varepsilon$

$$|P_{m,\theta}^\lambda(x) - P_{n,\theta}^\lambda(x)| < \varepsilon.$$

On the other hand, $R^{\lambda}_{m,\theta}(x) \to 0$ and $R^{\lambda}_{n,\theta}(x) \to 0$ for $x \to +\infty$, which means that

$$\varlimsup_{x \to +\infty} |Q^{\lambda}_{m,\theta}(x) - Q^{\lambda}_{n,\theta}(x)| \leqslant \varepsilon \qquad (n, m > N_{\varepsilon}).$$

Hence for the almost-periodic polynomials $Q^{\lambda}_{m,\theta}(x)$ and $Q^{\lambda}_{n,\theta}(x)$ we have the following inequality, valid on the entire real axis:

$$|Q^{\lambda}_{m,\theta}(x) - Q^{\lambda}_{n,\theta}(x)| \leqslant \varepsilon \qquad (-\infty < x < \infty).$$

Thus the sequence of almost-periodic polynomials $Q^{\lambda}_{n,\theta}(z)$ converges uniformly to an almost-periodic function $\psi^{\lambda}_{\theta}(x)$ and we have obtained the following lemma.

LEMMA 12. *Let $F(z)$ be an entire function to which a sequence of exponential polynomials $P_n(z)$ converges in the norm of B_I, the polynomials $P_n(z)$ having exponents belonging to the boundary of $\bar{I}$. Let $F^{\lambda}_{\theta}(z)$ be the function defined by equation* (6.61). *Then $F^{\lambda}_{\theta}(z)$ may be represented in the form*

$$F^{\lambda}_{\theta}(z) = \Phi^{\lambda}_{\theta}(z) + \psi^{\lambda}_{\theta}(z), \tag{6.63}$$

where $\psi^{\lambda}_{\theta}(z)$ is an almost-periodic function whose spectrum lies on a portion of the boundary of the region I^{λ}_{θ}, lying on the imaginary axis, and $\Phi^{\lambda}_{\theta}(z)$ converges uniformly to zero as $\operatorname{Re} z \to +\infty$ *and* $|\operatorname{Im} z| \leqslant M$ (*M any positive number*).

In particular, it may turn out that $\psi^{\lambda}_{\theta}(z) \equiv 0$.

REMARK 1. *The representation* (6.63) *is unique.*

Indeed, from the equality of the two representations

$$\Phi^{\lambda}_{1\theta}(z) + \psi^{\lambda}_{1\theta}(z) = \Phi^{\lambda}_{2\theta}(z) + \psi^{\lambda}_{2\theta}(z)$$

it follows that

$$\lim_{x \to \infty} [\psi^{\lambda}_{1\theta}(x) - \psi^{\lambda}_{2\theta}(x)] = 0,$$

and in view of the almost-periodicity of the functions $\psi^{\lambda}_{1\theta}(x)$ and $\psi^{\lambda}_{2\theta}(x)$ we obtain

$$\psi^{\lambda}_{1\theta}(x) = \psi^{\lambda}_{2\theta}(x).$$

REMARK 2. If λ is an extreme point of the region I, then the function $\psi^{\lambda}_{\theta}(z)$ is equal to a constant $a(\lambda)$. The function $F(z)$ may be represented in this case in the form

$$F(z) = e^{\lambda z}[a(\lambda) + \chi_{\lambda}(z)], \tag{6.64}$$

while $\lim_{r \to \infty} \chi_{\lambda}[(r + i\tau)e^{i\theta}] = 0$ and the convergence is uniform in τ for $|\tau| \leqslant M$, if θ is the direction corresponding to the point λ of the boundary of the region $\bar{I}$, and such that the corresponding supporting line has only the one point λ of contact.

Now we shall define the important concept of the *spectrum of a function of class E_I*.

DEFINITION. The point λ of the boundary of the bounded region $\bar{I}$ is said to be *a point of the spectrum* of the function $F(z)$ of class E_I if zero is a point of the spectrum of the almost-periodic function ψ_θ^λ in the representation

$$F(ze^{i\theta})\exp(-\lambda ze^{i\theta}) = \psi_\theta^\lambda(z) + \Phi_\theta^\lambda(z)$$

in which θ is the direction corresponding to the point λ with $\Phi_\theta^\lambda(x) \to 0$ when $x \to +\infty$.

The collection of all points of the spectrum of the function $F(z)$ is called its *spectrum*. From the uniqueness of the representation (6.63) it follows that the spectrum of the function $F(z)$ of class E_I does not depend on the choice of the approximating sequence of exponential polynomials.

We shall show that if λ is a point of the spectrum of the function $F(z)$, then for some $N_\lambda > 0$ and $n > N_\lambda$ all the exponential polynomials $P_n(z)$ contain λ among their exponents and

$$\lim_{n\to\infty} a_n(\lambda) = a(\lambda), \tag{6.65}$$

where $a_n(\lambda)$ is the coefficient of $e^{\lambda z}$ in the exponential polynomial $P_n(z)$, and $a(\lambda)$ is the free term in the Fourier series of the almost-periodic function $\psi_\theta^\lambda(z)$. Indeed, if λ is an extreme point of $\bar{I}$, then from (6.64) and the inequality

$$|F(re^{i\theta}) - P_n(re^{i\theta})|\, e^{-h(\theta)r} < \varepsilon,$$

which is equivalent to the inequality

$$|\exp(-\lambda re^{i\theta})[F(re^{i\theta}) - P_n(re^{i\theta})]| < \varepsilon,$$

one obtains on passing to the limit as $r \to \infty$

$$|a(\lambda) - a_n(\lambda)| \leqslant \varepsilon.$$

Therefore (6.65) follows immediately. If now λ is a point of a rectilinear segment of the boundary of $\bar{I}$, then (6.65) follows from the representation (6.63) of the function $F_\theta^\lambda(z)$ and the known property (f) of almost-periodic functions. Conversely, if (6.65) is satisfied, then the point λ enters into the spectrum of the function $F(z)$. Thus we obtain a new definition of the spectrum.

The spectrum of the entire function $F(z)$, approximated in the norm of B_I by exponential polynomials whose exponents belong to the boundary of $\bar{I}$, is *the set of those points λ of the boundary of $\bar{I}$ for which the quantity*

$$a(\lambda) = \lim_{n\to\infty} a_n(\lambda)$$

is not equal to zero.

Thus the spectrum of the function $F(z)$ does not depend on the choice of the norming region $\bar{I}$, whose boundary contains all the exponents $\{\lambda_k\}$ of the approximating sequence of exponential polynomials. In particular, the spectrum of the entire function $f(z)$, represented by the series

$$f(z) = \sum_{k=1}^{\infty} a_k e^{\lambda_k z} \qquad \left(\sum_{k=1}^{\infty} |a_k| < \infty\right) \tag{6.01}$$

in which the numbers λ_k belong to the boundary of a convex region, consists of the set of all exponents λ_k.

From the representation (6.63) it is easy to obtain an expression for the "Fourier coefficient" of the function $F(z)$, and indeed to obtain the formula

$$a(\lambda) = \lim_{T \to +\infty} \frac{1}{T} \int_0^T F(re^{i\theta}) \exp(-\lambda e^{i\theta} r)\, dr, \tag{6.66}$$

where λ and θ have the same meaning as in (6.61).

Thus the spectrum of the function $F(z)$ may also be defined as the set of values λ for which $a(\lambda)$, defined by formula (6.66), is distinct from zero.

From the representation (6.63) it also follows that if λ is a point of the spectrum of the function $F(z)$, then the point $\bar{\lambda}$ enters into the boundary of the indicator diagram of the function $F(z)$, and accordingly the conjugate diagram of the function $F(z)$ contains all the points of its spectrum. Observe that the conjugate diagram of a function $F(z)$ of class E_I does not necessarily coincide with the smallest convex region containing the whole spectrum of the function. To see this, we construct an entire function of the class E_I which has no spectrum. Let $f(z)$ be a function holomorphic outside the unit circle, not continuable through the unit circumference, equal to zero at infinity and continuous for $|z| \geqslant 1$.[9] Put

$$F(z) = \int_{(C)} e^{\lambda z} f(\lambda)\, d\lambda,$$

where C is the unit circumference. Evidently $F(z)$ is an entire function of exponential type and its conjugate diagram is the unit circle. Put

$$P_n(z) = \sum_{k=1}^{n} e^{\lambda_k z} f(\lambda_k)(\lambda_{k+1} - \lambda_k) \qquad (\lambda_k \in C).$$

We shall show that if $|\lambda_{k+1} - \lambda_k| < \delta$ $(k = 1, 2, \ldots, n)$ and $\delta \to 0$ the function $P_n(z)$ tends to $F(z)$ in the norm of E_I, if as $\bar{I}$ we choose the unit circle C. We may without loss of generality assume that $z = r$, and consider the quantity

$$|F(r) - P_n(r)|\, e^{-r} = \left| \sum_{k=1}^{n} \int_{s_k} [e^{(\lambda - 1) r} f(\lambda) - e^{(\lambda_k - 1) r} f(\lambda_k)]\, d\lambda \right|,$$

[9] For example one may take

$$f(z) = \sum_{n=1}^{\infty} \frac{1}{n^2 z^{n!}}.$$

where s_k is an arc of the circumference of C between the points λ_k and λ_{k+1} $(k = 1, 2, \ldots, n)$. We have

$$|F(r) - P_n(r)| e^{-r} \leqslant \sum_{k=1}^{n} \int_{s_k} |e^{(\lambda-1)r} [f(\lambda) - f(\lambda_k)]| |d\lambda|$$

$$+ \sum_k \int_{s_k'} |e^{(\lambda-1)r} - e^{(\lambda_k-1)r}| |f(\lambda_k)| |d\lambda| + \sum_k \int_{s_k''} |f(\lambda_k)| |d\lambda|,$$

where s_k' is the arc of the circumference on which $|\arg \lambda| \geqslant \eta > 0$, and s_k'' is the remainder of the arc. We choose the arcs s_k so small that the oscillation of the function $f(\lambda)$ on each of these arcs will be less than $\varepsilon/6\pi$ and so that if $\lambda \in s_k'$ and $r > 0$ we will have

$$|e^{(1-\lambda)r} - e^{(\lambda_k-1)r}| < \frac{\varepsilon}{6\pi M},$$

where $M = \max |f(\lambda)|$ for $\lambda \in C$. Choosing in addition $\eta < \varepsilon/6\pi$, we obtain

$$|F(r) - P_n(r)| e^{-r} < \varepsilon$$

and in view of the equivalence of all directions

$$|F(re^{i\theta}) - P_n(re^{i\theta})| e^{-r} < \varepsilon.$$

Thus $F(z)$ is of class E_I. On the other hand, it is easy to show that for any value of θ

$$\lim_{r\to\infty} F(re^{i\theta}) e^{-r} = 0,$$

i.e., the spectrum of the function $F(z)$ is empty.

In connection with this example the following theorem takes on particular interest.

THEOREM 7. *If the norming region $\bar{I}$ is a polygon, then a function $F(z)$ of class E_I may be represented in the form*

$$F(z) = \sum_{j=1}^{p} e^{\lambda_j z} \psi_j(ze^{i\varphi_j}), \tag{6.67}$$

where λ_j is a point on the side d_j of the polygon $\bar{I}$, the angle φ_j defines the direction of the normal to that side, and $\psi_j(z)$ is an almost-periodic function, while the whole spectrum of the function $\psi_j(z)$ lies on the segment into which the side d_j of the polygon $\bar{I}$ goes on displacing it by the vector $-\bar{\lambda}_j$ and rotating it by the angle $-\varphi_j$. In addition the set of singularities of the associated function $f(z)$ coincides with the closure of the spectrum of the function $F(z)$.

PROOF. The exponential polynomials $P_n(z)$ approximating the function $F(z)$ may be represented in the form

$$P_n(z) = \sum_{j=1}^{p} e^{\lambda_j z} \psi_{jn}(ze^{i\varphi_j}),$$

where λ_j and φ_j have the same meaning as in the formula (6.67) and $\psi_{jn}(z)$ are almost-periodic polynomials, while the entire spectrum of the function $\psi_{jn}(z)$ lies on the segment into which the side d_j of the polygon $\bar{I}$ goes under displacement by the vector $-\bar{\lambda}_j$ and rotation by the angle $-\varphi_j$. Since the sequence $P_n(z)$ converges in the norm of B_I, therefore for any $\varepsilon > 0$, some N_ε and $n, m > N_\varepsilon$

$$|P_n(re^{i\theta}) - P_m(re^{i\theta})| e^{-h(\theta) r} < \varepsilon \qquad (0 \leqslant \theta < 2\pi,\ r > 0).$$

Hence we immediately obtain

$$\overline{\lim_{r\to\infty}} |\psi_{jn}(r) - \psi_{jm}(r)| = \overline{\lim_{r\to\infty}} |P_n(re^{-i\varphi_j}) - P_m(re^{-i\varphi_j})| e^{-h(-\varphi_j) r} < \varepsilon,$$

and accordingly

$$|\psi_{jn}(x) - \psi_{jm}(x)| < \varepsilon \qquad (-\infty < x < \infty).$$

On passing to the limit we obtain (6.67).

In order to prove the second part of the theorem it is sufficient to show that the set of singularities of the function associated to the almost-periodic function $\psi(z)$ with a bounded spectrum coincides with the closure of the set of its Fourier exponents $\{i\lambda_k\}$. Denote by $\chi(z)$ the function associated to the function $\psi(z)$ and by

$$\chi_n(z) = \sum_{k=1}^{n} \frac{a_n(\lambda_k)}{z - i\lambda_k}$$

the function associated to the almost-periodic polynomial $\psi_n(z)$ of a sequence uniformly converging to $\psi(z)$. From the inequality

$$|\psi(x) - \psi_n(x)| < \varepsilon \qquad (-\infty < x < \infty)$$

it follows that

$$|\chi(z) - \chi_n(z)| < \frac{\varepsilon}{|x|} \qquad (z = x + iy).$$

On the other hand,

$$\lim_{x\to 0} x\chi_n(x + i\lambda_k) = a_n(\lambda_k),$$

and accordingly if λ_k is a point of the spectrum of the function $\psi(z)$, then

$$\lim_{x\to 0} x\chi(x + i\lambda_k) = a(\lambda_k) \neq 0.$$

Thus, if λ_k is a point of the spectrum of $\psi(z)$, then $i\lambda_k$ is a singular point of the function $\chi(z)$. Suppose conversely that the point iy_0 lies at a positive distance d from the set $\{i\lambda_k\}$. We shall show that iy_0 is a regular point of the function $\chi(z)$. Indeed, from the inequality

$$|\psi_n(x) - \psi_m(x)| < \varepsilon \qquad (-\infty < x < \infty)$$

it follows that

$$|\chi_n(z) - \chi_m(z)| < \frac{\varepsilon}{|x|},$$

and applying the Poisson formula to the function $\ln|\chi_n(z) - \chi_m(z)|$ in a circle with center at the point iy_0 and of radius $R < d$ we obtain

$$\ln|\chi_n(z)-\chi_m(z)| < \frac{1}{2\pi}\int_0^{2\pi} \ln\frac{\varepsilon}{R|\cos\psi|}\frac{R^2-\rho^2}{R^2+\rho^2-2R\rho\cos(\psi-\theta)}\,d\psi.$$

$$(z = iy_0 + \rho e^{i\theta}),$$

From this inequality follows the estimate

$$\ln|\chi_n(z)-\chi_m(z)| < \left(\ln\frac{\varepsilon}{R}+\frac{R+\rho}{R-\rho}\ln 2\right),$$

from which one immediately finds that the sequence $\{\chi_n(z)\}$ converges uniformly in the neighborhood of the point iy_0, so that there is no singularity of the function $\chi(z)$ at the point iy_0. The theorem is proved.

Another example of an entire function for which the closure of the spectrum coincides with the set of singularities of the associated function, so that its conjugate diagram coincides with the smallest convex region containing the spectrum, is given by a function represented by the series

$$f(z) = \sum_{k=1}^{\infty} a_k e^{\lambda_k z} \tag{6.01}$$

in which $\Sigma_{k=1}^{\infty} |a_k| < \infty$ and the λ_k belong to the boundary of the convex region I (Chapter I, § 21).

6. Growth of a Function of Class E_I for a Norming Region in the Form of a Polygon

In this section we consider the case mentioned at the end of the preceding section.

Namely, we suppose that a sequence of exponential polynomials converges in the norm (6.60), while the norming region I is a polygon and the exponents of the exponential polynomials approximating the function $F(z)$ lie on the boundary of that polygon. Our fundamental result consists in the fact that a function $F(z)$ defined in this way is a function of completely regular growth.

First we prove the following lemma.

Lemma 13. *Suppose that $\varphi_1(z)$ and $\varphi_2(z)$ are entire almost-periodic functions of exponential type with non-negative spectra. Then the entire function of exponential type*

$$F(z) = \varphi_1(ze^{-i\alpha_1}) + \varphi_2(ze^{-i\alpha_2}) \qquad (\alpha_1 < \alpha_2) \tag{6.68}$$

is a function of completely regular growth.

Proof. Without loss of generality we may assume that the point $z = 0$ does not enter simultaneously in the spectra of both functions $\varphi_1(x)$ and $\varphi_2(x)$.

Since the functions associated to the functions $\varphi_1(ze^{-i\alpha_1})$ and $\varphi_2(ze^{-i\alpha_2})$ have no common singularities other than perhaps the origin, the conjugate diagram of the function $F(z)$ is the smallest convex region containing two rectilinear segments d_1 and d_2 on the rays $\arg z = \pi/2 - \alpha_1$ and $\arg z = \pi/2 - \alpha_2$. These rays are obtained by rotating the smallest segment of the imaginary axis containing the spectra of the functions $\varphi_1(z)$ and $\varphi_2(z)$, by the angles α_1 and α_2.

The functions $\varphi_1(z)$ and $\varphi_2(z)$ are bounded on the real axis and therefore (Chapter V, § 4, Theorem 11) are functions of completely regular growth. Thus the following limits exist:

$$\lim_{r\to\infty}{}^{*} \frac{\ln|\varphi_j(re^{i(\theta-\alpha_j)})|}{r} = h_j(\theta) \qquad (j = 1,\ 2).$$

Evidently if for some value of the angle θ we have $h_1(\theta) \neq h_2(\theta)$, then the limit

$$\lim_{r\to\infty}{}^{*} \frac{\ln|F(re^{i\theta})|}{r} = \max\{h_1(\theta),\ h_2(\theta)\}$$

exists.

If the point $z = 0$ is not a common end of the segments d_1 and d_2, then there exist only two values of the angle θ for which $h_1(\theta) = h_2(\theta)$. These values of θ correspond to the directions of the normals to those sides of the indicator diagram of the function $F(z)$ which join the ends of the segments d_1 and d_2. The function $F(z)$ is of completely regular growth in all directions except possibly two, and, by the theorem that the set of directions of completely regular growth is closed, it is a function of completely regular growth in the entire plane (Chapter III, corollary to Theorem 1).

If the point $z = 0$ is a common end of the segments d_1 and d_2, then $h_1(\theta) = h_2(\theta)$ for $\alpha_1 \leqslant \theta \leqslant \pi + \alpha_2$. We must prove that $F(z)$ is of completely regular growth inside this angle.

Suppose that the point $z = 0$ is a point of the spectrum of the function $F(z)$ with Fourier coefficient a_0. From the representation (6.64) of this function it follows that for $\alpha_1 < \theta < \pi + \alpha_2$

$$\lim_{r\to\infty} F(re^{i\theta}) = a_0.$$

Hence it follows that the function $F(z)$ is of completely regular growth inside that angle, and thus in the entire plane.

The case when the point $z = 0$ is a common end point of the two segments d_1 and d_2 and does not enter into the spectrum of the function $F(z)$ is somewhat more complicated.

Suppose, without loss of generality, that $\alpha_1 = 0$, so that

$$F(z) = \varphi_1(z) + \varphi_2(ze^{-i\alpha_2}). \tag{6.69}$$

It is easy to see, using the approximation of the function $\varphi_2(z)$ by generalized trigonometric polynomials, that as $x \to +\infty$ $\varphi_2(ze^{-i\alpha_2})$ converges uniformly to zero.

From Lemma 1 of this chapter, in any horizontal strip $|y| \leqslant h$, but outside circles of radius δ with centers at the roots of the function $\varphi_1(z)$, the inequality

$$|\varphi_1(z)| > m(\delta, h) > 0$$

is satisfied.

Since for some $x_0 > \chi(\delta, h)$, $x > x_0$ and $|y| \leqslant h$ the inequality $|\varphi_2(ze^{-i\alpha_2})| < m(\delta, h)$ holds, therefore in the strip $|y| \leqslant h$ for $x > x_0$ there are no roots of the function $F(z)$ outside the regions formed by the exceptional circles.

By Lemma 2, the number of roots of the function $\varphi_1(z)$ in the rectangle $|y| \leqslant h$, $t \leqslant x < t + 1$ (t arbitrary) does not exceed some number k_h. For $\delta < [2(k_h + 1)]^{-1}$ the diameters of the regions formed by the "clusters" of exceptional circles do not exceed the quantity $2k_h\delta$, so that for a sufficiently small δ they will be arbitrarily small. By the well-known Rouché theorem the function $F(z)$ has, in each such region, the same number of roots as the function $\varphi_1(z)$.

Denote by $n_F(h, t)$ the number of roots of the function $F(z)$ in the rectangle $0 < x < t$, $|y| \leqslant h$, and by $n_\varphi^{(1)}(u, t)$ the number of roots of the function $\varphi_1(z)$ in the rectangle $|y| \leqslant u$, $0 < x \leqslant t$. From the above we have for $\eta > 0$ the following inequality:

$$\lim_{t\to\infty} \frac{n_\varphi^{(1)}(h - \eta, t)}{t} \leqslant \underline{\lim}_{t\to\infty} \frac{n_F(h, t)}{t} \leqslant \overline{\lim}_{t\to\infty} \frac{n_F(h, t)}{t}$$
$$\leqslant \lim_{t\to\infty} \frac{n_\varphi^{(1)}(h + \eta, t)}{t}. \tag{6.70}$$

The limits referred to in the inequality (6.70), which exist in view of Theorem 4 of this chapter, are continuous functions of $h + \eta$ and $h - \eta$ for all values of these quantities with the possible exclusion of a countable set. Thus, for almost all values of the quantity h there exists the limit

$$\lim_{t\to\infty} \frac{n_F(h, t)}{t} = \lim_{t\to\infty} \frac{n_\varphi^{(1)}(h, t)}{t}.$$

Making use of the corollary to Theorem 6 of this chapter, we may assert that for any $\varepsilon > 0$ and $h > h_\varepsilon$

$$\lim_{t\to+\infty} \frac{n_F(h, t)}{t} > \frac{l_1 - \varepsilon}{2\pi}, \tag{6.71}$$

where l_1 is the length of the spectrum of the function $\varphi_1(z)$.

Thus, the lower density of the set of roots inside the angle $(-\varepsilon, \varepsilon)$ $(\varepsilon > 0)$, containing the positive half-axis, satisfies the inequality

$$\underline{\lim}_{r\to\infty} \frac{n_F(r, -\varepsilon, \varepsilon)}{r} \geqslant \frac{l_1}{2\pi}.$$

Analogously, for an angle containing the normal to the second side of the indicator diagram.

$$\varliminf_{r\to\infty} \frac{n_F(r,\ \pi-\alpha_2-\varepsilon,\ \pi-\alpha_2+\varepsilon)}{r} \geqslant \frac{l_2}{2\pi}, \tag{6.72}$$

where l_2 is the length of the spectrum $\varphi_2(z)$. Outside the angle $(0, \pi - \alpha_2)$ the function $F(z)$ is of completely regular growth. Therefore inside a small angle, containing the normal to the third side of the indicator diagram, there exists a density for the set of roots of the function $F(z)$ and it is equal to the length of the third side l_3, divided by 2π (Chapter III). Thus,

$$\varliminf_{r\to\infty} \frac{n_F(r)}{r} \geqslant \frac{s_F}{2\pi}, \tag{6.73}$$

where s_F is the length of the entire indicator diagram of the function $F(z)$. By Theorem 3 of Chapter IV and the remark to it, $F(z)$ is a function of completely regular growth.

REMARK. In (6.71) we have also established the following fact relative to the asymptotic distribution of the roots of the function $F(z)$. In every strip parallel to the normals d_j, $|\mathrm{Im}\,(ze^{-i\alpha_j})| \leqslant h$, there exists a density for the set of roots of the function $F(z)$ for all values of h with the possible exception of a countable set, while for an unbounded expansion of the strip ($h \to +\infty$) this density converges to the quantity $l_j/2\pi$, where l_j is the length of the corresponding segment.

THEOREM 8. *If I is a polygon, then the entire function $F(z)$ of class E_I is a function of completely regular growth.*

PROOF. Consider the angle between the normals $\arg z = \theta_j$ and $\arg z = \theta_{j+1}$ to the two successive sides d_j and d_{j+1} of the indicator diagram I_F of the function $F(z)$. Without loss of generality one may assume that the common end point of the segments d_j and d_{j+1} lies at the point $z = 0$. The function $F(z)$ may be represented in the form

$$F(z) = \psi_j(ze^{-i\theta_j}) + \psi_{j+1}(ze^{-i\theta_{j+1}}) + F_{j,j+1}(z), \tag{6.74}$$

where $\psi_j(z)$ and $\psi_{j+1}(z)$ are almost-periodic functions with positive spectra (one of them may be equal to zero), and $F_{j,j+1}(z)$ is the sum of the remaining terms of (6.67).

The first two terms form, by Lemma 13, a function of completely regular growth with an indicator equal to zero inside the angle $\theta_j < \arg z < \theta_{j+1}$. The conjugate diagram of the function $F_{j,j+1}(z)$ is a polygon lying in the smallest convex polygon containing all the sides of the polygon I other than d_j and d_{j+1}. Therefore the indicator of $F_{j,j+1}(z)$ is a negative quantity inside the angle $\theta_j < \arg z < \theta_{j+1}$, and accordingly $F(z)$ is of completely regular growth inside that angle. Thus $F(z)$ is of completely regular growth inside any angle between two successive normals. Since the set of rays of completely regular growth is closed (Chapter III, Theorem 1), $F(z)$ is of completely regular growth in the whole plane. The theorem is proved.

Comparing this theorem with Theorem 2 of Chapter III, we find that *all the roots of the function $F(z)$ except possibly for a set of zero density lie inside arbitrarily small angles containing the normals to the sides of the indicator diagram, and the density of the set of roots inside each of these angles is equal to $l_j/2\pi$, where l_j is the length of the corresponding side.*

In a special case Theorem 8 may be substantially sharpened. This sharpening may be formulated as a separate theorem.

THEOREM 9. *If $F(z)$ is the entire function defined in Theorem 8, and all the vertices of the polygon I enter into its spectrum, then all of its roots lie in the half-planes $|\operatorname{Im}(ze^{-i\theta_j})| < h$, $\operatorname{Re}(ze^{-i\theta_j}) \geqslant 0$, where θ_j defines the direction of the normal to the side of the polygon I and h is some positive number. Moreover, if one denotes by $n_j(t)$ the number of roots of the function $F(z)$ in the rectangle*

$$|\operatorname{Im}(ze^{-i\theta_j})| \leqslant h \ u \ 0 \leqslant \operatorname{Re}(ze^{\ \ i\theta_j}) \leqslant t, \text{then } n_j(t) = \frac{d_j}{2\pi}t + O(1).$$

PROOF. This theorem may be proved independently of the general Theorem 8. Without loss of generality one may assume that the vertex λ_j, in which the sides d_j and d_{j+1} of the polygon come together, lies at the point 0.

From the representation

$$F(z) = a_0 + \psi_j(ze^{-i\theta_j}) + \psi_{j+1}(ze^{\ \ i\theta_{j+1}}) + F_{j,j+1}(z) \tag{6.75}$$

(in which $\psi_j(z)$, $\psi_{j+1}(z)$ and $F_{j,j+1}(z)$ have the same meaning as in (6.74)), it follows that for some $h > 0$ there are no roots of the function $F(z)$ inside the angle $\operatorname{Im}(ze^{-i\theta_j}) \geqslant h$, $\operatorname{Im}(ze^{-i\theta_{j+1}}) \geqslant h$ $(\theta_{j+1} > \theta_j)$.

Indeed, the spectrum of the almost-periodic function $\psi_j(z)$ is positive, and accordingly the function uniformly converges to zero as $\operatorname{Im} z \to +\infty$. Thus, the second term in (6.75) tends to zero as $\operatorname{Im}(e^{i\theta_j}z) \to +\infty$. Analogously one proves that the third term in (6.75) converges uniformly to zero as $\operatorname{Im}(e^{i\theta_{j+1}}z) \to +\infty$. In exactly the same way one easily proves that $F_{j,j+1}(z) \to 0$ uniformly for $\operatorname{Im}(e^{i\theta_j}z) \geqslant h$, $\operatorname{Im}(e^{i\theta_{j+1}}z) \geqslant h$, and $h \to +\infty$. Thus, in the indicated region for sufficiently small h we have

$$|F(z) - a_0| < |a_0|, \tag{6.76}$$

and thus $F(z)$ has no roots.

In order to obtain an asymptotic estimate of the number of roots we shall suppose without loss of generality that

$$F(z) = \psi_j(z) + F_1(z),$$

where $\psi_j(z)$ is an almost-periodic function of class $[\Delta]$, and $F_1(z) \to 0$ uniformly as $\operatorname{Re} z \to +\infty$ in any strip $|\operatorname{Im} z| \leqslant h$. From Theorem 3 of this chapter, all the roots of the function $\psi_j(z)$ lie in some strip $|\text{Im } z| \leqslant h$ and $n_j(t) = (l_j/2\pi)t + O(1)$, where l_j is the "length" of the spectrum $\psi_j(z)$, and $n_j(t)$ is the number of its roots in the rectangle $0 < \text{Re } z \leqslant t$, $|\text{Im } z| \leqslant h$. Moreover, outside regions covered by circles of radius δ with centers at the roots of the function $\psi_j(z)$, it is true that $|\psi_j(z)| \geqslant m(\delta) > 0$. For some $x_0 = x(\delta)$, $\text{Re } z > x_0$ and $|\text{Im } z| < h$ we obtain $|F_1(z)| < m(\delta)$, and accordingly the function $F(z)$ has no roots outside these regions. By Rouché's theorem the function $F(z)$ has in each of these regions the same number of roots as the function $\psi_j(z)$. Thus,

$$n_j(t) = \frac{l_j}{2\pi} t + O(1). \tag{6.77}$$

REMARK. Using Rouché's theorem it is easy to prove that the asymptotic picture of the distribution of the roots of the function

$$\tilde{F}(z) = F(z) + \Phi(z), \tag{6.78}$$

where $F(z)$ satisfies the conditions of Theorem 9, and $\Phi(z)$ is such that $|\Phi(re^{i\theta})| \, e^{-h(\theta)r} \to 0$ uniformly as $r \to \infty$ ($h(\theta)$ is the supporting function for the polygon I), is the same as for the function $F(z)$. In particular, one may represent any exponential polynomial in that form, if as $\bar{I}$ one takes the smallest convex region containing all the exponents of that exponential polynomial. The role of the function $\Phi(z)$ is played by the sum of the exponents of the functions with exponents lying strictly inside the polygon.

We shall construct an example showing that if a vertex of the polygon does not enter into the spectrum, then inside the angle corresponding to that vertex there may be an infinite set of roots.

EXAMPLE. Consider a function $F(z)$ represented by the series

$$F(z) = \sum_{n=1}^{\infty} a_n e^{\alpha_n z} \sin(1 - \alpha_n) z$$

$$(0 = \alpha_1 < \alpha_2 < \alpha_3 < \dots < \alpha_n < \dots < 1, \ \alpha_n \to 1), \tag{6.79}$$

where $\Sigma\, |a_n| < \infty$. As the region I one may employ the triangle $|y| \leqslant 1 - x$, $x \geqslant 0$ and the spectrum of $F(z)$ consists of the points $\alpha_n \pm i\alpha_n$. The point $z = 1$ does not enter into the spectrum of $F(z)$. We shall show that this function can have an infinite set of positive roots. To this end we shall select the coefficients a_n in the following way. Suppose that the numbers $a_1, a_2, \dots, a_n$ have already been chosen. Consider the function

$$P_m(z) = \sum_{n=1}^{m} a_n e^{\alpha_n z} \sin(1 - \alpha_n) z \tag{6.80}$$

in the strip $|y| \leqslant 1$. It is easy to see that there exists a number N_m such that for $x > N_m$ and outside the exceptional circles of radius $\delta < \frac{1}{2}$ with centers at the roots of the function $\sin(i - \alpha_m) z$ the modulus of the last term of the sum (6.80) exceeds twice the modulus of the sum of the remaining terms. Moreover, one may suppose that the modulus of the last term in that region is larger than unity. Define a positive number $a_{m+1} < 2^{-m-1}e^{-2N_m}$ and then an abscissa N_{m+1} such that for $x > N_{m+1}$ and outside the exceptional circles of radius δ with centers at the roots of the function $\sin(1 - \alpha_{m+1}) z$ the modulus of the last term in the sum $P_{m+1}(z)$ exceeds the doubled modulus of the sum of all the remaining terms, and moreover is larger than unity. We put $a_{m+2} < 2^{-m-2}e^{-2Nm+1}$ and so forth.

For $N_m < x < 2N_m$ and $|y| \leqslant 1$, but outside the exceptional circles, we will have

$$\frac{1}{2}\left|e^{\alpha_m z}\sin(1-\alpha_m)z\right| > |P_{m-1}(z)|$$

and

$$\left|\sum_{n=m+1}^{\infty} a_n e^{\alpha_n z}\sin(1-\alpha_n)z\right| \leqslant e^{2N_m}\sum_{n=m+1}^{\infty}|a_n| < \frac{1}{2^m}.$$

From the last inequality it follows that in the indicated region the term $a_m e^{\alpha_m z}\sin(1-\alpha_m)z$ of the sum (6.80) exceeds in modulus the entire remaining sum. From Rouché's theorem one finds that the function $F(z)$ has in each exceptional circle of the rectangle $N_m < x < 2N_m$, $|y| \leqslant 1$ exactly one root. Since $F(z)$ is real it follows that that root is real.

This example shows that if the conjugate diagram $\bar{I}$ of the function $F(z)$, with which Theorem 8 is concerned, is a polygon and not all the vertices of this polygon enter into the spectrum of the function $F(z)$, then that function may have an infinite set of roots outside of angles containing the normals to the sides of $\bar{I}$. In view of Theorem 8 this set must have zero density.[10]

---

[10] If one denotes by γ_j $(j = 1, 2, \ldots)$ the positive roots of the function $F(z)$ so constructed, then

$$\sum_{\gamma_j < 2N_m}\frac{1}{\gamma_j} > \frac{1}{2\pi}\sum_{k=1}^{m}(1-\alpha_k),$$

and if one chooses the points α_k so that the series

$$\sum_{k=1}^{\infty}(1-\alpha_k)$$

diverges, then also the series

$$\sum_{j=1}^{\infty}\frac{1}{\gamma_j}$$

will diverge. This shows that in spite of the zero density, the set of roots outside angles con taining the normals to the sides of $\bar{I}$ may be quite "thick."

7. Growth of a Function of Class E_I Outside an Arbitrary Norming Region $\bar{I}$

In this section we study the general case of an entire function of class E_I, for which the conjugate diagram coincides with the smallest convex region containing the entire spectrum. We have shown in § 5 that this always takes place if the norming region $\bar{I}$ is a polygon or if the function is represented by the series (6.01). Moreover, the example of § 5 shows that in the general case this is not the case. The growth of a function of class E_I, having that property is characterized by the following theorem.

THEOREM 10. *If the conjugate diagram of an entire function of class E_I coincides with the smallest convex region containing the entire spectrum of the function, then that function has completely regular growth.*

PROOF. From the representation (6.63) it follows that a direction θ which corresponds to a point of the spectrum of the function $F(z)$ is either a point of the spectrum or a condensation point for points of the spectrum.

Using this fact we shall show that a direction corresponding to any extreme point of the conjugate diagram is a direction of completely regular growth.

Suppose that the extreme point λ_0 of the conjugate diagram $\bar{I}_F$ is not a corner point of the boundary of $\bar{I}$. There corresponds to it a unique direction θ_0. If λ_0 is a point of the spectrum, then its direction is a direction of completely regular growth. Now if λ_0 does not enter into the spectrum, then it is the limit point for a sequence of points λ_j of the spectrum, and the direction θ_0 is a limiting value for directions θ_j corresponding to the points λ_j. The directions θ_j are directions of completely regular growth, and since such directions form a closed set (Chapter III, corollary to Theorem 1), then also θ_0 is a direction of completely regular growth.

Now suppose that λ_0 is an angular point of $\bar{I}$. If λ_0 enters into the spectrum of the function $F(z)$, then all the directions corresponding to it are directions of completely regular growth. It remains only to consider angular points not entering into the spectrum of the function. First we consider angular points which are not common end points for two rectilinear segments of the boundary of the conjugate diagram. In this case there exists a sequence $\{\lambda_j\}$ of points of the spectrum not belonging to a rectilinear segment of the boundary of $\bar{I}$, which converges to λ_0 as a limit. In addition, the sequence of corresponding directions $\{\theta_j\}$ converges to a direction θ_0 defining one of the sides of the angle (θ_1, θ_2) corresponding to the angular point λ_0. The directions θ_j are directions of completely regular growth, and accordingly the same is true for the direction θ_0. This direction is not perpendicular to a rectilinear segment of the boundary, and accordingly $h'_+(\theta_0) = h'_-(\theta_0)$. Moreover, within the angle (θ_1, θ_2) corresponding to the angular point λ_0 of the boundary of $\bar{I}$, the indicator $h(\theta)$ is trigonometric. According to Theorem 7 of Chapter III, $F(z)$ has completely regular growth within this angle.

Now suppose that the angular point λ_0 of the conjugate diagram $\bar{I}$ is a common end point of two rectilinear segments of the boundary of $\bar{I}$. Then the function $F(z)$ may be represented in the form

$$F(z) = e^{\lambda_0 z}[\psi_1(ze^{-i\theta_1}) + \psi_2(ze^{-i\theta_2}) + F_{1,2}(z)],$$

where $\psi_1(z)$ and $\psi_2(z)$ are almost-periodic functions with positive spectra, $\theta_1 \leqslant \arg z \leqslant \theta_2$ is the angle corresponding to the point λ_0, and $F_{1,2}(z)$ is an entire function of exponential type, whose indicator is negative inside that angle. Applying Lemma 13, we find from this representation that $F(z)$ is of completely regular growth inside that angle, corresponding to the point λ_0. Thus, $F(z)$ is of completely regular growth in all directions corresponding to the extreme points of $\bar{I}$. The directions corresponding to rectilinear segments of the indicator diagram form no more than a countable set. But the set of directions of completely regular growth is closed (Chapter III, § 1, corollary to Theorem 1), and accordingly $F(z)$ is of completely regular growth in the entire plane. The theorem is proved.

REMARK 1. From this theorem and Theorem 2 of Chapter III it follows that the set of roots of the function $F(z)$ has a density within the angle (ϑ, θ) for all values of ϑ and θ with the exclusion of at most a countable set, and this density $\Delta(\theta) - \Delta(\vartheta)$ is equal to the arc length of the boundary of I between the supporting points of the lines l_ϑ and l_θ.[11]

REMARK 2. If the function $F(z)$ satisfies all the conditions of Theorem 10 and in addition all the angular points of $\bar{I}$ enter into its spectrum, and if the function $F_1(z)$ satisfies the condition

$$|F_1(re^{i\theta})|\, e^{-h_F(\theta)r} \to 0$$

uniformly in θ as $r \to \infty$, then

$$\Phi(z) = F(z) + F_1(z)$$

is a function of completely regular growth with indicator $h_F(\theta)$.

Indeed, if θ_0 is a direction corresponding to an angular point λ_0 of the spectrum of the function $F(z)$, then

$$\lim_{r\to\infty} \Phi(re^{i\theta}) \exp(-\lambda_0 re^{i\theta}) = \lim_{r\to\infty} F(re^{i\theta}) \exp(-\lambda_0 re^{i\theta}) = a(\lambda_0) \neq 0.$$

Therefore each such direction is a direction of completely regular growth. These directions (values of the angle θ) are densely distributed on the interval $(0, 2\pi)$ and accordingly $\Phi(z)$ is a function of completely regular growth with indicator $h_\Phi(\theta) = h_F(\theta)$.

---

[11] Regensburger **[1]** obtained for sums of the form (6.01) the following result. If the points λ_k are densely distributed on some convex arc and $r_k e^{i\theta_k}$ are the roots of the function $f(z)$ lying between the normals to the end points of the arc, then $\sum_{k=1}^{\infty} r_k^{-1} = \infty$. This result of course follows from Theorem 10.

In conclusion we consider one further general example of a function of class E_I, where I is the unit circle, for which the smallest convex region containing the spectrum coincides with the conjugate diagram of the function.

EXAMPLE. Suppose that

$$F(z) = \sum_{n=1}^{\infty} \frac{\varphi(n)}{n!} z^n, \tag{6.81}$$

where $\varphi(n)$ is an almost-periodic function on the group of integers. It is known[12] that such a function may be approximated by almost-periodic functions of the form

$$P_m(n) = \sum_{k=1}^{m} a_k^{(m)} e^{i\lambda_k n}. \tag{6.82}$$

Thus, the function associated with the function $F(z)$ in the sense of Borel (the Borel transform) has the form

$$\psi(z) = \sum_{n=1}^{\infty} \frac{\varphi(n)}{z^{n+1}} = \sum_{k=1}^{m} \frac{a_k^{(m)}}{z - e^{i\lambda_k}} + \sum_{n=1}^{\infty} \frac{\varepsilon^{(m)}(n)}{z^{n+1}} \qquad (|\varepsilon^{(m)}(n)| < \varepsilon).$$

Therefore it easily follows that all the points $e^{i\lambda_k}$, where λ_k is a point of the spectrum of $\varphi(n)$, are singular points of the function $\psi(z)$. The points $e^{i\lambda_k}$ are points of the spectrum of the function $F(z)$ if one takes as the region I the unit circle.

We shall prove that the set of singularities of the function $\psi(z)$ coincides with the closure of the set of points $e^{i\lambda_k}$. This is evident if the points $e^{i\lambda_k}$ are densely distributed on the unit circumference.

Now suppose that some arc of the unit circumference is free from points $e^{i\lambda_k}$. Without loss of generality one may suppose that the middle of this arc lies at the point -1, i.e., that the arc is supplementary to the arc $|\arg z| \leqslant q < \pi$. The exponents λ_k in (6.82) are defined up to an integer multiple of 2π, so that one may suppose that $|\lambda_k| \leqslant q < \pi$. The polynomials $P_m(z) = \sum_{k=1}^{m} a_k^{(m)} e^{i\lambda_k z}$ form a sequence uniformly converging on the set of integers, i.e.,

$$|P_s(z) - P_t(z)| < \varepsilon \text{ for } s, t > N_2 \text{ and } \pm z = 0, 1, 2, \ldots. \tag{6.83}$$

For the proof of uniform convergence of this sequence on the whole real axis we refer to the theorem of M. Cartwright, proved in Chapter IV. Since the almost-periodic polynomials $P_s(z)$ are entire functions of degree not larger than $q < \pi$, by that theorem it follows from (6.83) that

$$|P_s(x) - P_t(x)| < c(q)\varepsilon \qquad (-\infty < x < \infty), \tag{6.84}$$

[12] See, for example, B. M. Levitan [1].

where $c(q)$ is a constant depending only on q. From (6.84) and from the uniform convergence of the sequence $P_m(z)$ on the set of real points it follows that this sequence converges uniformly on the entire real axis, which means that it also converges uniformly in any strip parallel to the real axis. The limit function $\varphi(z)$ is evidently an entire function of degree less than or equal to q. We represent the function $\varphi(z)$ in the form (Chapter I, § 21)

$$\varphi(z) = \frac{1}{2\pi i} \int_{\Delta_\varepsilon} \sigma(u) e^{uz} \, du, \tag{6.85}$$

where $\sigma(u)$ is the Borel transform of $\varphi(z)$, and Δ_ε is the rectangle formed by the lines $\operatorname{Re} u = \pm\varepsilon$ and $\operatorname{Im} u = \pm(q + \varepsilon)$ ($\varepsilon > 0$). For $|z| > 1 + \varepsilon$ the function $\psi(z)$ may be represented in the form

$$\psi(z) = \sum_{n=0}^{\infty} \left[\frac{1}{2\pi i} \int_{\Delta_\varepsilon} \sigma(u) e^{nu} z^{-n-1} \, du \right] = \frac{1}{2\pi i} \int_{\Delta_\varepsilon} \frac{\sigma(u)\, du}{z - e^u}.$$

From this representation it is clear that $\psi(z)$ has no singularities outside the region into which the rectangle Δ_ε goes under the mapping $z = e^u$. Since $\varepsilon > 0$ is arbitrary, the function $\psi(z)$ may have singularities only on the arc $|z| = 1$, $|\arg z| \leqslant q < \pi$. Thus the set of singularities of $\psi(z)$ coincides with the closure of the point spectrum of $F(z)$, from which it follows that the conjugate diagram of the function $F(z)$ is the smallest convex region containing all its point spectrum. By Theorem 10, $F(z)$ is of completely regular growth.[13]

[13] This example was considered in the paper of H. Mattison[1]. He obtained the following result: *if the points $e^{i\lambda_k}$ are distributed densely on some arc $\theta_1 \leqslant \arg z \leqslant \theta_2$ and if z_j are the roots of $F(z)$ inside the angle $(-\theta_2, -\theta_1)$, then* $\sum_1^\infty 1/|z_j| = \infty$.

CHAPTER VII

THE THEOREM OF HERMITE-BIEHLER FOR ENTIRE FUNCTIONS

In the preceding chapters we were occupied with the fundamental question of the regularity of the growth of entire functions and the connection of this regularity with the distribution of the roots of the entire function. In a somewhat different section lie other investigations also connected with the distribution of roots and the establishment of conditions for all the roots of an entire function to belong to some region G. In what follows we shall take as the region G the open or closed upper half-plane.

In the exposition of these investigations we shall make use of certain methods and theorems developed in the preceding chapters, in particular, in Chapters III and V.

The basic algebraic fact in this domain is the well-known theorem of Hermite-Biehler **[1]** for polynomials.

In order that the polynomial

$$\omega(z) = P(z) + iQ(z), \tag{7.00}$$

where $P(z)$ and $Q(z)$ are real polynomials,[1] *not have any roots in the closed lower half-plane* $\mathrm{Im}\, z \leqslant 0$, *it is necessary and sufficient that the following condition be satisfied*:

(1) *the polynomials $P(z)$ and $Q(z)$ have only simple real roots, while these roots separate one another, i.e., between two successive roots of one of these polynomials there lies exactly one root of the other*;

(2) *at some point x_0 of the real axis*

$$Q'(x_0)P(x_0) - Q(x_0)P'(x_0) > 0. \tag{7.01}$$

The fact that the two polynomials satisfy condition (1) will be expressed by saying that "the roots of the polynomials $P(z)$ and $Q(z)$ are interlaced," or that "the polynomials $P(z)$ and $Q(z)$ satisfy the interlaced roots condition." The same terminology will be used for entire functions also.

[1] An analytic function is said to be real if all the coefficients of its expansion into a power series in the independent variable are real, or, put differently, if it takes on real values on the real axis. Evidently every entire function can be represented in the form (7.00), where $P(z)$ and $Q(z)$ are real entire functions.

We observe that if the roots of the polynomials $P(z)$ and $Q(z)$ are interlaced, then the expression $Q'(x)P(x) - Q(x)P'(x)$ has the same sign on the whole real axis.

It is easy to see that in its direct form the Hermite-Biehler criterion does not carry over to arbitrary entire functions.

One may show with simple examples that, for the entire function $\omega(z) = P(z) + iQ(z)$, the two conditions of reality and interlacing of the roots of the functions $P(z)$ and $Q(z)$ are neither necessary nor sufficient for the function $\omega(z)$ to have no roots in the lower or in the upper half-plane. Indeed, suppose for example that $\varphi(z)$ and $\psi(z)$ are real polynomials whose roots interlace. Then the roots of the real and imaginary parts of the function

$$\omega(z) = e^{z}\varphi(z) + i\psi(z)$$

are also real and interlace. On the other hand, this function is of completely regular growth and its indicator diagram is the segment $0 \leqslant x \leqslant 1$ of the real axis (Chapter I, § 20). Thus, inside arbitrarily small angles containing rays of the imaginary axis, there is an infinite set of roots (with density $\Delta = (2\pi)^{-1}$) and accordingly the condition of interlacing of the roots of $P(x)$ and $Q(x)$ is not sufficient for all the roots of the function $\omega(z)$ to lie in one of the half-planes $\operatorname{Im} z \geqslant 0$ or $\operatorname{Im} z \leqslant 0$. In order to show that it also is not necessary, we introduce the following example. The function

$$\omega(z) = e^{-i\lambda z}(z - i)$$

has only one root i, and it is easy to verify that the real part $z \cos \lambda z - \sin \lambda z$ has for $\lambda > 1$ a pair of complex roots, and for $\lambda < 1$ three roots between the two roots of the imaginary part $-z \sin \lambda z - \cos \lambda z$ closest to zero.

For the solution of certain questions of the theory of automation one needs effective criteria for all the roots of the function

$$F(z) = \sum_{k,\, j=0}^{n,\, m} a_{kj} z^{i} e^{\lambda_k z} \tag{7.02}$$

to lie in the left half-plane. N. G. Čebotarev called functions of type (7.02) quasipolynomials. Quasipolynomials depend only on a finite number of parameters, and therefore it is natural to suppose that for them there exists an effective method of solving this problem.

In 1942 N. G. Čebotarev found such an effective criterion for a very special case of polynomials [3]. In another paper N. G. Čebotarev [5], generalizing the Sturm algorithm to quasipolynomials, gave a general principle for solving this problem for arbitrary quasipolynomials. However the application of this general principle required a generalization of the Hermite-Biehler theorem to quasipolynomials.

L. S. Pontrjagin [1] in 1942 generalized the Hermite-Biehler theorem to quasipolynomials of type $P(z, e^{z})$, where $P(z, u)$ is a polynomial in two variables.

The papers of N. G. Čebotarev [2; 3; 4; 5; 6] and of L. S. Pontrjagin [1] awakened interest in this group of questions, and there shortly after appeared a series of papers by Soviet mathematicians in this direction.[2] In these papers the question of carrying the Hermite-Biehler criterion to arbitrary entire functions was completely cleared up.

The investigations connected with the Hermite-Biehler theorem later found a new application. They play an essential role in the study of certain extremal properties of entire functions. These applications will be presented in Chapter IX.

1. Representation of a Real Meromorphic Function Mapping the Upper Half-Plane onto the Upper Half-Plane

In carrying over the Hermite-Biehler criterion to arbitrary entire functions, an essential role is played by a particular class of entire functions. This class was introduced and studied by M. G. Kreĭn in his paper "On a class of entire and meromorphic functions," which was devoted to the extension of the Hurwitz criterion to entire functions.[3]

DEFINITION. The entire function $\omega(z)$ is said to be a *function of class HB* if it has no roots in the closed lower half-plane $\operatorname{Im} z \leqslant 0$, and if

$$\left|\frac{\omega(z)}{\overline{\omega}(z)}\right| < 1 \quad \text{for} \quad \operatorname{Im} z > 0.^{4} \tag{7.03}$$

If $\omega(z)$ is a polynomial, then condition (7.03) evidently follows from the fact that all the roots of the polynomial lie in the upper half-plane, i.e., in this case condition (7.03) is superfluous.

It follows from the Phragmén-Lindelöf theorem that the same circumstance also holds for entire functions of zero degree. For entire functions of arbitrary growth the conditions defining the class HB are independent. We note further that it follows from (7.03) that the function $\omega(z)/\overline{\omega}(z)$ belongs in the upper half-plane to the class A, and since the roots of this function in the upper half-plane coincide with the roots of $\omega(z)$, therefore every function of class HB is an entire function of class A.

In what follows we will frequently need to use the equivalence of condition (7.03) for the function

$$\omega(z) = P(z) + iQ(z),$$

---

[2] See the papers of B. Ja. Levin [5], N. N. Meĭman [1], Ju. I. Neĭmark [1]. The greater part of these results are presented in the book of N. G. Čebotarev and N. N. Meĭman [1].

[3] The definition of this class presented here is due to N. N. Meĭman [1]. It is equivalent to the definition given earlier by M. G. Kreĭn (see the book of N. I. Ahiezer and M. G. Kreĭn [1]).

[4] Here, as usual, we understand by $\overline{\omega}(z)$ the entire function obtained from $\omega(z)$ by replacing all the coefficients in its Taylor series by their conjugates.

where $P(z)$ and $Q(z)$ are real functions, to the condition that the function

$$\theta(z) = \frac{Q(z)}{P(z)}$$

maps the upper half-plane into the upper half-plane.

Indeed, if we put

$$w = \frac{\omega(z)}{\bar{\omega}(z)}$$

and $\theta = \theta(z)$, then we have

$$w = \frac{1 + i\theta}{1 - i\theta},$$

and the unit circle in the plane of w corresponds to the upper half-plane in the plane of θ.

In Theorem 1 we give a representation of real meromorphic functions which map the upper half-plane onto the upper half-plane.

THEOREM 1. *In order that the real meromorphic function* $\psi(z)$ *map the upper half-plane onto the upper half-plane, i.e., in order that the inequality* $\operatorname{Im} \psi(z) \operatorname{Im} z > 0$ *be satisfied for* $\operatorname{Im} z \neq 0$, *it is necessary and sufficient that this function be represented in the form*[5]

$$\psi(z) = c\frac{z - a_0}{z - b_0} \prod{}' \left(1 - \frac{z}{a_k}\right)\left(1 - \frac{z}{b_k}\right)^{-1}, \tag{7.04}$$

where $b_k < a_k < b_{k+1}$, $\pm k = 0, 1, 2, \ldots$, $a_{-1} < 0 < b_1$ *and* $c > 0$.

The prime on the infinite product means that the index k takes on all integer values except zero.

PROOF. First we shall prove the sufficiency. From the interlacing of the quantities a_k and b_k we easily obtain the convergence of the series

$$\sum{}' \left(\frac{1}{b_k} - \frac{1}{a_k}\right),$$

and accordingly also the convergence of the series

$$\sum{}' \left[\left(1 - \frac{z}{a_k}\right)\left(1 - \frac{z}{b_k}\right)^{-1} - 1\right] = z\sum{}' \left(\frac{1}{b_k} - \frac{1}{a_k}\right)\left(1 - \frac{z}{b_k}\right)^{-1}.$$

Thus the product (7.04) converges uniformly on any bounded closed set not containing the points b_k ($\pm k = 0, 1, \ldots$).

For all values of the index k we have

$$\arg \frac{1 - \frac{z}{a_k}}{1 - \frac{z}{b_k}} = \arg(z - a_k) - \arg(z - b_k),$$

[5] This theorem was proved by M. G. Kreĭn in an unpublished paper. It is also implicitly contained in a paper of N. N. Meĭman [2].

so that this argument is equal to the angle subtended by the segment $[b_k, a_k]$ of the real axis from the point z. Thus, from the equation

$$\arg \psi(z) = \sum_{-\infty}^{\infty} [\arg(z - a_k) - \arg(z - b_k)]$$

it follows that

$$0 < \arg \psi(z) < \pi.$$

Now we shall prove the necessity. Suppose that the function $\psi(x)$ maps the upper half-plane into the upper half-plane, i.e., for $\operatorname{Im} z > 0$

$$0 \leqslant \arg \psi(z) \leqslant \pi.$$

Then the function $\psi(z)$ has neither roots nor poles in the upper half-plane.

Indeed, in the opposite case, on going around a certain contour lying in the upper half-plane and containing a pole or root of the function $\psi(z)$, the increment of the function $\arg \psi(z)$ will be larger or equal to 2π, which is impossible. By the symmetry principle the function $\psi(z)$ also has neither roots nor poles in the lower half-plane.

Thus all the roots and poles of the function $\psi(z)$ lie on the real axis.

For $\operatorname{Im} z < 0$ we have

$$-\pi < \arg \psi(z) < 0,$$

and when the point z makes one circuit of any circumference, the increment of the function $\arg \psi(z)$ does not exceed the number 2π in absolute value.

Therefore it follows that all roots and poles of the function $\psi(z)$ are simple and that on any interval of the real axis the number of roots of the function $\psi(z)$ differs from the number of its poles by no more than unity, i.e., that the roots and poles of the function interlace.

The infinite product

$$\Phi(z) = \frac{z - a_0}{z - b_0} \prod{}' \left(1 - \frac{z}{a_k}\right)\left(1 - \frac{z}{b_k}\right)^{-1}, \tag{7.04'}$$

made up over the roots $\{a_k\}$ and the poles $\{b_k\}$ of the function $\psi(z)$, converges uniformly on any bounded closed set not containing the points of $\{b_k\}$. The function $\Phi = \Phi(z)$ maps the upper half-plane of the plane of z into the upper half-plane of the plane of Φ.

The function

$$\chi(z) = \frac{\psi(z)}{\Phi(z)}$$

is entire and has no roots in the entire plane, and $|\arg \chi(z)| \leqslant 2\pi$ for all values of the variable z.

Thus the entire function $u = \ln \chi(z)$ maps the entire plane of z into the strip $|\operatorname{Im} u| \leqslant 2\pi$, so that $u = \text{const}$.

From this theorem we can at once make the following remark.

REMARK. *In order that the real meromorphic function $w = \psi(z)$ may be uniformly approximated in any bounded region by real rational functions with real interlacing roots and poles, it is necessary and sufficient that it map the upper half-plane of the plane of z onto the upper half-plane of the plane[6] of w.*

Another representation of real meromorphic functions mapping the upper half-plane into the upper half-plane was given considerably earlier by N. G. Čebotarev [**6**].

THEOREM 2 (N. G. ČEBOTAREV). *If the real meromorphic function $\psi(x)$ maps the upper half-plane onto the upper half-plane, then its poles a_k $(\pm k = 0, 1, 2, \ldots)$ are all real and simple, and it may be represented in the form*

$$\psi(z) = az + b + \sum_{k=\alpha}^{\omega} A_k\left(\frac{1}{a_k - z} - \frac{1}{a_k}\right) \tag{7.05}$$

$$(-\infty \leqslant \alpha < \omega \leqslant \infty),$$

where $a \geqslant 0$, b is real, $A_k \geqslant 0$ $(\pm k = 0, 1, 2, \ldots)$ and the series

$$\sum_{k=\alpha}^{\omega} \frac{A_k}{a_k^2}$$

converges.

PROOF. By Theorem 1, the function $\psi(z)$ may be represented up to a constant multiplier in the form (7.04), and so is the uniform limit of the sequence

$$\psi_n(z) = \frac{z - a_0}{z - b_0} \prod_{-n}^{n} \left(1 - \frac{z}{a_k}\right)\left(1 - \frac{z}{b_k}\right)^{-1},$$

which may be represented in the form

$$\psi_n(z) = \sum_{k=-n}^{n} \frac{A_{k,n}}{a_k - z} + c_n$$

(for $k < \alpha$ and $k > \omega$ we take $a_k = \infty$), while $A_{k,n} > 0$ and c_n is a real constant. Taking the derivative at the point zero, we find that the sequence of sums

$$\sum_{k=-n}^{n} \frac{A_{k,n}}{a_k^2}$$

converges to $\psi'(0)$. On the other hand, it is evident that $A_{k,n}$ tends to the residue A_k of the function $\psi(z)$ at the point a_k. Passing to the limit as $n \to \infty$, we find from the inequality

$$\sum_{k=-N}^{N} \frac{A_{k,n}}{a_k^2} \leqslant \psi_n'(0) \qquad (n \geqslant N)$$

[6] This fact was first established by N. G. Čebotarev [**6**].

that

$$\sum_{k=-N}^{N} \frac{A_k}{a_k^2} \leqslant \psi'(0),$$

and accordingly the series

$$\sum_{k=\alpha}^{\omega} \frac{A_k}{a_k^2}$$

converges. Thus we obtain the convergence of the series on the right side of formula (7.05).

Now we shall show that the difference between $\psi(z)$ and the sum of that series is a linear function. To this end we fix a number N and consider for $n > N$ the difference

$$\psi_n(z) - \sum_{k=-N}^{N} A_{k,n}\left(\frac{1}{a_k - z} - \frac{1}{a_k}\right) = \sum_{N<|k|\leqslant n} \frac{A_{k,n}}{a_k - z} + c_n'.$$

Evidently the imaginary part of this function is non-negative for $\operatorname{Im} z > 0$. Passing to the limit we have

$$\operatorname{Im}\left[\psi(z) - \sum_{k=-N}^{N} A_k\left(\frac{1}{a_k - z} - \frac{1}{a_k}\right)\right] \geqslant 0 \qquad (\operatorname{Im} z \geqslant 0).$$

Thus, the entire real function

$$\psi(z) - \sum_{k=\alpha}^{\omega} A_k\left(\frac{1}{a_k - z} - \frac{1}{a_k}\right)$$

maps the upper half-plane onto the upper half-plane, and so, from the lemma of N. G. Čebotarev (Chapter V), it is equal to $az + b$ ($a \geqslant 0$, $\operatorname{Im} b = 0$). The theorem is proved.[7]

2. Generalization of the Theorem of Hermite-Biehler to Arbitrary Entire Functions

From Theorem 1 it is very simple to obtain a criterion for an entire function to belong to the class HB, which is analogous to the Hermite-Biehler criterion. This criterion, due to N. N. Meĭman, is given by the following theorem.

THEOREM 3. *Suppose that*

$$\omega(z) = P(z) + iQ(z),$$

[7] Another proof of this theorem is obtained immediately from the following representation of a function mapping the upper half-plane into the upper half-plane, due to R. Nevanlinna:

$$\psi(z) = az + b + \int_{-\infty}^{\infty} \frac{1 + tz}{t - z}\, d\sigma(t),$$

where $\sigma(t)$ is a nondecreasing function.

where $P(z)$ and $Q(z)$ are real entire functions, and suppose that

$$P(z) = Ae^{u(z)}(z-a_0)\prod_{k=-\infty}^{\infty}{}'\left(1-\frac{z}{a_k}\right)e^{P_k\left(\frac{z}{a_k}\right)} \quad (u(0)=0)$$

and (7.06)

$$Q(z) = Be^{v(z)}(z-b_0)\prod_{k=-\infty}^{\infty}{}'\left(1-\frac{z}{b_k}\right)e^{P_k\left(\frac{z}{b_k}\right)} \quad (v(0)=0)$$

are their expansions into infinite products.

For the function $\omega(z)$ to belong to the class HB it is necessary and sufficient that the following conditions be satisfied:

(α) *the roots a_k and b_k of the functions $P(z)$ and $Q(z)$ interlace;*

(β) *the entire real functions $u(z)$ and $v(z)$ and the exponents $P_k(z/a_k)$ and $P_k(z/b_k)$ satisfy the condition*

$$u(z)-v(z)+\sum_{k=-\infty}^{\infty}\left[P_k\left(\frac{z}{a_k}\right)-P_k\left(\frac{z}{t_k}\right)\right]=0; \tag{7.07}$$

(γ) *the constants A and B are of the same sign.*

Proof of Necessity. It follows from (7.03) that the function

$$F(z) = \frac{\omega(z)}{\bar{\omega}(z)}$$

maps the upper half-plane onto the unit circle, and accordingly that the real function $\theta(z)$ defined by the equation

$$\theta(z) = \frac{Q(z)}{P(z)} = i\,\frac{F(z)-1}{F(z)+1}$$

maps the upper half-plane onto the upper half-plane. By Theorem 1 its roots and poles are real and simple, and interlace. Since the functions $\omega(z)$ and $\bar{\omega}(z)$ have no common roots, the functions $P(z)$ and $Q(z)$ also have no common roots. Thus we have proved that the roots of the functions $P(z)$ and $Q(z)$ satisfy condition (α).

From the representation (7.06) of the functions $P(z)$ and $Q(z)$ it follows that

$$\theta(z) = \frac{B}{A}\,e^{w(z)}\Phi(z),$$

$$w(z) = u(z)-v(z)+\sum_{k=-\infty}^{\infty}\left[P_k\left(\frac{z}{b_k}\right)-P_k\left(\frac{z}{a_k}\right)\right],$$

where $\Phi(z)$ is defined by equation (7.04').

From Theorem 1, the function $\tau = \Phi(z)$ maps the upper half-plane onto the upper half-plane. Thus we find that in the entire plane of z

$$\left|\arg\frac{\theta(z)}{\Phi(z)}\right| \leqslant \pi,$$

so that the entire real function

$$w=\frac{A}{B}\ln\frac{\theta(z)}{\Phi(z)}=v(z)-u(z)+\sum_{k=-\infty}^{\infty}\left[P_k\left(\frac{z}{b_k}\right)-P_k\left(\frac{z}{a_k}\right)\right]$$

maps the entire plane of z onto the strip $|\operatorname{Im} w| \leqslant \pi$, i.e., $w = \text{const.}$ One obtains condition (β) if one observes that $w(0) = 0$. Since the imaginary parts of $\theta(z)$ and of $\Phi(z)$ are of the same sign, therefore A and B are also of the same sign.

PROOF OF SUFFICIENCY. If conditions (α), (β) and (γ) are satisfied, the function $\theta(z)$ coincides up to a positive constant multiplier with the function $\Phi(z)$, i.e., it maps the upper half-plane onto the upper half-plane.

The function

$$F(z)=\frac{1+i\theta(z)}{1-i\theta(z)}$$

maps the upper half-plane onto the unit circle, i.e., inequality (7.03) is satisfied. The functions $P(z)$ and $Q(z)$, and therefore $\omega(z)$ and $\bar{\omega}(z)$, do not have common roots, and from inequality (7.03) the function $\omega(z)$ has no roots in the half-plane $\operatorname{Im} z \leqslant 0$. The theorem is proved.

Observe that the condition $AB > 0$ may be replaced by the condition

$$Q'(x)P(x)-Q(x)P'(x)>0 \qquad (-\infty<x<\infty).$$

Indeed, if conditions (α) and (β) are satisfied, the function $\theta(z)$ maps the upper half-plane either onto the upper half-plane or onto the lower half-plane. In the first case the function $\theta(x)$ increases monotonically, and in the second it decreases monotonically. It is easy to see that $\theta'(x) \neq 0$ and that the sign of $\theta'(x)$ coincides with the sign of the expression $Q'(x)P(x) - Q(x)P'(x)$.

DEFINITION. An entire function $\omega(z)$ having no roots in the open lower half-plane ($\operatorname{Im} z < 0$) and satisfying the condition

$$\left|\frac{\omega(z)}{\bar{\omega}(z)}\right| \leqslant 1 \quad \text{for} \quad \operatorname{Im} z > 0$$

will be said to be *a function of class* $\overline{HB}$.

If $\omega(z) \in \overline{HB}$, then all the common roots of the functions $\omega(z)$ and $\bar{\omega}(z)$ are real. Setting up the canonical product $R(z)$ over these common roots, we find that

$$\omega(z)=R(z)\,\omega_1(z), \tag{7.08}$$

where $\omega_1(z) \in HB$. Conversely, if the function $\omega(z)$ is represented in the form (7.08), where $\omega_1(z) \in HB$, and $R(z)$ is a real function having only real roots, then $\omega(z) \in \overline{HB}$. It is worth noting that if the function $\omega(z)$ is of integral order, then the functions $R(z)$ and $\omega_1(z)$ may turn out to have greater growth than $\omega(z)$. In particular, if $\omega(z)$ is of normal type, then the functions $R(z)$ and $\omega_1(z)$ may turn out to be functions of maximal type.

We also note that if a sequence of functions of class $\overline{HB}$ converges uniformly in each finite portion of the plane, then the limit function is also a function of class $\overline{HB}$.

From Theorem 3 it immediately follows that:

THEOREM 3′. *In order that the entire function*

$$\omega(z) = P(z) + iQ(z)$$

be a nontrivial function of class $\overline{HB}$, it is necessary and sufficient that the following conditions be satisfied:

(α') *the functions $P(z)$ and $Q(z)$ may be represented in the form*

$$P(z) = R(z)P_1(z) \quad and \quad Q(z) = R(z)Q_1(z),$$

while the roots of the entire functions $R(z)$, $P_1(z)$ and $Q_1(z)$ are all real and the roots of $P_1(z)$ and $Q_1(z)$ interlace;

(β) *in the representations of the functions $P_1(z)$ and $Q_1(z)$*

$$P_1(z) = Ae^{u(z)}(z - a_0)\prod_{k=-\infty}^{\infty}{}'\left(1 - \frac{z}{a_k}\right)e^{P_k\left(\frac{z}{a_k}\right)} \qquad (u(0) = 0),$$

$$Q_1(z) = Be^{v(z)}(z - b_0)\prod_{k=-\infty}^{\infty}{}'\left(1 - \frac{z}{b_k}\right)e^{P_k\left(\frac{z}{b_k}\right)} \qquad (v(0) = 0)$$

($a_{-1} < 0 < b_1$) the entire real functions $u(z)$ and $v(z)$ and the Weierstrass exponents $P_k(z/a_k)$ and $P_k(z/b_k)$ satisfy the condition

$$u(z) - v(z) + \sum_{k=-\infty}^{\infty}\left[P_k\left(\frac{z}{a_k}\right) - P_k\left(\frac{z}{b_k}\right)\right] = 0;$$

(γ) *the constants A and B have the same sign.*

Condition (γ) may here be replaced by the condition

$$Q'(x)P(x) - Q(x)P'(x) \geqslant 0$$

at any point of the real axis.

Sometimes it is more convenient to use other conditions for an entire function to belong to the class HB or $\overline{HB}$. Before formulating them we introduce the following concept, due to N. G. Čebotarev [3].

DEFINITION. Two entire real functions $P(z)$ and $Q(z)$ are said to be a *real pair*, if they have no common roots and if any linear combination $\mu P(z) + \nu Q(z)$ of them with real coefficients μ and ν has no complex roots.[8]

[8] We refer to roots not lying on the real axis as complex roots.

THEOREM 4. *In order that the entire function* $\omega(z) = P(z) + iQ(z)$ *belong to the class HB, it is necessary and sufficient that the functions* $P(z)$ *and* $Q(z)$ *form a real pair and that at any point* x_0 *of the real axis the inequality*

$$Q'(x_0)P(x_0) - Q(x_0)P'(x_0) > 0$$

hold.

PROOF. From the fact that $P(z)$ and $Q(z)$ form a real pair, it follows that the real function

$$\theta(z) = \frac{Q(z)}{P(z)} \tag{7.09}$$

does not take on real values for $\operatorname{Im} z \neq 0$, so that it maps the upper half-plane either onto the upper half-plane or onto the lower half-plane.

If in addition $Q'(x_0)P(x_0) - Q(x_0)P'(x_0) > 0$, then $\theta'(x_0) > 0$, and $\theta(z)$ maps the upper half-plane onto the upper half-plane. Hence it follows that the function

$$F(z) = \frac{\omega(z)}{\bar{\omega}(z)} \tag{7.10}$$

maps the upper half-plane onto the unit circle, and we obtain inequality (7.03).

Since the functions $P(z)$ and $Q(z)$ have no common roots, $\omega(z)$ and $\bar{\omega}(z)$ also have no common roots, and it follows from the inequality (7.03) that $\bar{\omega}(z)$ has no roots in the half-plane $\operatorname{Im} z \leqslant 0$. Thus, $\omega(z) \in HB$.

Conversely, if $\omega(z) \in HB$, then $\omega(z)$ and $\bar{\omega}(z)$ have no common roots, and inequality (7.03) is satisfied. But then the real functions $P(z)$ and $Q(z)$ also have no common roots, the function $\theta(z)$ is holomorphic in the upper half-plane and $\operatorname{Im} \theta(z) > 0$ for $\operatorname{Im} z > 0$. Thus the function $\theta(z)$ takes on real values only on the real axis, and, accordingly, the linear combination $\mu P(z) + \nu Q(z)$ with real coefficients can have only real roots. The theorem is proved.

DEFINITION. Two entire real functions $P(z)$ and $Q(z)$ will be called *a generalized real pair* if an arbitrary linear combination of them, $\mu P(z) + \nu Q(z)$, with real coefficients has no complex roots. In this connection, the functions $P(z)$ and $Q(z)$ are permitted to have common roots.

From Theorem 4 and representation (7.08) there immediately follows

THEOREM 4'. *In order that the entire function*

$$\omega(z) = P(z) + iQ(z)$$

be a nontrivial function of class $\overline{HB}$, *it is necessary and sufficient that the functions* $P(z)$ *and* $Q(z)$ *form a generalized real pair and that at some real point the inequality*

$$Q'(x_0)P(x_0) - Q(x_0)P'(x_0) > 0$$

be satisfied.

Now we shall prove that it is impossible to generalize the Hermite-Biehler theorem to a wider class of entire functions.[9] Here we restrict ourselves for simplicity to entire functions of finite order.

THEOREM 5. *If the entire function of finite order*

$$\omega(z) = P(z) + iQ(z)$$

has roots and all of its roots lie in the open upper half-plane, and the roots of the functions $P(z)$ and $Q(z)$ interlace, then $\omega(z) \in HB$.

PROOF. From the estimate of the entire function $\bar{\omega}(z)$ from below (Chapter I, § 8), it follows that *outside the exceptional circles* the function $F(z) = \omega(z)/\bar{\omega}(z)$ satisfies the inequality

$$\ln |F(z)| < \ln M_\varepsilon + r^{\rho+\varepsilon} \qquad (\operatorname{Im} z > 0), \tag{7.11}$$

where ρ is the order of $\omega(z)$ ($\varepsilon > 0$), and M_ε is some constant. In the half-plane $\operatorname{Im} z > 0$ the function $F(z)$ has no singularities, and since on the real axis $|F(x)| = 1$, by the maximum principle the inequality (7.11) will be satisfied *everywhere* in the upper half-plane.

On the other hand, the function $\theta(z) = P(z)/Q(z)$ may be represented in the form

$$\theta(z) = e^{g(z)}\psi(z), \tag{7.12}$$

where $g(z) = a_0 z^n + a_1 z^{n-1} + \ldots + a_n$, $n \leqslant [\rho]$ and all of the coefficients a_j ($j = 0, 1, \ldots, n-1$) are real, and the function $\psi(z)$ is defined by equation (7.04′) and, from Theorem 1, maps the upper half-plane onto the upper half-plane.

If $n = 0$, i.e., $g(z) \equiv c$, then $\theta(z)$ maps the upper half-plane onto the upper half-plane, and since $F(z) = \omega(z)/\bar{\omega}(z)$ may be represented in the form

$$F(z) = \frac{1 + i\theta(z)}{1 - i\theta(z)},$$

then $|F(z)| < 1$ for $\operatorname{Im} z > 0$, and so $\omega(z) \in HB$.

Now suppose that $n > 0$. We have

$$F(z) = \frac{1 + ie^{g(z)}\psi(z)}{1 - ie^{g(z)}\psi(z)}. \tag{7.13}$$

From the Carathéodory inequality (Chapter I, § 6, Theorem 8′), it follows for $\operatorname{Im} z > 0$ that

$$c_1 \frac{\sin\theta}{r} < |\psi(z)| < c_2 \frac{r}{\sin\theta}, \tag{7.14}$$

where c_1 and c_2 are constants and $z = re^{i\theta}$ ($0 < \theta < \pi$).

On the other hand, we have

$$\operatorname{Re} g(re^{i\theta}) \approx a_0 r^n \cos n\theta. \tag{7.15}$$

[9] This result is due to N. N. Meĭman (see N. G. Čebotarev and N. N. Meĭman [1]).

Comparing (7.13), (7.14) and (7.15) we find that outside arbitrarily small angles containing the rays $\arg z = (k + \frac{1}{2})\pi/n$, the function $F(z)$ is bounded. Moreover, $F(z)$ is bounded on the real axis.

Taking into account estimate (7.11), we find from the Phragmén-Lindelöf principle that $F(z)$ is bounded in the upper half-plane and, applying once again the same principle, we find that $|F(z)| < 1$ for $\operatorname{Im} z > 0$, i.e., $\omega(z) \in HB$.

3. Representation of a Function of Class *HB*

In this section we shall obtain a representation of functions of class *HB* using special infinite products, which is due to M. G. Kreĭn **[1]**. In this connection we shall use the following notations.

If $f(z) = \sum_{k=0}^{\infty}(a_k + ib_k)z^k$, where a_k and b_k are real numbers, then

$$Rf(z) = \sum_{k=0}^{\infty} a_k z^k \quad \text{and} \quad If(z) = \sum_{k=0}^{\infty} b_k z^k \tag{7.16}$$

or, put differently, $2Rf(z) = f(z) + \bar{f}(z)$ and $2iIf(z) = f(z) - \bar{f}(z)$. Suppose

$$\omega(z) = z^m e^{g(z)} \prod_{k=1}^{\infty} \left(1 - \frac{z}{\alpha_k}\right) e^{P_k\left(\frac{z}{\alpha_k}\right)} \tag{7.17}$$

is a decomposition of the function $\omega(z)$ into a canonical product. If $\omega(z) \in HB$, then the function

$$F(z) = \frac{\omega(z)}{\bar{\omega}(z)}$$

is bounded in the upper half-plane and therefore, by the remark to Theorem 2 of Chapter V, belongs in this half-plane to the class A, i.e., the series $\sum_{k=-\infty}^{\infty} |\operatorname{Im} 1/\alpha_k|$ converges. By Lemma 4 of Chapter V the product

$$\chi(z) = \prod_{k=1}^{\infty} \left(1 - \frac{z}{\alpha_k}\right)\left(1 - \frac{z}{\bar{\alpha}_k}\right)^{-1} \tag{7.18}$$

converges uniformly on every closed bounded set not containing any of the points α_k. We write

$$\chi_n(z) = \prod_{k=1}^{n} \left(1 - \frac{z}{\alpha_k}\right)\left(1 - \frac{z}{\bar{\alpha}_k}\right)^{-1}$$

and

$$\omega_n(z) = z^m e^{g(z)} \prod_{k=1}^{n} \left(1 - \frac{z}{\alpha_k}\right) e^{P_k\left(\frac{z}{\alpha_k}\right)}. \tag{7.19}$$

Then we will have

$$\frac{\omega_n(z)}{\bar{\omega}_n(z)} [\chi_n(z)]^{-1} = \exp 2i\left[Ig(z) + \sum_{k=1}^{n} IP_k\left(\frac{z}{\alpha_k}\right)\right]. \tag{7.20}$$

From the convergence of the products (7.17) and (7.18) we find that the series $\Sigma_{k=1}^{\infty} IP_k(z/\alpha_k)$ converges uniformly in any finite portion of the plane. Writing $v(z) = Ig(z) + \Sigma_{k=1}^{\infty} IP_k(z/\alpha_k)$ and $u(z) = Rg(z)$, we represent the function $\omega(z)$ in the form

$$\omega(z) = z^m e^{u(z)+iv(z)} \prod_{k=1}^{\infty} \left(1 - \frac{z}{\alpha_k}\right) e^{RP_k\left(\frac{z}{\alpha_k}\right)}. \tag{7.21}$$

From the Phragmén-Lindelöf principle for the half-plane (Chapter I, § 14) one may easily find that

$$\left| \frac{\omega(z)}{\bar{\omega}(z)} [\chi_n(z)]^{-1} \right| \leqslant 1 \quad \text{for} \quad \operatorname{Im} z > 0. \tag{7.22}$$

Passing in (7.22) to the limit as $n \to \infty$ and comparing with (7.21), we see that the entire real function $v(z)$ satisfies the inequality $|e^{iv(z)}| \leqslant 1$ for $\operatorname{Im} z > 0$, or, what is the same thing, maps the upper half-plane onto the upper half-plane. From a lemma of N. G. Čebotarev (Chapter V, § 3, Lemma 3) we will have $v(z) = \nu z + \delta$, where $\nu \geqslant 0$. We finally obtain

$$\omega(z) = z^m e^{u(z)+i(\nu z+\delta)} \prod_{k=1}^{\infty} \left(1 - \frac{z}{\alpha_k}\right) e^{RP_k\left(\frac{z}{\alpha_k}\right)}$$
$$\left(\sum_1^{\infty} \left| \operatorname{Im} \frac{1}{\alpha_k} \right| < \infty, \qquad \operatorname{Im} \alpha_k > 0\right), \tag{7.23}$$

with $\nu \geqslant 0$ and $u(z)$ a real entire function.

Conversely, if the function $\omega(z)$ is represented in the form (7.23), then

$$\frac{\omega(z)}{\bar{\omega}(z)} = e^{2i(\nu z+\delta)} \chi(z) \tag{7.24}$$

and $\omega(z) \in \overline{HB}$. Thus, we have obtained the following theorem of M. G. Kreĭn.[10]

THEOREM 6. *In order that the entire function $\omega(z)$ belong to the class HB, it is necessary and sufficient that it can be represented in the form* (7.23).

Functions of class $\overline{HB}$ have evidently the same representation, the only difference being that $\operatorname{Im} \alpha_k \geqslant 0$ and the equality sign is possible.

4. The Hermite-Biehler Theorem for Entire Functions of Exponential Type

The generalized criterion of Hermite-Biehler is usually necessary in applications to entire functions of exponential type. For this case it may be given a far

[10] See N. I. Ahiezer and M. G. Kreĭn [1].

more complete formulation than in the general case of an arbitrary function of class HB. In this case the conditions (7.03) and (7.06), which are difficult to verify, may be replaced by conditions imposed on the indicator diagrams of the functions $\omega(z)$, $P(z)$, and $Q(z)$, which in many cases are easily seen to be satisfied.

We shall define the *defect* d_ω of the function $\omega(z)$ of exponential type by the equation

$$2d_\omega = h\left(-\frac{\pi}{2}\right) - h\left(\frac{\pi}{2}\right).$$

For a function of exponential type and in the class A (and thus in HB) the indicator diagram is symmetric relative to an axis parallel to the real axis (Chapter V, § 5). Evidently the defect is equal to length of the segment which this axis of symmetry cuts off from the imaginary axis, taken with the opposite sign.

The function $\omega(z)$ of exponential type will be called *a function of class P* if

(a) $\omega(z)$ has no roots in the open lower half-plane;

(b) the defect of the function $\omega(z)$ is not negative.

Functions of class P and with degrees not exceeding $\sigma \geqslant 0$ form a subclass which we shall denote by P_σ. A function which is real up to a constant multiplier will be called a trivial function of class P.

The following lemma is very important.

Lemma 1. *In order that the entire function $\omega(z)$ belong to the class P, it is necessary and sufficient that it be of the class $\overline{HB}$ and of exponential type.*

Proof. Suppose that $\omega(z) \in P$ and that its decomposition is

$$\omega(z) = z^m e^{uz+b} \prod_{k=1}^{\infty} \left(1 - \frac{z}{a_k}\right) e^{\frac{z}{a_k}}$$
$$\left(\operatorname{Im} a \geqslant 0 \text{ and } \sum_{k=1}^{\infty} \left|\operatorname{Im} \frac{1}{a_k}\right| < \infty\right). \qquad (7.25)$$

For the function $F(z) = \omega(z)/\bar{\omega}(z)$ we obtain the decomposition

$$F(z) = e^{i(pz+q)} \chi(z), \qquad (7.26)$$

where

$$\chi(z) = \prod_{k=1}^{\infty} \left(1 - \frac{z}{a_k}\right)\left(1 - \frac{z}{\bar{a}_k}\right)^{-1},$$

and p and q are real numbers.

From Lemma 5 of Chapter V the function $\chi(z)$ is of completely regular growth in the half-plane $0 < \theta < \pi$ and

$$\lim_{r\to\infty}{}^* \frac{\ln|\chi(re^{i\theta})|}{r} = 0. \qquad (7.27)$$

Therefore it follows that $F(z)$ is also of completely regular growth in the same half-plane and

$$h_F(\theta) = -p \sin \theta \qquad (0 < \theta < \pi). \tag{7.28}$$

When two functions, one of which is of completely regular growth, are multiplied, their indicators are added (Chapter III, Theorem 5); therefore

$$h_\omega(\theta) = h_\omega(-\theta) - p \sin \theta. \tag{7.29}$$

Putting $\theta = \pi/2$, we see that $p = 2d_\omega$.

Since by hypothesis the defect of the function $\omega(z)$ is non-negative, therefore $p \geqslant 0$ and we obtain from (7.26)

$$\left| \frac{\omega(z)}{\bar{\omega}(z)} \right| \leqslant 1 \quad \text{for} \quad \operatorname{Im} z > 0.$$

Conversely, if $\omega(z) \in \overline{HB}$, then $|\omega(iy)| \leqslant |\omega(-iy)|$ $(y \geqslant 0)$. Accordingly, $h_\omega(\pi/2) \leqslant h_\omega(-\pi/2)$, and if $\omega(z)$ is of exponential type, then it is of class P. The lemma is proved.

REMARK. *The product of two functions $\omega_1(z)\omega_2(z)$, each of which is of class P, is a function of class P, while the defect of the product is equal to the sum of the defects of the factors.*

The first assertion immediately follows from the lemma. In order to prove the second we observe that the functions

$$F_i(z) = \frac{\omega_i(z)}{\bar{\omega}_i(z)} \qquad (i = 1, 2)$$

are of completely regular growth and that their indicators are $h_{F_i}(\theta) = -2d_{\omega_i} \sin \theta$. It is sufficient to recall that on multiplication of functions of completely regular growth their indicators are added.

LEMMA 2. *Suppose that $P(z)$ and $Q(z)$ are entire functions of exponential type, whose roots are real and interlacing, and suppose that*

$$\left.\begin{aligned} P(z) &= z^m e^{az+c} \prod_{-\infty}^{\infty} \left(1 - \frac{z}{a_k}\right) e^{\frac{z}{a_k}} \\ &\textit{and} \\ Q(z) &= z^n e^{bz+d} \prod_{-\infty}^{\infty} \left(1 - \frac{z}{b_k}\right) e^{\frac{z}{b_k}} \\ &(a_{-1} < 0 < b_1) \end{aligned}\right\} \tag{7.30}$$

are their canonical representations.

In order that the equation

$$b - a + \sum_{-\infty}^{\infty} \left(\frac{1}{b_k} - \frac{1}{a_k}\right) = 0 \tag{7.31}$$

hold, it is necessary and sufficient that the indicator diagrams of these functions coincide.

PROOF. From the representation (7.30) it follows that

$$\frac{Q(z)}{P(z)} = ce^{\delta z}\frac{z-b_0}{z-a_0}\prod{}'\left(1-\frac{z}{b_k}\right)\left(1-\frac{z}{a_k}\right)^{-1}, \tag{7.32}$$

where c is a constant and $\delta = b - a + \Sigma_{-\infty}^{\infty}[(1/b_k) - (1/a_k)]$.

Denote by $\Phi(z)$ the infinite product on the right side of the equation, whose convergence has already been proved (Theorem 1). The real function $w = \Phi(z)$ maps the upper half-plane of the plane of z either onto the upper or onto the lower half-plane of the plane of w (§ 1).

By Carathéodory's theorem for the half-plane (Chapter I, § 6, Theorem 8′), the function $\Phi(z)$ satisfies the inequalities

$$c_1\frac{|\sin\theta|}{r} < |\Phi(re^{i\theta})| < c_2\frac{r}{|\sin\theta|} \qquad (\theta \neq 0,\ \pi), \tag{7.33}$$

in which c_1 and c_2 are constants depending on the function $\Phi(z)$. It follows from these inequalities that if $\theta \neq 0, \pi$

$$\lim_{r\to\infty}\frac{\ln|\Phi(re^{i\theta})|}{r} = 0.$$

Thus, we find that if $\theta \neq 0, \pi$

$$h_Q(\theta) = h_P(\theta) + \delta\cos\theta, \tag{7.34}$$

and in view of the continuity of the indicators the equation (7.34) holds for all values of θ.

Evidently the condition $\delta = 0$ is equivalent to the identity equation $h_P(\theta) \equiv h_Q(\theta)$. The lemma is proved.

We make the following observation, which follows immediately from the equation (7.34).

REMARK. *If the roots of the functions $P(z)$ and $Q(z)$ of exponential type interlace, then the indicator diagram of the function $Q(z)$ is obtained from the indicator diagram of the function $P(z)$ by a shift through a vector δ. Evidently if $P(z)$ and $Q(z)$ are real functions, then the vector δ is real as well.*

Note that if the roots of a real function of exponential type are real, then it is a trivial function of class P.

Now we formulate the fundamental theorem of this section.[11]

THEOREM 7. *In order that the entire function $\omega(z) = P(z) + iQ(z)$ be a nontrivial function of class P, it is necessary and sufficient that the real functions $P(z)$ and $Q(z)$ satisfy the conditions:*

(A) *the functions $P(z)$ and $Q(z)$ are representable in the form*

$$P(z) = R(z)P_1(z) \quad and \quad Q(z) = R(z)Q_1(z), \tag{7.35}$$

[11] This theorem was proved simultaneously and independently by me [5] and by M. G. Kreĭn, but the paper of M. G. Kreĭn was not published.

while all the roots of the functions $R(z)$, $P_1(z)$ and $Q_1(z)$ are real and the roots of $P_1(z)$ and $Q_1(z)$ interlace;

(B) *the functions $P(z)$ and $Q(z)$ are of exponential type and their indicator diagrams coincide;*

(C) *at some real point*

$$Q'(x_0)P(x_0) - Q(x_0)P'(x_0) > 0. \tag{7.36}$$

PROOF. By Lemma 1, the property of $\omega(z)$ belonging to the class P is equivalent to $\omega(z) \in \overline{HB}$ and its being of exponential type. The conditions for the function $\omega(z)$ to belong to the class $\overline{HB}$ were given in Theorem 3′.

Conditions (α') and (γ) of Theorem 3′ coincide with conditions (A) and (C) of Theorem 5. If $\omega(z)$ is of exponential type, then the functions $P(z)$ and $Q(z)$ are also of exponential type, while the exponential type of at least one of them is equal to that of the function $\omega(z)$. Conversely, if $P(z)$ and $Q(z)$ are of exponential type, then $\omega(z)$ is also of exponential type and the exponential type of $\omega(z)$ is equal to the larger of the types of the functions $P(z)$ and $Q(z)$. This deduction follows immediately from the expression of the exponential type of a function in terms of the coefficients of its expansion into power series (Chapter I, § 2).

By Lemma 2, the condition of coincidence of the indicator diagrams for the functions $P(z)$ and $Q(z)$ of exponential type with real interlacing roots is equivalent to the equation

$$b - a + \sum\left(\frac{1}{b_k} - \frac{1}{a_k}\right) = 0,$$

or, what is the same thing, to condition (β) of Theorem 3′. Thus, if conditions (A), (B) and (C) are satisfied, then the function $\omega(z)$ is of exponential type and belongs to the class $\overline{HB}$, i.e., it is of class P.

Conversely, if $\omega(z)$ is of class P, then by Lemma 1 it belongs to $\overline{HB}$ and, accordingly satisfies conditions (α'), (β) and (γ). Since $P(z)$ and $Q(z)$ are also of exponential type, condition (β) is equivalent to condition (B) and thus conditions (A), (B) and (C) are satisfied. The theorem is proved.

We shall now analyze more in detail the connection between conditions (A), (B) and (C) and conditions (a) and (b) which define the class P. First of all we make the following remark.

If condition (A) is satisfied but the indicator diagrams of the functions $P(z)$ and $Q(z)$ do not coincide, then the function $\omega(z)$ has roots in both half-planes $\operatorname{Im} z > 0$ and $\operatorname{Im} z < 0$.

Indeed, from Theorem 5 it is easy to show that if, with condition (A) satisfied, all the roots of the function $\omega(z)$ belong to one of the half-planes $\operatorname{Im} z \geqslant 0$ or $\operatorname{Im} z \leqslant 0$, then either $\omega(z)$ or $\bar{\omega}(z)$ belongs to class $\overline{HB}$, or else, what is the same thing, to the class P. But in this case condition (B) must be satisfied, which contradicts the hypothesis. So the function $\omega(z)$ must have roots in both half-planes.

Thus the requirements that all the roots of the function $\omega(z)$ should lie in the upper half-plane and that the roots of $P(z)$ and $Q(z)$ should all be real and interlacing are compatible only in the case when the indicator diagrams of the functions $P(z)$ and $Q(z)$ coincide. If now this condition is satisfied, then from the reality and interlacing of the roots of $P(z)$ and $Q(z)$ it follows that either $\omega(z)$ or $\bar{\omega}(z)$ belongs to the class P.

REMARK. If the function $\omega(z)$ is of zero order, then the indicator diagrams of the functions $\omega(z)$, $P(z)$ and $Q(z)$ reduce to the single point zero. Thus in this case the property of $\omega(z)$ belonging to the class P is defined by the requirement that $\omega(z)$ should not have roots in the open lower half-plane.

The indicator diagrams of the functions $P(z)$ and $Q(z)$ in this case coincide, i.e., condition (B) is satisfied.

Thus it follows from Theorem 7 that for functions of zero exponential type, reality and interlacing of the roots of the functions $P(z)$ and $Q(z)$, and the condition (C), namely that $Q'(x)P(x) - Q(x)P'(x) > 0$, are necessary and sufficient conditions for the function $\omega(z) = P(z) + iQ(z)$ not to have roots in the half-plane $\operatorname{Im} z \leqslant 0$.

THEOREM 8. *If the defect d_ω of the entire function $\omega(z)$ of exponential type is positive, then condition* (A) *is necessary and sufficient for the function $\omega(z)$ to belong to the class P.*

PROOF. If condition (A) is satisfied but condition (B) is not satisfied, then, as stated above, the indicator diagrams of the functions $P(z)$ and $Q(z)$ can be obtained one from the other by a shift by a nonzero vector. But since the indicator diagram of the function $\omega(z)$ is the smallest convex region containing the indicator diagrams of the functions $P(z)$ and $Q(z)$ (Chapter I, § 21), it must be symmetric with respect to the real axis, since the indicator diagrams of the real functions $P(z)$ and $Q(z)$ are symmetric with respect to the real axis, so that $d_\omega = 0$. Thus, if condition (A) is satisfied and $d_\omega > 0$, then condition (B) is also satisfied, from which it follows that $\omega(z) \in P$.

We consider two examples of the application of the Hermite-Biehler theorem.

EXAMPLE 1. Let

$$\omega(z) = \sum_{k=1}^{n} e^{i\lambda_k z} P_k(z), \tag{7.37}$$

where $\lambda_1 < \lambda_2 < \ldots < \lambda_n$, $|\lambda_1| < \lambda_n$ and $P_k(z)$ are polynomials.

The indicator diagram of this quasipolynomial is a segment of the imaginary axis between the points $-i\lambda_n$ and $-i\lambda_1$, so that

$$d_\omega = \frac{1}{2}(\lambda_n + \lambda_1) > 0.$$

Let

$$\omega(z) = P(z) + iQ(z).$$

It follows from Theorem 8 that the condition that the roots of $P(z)$ and $Q(z)$ interlace is equivalent to the condition that all the roots of the quasi-polynomial $\omega(z)$ lie in the open upper half-plane.

EXAMPLE 2. Suppose given the differential equation

$$y'' + q(x)y + \lambda\rho(x)y = 0 \qquad (\rho(x) \geqslant 0,\ \operatorname{Im} q(x) = 0). \tag{7.38}$$

Determine two solutions $\varphi(x, \lambda)$ and $\psi(x, \lambda)$, satisfying the conditions

$$\varphi(0, \lambda) = \cos\alpha, \qquad \varphi'(0, \lambda) = \sin\alpha,$$
$$\psi(0, \lambda) = -\sin\alpha, \qquad \psi'(0, \lambda) = \cos\alpha,$$

where α is any real number and the functions $\varphi(x, \lambda)$ are entire functions of λ, since the coefficients of the equation (7.38) are entire functions of λ.

Let $\chi(x, \lambda)$ be some solution of the same equation, satisfying the condition

$$\chi'(l, \lambda) + h\chi(l, \lambda) = 0 \qquad (h \text{ real}). \tag{7.39}$$

Then under an appropriate normalization

$$\chi(x, \lambda) = \varphi(x, \lambda) - \omega(\lambda)\psi(x, \lambda).$$

Writing out in detail the boundary condition (7.39), we obtain

$$\varphi'(l, \lambda) - \omega(\lambda)\psi'(l, \lambda) + h[\varphi(l, \lambda) - \omega(\lambda)\psi(x, \lambda)] = 0$$

or

$$\omega(\lambda) = \frac{\varphi'(l, \lambda) + h\varphi(l, \lambda)}{\psi'(l, \lambda) + h\psi(l, \lambda)}. \tag{7.40}$$

By a well-known Abel formula, the Wronskian formed from the solutions $\varphi(x, \lambda)$ and $\psi(x, \lambda)$ does not depend on x, and thus

$$\varphi'(x, \lambda)\psi(x, \lambda) - \varphi(x, \lambda)\psi'(x, \lambda) = 1.$$

Moreover, multiplying the first of the relations

$$\chi'' + q(x)\chi + \lambda\rho(x)\chi = 0,$$
$$\bar{\chi}'' + q(x)\bar{\chi} + \bar{\lambda}\rho(x)\bar{\chi} = 0$$

by $\bar{\chi}$ and the second by χ, and subtracting and integrating from zero to l, we obtain

$$\int_0^l (\chi''\bar{\chi} - \bar{\chi}''\chi)\,dx + (\lambda - \bar{\lambda})\int_0^l |\chi(x, \lambda)|^2\rho(x)\,dx = 0,$$

or

$$\chi'(0, \lambda)\bar{\chi}(0, \lambda) - \bar{\chi}'(0, \lambda)\chi(0, \lambda) = (\lambda - \bar{\lambda})\int_0^l |\chi(x, \lambda)|^2\rho(x)\,dx$$

and

$$\frac{\omega(\lambda)-\overline{\omega}(\lambda)}{\lambda-\overline{\lambda}}=\int_0^l|\chi(x,\lambda)|^2\rho(x)\,dx>0. \tag{7.41}$$

Thus $\omega(\lambda)$ is a real meromorphic function, mapping the upper half-plane into the upper half-plane. This function can have only real poles.

The roots of the equation

$$\psi'(l,\lambda)+h\psi(l,\lambda)=0$$

are the characteristic numbers of the problem under the boundary conditions

$$y'+\operatorname{ctg}\alpha y|_{x=0}=0,\qquad y'+hy|_{x=l}=0. \tag{7.42}$$

Thus in this case (for real h) all the characteristic numbers are real and simple. Since h is any real number, $\psi(l,\lambda)$ and $\psi'(l,\lambda)$ form a real pair. Evidently, if $h_1\neq h_2$ the functions $\psi(l,\lambda)+h_1\psi'(l,\lambda)$ and $\psi(l,\lambda)+h_2\psi'(l,\lambda)$ also form a real pair. For $h=p+iq$ we must have

$$\varphi(\lambda)=\psi(l,\lambda)+h\psi'(l,\lambda)=\psi(l,\lambda)+p\psi'(l,\lambda)+iq\psi'(l,\lambda),$$

and from Theorem 4 we find that all the characteristic numbers lie in one of the half-planes Im >0 or Im $\lambda<0$.

If for complex values of λ the general solution of equation (7.38) has an integrable square on the half-axis $(0,\infty)$ with the weighting function $\rho(x)$, then we may put $l=\infty$ in the above discussion. From a theorem of M. G. Kreĭn (Chapter V, Theorem 12), we can obtain more complete information on the distribution of eigenvalues of this boundary problem.

From (7.40) for $h=0$ and $h=\infty$ we find that $\varphi(l,\lambda)/\psi(l,\lambda)$ and $\varphi'(l,\lambda)/\psi'(l,\lambda)$ are meromorphic functions mapping the upper half-plane into the upper half-plane. We have

$$\frac{1}{\psi(l,\lambda)\psi'(l,\lambda)}=\frac{\varphi(l,\lambda)}{\psi(l,\lambda)}-\frac{\varphi'(l,\lambda)}{\psi'(l,\lambda)},$$

and by a theorem of N. G. Čebotarev (Theorem 2) we have the decomposition

$$\frac{1}{\psi(l,\lambda)\psi'(l,\lambda)}=\sum A_k\left(\frac{1}{\lambda_k-\lambda}-\frac{1}{\lambda_k}\right)-\sum B_k\left(\frac{1}{\mu_k-\lambda}-\frac{1}{\mu_k}\right), \tag{7.43}$$

where λ_k and μ_k are roots of the functions $\psi(l,\lambda)$ and $\psi'(l,\lambda)$ and A_k and B_k are positive numbers.

By the theorem of M. G. Kreĭn, it follows from the existence of this decomposition that $\psi(l,\lambda)\psi'(l,\lambda)$ is an entire function of exponential type. It is known[12] that almost all the characteristic numbers of the equation (7.38) under conditions (7.42) and real h are positive. Putting $\lambda = \mu^2$ in (7.43) and expanding each term into elementary divisors, we obtain for the entire function $\psi(l, \mu^2)\psi'(l, \mu^2)$ an expansion analogous to (7.43). Again applying the Kreĭn theorem, we conclude that the entire function $\psi(l, \lambda)\psi'(l, \lambda)$ is of order not greater than $\frac{1}{2}$ and is of normal type. Denoting the numbers of roots of the functions $\psi(l, \lambda)\psi'(l, \lambda)$ and $\psi(l, \lambda) + h\psi'(l, \lambda)$ in the circle $|\lambda| \leqslant r$ by $m(r)$, $n_h(r)$, we find that since $\psi(l, \lambda)\psi'(l, \lambda)$ is of completely regular growth there exists a finite limit

$$\Delta = \lim_{r \to \infty} \frac{m(r)}{r^{\frac{1}{2}}} \qquad (\Delta > 0).$$

From the interlacing of the roots of $\psi(l, \lambda)$ and $\psi'(l, \lambda)$ we may conclude that there exists a finite limit

$$\lim_{r \to \infty} \frac{n_h(r)}{r^{\frac{1}{2}}} = \frac{\Delta}{2}.$$ [13]

[12] See, for example, G. Sansone [1].

[13] The discussion presented here is taken from the paper [4] of M. G. Kreĭn. On the basis of these results, he obtains in his paper a very precise characterization of the spectrum of a boundary problem.

CHAPTER VIII

APPROXIMATION OF ENTIRE FUNCTIONS BY POLYNOMIALS WHOSE ZEROS LIE IN A GIVEN REGION[1]

The representation of an entire function by a power series shows that any entire function is the limit of a sequence of polynomials which converges uniformly in every bounded domain. If we impose on the polynomials which are to approximate the function uniformly the additional requirement that their zeros belong to a certain set, then the limit functions will form a special class, depending on the set.

The first results in this direction are due to Laguerre [2], who gave a complete characterization of the entire functions that can be uniformly approximated by polynomials:

I. whose zeros are all positive;

II. whose zeros are all real.

In case II Laguerre formulated the theorem without proof. A proof of his theorem was later given by Pólya [2]. A more complete investigation of the convergence of sequences of such polynomials was carried out by Lindwart and Pólya [1]. They showed, in particular, that in cases I and II (as well as in more general cases) the uniform convergence of a sequence of polynomials in some disk $|z| < R$ implies its uniform convergence on any bounded subset of the plane.

We will follow the method of Lindwart and Pólya in the presentation of these questions.

In the third section of this chapter we give an exposition of the theorems on "multiplier sequences" which are closely related to the above set of problems.

1. Functions which can be Approximated by Polynomials, all of Whose Zeros Lie in an Angle

LEMMA 1. *A set of polynomials of the form*

$$P_m(z) = \prod_{k=1}^{n_m}\left(1 - \frac{z}{a_{k,m}}\right) = 1 + c_{1,m}z + \dots + c_{n_m,m}z^{n_m} \quad (8.00)$$

[1] *Translator's Note*: For a survey of recent work in this area see the article by J. Korevaar, "Limits of polynomials with restricted zeros," in "Essays in honor of George Pólya," Stanford, 1962.

satisfying the conditions

$$\sum_{k=1}^{n_m} |a_{k,m}|^{-\rho} < M \tag{8.01}$$

and

$$|c_{k,m}| < M \qquad (k = 1, 2, \ldots, p, \rho - 1 \leqslant p < \rho) \tag{8.02}$$

constitutes a normal family in the plane.[2]

PROOF. In Chapter I (§ 4, Lemma 2) we proved the existence of a constant A_ρ such that for all values of the variable u one has the inequality

$$|(1-u)e^{u+\frac{u^2}{2}+\cdots+\frac{u^p}{p}}| \leqslant e^{A_\rho |u|^\rho} \qquad (0 \leqslant p < \rho \leqslant p+1).$$

From this inequality it is easily seen that all polynomials of the set under consideration satisfy the inequality

$$|P_m(z)e^{s_{1,m}z+\frac{1}{2}s_{2,m}z^2+\cdots+\frac{1}{p}s_{p,m}z^p}| \leqslant e^{A_\rho M|z|^\rho},$$

where

$$s_{j,m} = \sum_{k=1}^{n_m} a_{k,m}^{-j} \qquad (j = 1, 2, \ldots, p)$$

and z is an arbitrary complex number. Expressing the quantities $s_{j,m}$ in terms of the coefficients of the polynomials by means of Newton's formulas, and making use of condition (8.02), we find that for certain constants $\sigma_j > 0$ and for all polynomials of the set considered one will have the inequality

$$|s_{j,m}| < \sigma_j \qquad (j = 1, 2, \ldots, p).$$

Hence, throughout the plane,

$$|P_m(z)| \leqslant e^{A_\rho M|z|^\rho + \sigma_1|z| + \cdots + \frac{1}{p}\sigma_p|z|^p} \tag{8.03}$$

The boundedness of the set of polynomials $\{P_m(z)\}$ on every bounded subset of the plane implies that it constitutes a normal family.[3]

LEMMA 2. *If a sequence of polynomials $\{P_m(z)\}$ of the form* (8.00) *satisfies condition* (8.01) *and converges uniformly in some neighborhood of the point* 0, *then the sequence converges uniformly in every bounded domain. If for such a sequence the number ρ in condition* (8.01) *is not an integer, the genus p of the limit entire function does not exceed* $[\rho]$. *If the number ρ is an integer, the limit function has the form*

$$F(z) = e^{\gamma z^\rho}\Phi(z), \tag{8.04}$$

where γ is a constant and $\Phi(z)$ an entire function of genus not larger than $\rho - 1$.

---

[2] A family of holomorphic functions in a given domain is called normal if every sequence of functions of the family contains a subsequence which converges uniformly on every bounded closed subset of the domain.

[3] See, for example, A. I. Markuševič [1].

PROOF. It follows from the convergence of the sequence of polynomials $P_m(z)$ in a neighborhood of zero that the coefficients $c_{j,m}$ $(j = 1, 2, \ldots)$ of these polynomials converge to the corresponding coefficients c_j of the limit function as $m \to \infty$, and hence condition (8.02) is satisfied. By Lemma 1 the sequence $\{P_m(z)\}$ forms a normal family and by a well-known theorem of Stieltjes[4] its uniform convergence in a neighborhood of zero implies its uniform convergence in every bounded domain.

It follows from the estimate (8.03) that the limit entire function is at most of order ρ and normal type.

Furthermore, the zeros $a_{k,m}$ of the polynomial $P_m(z)$ in an arbitrary circle tend to the zeros a_k of the limit function. Thus

$$\sum_{|a_k|<r} |a_k|^{-\rho} \leqslant \lim_{m\to\infty} \sum_{|a_{k,m}|<r} |a_{k,m}|^{-\rho} \leqslant M$$

or

$$\sum_{k=1}^{\infty} |a_k|^{-\rho} \leqslant M,$$

and hence the genus of the canonical product, constructed with the zeros of the limit function, does not exceed the number p such that $p < \rho \leqslant p + 1$.

When ρ is not an integer it follows therefore that the limit function $F(z)$ is at most of order ρ and minimal type.

When ρ is an integer the genus of the canonical product is not larger than $\rho - 1$. By Hadamard's theorem the function $F(z)$ differs from its canonical product only by a factor of the form

$$e^{\gamma z^\rho + \gamma_1 z^{\rho-1} + \ldots + \gamma_\rho},$$

that is, it has the form (8.04).

THEOREM 1. *If a sequence of polynomials $\{P_m(z)\}$ converges uniformly in a neighborhood of the origin to a function not identically equal to zero, and all zeros of the polynomials $P_m(z)$ lie in the angle $\theta_1 \leqslant \arg z \leqslant \theta_2$ $(|\theta_2 - \theta_1| < \pi)$, then the sequence converges uniformly on every finite subset of the plane, and the limit function has the form*

$$F(z) = cz^q e^{\sigma z} \prod_{k=1}^{\infty} \left(1 - \frac{z}{a_k}\right), \tag{8.05}$$

where

$$\sum_{k=1}^{\infty} |a_k|^{-1} < \infty, \quad \theta_1 \leqslant \arg a_k \leqslant \theta_2 \ \textit{and} \ \theta_1 \leqslant -\arg \sigma \leqslant \theta_2.$$

PROOF. We may assume without loss of generality that the polynomials $P_m(z)$ have no zeros in some disk $|z| < \delta$ and that $P_m(0) = 1$. Otherwise we

[4] See, for example, A. I. Markuševič [1].

could consider, instead of the polynomials $P_m(z)$, the polynomials

$$\tilde{P}_m(z) = \frac{1}{\prod_{|a_{k,m}|<\delta} (z - a_{k,m})} P_m(z),$$

where $a_{k,m}$ runs through the zeros of $P_m(z)$. Since the zeros of $P_m(z)$ tend to the zeros of the limit function, the convergence of the sequence $\{\tilde{P}_m(z)\}$ is equivalent to the convergence of the sequence $\{P_m(z)\}$. Noting that $\lim \tilde{P}_m(0) \neq 0$ we find that the sequence $\{\tilde{P}_m(z)/\tilde{P}_m(0)\}$ converges uniformly in a neighborhood of zero. We will therefore assume in the following that $P_m(z) \neq 0$ in some neighborhood of zero and that $P_m(0) = 1$.

We may also suppose that

$$\theta_1 = -\frac{\pi}{2} + \delta_1 \text{ and } \theta_2 = \frac{\pi}{2} - \delta_1 \qquad (\delta_1 > 0).$$

It follows from the convergence of the sequence of polynomials

$$P_m(z) = 1 + c_{1,m}z + c_{2,m}z^2 + \ldots + c_{n_m, m}z^{n_m}$$

in the disk $|z| < R$ that $c_{1,m} \to c_1$, and consequently $|c_{1,m}| < M$. Thus we obtain that

$$\sum_{k=1}^{n_m} \operatorname{Re} (a_{k,m})^{-1} < \Big|\sum_{k=1}^{n_m} (a_{k,m})^{-1}\Big| < M,$$

and by virtue of the inequality $|a_{k,m}|^{-1} \sin \delta_1 \leqslant \operatorname{Re} (a_{k,m})^{-1}$

$$\sum_{k=1}^{n_m} |a_{k,m}|^{-1} < \frac{M}{\sin \delta_1}.$$

Lemma 2 therefore shows that the sequence $\{P_m(z)\}$ converges uniformly in every bounded domain, and that the limit function $F(z)$ has the form

$$F(z) = ce^{-\sigma z} \prod_{k=1}^{\infty} \left(1 - \frac{z}{a_k}\right).$$

The zeros a_k are limit points of the zeros of the polynomials $P_m(z)$, and hence $|\arg a_k| \leqslant \frac{1}{2}\pi - \delta_1$. We still have to prove that $|\arg \sigma| \leqslant \frac{1}{2}\pi - \delta_1$.

From the expansions

$$-\frac{F'(z)}{F(z)} = s_1 + s_2 z + s_3 z^2 + \ldots$$

and

$$-\frac{P'_m(z)}{P_m(z)} = s_{1,m} + s_{2,m}z + s_{3,m}z^2 + \ldots$$

we obtain that

$$s_1 = \sigma + \sum_{k=1}^{\infty} a_k^{-1} \text{ and } s_{1,m} = \sum_{k=1}^{n_m} a_{k,m}^{-1}.$$

Here $s_1 = \lim_{m\to\infty} s_{1,m}$ and $\lim_{m\to\infty} a_{k,m} = a_k$. Observing that for $-\delta_1 \leqslant \tau \leqslant \delta_1$ one has $\operatorname{Re}\{e^{i\tau}a_{k,m}^{-1}\} \geqslant 0$, we find that

$$\operatorname{Re}\left\{e^{i\tau}\sum_{k=1}^{N} a_k^{-1}\right\} = \lim_{m\to\infty} \operatorname{Re}\left\{e^{i\tau}\sum_{k=1}^{N} a_{k,m}^{-1}\right\}$$
$$\leqslant \lim_{m\to\infty} \operatorname{Re}\left\{e^{i\tau}\sum_{k=1}^{n_m} a_{k,m}^{-1}\right\} = \operatorname{Re}\{e^{i\tau}s_1\},$$

and taking the limit as $N \to \infty$ we obtain

$$\operatorname{Re}\{e^{i\tau}\sigma\} = \operatorname{Re}\{e^{i\tau}s_1\} - \operatorname{Re}\left\{e^{i\tau}\sum_{k=1}^{\infty} a_k^{-1}\right\} \geqslant 0 \qquad (-\delta_1 \leqslant \tau \leqslant \delta_1).$$

Thus we conclude that $|\arg \sigma| \leqslant \frac{1}{2}\pi - \delta_1$, and this completes the proof.

We remark that if an entire function has the form (8.05), then it can be represented in the form

$$F(z) = \lim_{m\to\infty} cz^q\left(1 - \frac{\sigma z}{m}\right)^m \prod_{k=1}^{m}\left(1 - \frac{z}{a_k}\right).$$

Thus we obtain the following theorem.

THEOREM 2. *In order that an entire function $F(z)$ can be uniformly approximated in every bounded domain by polynomials, all of whose zeros lie in the angle $\theta_1 \leqslant \arg z \leqslant \theta_2$ ($|\theta_2 - \theta_1| < \pi$), it is necessary and sufficient that the function have the form* (8.05).

We will now investigate the more delicate case $|\theta_2 - \theta_1| = \pi$. A polynomial which has no zeros in the open lower half-plane will be called an *H-polynomial.*

THEOREM 3. *If a sequence $\{P_m(z)\}$ of H-polynomials converges uniformly in a δ-neighborhood of the origin to a function $F(z)$ not identically equal to zero, then it converges uniformly in every bounded domain. The limit function $F(z)$ belongs to the class $\overline{HB}$, and can be represented in the form*

$$F(z) = cz^q e^{-\gamma z^2 + \beta z} \prod_{k=1}^{\infty}\left(1 - \frac{z}{a_k}\right)e^{\frac{z}{a_k}}, \tag{8.06}$$

where

$$\sum_{k=1}^{\infty} |a_k|^{-2} < \infty \text{ and } \gamma \geqslant 0.$$

If all zeros of the polynomials $P_m(z)$ are real, then all zeros of $F(z)$ are real[5] *and* $\operatorname{Im}\beta = 0$.

[5] In the case of real zeros this is the theorem of Laguerre and Pólya. For the case of complex zeros the theorem was proved by N. Obrechkoff [1]. *Translator's Note*: This theorem was stated by Pólya.

PROOF. Taking $0 < a < \delta$ we construct the sequence of polynomials

$$Q_m(z) = \frac{P_m(z - ia)}{P_m(-ia)} = 1 + c_{1,m}z + c_{2,m}z^2 + \ldots + c_{n_m, m}z^{n_m}.$$

Since $F(-ia) \neq 0$ the sequence $\{Q_m(z)\}$ converges uniformly in the disk $|z| < \delta - a$. Hence the coefficient $c_{1,m}$ tends to a limit as $m \to \infty$, and we will have an inequality $|c_{1,m}| < M$ for some $M > 0$ and all $m = 1, 2, \ldots$.

It follows from the equation

$$\sum_{k=1}^{n_m} (a_{k,m} + ia)^{-1} = -c_{1,m}$$

that

$$\sum_{k=1}^{n_m} \frac{a}{|a_{k,m} + ia|^2} \leqslant \sum_{k=1}^{n_m} \frac{a + \operatorname{Im} a_{k,m}}{|a_{k,m} + ia|^2} = \operatorname{Im} c_{1,m} < M.$$

Thus the polynomials $Q_m(z)$ satisfy condition (8.01) with $\rho = 2$, and by Lemma 2 the sequence of these polynomials will converge uniformly in every bounded domain. The limit function of this sequence is $F(z - ia)/F(-ia)$, and hence $F(z)$ must have the form (8.04) with $\rho = 2$.

The H-polynomials $P_m(z)$ belong to the class $\overline{HB}$, and hence the limit function $F(z)$ also belongs to $\overline{HB}$. It remains to show that $\gamma \geqslant 0$. To prove this we suppose first that all zeros of the polynomials $P_m(z)$ are real. We may assume without loss of generality that $q = 0$ in (8.06). Using the notation

$$s_2 = -\frac{F''(0)F(0) - F'^2(0)}{F^2(0)} = 2\gamma + \sum_{k=1}^{\infty} a_k^{-2}$$

and

$$s_{2,m} = -\frac{P''_m(0)P_m(0) - P'^2_m(0)}{P^2_m(0)} = \sum_{k=1}^{n_m} a_{k,m}^{-2},$$

we will have $\lim_{m\to\infty} s_{2,m} = s_2$. For any positive integer N

$$\sum_{k=1}^{N} a_k^{-2} = \lim_{m\to\infty} \sum_{k=1}^{N} a_{k,m}^{-2} \leqslant \lim_{m\to\infty} s_{2,m} = s_2,$$

and hence

$$\sum_{k=1}^{\infty} a_k^{-2} \leqslant s_2.$$

Thus in this case $\gamma \geqslant 0$.

We now consider the general case. We set

$$P_m(z) = Q_m(z) + iR_m(z).$$

The uniform convergence of the sequence $\{P_m(z)\}$ implies the uniform convergence of the sequence $\{\bar{P}_m(z)\}$, since $\bar{P}_m(z) = \overline{P_m(\bar{z})}$.

The equations

$$2Q_m(z) = P_m(z) + \bar{P}_m(z) \text{ and } 2iR_m(z) = P_m(z) - \bar{P}_m(z)$$

now imply the uniform convergence of the sequences of polynomials $\{Q_m(z)\}$ and $\{R_m(z)\}$, to functions $Q(z)$ and $R(z)$, say. By the theorem of Hermite and Biehler the zeros of $Q_m(z)$ and $R_m(z)$ are real, and hence by what has been proved already $Q(z)$ and $R(z)$ have the representations

$$Q(z) = c_1 z^{q_1} e^{-\gamma_1 z^2 + \beta_1 z} \prod_{k=1}^{\infty} \left(1 - \frac{z}{a_k^{(1)}}\right) e^{\frac{z}{a_k^{(1)}}}$$

and

$$R(z) = c_2 z^{q_2} e^{-\gamma_2 z^2 + \beta_2 z} \prod_{k=1}^{\infty} \left(1 - \frac{z}{a_k^{(2)}}\right) e^{\frac{z}{a_k^{(2)}}}.$$

Since the function

$$F(z) = Q(z) + iR(z)$$

belongs to the class $\overline{HB}$, condition (β) of Theorem 3 in Chapter VII shows that $\gamma_1 = \gamma_2$.

Thus the function $F(z)$ can be represented in the form

$$F(z) = e^{-\gamma_1 z^2} \Phi_1(z),$$

where $\gamma_1 \geqslant 0$ and $\Phi_1(z)$ is a function of at most order two and of minimal type. Such a representation is unique since it implies that $h_F(\theta) = -\gamma_1 \cos 2\theta$, and hence γ_1 is uniquely determined.[5a]

REMARK. *A set of H-polynomials*

$$P_m(z) = Q_m(z) + iR_m(z),$$

satisfying the conditions $|P_m(0)| \geqslant c_0 > 0$, $|P'_m(0)| \leqslant c_1$, $|P''_m(0)| \leqslant c_2$ (*where* c_0, c_1, c_2 *are arbitrary positive constants*), *forms a normal family.*

PROOF. Denoting the zeros of the polynomials $Q_m(z)$ and $R_m(z)$ by x_{km} and y_{km} we obtain

$$\sum_1^{n_m} x_{km}^{-2} = \frac{Q_m'^2(0) - Q_m(0)\, Q_m''(0)}{Q_m^2(0)},$$

$$\sum_1^{n_m} y_{km}^{-2} = \frac{R_m'^2(0) - R_m(0)\, R_m''(0)}{R_m^2(0)}.$$

[5a] The requirement of convergence of the sequence of H-polynomials $\{P_n(z)\}$ in a neighborhood of zero can be replaced by the weaker requirement of the convergence of this sequence on a "sufficiently large set" (see Levin [17]).

At least one of these sums will be bounded by the number

$$M = \frac{2(c_1^2 + c_0 c_2)}{c_0^2}.$$

Hence, since the zeros of $Q_m(z)$ and $R_m(z)$ alternate, we obtain

$$\sum_3^{n_m} x_{km}^{-2} \leqslant M, \quad \sum_3^{n_m} y_{km}^{-2} \leqslant M.$$

It follows from these estimates and Lemma 1 that the polynomials

$$Q_m(z)(z - x_{1m})^{-1}(z - x_{2m})^{-1}, \qquad R_m(z)(z - y_{1m})^{-1}(z - y_{2m})^{-1}$$

form normal families. From this one immediately obtains the normality of the family $\{P_m(z)\}$.[6]

Our conclusion remains valid if $\lim_{m\to\infty} P_m^{(k)}(0) = 0$ $(k = 0, 1, 2, \ldots, p-1)$, $\liminf_{m\to\infty} |P_m^{(p)}(0)| > 0$, and $|P_m^{(p+1)}(0)| \leqslant c_1$, $|P_m^{(p+2)}(0)| \leqslant c_2$. Indeed, this result follows from the preceding if instead of the polynomials $P_m(z)$ we consider the polynomials

$$s_m(z) = \frac{P_m(z)}{\prod_1^p (z - z_j)},$$

where the z_j $(j = 1, 2, \ldots, p)$ are the zeros of $P_m(z)$ closest to 0. The requirement $\liminf_{m\to\infty} |P_m^{(p)}(0)| > 0$ is essential, as is shown by the example of the sequence of H-polynomials $\{z^m\}$, which does not form a normal family in the plane.

We remark also that the result can be generalized by replacing the H-polynomials with entire functions $F(z)$ of the form (8.06). This follows immediately if one makes use of the formula

$$2\gamma + \sum_1^\infty \frac{1}{a_k^2} = \frac{F'^2(0) - F(0)F''(0)}{F^2(0)}.$$

DEFINITION. We will define *the class P^** as the class of the entire functions which belong to $\overline{HB}$ and have a representation of the form

$$F(z) = e^{-\gamma z^2}\Phi(z),$$

with $\gamma \geqslant 0$ and $\Phi(z)$ an entire function of genus not larger than one.

THEOREM 4. *In order that an entire function $F(z)$ be the limit of a uniformly convergent sequence of H-polynomials, it is necessary and sufficient that it belong to the class P^*.*

[6] *Translator's Note*: The proof becomes simpler if one assumes at first that $P_m(0) = 1$. For a different proof see O. Szász, *On sequences of polynomials and the distribution of their zeros*, Bull. Amer. Math. Soc. **49** (1943), 377–383.

PROOF. The necessity follows immediately from Theorem 3. In order to prove the sufficiency we will construct, for a given function $F(z)$ of class P^* and prescribed $\varepsilon > 0$ and $R > 0$, an H-polynomial $P(z)$ such that $|F(z) - P(z)| < \varepsilon$ for $|z| \leqslant R$.

The representation given in Theorem 6 of Chapter VII for an arbitrary entire function of class $\overline{HB}$ shows that $F(z)$ can be represented in the form

$$F(z) = cz^q e^{-\gamma z^2 + \delta z + i\nu z} \prod_{n=1}^{\infty} \left(1 - \frac{z}{a_n}\right) e^{\operatorname{Re} \frac{1}{a_n} z},$$

with $\gamma \geqslant 0$, δ real and $\nu \geqslant 0$.

It follows from this representation that for $|z| \leqslant R$ and sufficiently large n

$$\left| F(z) - cz^q e^{-\gamma z^2 + \sigma_n z} \prod_{k=1}^{n} \left(1 - \frac{z}{a_k}\right) \right| < \frac{\varepsilon}{2},$$

where

$$\sigma_n = \delta + i\nu + \sum_{k=1}^{n} \operatorname{Re} a_k^{-1}.$$

Every polynomial of the form

$$P_n(z) = cz^q \left(1 - \frac{\gamma z^2}{p_n}\right)^{p_n} \left(1 + \frac{\sigma_n z}{p_n}\right)^{p_n} \prod_{k=1}^{n} \left(1 - \frac{z}{a_k}\right),$$

with p_n a positive integer, is obviously an H-polynomial, and for p_n sufficiently large it will satisfy

$$\left| cz^q e^{-\gamma z^2 + \sigma_n z} \prod_{k=1}^{n} \left(1 - \frac{z}{a_k}\right) - P_n(z) \right| < \frac{\varepsilon}{2} \qquad (|z| \leqslant R).$$

Combining this inequality with the preceding one we obtain

$$|F(z) - P_n(z)| < \varepsilon \qquad (|z| \leqslant R).$$

This proves the theorem.

In the next theorem we establish a relation between membership in the class P^* for an entire function and membership for its real and imaginary component.

THEOREM 5. *In order that the function*

$$\omega(z) = P(z) + iQ(z)$$

belong to P^*, it is necessary and sufficient that $\omega(z)$ belong to the class $\overline{HB}$ and that $P(z)$ and $Q(z)$ be functions of class P^*.

PROOF. If $\omega(z) \in P^*$, there exists a sequence of H-polynomials

$$\omega_n(z) = P_n(z) + iQ_n(z)$$

which converges uniformly to $\omega(z)$. It is clear that the sequences of real H-polynomials $\{P_n(z)\}$ and $\{Q_n(z)\}$ will converge uniformly to the entire functions

$P(z)$ and $Q(z)$ which, as a consequence, must belong to P^*. Conversely, let $\omega(z) \in \overline{HB}$ and let

$$P(z) = c_1 e^{-\gamma_1 z^2 + \beta_1 z} \prod_{k=1}^{\infty} \left(1 - \frac{z}{a_k}\right) e^{\frac{z}{a_k}},$$

$$Q(z) = c_2 e^{-\gamma_2 z^2 + \beta_2 z} \prod_{k=1}^{\infty} \left(1 - \frac{z}{b_k}\right) e^{\frac{z}{b_k}}$$

be functions of class P^*.

It follows from Theorem 1 in § 1 of Chapter VII that the zeros a_k and b_k alternate and that

$$\gamma_1 = \gamma_2, \quad \beta_1 - \beta_2 + \sum_1^{\infty} \left(\frac{1}{a_k} - \frac{1}{b_k}\right) = 0.$$

The polynomials

$$P_{m,n}(z) = \left(1 - \frac{\gamma z^2}{n}\right)^n \left[1 + zn^{-1}\left(\beta_1 + \sum_1^m a_k^{-1}\right)\right]^n \prod_{k=1}^{m} \left(1 - \frac{z}{a_k}\right)$$

and

$$Q_{m,n}(z) = \left(1 - \frac{\gamma z^2}{n}\right)^n \left[1 + zn^{-1}\left(\beta_1 + \sum_1^m a_k^{-1}\right)\right]^n \prod_{k=1}^{m} \left(1 - \frac{z}{b_k}\right)$$

form a generalized real pair, and for $b_k > a_k$ the polynomial $\omega_{m,n}(z) = P_{m,n}(z) + iQ_{m,n}(z)$ is an H-polynomial. It is obvious that one can select a sequence of integers $n = n_m$ such that the sequences of polynomials $\{P_{m,n_m}(z)\}$ and $\{Q_{m,n_m}(z)\}$ converge uniformly to $P(z)$ and $Q(z)$ as $m \to \infty$. The corresponding sequence $\{\omega_{m,n_m}(z)\}$ of H-polynomials will converge uniformly to $\omega(z)$. Thus $\omega(z) \in P^*$.[6a]

Additional properties of functions of class P^* will be established in the next chapter.

2. Theorems on Composition of Polynomials

Later on we will need transformations which take H-polynomials into H-polynomials. Theorems on composition of polynomials and on multiplier sequences will play an essential part there. We present these theorems in the next two sections.

THEOREM 6 (SCHUR [1]). *If all zeros of the polynomial*

$$f(x) = a_0 + a_1 x + a_2 x^2 + \ldots + a_m x^m \qquad (a_m \neq 0)$$

are real, and all zeros of the polynomial

$$g(x) = b_0 + b_1 x + b_2 x^2 + \ldots + b_n x^n \qquad (b_n \neq 0)$$

[6a] Korevaar [1; 2] has studied functions that are uniform limits of polynomials with zeros in domains composed of rays issuing from the origin. He has obtained very significant results. In particular, he has established that an entire function which is a uniform limit of polynomials that have no zeros in an angle of opening less than π can have arbitrarily rapid growth.

are real and of the same sign, then all zeros of the polynomial

$$P(x) = a_0 b_0 + 1!\, a_1 b_1 x + 2!\, a_2 b_2 x^2 + \ldots + k!\, a_k b_k x^k$$
$$(k = \min(m, n))$$

are real. If $m \leqslant n$ *here and* $a_0 b_0 \neq 0$, *then all zeros of* $P(x)$ *are distinct.*

For the proof of this theorem we follow the article of I. Schur. As a preliminary we prove two lemmas.

LEMMA 3. *If all zeros of the real polynomial*

$$h(x) = c_0 + c_1 x + \ldots + c_n x^n \ (c_n \neq 0)$$

are real, $c_0 \neq 0$ *and* $c_p = 0$ $(0 < p < n)$, *then* $c_{p-1} c_{p+1} < 0$.

PROOF. Denoting the zeros of $h(x)$ by x_k we will have

$$c_1^2 - 2c_0 c_2 = c_0^2 \sum_{k=1}^{n} x_k^{-2}. \tag{8.07}$$

Hence it follows that for $c_0 \neq 0$

$$c_1^2 - 2c_0 c_2 > 0.$$

Rolle's theorem shows that all zeros of the derivative of order $p - 1$ of the polynomial $h(x)$ are real; one has

$$h^{(p-1)}(x) = (p-1)!\,c_{p-1} + \frac{p!}{1!} c_p x + \frac{(p+1)!}{2!} c_{p+1} x^2 + \ldots + \frac{n!}{(n-p+1)!} c_n x^{n-p+1}.$$

Forming the expression (8.07) for this polynomial we obtain the inequality

$$p c_p^2 - (p+1) c_{p-1} c_{p+1} > 0,$$

and the assertion of the lemma is an immediate consequence.

In particular, it follows from this lemma that in a polynomial $h(x)$ with $c_0 \neq 0$ no two consecutive coefficients can be equal to zero.

LEMMA 4. *If all zeros of the polynomials*

$$f(x) = a_0 + a_1 x + \ldots + a_n x^n$$

and

$$h(x) = c_0 + c_1 x + \ldots + c_n x^n \qquad (c_0 \neq 0,\ c_n \neq 0)$$

are real, then all zeros of the polynomial

$$F(x) = h(D) f(x) = c_0 f(x) + c_1 f'(x) + \ldots + c_n f^{(n)}(x)$$

are real. Here every multiple zero of the polynomial $F(x)$ *is also a multiple zero of the polynomial* $f(x)$.

PROOF. The differential operator $h(D)$ can be written in factored form

$$h(D) = c_n \prod_{k=1}^{n} (D - \alpha_k),$$

where all zeros α_k are real and different from zero. For every polynomial $\varphi(x)$ and real $\alpha \neq 0$ the function $e^{-\alpha x}\varphi(x)$ tends to zero either as $x \to +\infty$ or as $x \to -\infty$.

By Rolle's theorem the number of real zeros of the function

$$\varphi_1(x) = (D - \alpha)\varphi(x) = e^{\alpha x} D[e^{-\alpha x}\varphi(x)]$$

is not less than that of the function $\varphi(x)$. Since the polynomial $f(x)$ has n real zeros the corresponding polynomial $\varphi_1(x)$ of degree n also has n real zeros. Consequently $F(x) = h(D)f(x)$ also has only real zeros.

To complete the proof of the lemma we still have to show that if ξ is a multiple zero of the polynomial

$$\Phi(x) = f'(x) - \alpha f(x),$$

then it will be a multiple zero of $f(x)$. We expand $f(x)$ and $\Phi(x)$ in powers of $x - \xi$:

$$f(x) = d_0 + d_1(x - \xi) + \ldots + d_n(x - \xi)^n,$$
$$\Phi(x) = d_1 - \alpha d_0 + (2d_2 - \alpha d_1)(x - \xi) + \ldots - \alpha d_n (x - \xi)^n.$$

If ξ is a multiple zero of the polynomial $\Phi(x)$, then $d_1 - \alpha d_0 = 0$ and $2d_2 - \alpha d_1 = 0$, and hence

$$d_1^2 - 2d_0 d_2 = 0.$$

By Lemma 3 this relation can not hold when $d_0 \neq 0$. Thus we have $d_0 = 0$, $d_1 = 0$, $d_2 = 0$, that is, ξ is a multiple zero of the polynomial $f(x)$.

PROOF OF THEOREM 6. We first consider the case $a_0 \neq 0$, $b_0 \neq 0$ and $m \leqslant n$. We may assume without loss of generality that $a_m > 0$, $b_n > 0$, and that all coefficients b_k $(k = 0, 1, \ldots, n)$ are positive. For arbitrary real z we form the polynomial

$$F(x) = b_0 f(x) + b_1 z f'(x) + \ldots + b_m z^m f^{(m)}(x)$$

and we write it in the form

$$F(x) = P_0(z) + \frac{P_1(z)}{1!}x + \frac{P_2(z)}{2!}x^2 + \ldots + \frac{P_m(z)}{m!}x^m,$$

where

$$P_0(z) = a_0 b_0 + 1!a_1 b_1 z + 2!a_2 b_2 z^2 + \ldots + m!a_m b_m z^m$$

and

$$P_\mu(z) = \mu!a_\mu b_0 + (\mu + 1)!\,a_{\mu+1} b_1 z + \ldots + m!a_m b_{m-\mu} z^{m-\mu}$$
$$(\mu = 1, 2, \ldots, m).$$

By Lemma 4 all zeros of $F(x)$ are real.

The polynomials $P_0(z)$ and $P_1(z)$ have no common zero, for if they did, the origin would be a multiple zero of the corresponding polynomial $F(x)$ and, by Lemma 4, a multiple zero of $f(x)$, contrary to the assumption that $a_0 \neq 0$.

Thus we have a sequence of polynomials

$$P_0(z), \quad P_1(z), \quad \ldots, \quad P_m(z)$$

with the following properties:

(1) The last element $P_m(z)$ of the sequence is a constant different from zero.

(2) Two consecutive polynomials of the sequence never vanish for the same real value of z (Lemma 3).

(3) If one of the polynomials of the sequence vanishes for a real value of z, then the polynomials on either side of it have values of opposite sign for this particular z (Lemma 3).

Thus our sequence of polynomials forms a generalized Sturm sequence. If we denote by $v(z)$ the number of sign changes in the sequence of these polynomials for a given value of z, and by r the number of real zeros of the polynomial $P_0(z)$, then we will have

$$r \geqslant v(-\infty) - v(\infty).$$

When $a_m > 0$ and $b_k > 0$ ($k = 0, 1, \ldots, m$) we find that $v(\infty) = 0$ and $v(-\infty) = m$, that is, all zeros of the polynomial $P_0(z)$ are real. As we pass through a zero of the polynomial $P_0(z)$ the number of sign changes $v(z)$ must, in this case, increase by the multiplicity of the zero. Since the polynomials $P_0(z)$ and $P_1(z)$ have no common zero the number of sign changes can not increase by more than one, that is, $P_0(z)$ has no multiple zeros.

We now consider the case $a_0 b_0 \neq 0$ and $m > n$. If instead of the polynomial $g(x)$ we consider the polynomial

$$(1 - \varepsilon x)^{m-n} g(x) = b_0(\varepsilon) + b_1(\varepsilon) x + \ldots + b_m(\varepsilon) x^m,$$

we will have

$$\lim_{\varepsilon \to 0} b_k(\varepsilon) = \begin{cases} b_k & \text{for} \quad k = 0, 1, \ldots, n, \\ 0 & \text{for} \quad k > n. \end{cases}$$

All zeros of the polynomial

$$b_0(\varepsilon) a_0 + 1! b_1(\varepsilon) a_1 x + \ldots + n! b_n(\varepsilon) a_n x^n + \ldots + m! b_m(\varepsilon) a_m x^m$$

will be real, and, taking the limit as $\varepsilon \to 0$, we find that all zeros of the polynomial

$$b_0 a_0 + 1! b_1 a_1 x + \ldots + n! b_n a_n x^n$$

are real.

If one or both of the numbers a_0 and b_0 are equal to zero, that is, if

$$f(x) = x^{\mu} f_1(x) \text{ and } g(x) = x^{\nu} g_1(x),$$

we form the composition of the polynomials

$$f_\varepsilon(x) = (x - \varepsilon)^{\mu} f_1(x) \text{ and } g_\varepsilon(x) = (x - \varepsilon)^{\nu} g_1(x),$$

thus obtaining a polynomial all of whose zeros are real. Taking the limit as $\varepsilon \to 0$ we find that the composition of the polynomials $f(x)$ and $g(x)$ has all its zeros real.

From the composition theorem of Schur one easily obtains

THEOREM 7 (MALÓ).[7] *If all zeros of the polynomial*

$$f(x) = a_0 + a_1 x + \dots + a_n x^n$$

are real, and all zeros of the polynomial

$$g(x) = b_0 + b_1 x + \dots + b_m x^m$$

are real and of the same sign, then all zeros of the polynomial

$$P(x) = a_0 b_0 + a_1 b_1 x + \dots + a_k b_k x^k \quad (k = \min(m, n))$$

are real.

PROOF. If all zeros of the polynomial

$$f(x) = a_0 + a_1 x + \dots + a_n x^n$$

are real, then replacing x by $1/x$ we see that all zeros of the polynomial

$$f_1(x) = a_n + a_{n-1} x + \dots + a_0 x^n$$

are real also. Forming the Schur composition of this polynomial and the polynomial

$$(1 + x)^n = 1 + \binom{n}{1} x + \binom{n}{2} x^2 + \dots + x^n,$$

we find that the polynomial

$$f_2(x) = a_n + a_{n-1} n x + a_{n-2} n(n-1) x^2 + \dots + a_0 n! x^n$$

likewise has only real zeros. Substituting again for x the quantity $1/x$ and dividing by $n!$ we find that the polynomial

$$f_3(x) = a_0 + \frac{a_1}{1!} x + \frac{a_2}{2!} x^2 + \dots + \frac{a_n}{n!} x^n$$

has only real zeros. If we now form the composition of this polynomial and the polynomial $g(x)$, we find that the polynomial $P(x)$ has only real zeros.

3. Multiplier Sequences

Let

$$\gamma_0, \gamma_1, \dots, \gamma_n, \dots \tag{8.08}$$

be an arbitrary sequence of real numbers, and let $\Gamma^*[P(x)]$ be the operator which takes the arbitrary polynomial

$$P(x) = a_0 + a_1 x + a_2 x^2 + \dots + a_n x^n$$

---

[7] We will not use Maló's theorem in the sequel.

into the polynomial

$$\Gamma^*[P(x)] = \gamma_0 a_0 + \gamma_1 a_1 x + \gamma_2 a_2 x^2 + \ldots + \gamma_n a_n x^n.$$

The sequence (8.08) is called a *multiplier sequence of the first kind* if the corresponding Γ^*-operator takes every polynomial whose zeros are real into a polynomial of the same class. It is called a *multiplier sequence of the second kind* if the corresponding Γ^*-operator takes every polynomial whose zeros are positive into a polynomial whose zeros are real. These concepts were introduced by Pólya and Schur [3].

Laguerre [2] considered a number of sequences of this type. He proved, in particular, that the sequences

$$1, \frac{1}{\omega}, \frac{1}{\omega(\omega+1)}, \ldots \qquad (\omega > 0)$$

and

$$1, q, q^4, q^9, \ldots, q^{n^2}, \ldots \qquad (|q| \leqslant 1)$$

are multiplier sequences of the first kind, and that for arbitrary real λ and θ the sequence

$$\cos\lambda, \cos(\lambda+\theta), \cos(\lambda+2\theta), \ldots$$

is a multiplier sequence of the second kind.

If $\gamma_n = n$, then $\Gamma^*[P(x)] = xP'(x)$; hence Rolle's theorem shows that the sequence 0, 1, 2, ... is a multiplier sequence of the first kind.

We will establish various properties of multiplier sequences.

1. *If $\gamma_0, \gamma_1, \ldots, \gamma_n, \ldots$ is a multiplier sequence of the first or second kind, then $\gamma_k, \gamma_{k+1}, \gamma_{k+2}, \ldots, \gamma_{k+n}, \ldots$ $(k > 0)$ is a multiplier sequence of the same kind.*

Indeed, if all zeros of the polynomial

$$P(x) = a_0 + a_1 x + \ldots + a_n x^n$$

are real (or real and of the same sign), and $\gamma_0, \gamma_1, \ldots$ is a multiplier sequence of the first (or second) kind, then all zeros of the polynomial

$$x^{-k}\Gamma^*[x^k P(x)] = a_0\gamma_k + a_1\gamma_{k+1}x + \ldots + a_n\gamma_{n+k}x^n$$

are real. We will assume in what follows that $\gamma_0 \neq 0$.

2. *If a certain element in a multiplier sequence of the first kind is equal to zero, then all subsequent elements are equal to zero.*

Indeed, all zeros of the polynomial

$$\Gamma^*[(1+x)^n] = \gamma_0 + \gamma_1\binom{n}{1}x + \gamma_2\binom{n}{2}x^2 + \ldots + \gamma_n x^n$$

will be real. It follows from Lemma 3 that if $\gamma_n \neq 0$ and $\gamma_p = 0$ for some p $(0 < p < n)$, then $\gamma_{p-1}\gamma_{p+1} < 0$. On the other hand the zeros of the polynomial

$$\Gamma^*[x^{p+1} - x^{p-1}] = (\gamma_{p+1}x^2 - \gamma_{p-1})\,x^{p-1}$$

will be real. But when $\gamma_{p-1}\gamma_{p+1} < 0$, this is impossible.

Thus the relations $\gamma_n \neq 0$ and $\gamma_p = 0$ (where $0 < p < n$) can not hold at the same time.

3. *In a multiplier sequence of the first kind the elements either all have the same sign or they have alternating signs.*

Indeed, it follows from the reality of the zeros of the polynomial $\gamma_{p+1}x^{p+1} - \gamma_{p-1}x^{p-1}$ that γ_{p-1} and γ_{p+1} have the same sign.

A non-negative multiplier sequence of the first kind will be called a Γ-*sequence*, the corresponding operator a Γ-*operator*.

4. *A Γ-operator takes an H-polynomial into an H-polynomial.*

Indeed, let

$$\omega(z) = P(z) + iQ(z)$$

be an H-polynomial. We may assume without loss of generality that $\omega(0) \neq 0$.[8] By Theorem 2 in § 1 of Chapter VII the polynomials

$$P(z) = a_0 + a_1 z + \dots + a_n z^n$$

and

$$Q(z) = b_0 + b_1 z + \dots + b_n z^n$$

will form a generalized real pair, and

$$a_0 b_1 - a_1 b_0 > 0.$$

A Γ-operator takes the generalized real pair $P(z)$, $Q(z)$ into the generalized real pair

$$\Gamma[P(z)] = \gamma_0 a_0 + \gamma_1 a_1 z + \dots + \gamma_n a_n z^n,$$

$$\Gamma[Q(z)] = \gamma_0 b_0 + \gamma_1 b_1 z + \dots + \gamma_n b_n z^n,$$

where $\gamma_0\gamma_1(a_0 b_1 - a_1 b_0) > 0$.

Hence it follows that $\Gamma[\omega(z)]$ is an H-polynomial.

Any half-plane of the form $\alpha < \arg z < \pi + \alpha$ can be mapped into the lower half-plane ($\operatorname{Im} z < 0$) by a simple rotation, and hence, if all zeros of a polynomial $R(z)$ belong to the angle $\alpha \leqslant \arg z \leqslant \beta$ $(0 \leqslant \beta - \alpha \leqslant \pi)$, then the zeros of the polynomial $\Gamma[R(z)]$ will all belong to the same angle.

In particular, if all zeros of $R(z)$ lie on a ray $\arg z = c$, then the zeros of the polynomial $\Gamma[R(z)]$ will lie on the same ray.

A Γ-operator can be extended to a larger class of functions with the aid of the following definition. Let

$$f(z) = \sum_{k=0}^{\infty} a_k z^k$$

be an entire function. Then we set

$$\Gamma[f(z)] = \sum_{k=0}^{\infty} a_k \gamma_k z^k$$

whenever the latter series converges.

---

[8] If this is not the case, we can consider the function $\omega(z + \varepsilon)$ and later take the limit as $\varepsilon \to 0$.

THEOREM 8. *A Γ-operator can be applied to all functions of class P^* and it takes functions of class P^* into functions of the same class.*

PROOF. A function

$$\omega(z) = \sum_{k=0}^{\infty} a_k z^k$$

of class P^* is the limit of a uniformly convergent sequence of H-polynomials

$$\omega_n(z) = \sum_{k=0}^{n} a_{k,n} z^k.$$

Applying a Γ-operator to this sequence we obtain a sequence of H-polynomials

$$\Gamma[\omega_n(z)] = \sum_{k=0}^{n} \gamma_k a_{k,n} z^k,$$

where $\lim_{n\to\infty} \gamma_k a_{k,n} = \gamma_k a_k$. It follows from the remark made after Theorem 3 that the polynomials $\Gamma[\omega_n(z)]$ form a normal family. The limit function $F(z)$ of a uniformly convergent subsequence of polynomials $\Gamma[\omega_n(z)]$ will belong to the class P^*. The coefficients in the series expansion of such a function can be obtained by passage to the limit; that is, we have

$$F(z) = \sum_{k=0}^{\infty} \gamma_k a_k z^k.$$

Since the limit is unique, the whole sequence $\Gamma[\omega_n(z)]$ will converge and

$$F(z) = \Gamma[\omega(z)].$$

An important example of a Γ-operator was constructed by Jensen.

MULTIPLIERS OF JENSEN [1]. Suppose that all zeros of the polynomial

$$P(x) = a_0 + a_1 x + \dots + a_k x^k$$

are real. Forming the composition of this polynomial and the polynomial $(1 + x)^n$ we obtain from Schur's theorem that all zeros of the polynomial

$$\tilde{P}_1(x) = a_0 + a_1 n x + a_2 n(n-1)x^2 + \dots + a_k n(n-1)\dots(n-k+1)x^k$$

are real. Replacing x by x/n we find that the polynomial

$$P_n(x) = a_0 + a_1 x + \left(1 - \frac{1}{n}\right) a_2 x^2 + \dots + \left(1 - \frac{1}{n}\right)\left(1 - \frac{2}{n}\right)\dots\left(1 - \frac{k-1}{n}\right) a_k x^k$$

has all its zeros real. Thus for any positive integer n the sequence

$$1,\ 1,\ 1 - \frac{1}{n},\ \left(1 - \frac{1}{n}\right)\left(1 - \frac{2}{n}\right),\ \dots,\ \left(1 - \frac{1}{n}\right)\left(1 - \frac{2}{n}\right)\dots\left(1 - \frac{n-1}{n}\right),\ 0,\ 0,\ \dots$$

is a Γ-sequence. The corresponding Γ-operator will be denoted by $I_n[P(x)]$. We will sometimes use the notation $\tilde{P}_n(x) = I_n[P(x)]$.

The operator $I_n[f]$ associates a polynomial of degree n with every entire function. It is easy to see that

$$\lim_{n \to \infty} I_n[f(z)] = f(z).$$

Indeed, let

$$f(z) = \sum_{k=0}^{\infty} a_k z^k$$

and

$$I_n[f(z)] = a_0 + a_1 z + \sum_{k=2}^{n} \left(1 - \frac{1}{n}\right) \ldots \left(1 - \frac{k-1}{n}\right) a_k z^k.$$

From the inequality for the coefficients of a power series

$$|a_k| \leqslant \frac{M(R)}{R^k} \qquad (M(R) = \max_{|z|=R} |f(z)|, \ k = 1, 2, \ldots)$$

we obtain the inequalities

$$\left| \sum_{k=m+1}^{n} \left(1 - \frac{1}{n}\right)\left(1 - \frac{2}{n}\right) \ldots \left(1 - \frac{k-1}{n}\right) a_k z^k \right| \leqslant \frac{M(R)\, r^{m+1}}{R^m (R-r)}$$

and

$$\left| \sum_{k=m+1}^{\infty} a_k z^k \right| \leqslant \frac{M(R)\, r^{m+1}}{R^m (R-r)}$$

which are valid for $|z| \leqslant r$. Choosing $R = 2r$ and taking m sufficiently large we will have

$$\left| \sum_{k=m+1}^{n} \left(1 - \frac{1}{n}\right)\left(1 - \frac{2}{n}\right) \ldots \left(1 - \frac{k-1}{n}\right) a_k z^k \right| < \frac{\varepsilon}{3}$$

and

$$\left| \sum_{k=m+1}^{\infty} a_k z^k \right| \leqslant \frac{\varepsilon}{3} \qquad (|z| \leqslant r).$$

Furthermore, for sufficiently large n

$$\left| \sum_{k=2}^{m} \left(1 - \frac{1}{n}\right)\left(1 - \frac{2}{n}\right) \ldots \left(1 - \frac{k-1}{n}\right) a_k z^k - \sum_{k=2}^{m} a_k z^k \right| < \frac{\varepsilon}{3}$$
$$(|z| \leqslant r)$$

and hence

$$|f(z) - I_n[f(z)]| < \varepsilon \quad \text{for} \quad |z| \leqslant r.$$

If the function $\omega(z) \in P^*$, it can be written as the limit of a uniformly convergent sequence of H-polynomials

$$\omega_m(z) = \sum_{k=0}^{m} a_{k,m} z^k.$$

Applying the operator I_n we obtain the sequence of H-polynomials $I_n[\omega_m(z)]$. Taking the limit as $m \to \infty$ we conclude that the polynomials

$$I_n[\omega(z)] = a_0 + a_1 z + \left(1 - \frac{1}{n}\right) a_2 z^2 + \dots$$
$$\dots + \left(1 - \frac{1}{n}\right) \dots \left(1 - \frac{n-1}{n}\right) a_n z^n,$$

which converge uniformly to $\omega(z)$, are H-polynomials.

We have thus proved the following theorem.

THEOREM 9. *In order that a function*

$$\omega(z) = \sum_{k=0}^{\infty} a_k z^k$$

belong to the class P^*, it is necessary and sufficient that the sequence $\{I_n[\omega(z)]\}$,

$$I_n[\omega(z)] = a_0 + a_1 z + \sum_{k=2}^{n} \left(1 - \frac{1}{n}\right)\left(1 - \frac{2}{n}\right) \dots \left(1 - \frac{k-1}{n}\right) a_k z^k,$$

which converges uniformly to $\omega(z)$, consist entirely of H-polynomials.[9]

We conclude this chapter by proving the following theorem of G. Pólya and I. Schur [1].

---

[9] This theorem was proved by Jensen [1] for the case of entire functions whose zeros are real. Writing down the conditions of Hurwitz for the polynomials $I_n[\omega(z)]$ we obtain a necessary and sufficient condition for membership in the class P^* in the form of an infinite set of determinantal inequalities. It is interesting to compare this result with existing extensions of Hurwitz's theorem to entire functions. Grommer [1] proved that for real entire functions $f(z) = \Sigma_{k=0}^{\infty} a_k z^k$ of genus zero the positivity of all Hurwitz determinants, that is, the determinants of the consecutive principal minors in the infinite matrix

$$\left\| \begin{matrix} a_1 & a_0 & 0 & 0 & 0 & 0 & 0 \dots \\ a_3 & a_2 & a_1 & a_0 & 0 & 0 & 0 \dots \\ a_5 & a_4 & a_3 & a_2 & a_1 & a_0 & 0 \dots \\ \cdot & \cdot & \cdot & \cdot & \cdot & \cdot & \cdot \\ \cdot & \cdot & \cdot & \cdot & \cdot & \cdot & \cdot \end{matrix} \right\|,$$

is necessary and sufficient in order that all zeros of the function lie in the left half-plane. Fujiwara [1] extended this result to entire functions of arbitrary genus. However, it was discovered by M. G. Kreĭn [1] that the results of Grommer and Fujiwara are incorrect. Kreĭn proved that the positivity of all Hurwitz determinants is necessary and sufficient in order that a real function $f(z)$ can be represented in the form $f(z) = E(z^2)\,G(z)$, where $E(z)$ is an arbitrary real entire function and $G(iz)$ is an arbitrary entire function of class $\overline{HB}$.

THEOREM 10. *In order that the sequence of numbers*

$$\gamma_0, \gamma_1, \ldots, \gamma_n, \ldots \qquad (\gamma_0 \neq 0) \tag{8.08}$$

be a Γ*-sequence, it is necessary and sufficient that the series*

$$\varphi(z) = \sum_{k=0}^{\infty} \frac{\gamma_k}{k!} z^k \tag{8.09}$$

converge in the whole plane, and that the entire function $\varphi(z)$ *can be represented in the form*

$$\varphi(z) = ce^{\sigma z} \prod_{k=1}^{\infty} (1 + \varkappa_k z) \tag{8.10}$$

with $\sigma \geqslant 0$, $\kappa_k > 0$ *and* $\Sigma_{k=1}^{\infty} \kappa_k < \infty$.

PROOF. Let (8.08) be a Γ-sequence. Applying the corresponding Γ-operator to the polynomial

$$(1 + x)^n = \sum_{k=0}^{n} \binom{n}{k} x^k,$$

we find that the polynomial

$$Q_n(z) = \sum_{k=0}^{n} \gamma_k \binom{n}{k} x^k \tag{8.11}$$

has all its zeros real and negative. Putting $x = z/n$ we find that the polynomial

$$P_n(z) = \sum_{k=0}^{n} \frac{\gamma_k}{k!} \left(1 - \frac{1}{n}\right) \cdots \left(1 - \frac{k-1}{n}\right) z^k \tag{8.12}$$

has all its zeros real and negative.

We will show that the sequence of polynomials $\{P_n(z)\}$ converges uniformly to an entire function. Indeed, our polynomials will form a normal family since all their zeros lie in the half-plane $\operatorname{Im} z \geqslant 0$, while their first three coefficients are bounded and $\gamma_0 \neq 0$ (see the remark following Theorem 3). Since the various coefficients in the polynomials $P_n(z)$ tend to limits as $n \to \infty$ our normal family has the unique limit function

$$\varphi(z) = \sum_{0}^{\infty} \frac{\gamma_k}{k!} z^k,$$

and hence the sequence of polynomials $\{P_n(z)\}$ converges uniformly to $\varphi(z)$. Since the polynomials $\{P_n(z)\}$ have all their zeros negative real, Theorem 1 shows that the limit function $\varphi(z)$ can be written in the form (8.10).

We now prove the sufficiency. Let $\varphi(z)$ be an entire function of the form (8.10). We take a sequence of polynomials

$$R_m(z) = \sum_{k=0}^{m} \frac{\gamma_{k,m}}{k!} z^k$$

whose zeros are negative real and which converge uniformly to the function $\varphi(z)$. We form the Schur composition of the polynomials $R_m(z)$ and an arbitrary polynomial

$$P(z) = a_0 + a_1 z + a_2 z^2 + \ldots + a_n z^n$$

which has all its zeros real. By Schur's theorem all zeros of the polynomials

$$S_m(z) = \sum_{k=0}^{n} a_k \gamma_{k,m} z^k$$

will be real. Passing to the limit as $m \to \infty$ we find that all zeros of the polynomial

$$\Gamma[P(z)] = \gamma_0 a_0 + \gamma_1 a_1 z + \ldots + \gamma_n a_n z^n$$

will be real. Hence it follows that the sequence of numbers (8.08) is a Γ-sequence.

REMARK. All alternating multiplier sequences of the first kind can be obtained by forming the series expansions of the functions of the form $\varphi(-z)$.

The proof of the following theorem of G. Pólya and I. Schur [1] is completely analogous to that of Theorem 10.

THEOREM 11. *In order that the sequence of numbers*

$$\gamma_0, \gamma_1, \ldots, \gamma_n, \ldots$$

be a multiplier sequence of the second kind, it is necessary and sufficient that the series

$$\varphi(z) = \sum_{k=0}^{\infty} \frac{\gamma_k}{k!} z^k$$

converge in the whole plane, and that $\varphi(z)$ *be a (constant multiple of a) real function of class* P^*.

CHAPTER IX

OPERATORS PRESERVING INEQUALITIES AMONG ENTIRE FUNCTIONS

The classes of entire functions considered in Chapters VII and VIII play an important role in the study of inequalities among entire functions and extremal properties of entire functions.

S. N. Bernšteĭn discovered the following significant property of entire functions of exponential type **[7]**.

If the entire function $f(z)$ of exponential type σ is bounded on the real axis, i.e.,

$$|f(x)| \leqslant M \qquad (-\infty < x < \infty),$$

then

$$|f'(x)| \leqslant M\sigma \qquad (-\infty < x < \infty).$$

This theorem, which plays an important role in the theory of approximation, aroused great interest. It was proved again several times and was generalized in various directions.

The inequalities in the formulation of this theorem may be written in the form

$$|f(x)| \leqslant |Me^{i\sigma x}| \text{ and } |f'(x)| \leqslant |(Me^{i\sigma x})'|,$$

i.e., the Bernšteĭn theorem consists in the statement that under the indicated conditions an inequality between certain functions implies an inequality between their derivatives.

Bernšteĭn generalized his theorem **[8]**, replacing the function $e^{i\sigma z}$ by the more general functions $\omega(z) = e^{i\sigma z}\varphi(z)$, where $\varphi(z)$ is an entire function of genus zero, having no roots in one of the half-planes $\operatorname{Im} z < 0$, $\operatorname{Im} z > 0$, at which $\sum_1^\infty |1/\operatorname{Im} a_k| < \infty$.

N. I. Ahiezer [2] carried out a further generalization, replacing the function $\varphi(z)$ by functions of exponential type satisfying certain regularity conditions.

In the papers **[6; 7]**[1] these regularity requirements were removed, and it was proved that if $\omega(z) \in P$ is an entire function of exponential type not larger than that of $\omega(z)$, then from the inequality

$$|f(x)| \leqslant |\omega(x)| \qquad (-\infty < x < \infty)$$

[1] The bracketed numbers without further indication of an author refer to papers by the author of this book.

there follows the inequality

$$|f'(x)| \leqslant |\omega'(x)| \qquad (-\infty < x < \infty).$$

Essentially one can not further extend the class of functions $\omega(x)$ playing the same role as $e^{i\sigma x}$ in the Bernšteĭn inequality. It was proved [7] that if $\omega(z) \bar{\in} P$ and $\bar{\omega}(z) \bar{\in} P$, then there exists an entire function of lower genus than $\omega(z)$ which on the real axis satisfies the inequality $|f(x)| \leqslant |\omega(x)|$, and is also such that at some point of the real axis $|f'(x)| > |\omega'(x)|$.

In the proofs of these theorems essential use was made of the fact that, under the conditions of the theorem, from the inequality $|f(x)| \leqslant |\omega(x)|$ on the real axis and $\sigma_\omega \leqslant \sigma_f$ there follows the inequality in the half-plane:

$$|f(z)| \leqslant |\omega(z)| \qquad (\operatorname{Im} z \leqslant 0).$$

N. N. Meĭman [2] proposed a direct definition of the majorizing condition by means of the inequalities

$$|f(z)| \leqslant |\omega(z)|, \qquad |f(\bar{z})| \leqslant |\omega(z)| \qquad (\operatorname{Im} z \leqslant 0) \tag{9.00}$$

and proved that if $\omega(z) \in HB$ there follows from these inequalities the following inequality on the real axis:

$$|f'(x)| \leqslant |\omega'(x)| \qquad (-\infty < x < \infty).$$

It remains to note that in this case, in contrast to the preceding, the majorizing relation holding for functions does not necessarily hold also for the derivatives. Therefore in the formulation of this theorem one can not replace the operation of differentiation by multiple differentiation.

It is natural to pose the question as to the widest class of majorants in this new sense which preserve the majorizing relation (9.00) after differentiation. It turned out [9] that the widest class of majorants, invariant under multiplication by the function $e^{\tau x}$ for any real value of the number τ and preserving the majorizing condition (9.00) after differentiation, coincides with P^*.

In the papers [7; 8], it was established that the differentiation operator may be replaced in this range of questions by a set of other operators of considerably more general form. In order to obtain these theorems one introduces the concepts of admissible classes of functions and of $\mathfrak{B}$-operators. A subclass T of the class $\overline{HB}$ will be called admissible if, from the condition that $\omega(z)$ belongs to the class T and from inequality (9.00), it follows that $\omega(z) + f(z) \in T$ and $v\omega(z) \in T$ (v a constant). The function $\omega(z)$ here is called a *T-majorant* of the function $f(z)$.

The general idea of the proof of inequalities for entire functions consists of proving the following theorem. *If T is an admissible class and if for $|v| > 1$ the function $f(z) - v\omega(z) \in T$, then $\omega(z)$ is a T-majorant of the entire function $f(z)$.*

Now suppose that $\omega(z)$ is a T-majorant of the entire function $f(z)$. Let $K[f]$ be *a linear operator defined on the linear envelope of the class T and mapping*

T into itself. Since T is an admissible class, therefore $f(z) - v\omega(z) \in T$ for $|v| \geqslant 1$. Applying the operator $K[f]$, we obtain $K[f(z)] - vK[\omega(z)] \in T$ for $|v| \geqslant 1$, and, accordingly, $K|\omega(z)|$ is a T-majorant for $K[f(z)]$. Thus the application of such an operator preserves T-majorizing.

Linear operators mapping the admissible class T onto itself will be called $\mathfrak{B}_T$-*operators.* The $\mathfrak{B}_{\bar{p}}$- and $\mathfrak{B}_{\bar{p}^*}$-operators are of particular interest. These will be abbreviated by $\mathfrak{B}$ and $\mathfrak{B}^*$. In particular, the differentiation operator is a $\mathfrak{B}$- and a $\mathfrak{B}^*$-operator.

In § 7 of this chapter we shall find the general form of $\mathfrak{B}$- and $\mathfrak{B}^*$- operators under certain not very restrictive conditions relating to the continuity of the operator (see **[10; 15]**).

From this general form one obtains, in particular, the fact that any $\mathfrak{B}$-operator which commutes with the differentiation operator has the form

$$\mathfrak{B}[f(z)] = \sum_{k=0}^{\infty} \bar{a}_k f^{(k)}(z),$$

where

$$F(z) = \sum_{k=0}^{\infty} a_k z^k$$

is an entire function of class P^*. The operator $\Gamma[f]$ defined in Chapter VIII is the most general form of $\mathfrak{B}$ (and $\mathfrak{B}^*$)-operator commuting with the operator $z(d/dz)$.

The general expression for the operators $\mathfrak{B}$ and $\mathfrak{B}^*$ required the introduction of the classes P, P^*, and HB for functions of several variables[2] **[8]**; $\mathfrak{B}$-operators were defined for classes of functions of several variables and the theorems on the preservation of inequalities were extended. The extension of the theory to functions of several variables made it possible to generalize to such functions the well-known theorem of Laguerre-Pólya given in Chapter VIII (see **[8; 9; 15]**). These results are discussed in § 5.

1. Majorants and Admissible Classes

DEFINITION 1. The function $\omega(z)$ is called an $\overline{HB}$-*majorant* of the entire function $f(z)$, if $\omega(z) \in \overline{HB}$ and for $\operatorname{Im} z \leqslant 0$

$$|f(z)| \leqslant |\omega(z)|, \quad |f(\bar{z})| \leqslant |\omega(z)|. \tag{9.00}$$

If $\omega(z)$ belongs to some subclass T of the class $\overline{HB}$, admitting multiplication by a constant,[3] then we will call it a T-*majorant* of the function $f(z)$.

Evidently if $\omega(z)$ is a T-majorant of the function $f(z)$, then for $|u| \geq 1$ also $u\omega(z)$ is a T-majorant of the same function.

---

[2] The first Bernšteĭn inequality for entire functions was carried over to functions of several variables by Bernšteĭn himself **[3]**.

[3] I.e., if $\omega(z) \in T$ then also $c\omega(z) \in T$.

THEOREM 1.[4] *Suppose that $f(z)$ and $\omega(z)$ are entire functions. In order that the function*

$$\varphi_v(z) = f(z) - v\omega(z)$$

belong to the class $\overline{HB}$ for all v, taken outside the unit circle ($|v| \geqslant 1$), it is necessary and sufficient that $\omega(z)$ be an $\overline{HB}$-majorant for $f(z)$.

PROOF. First we prove the necessity. Suppose that $\varphi_v(z) \in \overline{HB}$. Then from the equation

$$\omega(z) = -\lim_{v\to\infty} v^{-1}\varphi_v(z)$$

it follows that $\omega(z) \in \overline{HB}$. Moreover, for $|v| > 1$ the function $\varphi_v(z)$ has no roots in the lower half-plane and accordingly

$$\frac{f(z)}{\omega(z)} - v \neq 0$$

for $\operatorname{Im} z < 0$ and $|v| > 1$. In other words,

$$\left|\frac{f(z)}{\omega(z)}\right| \leqslant 1 \tag{9.01}$$

for $\operatorname{Im} z \leqslant 0$. In order to obtain the second majorizing condition, we observe that if $\operatorname{Im} z \leqslant 0$

$$|f(\bar{z}) - v\omega(\bar{z})| \leqslant |f(z) - v\omega(z)|,$$

or, taking into account (9.01) and the fact that $\omega(z)$ belongs to the class $\overline{HB}$,

$$|f(\bar{z})| \leqslant (1 + 2|v|)|\omega(z)|.$$

Thus, in view of the equality $|\bar{f}(z)| = |f(\bar{z})|$ we shall have throughout the lower half-plane

$$\left|\frac{\bar{f}(z)}{\omega(z)}\right| \leqslant (1 + 2|v|) \qquad (\text{for } |v| \geqslant 1).$$

From (9.01) we have for $-\infty < x < \infty$

$$\left|\frac{\bar{f}(x)}{\omega(x)}\right| \leqslant 1$$

and by the Phragmén-Lindelöf theorem (Chapter I, § 14)

$$\left|\frac{\bar{f}(z)}{\omega(z)}\right| \leqslant 1 \qquad (\operatorname{Im} z \leqslant 0),$$

i.e.,

$$|f(\bar{z})| \leqslant |\omega(z)| \qquad (\operatorname{Im} z \leqslant 0).$$

Thus $\omega(z)$ is an $\overline{HB}$-majorant of the function $f(z)$.

---

[4] This theorem was proved by the author for functions of exponential type and formulated as a general principle for the proof of inequalities of the Bernšteĭn type (see [7], Lemma 3). The proof of the generalized inequality of Bernšteĭn with the use of this lemma is an improvement of a method due to N. I. Ahiezer [2]. For the class $\overline{HB}$ the sufficiency part of the theorem was proved by N. N. Meĭman [2].

Now we prove the sufficiency. If $\omega(z)$ is an $\overline{HB}$-majorant of the function $f(z)$, then for $|v| > 1$ the function $\varphi_v(z)$ can not vanish in the half-plane Im $z < 0$. Moreover, for $|v| > 1$ and Im $z < 0$

$$\left|\frac{\varphi_v(\bar{z})}{\varphi_v(z)}\right| = \left|\frac{f(\bar{z}) - v\omega(\bar{z})}{f(z) - v\omega(z)}\right| = \left|\frac{f(\bar{z})}{\omega(z)} - v\frac{\omega(\bar{z})}{\omega(z)}\right| : \left|\frac{f(z)}{\omega(z)} - v\right| \leqslant \frac{1 + |v|}{|v| - 1}.$$

For real z this ratio is equal to unity, and applying the Phragmén-Lindelöf theorem to the function

$$\frac{\bar{\varphi}_v(z)}{\varphi_v(z)},$$

we find that

$$|\varphi_v(\bar{z})| = |\bar{\varphi}_v(z)| \leqslant |\varphi_v(z)| \qquad (\text{Im } z \leqslant 0).$$

Thus, $\varphi_v(z) \in \overline{HB}$ for $|v| \geqslant 1$. The lemma is proved.

DEFINITION 2. We shall call a subclass T of the class $\overline{HB}$ *admissible* if from the fact that $\omega(z)$ is a T-majorant of the entire function $f(z)$ it follows that

$$f(z) + \omega(z) \in T.$$

From Theorem 1 we see that the class $\overline{HB}$ is admissible. *The class P is also an admissible class.* Indeed, if $\omega(z)$ is a P-majorant of the function $f(z)$, then $f(z)$ is of exponential type. The function $f(z) + \omega(z)$ is also of exponential type. By Theorem 1 this function belongs to $\overline{HB}$ and accordingly to P (Chapter VII, Lemma 1).

We shall prove the admissibility of the class P^* somewhat further on.

2. Some Properties of the Class P^*

Suppose that $P_n(z)$ is any H-polynomial. From the formula

$$\frac{P_n'(z)}{P_n(z)} = \sum_{k=1}^{n} \frac{1}{z - z_k}$$

in which the z_k are the roots of the polynomial $P_n(z)$ (Im $z_k \geqslant 0$), it follows that for Im $z < 0$

$$\text{Im}\,\frac{P_n'(z)}{P_n(z)} > 0.$$

Thus $P_n'(z)$ has no roots in the lower half-plane and accordingly is also an H-polynomial.[5]

[5] The following theorem of Gauss follows easily from the considerations just introduced: *the roots of the derivative of a polynomial are all situated in the smallest convex region containing the roots of the polynomial.*

Approximating a given function $\omega(z)$ of class P^* by a sequence $\{P_n(z)\}$ of H-polynomials, we find that $\omega'(z)$ is also a limit of a sequence $\{P_n'(z)\}$ of H-polynomials. Thus, the derivative of a function of class P^* also belongs to P^*.

The following theorem shows that the class P^* is in a certain sense the widest class of functions admitting differentiation.

THEOREM 2 [9]. *If $\omega(z)$ is an entire function and if for any real τ*

$$[e^{\tau z}\omega(z)]' \in \overline{HB}, \tag{9.02}$$

then $\omega(z)$ is a function of class P^*.

PROOF. Suppose that $\omega(z) = P(z) + iQ(z)$. From Theorem 4′ of Chapter VII it follows from (9.02) that for all real values of τ the functions

$$P'(z) + \tau P(z) \quad \text{and} \quad Q'(z) + \tau Q(z)$$

form a generalized real pair. Thus all the roots of the function $P'(z) + \tau P(z)$ are real, and accordingly the quantity

$$\operatorname{Im} \frac{P'(z)}{P(z)} \tag{9.03}$$

preserves its sign for $\operatorname{Im} z > 0$. In the upper semicircle of each pole this quantity is negative, and thus it is negative in the entire upper half-plane. By Theorem 2 of Chapter VII the meromorphic function $P'(z)/P(z)$ admits a representation

$$\frac{P'(z)}{P(z)} = -2\gamma z + \delta + \sum_{k=\alpha}^{\omega} \left[\frac{1}{z-a_k} - \frac{1}{a_k}\right],$$

in which δ and a_k $(k = 1, 2, \ldots)$ are real, $\gamma \geqslant 0$ and $\sum_{k=1}^{\infty} |a_k|^{-2} < \infty$. Thus $P(z)$ may be represented in the form

$$P(z) = ce^{-\gamma z^2 + \delta z} \prod_{k=\alpha}^{\omega} \left(1 - \frac{z}{a_k}\right) e^{\frac{z}{a_k}}$$

so that $P(z) \in P^*$. Analogously one shows that $Q(z) \in P^*$.

Since for any real μ, ν, and τ all the roots of the function

$$\mu[P'(z) + \tau P(z)] + \nu[Q'(z) + \tau Q(z)]$$

are real, by dividing by τ and passing to the limit as $\tau \to \infty$ we find that $P(z)$ and $Q(z)$ are a generalized real pair. Hence it follows that either $\omega(z) \in \overline{HB}$ or $\bar{\omega}(z) \in \overline{HB}$, and from Theorem 5 of § 1 of Chapter VIII, either $\omega(z) \in P^*$ or $\bar{\omega}(z) \in P^*$.

For a nonreal function $\omega(z)$ the second case is excluded, since $\omega'(z) \in \overline{HB}$.

Thus $\omega(z) \in P^*$ This theorem shows the close connection between the class P^* and the differentiation operation.[6]

In the following theorem we shall prove another *characteristic property of the class P^**.

THEOREM 3 [9]. *In order that the entire function $\omega(z)$ belong to the class P^*, it is necessary and sufficient that the condition*

$$\omega(z - i\sigma) \in \overline{HB} \tag{9.04}$$

be satisfied for any positive σ.

PROOF. The necessity of the condition (9.04) follows from the fact that H-polynomials have this property and that it carries through under the limit process.

For the proof of sufficiency we note first of all that condition (9.04) is satisfied if the number $-i\sigma$ is replaced by any number u of the lower half-plane. We choose an arbitrary number z_0 of the lower half-plane and a real number $h > 0$ and we put $u = z_0 - i(h/2)$. Then for $\operatorname{Im} z \leqslant 0$

$$\left|\omega\left(z + z_0 - i\frac{h}{2}\right)\right| \geqslant \left|\omega\left(\bar{z} + z_0 - i\frac{h}{2}\right)\right|,$$

and if we put $z = -ih/2$, then

$$|\omega(z_0 - ih)| \geqslant |\omega(z_0)|. \tag{9.05}$$

The function $\omega(z - ih)$ is an HB-majorant for the function $\omega(z)$. Indeed, in view of (9.05) we will have

$$|\omega(z - ih)| \geqslant |\omega(z)| \geqslant |\omega(\bar{z})| \qquad (\operatorname{Im} z \leqslant 0).$$

From Lemma 1 it follows that

$$\frac{\omega(z - ih) - \omega(z)}{ih} \in HB,$$

and passing to the limit as $h \to 0$ we obtain $\omega'(z) \in \overline{HB}$.

If $\omega(z)$ satisfies the condition (9.04), then $e^{\tau z}\omega(z)$ also satisfies this condition for any real τ. Thus from (9.04) it follows that

$$[e^{\tau z}\omega(z)]' \in \overline{HB},$$

and from Theorem 2 we obtain $\omega(z) \in P^*$.

---

[6] Pólya [4] set the following problem in 1914: Can we conclude that $f(z) \in P^*$ if $f(z)$ and all its derivatives have only real zeros? Ålander [1] remarked in his dissertation that a somewhat more general conjecture was made by Wiman in his lectures at Uppsala in 1911. These problems of Pólya and Wiman have inspired many investigations; references are given by Gol'dberg and Ostrovskiĭ [1]. Pólya's conjecture was recently established by Hellerstein and Williamson [1]. Their result can be stated as follows: *If a real entire function $f(z)$ and the two derivatives $f'(z)$ and $f''(z)$ have only real zeros, then $f(z) \in P^*$*. It follows that the conclusion of Theorem 2 remains valid if (9.02) is replaced by the requirement that $\omega^{(k)}(z) \in \overline{HB}$ for $k = 0, 1, 2$. In fact, it follows from the theorem of Hellerstein and Williamson that $\varphi(z) = \lambda P(z) + \mu Q(z) \in P^*$ for arbitrary λ and μ, and this is easily seen to imply that $\omega(z) \in P^*$.

THEOREM 4. *If $\omega(z)$ is a P^*-majorant of the entire function $f(z)$, then*

$$\omega(z)+f(z)\in P^*,$$

in other words, P^* is an admissible class.

PROOF. Suppose that $\omega(z)$ is a P^*-majorant of the function $f(z)$. We shall prove that for any $h>0$ the function $\omega(z-ih)$ is a P^*-majorant of the function $f(z-ih)$. Indeed, from the fact that $\omega(z)$ is a majorant of $f(z)$ it follows that

$$|\omega(z-ih)|\geqslant|f(z-ih)| \tag{9.06}$$

for $\operatorname{Im} z\leqslant h$.

In the proof of Theorem 3 we established in (9.05) that for $\operatorname{Im} z\leqslant 0$ the function $|\omega(z-ih)|$ increases monotonically[7] together with $h(h\geqslant 0)$. Therefore for $\operatorname{Im} z\leqslant -h$ we will have

$$|\omega(z-ih)|\geqslant|\omega(z+ih)|\geqslant|\overline{f}(z+ih)|=|f(\bar{z}-ih)|. \tag{9.07}$$

It remains to be verified that the inequality (9.07) is satisfied for $-h\leqslant \operatorname{Im} z\leqslant 0$. From the monotonicity of the function $|\omega(z-ih)|$ it follows that for $-h\leqslant\operatorname{Im} z\leqslant 0$

$$|\omega(z-ih)|\geqslant|\omega(\bar{z}-ih)|.$$

Moreover, in view of (9.06), for $\operatorname{Im}\bar{z}\leqslant h$

$$|\omega(\bar{z}-ih)|\geqslant|f(\bar{z}-ih)|$$

and in view of the last two inequalities and inequality (9.07)

$$|\omega(z-ih)|\geqslant|f(\bar{z}-ih)| \tag{9.08}$$

for $\operatorname{Im} z\leqslant 0$.

The inequalities (9.06) and (9.08) mean that the function $\omega(z-ih)$ $(h>0)$ is a P^*-majorant of the function $f(z-ih)$.

By Theorem 1 we find that the function

$$\omega(z-ih)+f(z-ih)\in\overline{HB}$$

for any $h\geqslant 0$, and in view of Theorem 3, $\omega(z)+f(z)\in P^*$. The theorem is proved.

3. Operators Preserving Subordination ($\mathfrak{B}_T$-Operators)

The following important concept, connected with the extremal properties of entire functions, deals with the so-called $\mathfrak{B}$-operators.

DEFINITION 3. *An additive, homogeneous operator,[8] defined on the linear envelope of the admissible class T, will be denoted by $\mathfrak{B}_T$ if it carries functions of the class T into functions of the class T.*

---

[7] This property may be obtained immediately. Indeed, it is obvious for an H-polynomial and is preserved through the limiting process.

[8] The operator $\mathfrak{B}$ is said to be additive if $\mathfrak{B}[f(z)+\varphi(z)]=\mathfrak{B}[f(z)]+\mathfrak{B}[\varphi(z)]$, and homogeneous, if $\mathfrak{B}[cf(z)]=c\mathfrak{B}[f(z)]$.

From this definition it follows immediately that every power of a $\mathfrak{B}_T$-operator, and more generally the products of $\mathfrak{B}_T$-operators, are $\mathfrak{B}_T$-operators.

We shall call a $\mathfrak{B}_P$-operator simply a $\mathfrak{B}$-operator, and a $\mathfrak{B}_{P*}$-operator simply a $\mathfrak{B}^*$-operator.[9]

THEOREM 5 [8]. *If T is an admissible class and $\omega(z)$ is a T-majorant of the entire function $f(z)$, then the function $\mathfrak{B}_T[\omega(z)]$ is a T-majorant of the function $\mathfrak{B}_T[f(z)]$.*

The proof is based on Theorem 1. Indeed, since the class T is admissible and $\omega(z)$ is a T-majorant of $f(z)$, therefore

$$\varphi_v(z) = f(z) - v\omega(z) \in T$$

for $|v| \geqslant 1$. By definition the $\mathfrak{B}_T$-operator

$$\mathfrak{B}_T[\varphi_v(z)] = \mathfrak{B}_T[f(z)] - v\mathfrak{B}_T[\omega(z)] \in T \qquad (|v| \geqslant 1).$$

It therefore follows from Lemma 1 that $\mathfrak{B}_T[\omega(z)]$ is an $\overline{HB}$-majorant of the function $\mathfrak{B}_T[f(z)]$, and since $\mathfrak{B}_T[\omega(z)] \in T$, therefore $\mathfrak{B}_T[\omega(z)]$ is a T-majorant of the function $\mathfrak{B}_T[f(z)]$.

Theorem 5 makes it possible to obtain various inequalities for entire functions.

In what follows we shall restrict ourselves to the study of $\mathfrak{B}$- and $\mathfrak{B}^*$-operators.

THEOREM 6. *If $\omega(z) \in P^*$ and if for* $\text{Im}\, z \leqslant 0$ *the inequalities*

$$|f(z)| \leqslant |\omega(z)|, \qquad |f(\bar{z})| \leqslant |\omega(z)| \tag{9.00}$$

are satisfied, then also

$$|f^{(k)}(z)| \leqslant |\omega^{(k)}(z)|, \qquad |f^{(k)}(\bar{z})| \leqslant |\omega^{(k)}(z)| \tag{9.09}$$

for $\text{Im}\, z \leqslant 0$ *and* $k = 1, 2, \dots,$ *while $\omega^{(k)}(z) \in P^*$ (more concisely, the relation of P^*-majorizing is preserved under differentiation). If $\omega(z)$ is not a polynomial and if, in addition, there is at least one point $z_0(\text{Im}\, z_0 \leqslant 0)$ at which $\omega(z)$ is not equal to zero, and if for some k in* (9.09) *the equality sign holds, then*

$$f(z) = c_1\omega(z) + c_2\bar{\omega}(z), \tag{9.10}$$

where c_1 and c_2 are complex numbers, while $|c_1| + |c_2| = 1$.[10]

PROOF. In § 2 it was shown that the class P^* is admissible and that the differentiation operator maps that class onto itself. Thus by Definition 2 *differentiation is a $\mathfrak{B}^*$-operator.*

In order to obtain the first part of the theorem, it suffices to refer to Theorem 5.

[9] No nontrivial example of a $\mathfrak{B}_T$-operator is known with T a wider admissible class than P^*.

[10] In [7] a somewhat more general theorem is proved.

For the proof of the second part of the theorem we observe that if some point $z_0(\mathrm{Im}\, z_0 \leqslant 0)$ $|\omega^{(k)}(z_0)| = |f^{(k)}(z_0)|$, then one may choose a real number α such that the derivative of order k of the function

$$\varphi_\alpha(z) = \omega(z) - e^{-i\alpha} f(z)$$

vanishes at the point z_0.

But $\varphi_\alpha^{(k)}(z) \in P^*$ or else is identically equal to zero. Therefore, if $\mathrm{Im}\, z_0 < 0$, then $\varphi_\alpha^{(k)}(z) \equiv 0$ and

$$f^{(k-1)}(z) = e^{i\alpha}\omega^{(k-1)}(z) + c.$$

The variation of the argument of $\omega^{k-1}(z)$ on the real axis is not less than π, and therefore, for $c \neq 0$ and at some points of the real axis, the impossible inequality

$$|f^{(k-1)}(x)| > |\omega^{(k-1)}(x)|$$

will be satisfied. Accordingly, $c = 0$. Integrating and repeating this discussion we find that for $\mathrm{Im}\, z_0 < 0$

$$f(z) = e^{i\alpha}\omega(z).$$

Now suppose that z_0 is real, i.e., the function $\varphi_\alpha^{(k)}(z)$ has a real root. We put $\psi(z) = \varphi_\alpha^{(k-1)}(z)$. Since $\psi(z) \in P^*$, then

$$\frac{\psi'(z)}{\psi(z)} = -2\gamma z + \delta + \sum_{j=1}^{\infty}\left[\frac{1}{z-a_j} - \mathrm{Re}\,\frac{1}{a_j}\right] \qquad (\gamma \geqslant 0,\ \mathrm{Im}\,\delta \geqslant 0)$$

and

$$\mathrm{Im}\,\frac{\psi'(x)}{\psi(x)} = \mathrm{Im}\,\delta + \sum_{j=1}^{\infty}\frac{\mathrm{Im}\, a_j}{|x-a_j|^2}. \tag{9.11}$$

This quantity can be equal to zero for $x = z_0$ only if δ is real and all the roots a_j are real. If $\psi(z_0) = 0$, then one must replace the functions $\varphi_\alpha^{(k)}(z)$ and $\varphi_\alpha^{(k-1)}(z)$ in the relation (9.11) by $\varphi_\alpha^{(k-1)}(z)$ and $\varphi_\alpha^{(k-2)}(z)$; if $\varphi_\alpha^{(k-2)}(z_0) = 0$, then one takes $\varphi_\alpha^{(k-2)}(z)$ and $\varphi_\alpha^{(k-3)}(z)$ and so forth. The case $\varphi_\alpha(z_0) = 0$ and $\varphi_\alpha'(z_0) = 0$ is impossible. Indeed, in this case we would have in the neighborhood of the point z_0 the following decomposition:

$$\frac{f(z)}{\omega(z)} = e^{i\alpha} + (z-z_0)^p \cdot \theta(z),$$

where $p \geqslant 2$, and $\theta(z)$ is holomorphic at the point z_0 and $\theta(z_0) \neq 0$. But such a representation is impossible, since in the lower half-plane the function $w = f(z)/\omega(z)$ takes on values lying in the unit disc $|w| < 1$. Thus, one may suppose without loss of generality that $\psi(z_0) \neq 0$ or $\psi'(z_0) \neq 0$.

First we consider the case $k = 1$. If $|f'(z_0)| = |\omega'(z_0)|$ for some real z_0, then one may choose a real number α such that $\varphi_\alpha'(z_0) = 0$, and, from (9.11), the function $\varphi_\alpha(z)$ is real up to a constant multiplier.

Therefore one may choose real constants γ_1 and γ_2 such that we have identically

$$e^{i\gamma_1}f(z) - e^{i\gamma_2}\omega(z) \equiv i\varphi(z),$$

where $\varphi(z)$ is a real function of the class P^*. We have

$$e^{i\gamma_1}f(z) = P(z) + iQ(z),$$

$$e^{i\gamma_2}\omega(z) = P(z) + iS(z),$$

while $P(z)$, $Q(z)$ and $S(z)$ are real functions.

Moreover,

$$|P(x) + iQ(x)| \leqslant |P(x) + iS(x)|,$$

from which it follows that on the real axis

$$|Q(x)| \leqslant |S(x)| \qquad (-\infty < x < \infty).$$

Since $S(z)$ has only real roots and all of these are roots of $Q(x)$, the function

$$\vartheta(z) = \frac{Q(z)}{S(z)} \tag{9.12}$$

is entire and $|\vartheta(x)| \leqslant 1$ $(-\infty < x < \infty)$. We shall prove that $\vartheta(z) \equiv$ const. To this end we observe first that for $-1 < t < 1$ we have $\omega(z) + tf(z) \in P^*$, and by the generalized theorem of Hermite-Biehler (Chapter VII, Theorem 3) the function $tQ(z) + S(z)$ has only real roots. From (9.12) one finds that the function $S(z)[\vartheta(z)t + 1]$ has only real roots for $-1 < t < 1$. But $|\vartheta(x)| \leqslant 1$, so that the second multiplier has no real roots. Thus more generally it has no roots for $-1 < t < 1$ and accordingly is equal to a constant. So we have

$$e^{i\gamma_1}f(z) = P(z) + icS(z),$$

where $P(z) + iS(z) = e^{i\gamma_2}\omega(z)$ and c is a real constant.

We have considered the case $k = 1$. We shall show that for $k > 1$ we still have the same representation of the function $f(z)$.

Suppose that

$$|f^{(k)}(z_0)| = |\omega^{(k)}(z_0)| \qquad (\operatorname{Im} z_0 \leqslant 0).$$

Then

$$e^{i\gamma_1}f^{(k-1)}(z) = P^{(k-1)}(z) + icS^{(k-1)}(z).$$

Integrating we have

$$e^{i\gamma_1}f^{(k-2)}(z) = P^{(k-2)}(z) + icS^{(k-2)}(z) + c_1.$$

Replacing z by the root x_j of the function $S^{(k-2)}(z)$, we obtain

$$e^{i\gamma_1}f^{(k-2)}(x_j) = P^{(k-2)}(x_j) + c_1.$$

Since at successive roots of $S^{(k-2)}(z)$ the function $P^{(k-2)}(z)$ takes on values with different signs, therefore for $c_1 \neq 0$ we can find a root x_j of the function $S^{(k-2)}(z)$ such that

$$|f^{(k-2)}(x_j)| > |P^{(k-2)}(x_j)| = |\omega^{(k-2)}(x_j)|.$$

This is impossible, so that $c_1 = 0$. Repeating the argument, we arrive again at the equality

$$e^{i\gamma_1} f(z) = P(z) + icS(z) \qquad (|c| \leqslant 1),$$

which we may rewrite in the form

$$f(z) = e^{i(\gamma_2-\gamma_1)} \frac{1+c}{2} \omega(z) + e^{-i(\gamma_2+\gamma_1)} \frac{1-c}{2} \bar{\omega}(z).$$

Denoting the coefficients on $\omega(z)$ and $\bar{\omega}(z)$ by c_1 and c_2, we find for $f(z)$ the representation (9.10).

REMARK. Theorem 6 remains true if in it the differentiation operator is replaced by the more general operator

$$(D + \tau I) f(z) = f'(z) + \tau f(z)$$

for some real τ. For the proof it suffices to remark that if $\omega(z) \in P^*$, then also $e^{\tau z}\omega(z) \in P^*$, and if $\omega(z)$ is a majorant of $f(z)$, then $e^{\tau z}\omega(z)$ is a majorant of $e^{\tau z}f(z)$. We note the following theorem, which in a certain sense may be considered as the converse of Theorem 6.

THEOREM 7. *If the entire function $\omega(z)$ is such that for any entire function not representable in the form* (9.10) *and for any real τ it follows from*

$$|\omega(z)| > |f(z)|, \qquad |\omega(z)| > |f(\bar{z})| \qquad (\operatorname{Im} z \leqslant 0)$$

that the inequalities

$$\left.\begin{aligned} |\omega'(z) + \tau\omega(z)| &> |f'(z) + \tau f(z)|, \\ |\omega'(z) + \tau\omega(z)| &> |f'(\bar{z}) + \tau f(\bar{z})| \end{aligned}\right\} \qquad (\operatorname{Im} z < 0) \qquad (9.13)$$

hold, then $\omega(z) \in P^*$.

PROOF. It follows from inequalities (9.13) that $\omega'(z) + \tau\omega(z)$ has no roots in the lower half-plane.

The function $f(z) = c\bar{\omega}(z)$ for $|c| < 1$ is not a function of the form (9.10), and since $\omega(z)$ is a majorant for it, therefore

$$|\omega'(z) + \tau\omega(z)| > |c||\bar{\omega}'(z) + \tau\bar{\omega}(z)| \qquad (\operatorname{Im} z < 0,\ |c| < 1),$$

i.e., $\omega'(z) + \tau\omega(z) \in \overline{HB}$. Thus, for any real τ

$$[e^{\tau z}\omega(z)]' \in \overline{HB}.$$

It therefore follows from Theorem 2 that $\omega(z) \in P^*$.

Let us consider some other examples of operators. Let $\omega_n(z)$ be a H-polynomial and $\operatorname{Im} \alpha \leqslant 0$. Then $e^{-\alpha z}\omega_n(z) \in P$, and we find that

$$(D-\alpha)[\omega_n(z)] = e^{\alpha z} D[e^{-\alpha z}\omega_n(z)]$$

is also an H-polynomial. If the function $\omega(z) \in P^*$, then, approximating it by H-polynomials $\omega_n(z)$, applying the operator $D - \alpha$ and passing to the limit, we find that $D - \alpha$ is a $\mathfrak{B}^*$-operator.

Now suppose that

$$F(z) = a_0 + a_1 z + \ldots + a_n z^n + \ldots$$

is an entire function. We define an operator $F(D)$ by the equality

$$F(D)[f(z)] = a_0 + a_1 f'(z) + \ldots + a_n f^{(n)}(z) + \ldots. \tag{9.14}$$

The operator is defined for all those functions $f(z)$ for which the series (9.14) converges.

If $F(z)$ is an H-polynomial, then evidently

$$\bar{F}(D)[f(z)] = \prod_{k=1}^{n} (D - \bar{\alpha}_k I)[f(z)]$$

and $F(-D)$ is a $\mathfrak{B}^*$-operator.

THEOREM 8. *If $F(z)$ and $\omega(z)$ are of class P^**

$$F(z) = e^{-\gamma_1 z^2} F_1(z) \text{ and } \omega(z) = e^{-\gamma_2 z^2}\omega_1(z), \tag{9.15}$$

where $F_1(z)$ and $\omega_1(z)$ are entire functions of genus one, then for $\gamma_1\gamma_2 < \frac{1}{4}$ the series $\bar{F}(D)[\omega(z)]$ converges and represents a function of class P^*.

PROOF. From the estimate for the coefficients of series expansion of an entire function of finite order 2 and of normal type (Chapter I, Theorem 2), we have

$$\overline{\lim_{n\to\infty}}\, n^{\frac{1}{2}} \sqrt[n]{|a_n|} = (2\gamma_1 e)^{\frac{1}{2}}. \tag{9.16}$$

Moreover, setting in the well-known Cauchy inequality

$$|\omega^{(n)}(z)| \leqslant n! \frac{M_\omega(R + |z|)}{R^n}$$

$R = \sqrt{n/2(\gamma_2 + \varepsilon)}$, we obtain for any $|z| \leqslant \rho (\rho > 0)$, any $\varepsilon > 0$ and sufficiently large $n > n_{\rho,\varepsilon}$:

$$|\omega^{(n)}(z)| < \frac{n!}{n^{\frac{n}{2}}} (2e)^{\frac{n}{2}} (\gamma_2 + \varepsilon)^{\frac{n}{2}} e^{(\gamma_2+\varepsilon)|z|^2}. \tag{9.17}$$

It follows from (9.16) and (9.17) that

$$\overline{\lim_{n\to\infty}} \sqrt[n]{|a_n \omega^{(n)}(z)|} \leqslant (4\gamma_1\gamma_2)^{\frac{1}{2}}.$$

Thus the uniform convergence of the series (9.14) in any bounded region is proved.

For the proof that a function $\bar{F}(D)[\omega(z)]$ belongs to the class P^* we approximate the function $\bar{F}(z)$ by the Jensen polynomials

$$\tilde{\bar{F}}_n(z) = a_0 + a_1 z + \dots + a_k\left(1 - \frac{1}{n}\right)\dots\left(1 - \frac{k-1}{n}\right)z^k + \dots + \frac{n!}{n^n}a_n z^n.$$

Evidently $\tilde{\bar{F}}_n(D)[\omega(z)] \in P^*$. From the estimates (9.16) and (9.17) it easily follows that the sequence $\{\tilde{\bar{F}}_n(D)[\omega(z)]\}$ converges uniformly in any bounded region to the function $\bar{F}(D)[\omega(z)]$, which accordingly belongs to P^*.

REMARK 1. The requirement $\gamma_1\gamma_2 < \frac{1}{4}$ is essential. In order to prove this, consider the example:[12] $F(D) = e^{-\gamma_1 D^2}$ and $\omega(z) = e^{-\gamma_2 z^2}$. We will have

$$\bar{F}(D)[\omega(z)] = e^{-\gamma_2 z^2}\sum_{n=0}^{\infty}\frac{(-\gamma_1\gamma_2)^n}{n!}H_{2n}(z),$$

where $H_{2n}(z)$ are Hermite polynomials. For $z = 0$

$$H_{2n}(0) = (-1)^n 2^n 1\cdot 3\dots(2n-1)$$

and

$$\bar{F}(D)[\omega(z)]\big|_{z=0} = \sum_{n=0}^{\infty}\frac{(2n)!}{(n!)^2}(\gamma_1\gamma_2)^n.$$

For $\gamma_1\gamma_2 \geqslant \frac{1}{4}$ the series converges.

REMARK 2. *For $\gamma_1 = 0$ the operator $\bar{F}(D)$ is a $\mathfrak{B}^*$-operator.*

It is also easy to prove, using analogous estimates, that for any $\gamma_1 > 0$ the operator $\bar{F}(D)$ is a $\mathfrak{B}$-operator [7].

Another example of a $\mathfrak{B}^*$-operator is given by the operator Γ, defined in § 3 of Chapter VIII. Indeed, by Theorem 9 of Chapter VIII a Γ-operator carries every function of class P^* into another function of the same class.

Thus, if the entire functions $\omega(z) \in P^*$ and $f(z)$ satisfy inequalities (9.00), then $\Gamma[\omega(z)]$ is also a majorant of $\Gamma[f(z)]$. The Γ-operator of Jensen is particularly interesting. When applied to the function

$$\omega(z) = \sum_{k=0}^{\infty} a_k z^k$$

it yields the polynomial

$$\tilde{\omega}_n(z) = a_0 + \sum_{k=1}^{n} a_k\left(1 - \frac{1}{n}\right)\dots\left(1 - \frac{k-1}{n}\right)z^k.$$

We formulate the result in the form of a theorem.

[12] See G. Pólya [3].

THEOREM 9. *If the entire functions* $\omega(z) \in P^*$ *and* $f(z)$ *satisfy inequalities* (9.00), *then*

$$|\tilde{f}_n(z)| \leqslant |\tilde{\omega}_n(z)| \quad \text{and} \quad |\tilde{f}_n(\bar{z})| \leqslant |\tilde{\omega}_n(z)| \qquad (\operatorname{Im} z \leqslant 0).$$

The meaning of this theorem is that the functions $f(z)$ and $\omega(z)$ are approximated by polynomials with preservation of the subordination relation, while the operation of passing from functions to polynomials is linear.

4. The Class P and Inequalities on the Real Axis

In the study of inequalities for entire functions a special role is played by functions of exponential type. This special role is determined by the fact that in this case the majorizing condition essentially comes down to the inequality

$$|f(x)| \leqslant |\omega(x)| \qquad (-\infty < x < \infty) \tag{9.18}$$

on the real axis. More precisely, the following lemma holds.

LEMMA 1. *In order that the entire function* $\omega(z)$ *of class* P *be a majorant for the entire function* $f(z)$, *it is necessary and sufficient that the inequality*

$$|f(x)| \leqslant |\omega(x)| \qquad (-\infty < x < \infty)$$

be satisfied and that the function $f(z)$ *be of exponential type* τ *not exceeding the type* σ *of the function* $\omega(z)$ ($\tau \leqslant \sigma$).

PROOF. The necessity of these conditions follows directly from the definition of majorization.

For the proof of sufficiency we consider the ratio

$$\psi(z) = \frac{f(z)}{\omega(z)}, \tag{9.19}$$

which is obviously a function of exponential type in the half-plane $\operatorname{Im} z \leqslant 0$. Moreover, $|\psi(x)| \leqslant 1$ for $-\infty < x < \infty$. This function is of exponential type in the half-plane $\operatorname{Im} z < 0$ and is bounded on the axis. Therefore it is of completely regular growth (Chapter V, Theorem 11) with the indicator $h_\psi(\theta) = k\,|\sin\theta|$ $(-\pi \leqslant \theta \leqslant 0)$. Thus,

$$h_f(\theta) = h_\omega(\theta) + k\,|\sin\theta|. \tag{9.20}$$

We shall show that $k \leqslant 0$. Since $\omega(z) \in P$, therefore $h_\omega(\theta) \geqslant h_\omega(-\theta)$ for $-\pi \leqslant \theta \leqslant 0$, and accordingly $h_\omega(\theta)$ takes on its largest value σ for some θ_0 $(-\pi \leqslant \theta_0 \leqslant 0)$. If $\theta \neq 0, \pi$, then

$$h_f(\theta_0) - \sigma = k\,|\sin\theta_0|,$$

and since $\max h_f(\theta) = \tau \leqslant \sigma$, therefore $k \leqslant 0$.

Suppose that $\theta_0 = 0$ (or π). It follows from (9.20) that for $\operatorname{Im} z \leqslant 0$ the indicator diagram of the function $f(z)$ is obtained from the indicator diagram of the function $\omega(z)$ by a shift by the vector $-ki$. If in addition $k \geqslant 0$, then we would have $\tau \geqslant \sqrt{\sigma^2 + k^2}$, which is impossible.

Thus $k \leqslant 0$, and by the Phragmén-Lindelöf principle $|\psi(z)| \leqslant 1$ on the entire lower half-plane. In other words,

$$|f(z)| \leqslant |\omega(z)| \qquad (\mathrm{Im}\, z \leqslant 0).$$

Carrying through the same argument for the function $\bar{f}(z)$, we find

$$|f(\bar{z})| \leqslant |\omega(z)|.$$

Thus $\omega(z)$ is a P-majorant for the function $f(z)$. The lemma is proved.

We have defined in § 3 the concept of a $\mathfrak{B}$-operator carrying the class P into itself. Since P is an admissible class, by Theorem 5 a $\mathfrak{B}$-operator preserves the subordination relation. From Lemma 1, we obtain the following theorem, which is fundamental in this group of questions.

THEOREM 10. *If $\omega(z)$ is a function of class P and of exponential type $\sigma \geqslant 0$, and $f(z)$ is an entire function of exponential type $\tau \leqslant \sigma$, and $\mathfrak{B}[f(z)]$ is any $\mathfrak{B}$-operator, then from the inequality*

$$|f(x)| \leqslant |\omega(x)| \qquad (-\infty < x < \infty)$$

there follows the inequality

$$|\mathfrak{B}[f(x)]| \leqslant |\mathfrak{B}[\omega(x)]| \qquad (-\infty < x < \infty),$$

while $\mathfrak{B}[f(z)]$ is a function of exponential type not exceeding that of the function $\mathfrak{B}[\omega(z)]$.

In § 2 we have proved that differentiation is a $\mathfrak{B}^*$-operator. Moreover, under differentiation a function of exponential type goes into a function of the same type.

Thus, differentiation is a $\mathfrak{B}$-operator.

THEOREM 11. *If $\omega(z)$ is a function of class P and of exponential type σ, and $f(z)$ an entire function of type $\tau \leqslant \sigma$, then from the inequality*

$$|f(x)| \leqslant |\omega(x)| \qquad (-\infty < x < \infty) \tag{9.21}$$

there follows the inequality

$$|f^{(k)}(x)| \leqslant |\omega^{(k)}(x)| \qquad (-\infty < x < \infty). \tag{9.22}$$

If at some point of the real axis at which $\omega(x) \neq 0$ we have the equality sign, then

$$f(z) = c_1\omega(z) + c_2\bar{\omega}(z), \tag{9.10}$$

where c_1 and c_2 are complex constants, while $|c_1| + |c_2| = 1$.

The first part of the theorem is a direct consequence of Theorem 10, and the second is obtained from Theorem 6 if one observes that $P \subset P^*$ and $\omega(z)$ is a P^*-majorant of the function $f(z)$.

We shall now show that the class P is the widest class of functions $\omega(z)$ for which the inequality (9.21) implies the inequality (9.22). Before formulating the theorem we note that if for the entire functions $f(z)$ and $\omega(z)$ inequalities (9.22) are satisfied, then the coefficients of the power expansion of $f(z)$ do not exceed the corresponding coefficients of the expansion of $\omega(z)$, and therefore the proximate order of $f(z)$ is not larger than the proximate order of $\omega(z)$, and if the proximate orders are equal the type of $f(z)$ is not larger than the type of $\omega(z)$.

THEOREM 12. *If neitner $\omega(z)$ nor $\bar{\omega}(z)$ belongs to the class P, then there exists a function $f(z)$ of the same proximate order and of the same type as $\omega(z)$, satisfying the equation*

$$|f(x)| = |\omega(x)| \qquad (-\infty < x < \infty),$$

and such that at some point x_0 of the real axis

$$|f'(x_0)| > |\omega'(x_0)|.$$

PROOF. Suppose first that $\omega(z)$ is of exponential type. If $\omega(x_0) \neq 0$, then the function

$$f(z) = e^{i\nu(z-x_0)}\omega(z),$$

where ν is a real number, is of the same proximate order and type as $\omega(z)$, and

$$|f(x)| = |\omega(x)| \qquad (-\infty < x < \infty)$$

and for sufficiently large ν

$$|f'(x_0)| > |\omega'(x_0)|.$$

Now suppose that $\omega(z)$ is of exponential type and that the roots of $\omega(z)$ lie in both half-planes $\operatorname{Im} z < 0$, $\operatorname{Im} z > 0$. Choose the point x_0 so that $\omega(x_0) \neq 0$ and $\omega'(x_0) \neq 0$. Suppose for definiteness that $\operatorname{Im}(\omega(x_0)/\omega'(x_0)) \leqslant 0$. Put

$$f(z) = \frac{z-\bar{\alpha}}{z-\alpha}\,\omega(z),$$

where α is a root of the function $\omega(z)$ in the lower half-plane.[13] Evidently we have

$$|f(x)| = |\omega(x)| \qquad (-\infty < x < \infty).$$

On the other hand,

$$|f'(x)| = |\omega'(x)|\left|1 - 2i\frac{\omega(x)}{\omega'(x)}\operatorname{Im}\frac{1}{x-\alpha}\right|.$$

Since

$$\operatorname{Im}\frac{1}{x-\alpha} < 0,$$

[13] If $\operatorname{Im}(\omega(x_0)/\omega'(x_0)) > 0$, then one has to choose the root α from the lower half-plane.

we find that

$$\operatorname{Re}\left[2i\,\frac{\omega(x_0)}{\omega'(x_0)}\operatorname{Im}\frac{1}{x_0-\alpha}\right] \geqslant 0.$$

Thus the vector $2i \operatorname{Im}\left(\frac{1}{x_0-\alpha}\right)\frac{\omega(x_0)}{\omega'(x_0)}$, is distinct from zero and forms with the positive direction of the real axis an angle not exceeding a right angle, so that

$$|f'(x_0)| > |\omega'(x_0)|.$$

Suppose that $\omega(z)$ has no roots in the lower half-plane. In order that neither $\omega(z)$ nor $\bar{\omega}(z)$ belong to the class P, it is necessary that $\omega(z)$ have at least one root in the upper half-plane and a negative defect $d_\omega < 0$. We put in this case

$$f(z) = e^{2i\delta z}\omega(z),$$

where $\delta = -d_\omega > 0$. Evidently, $|f(x)| = |\omega(x)|$ and

$$\operatorname{Re}(f'(x)/f(x)) = \operatorname{Re}(\omega'(x)/\omega(x)).$$

Moreover, we have $\omega(z) = e^{-i\delta z}\prod(1 - z/\alpha_k)\, e^{\operatorname{Re} z/\alpha_k}$ and therefore

$$\operatorname{Im}\frac{\omega'(x)}{\omega(x)} = -\delta + \sum_{k-1}^{\infty}\frac{\gamma_k}{(x-\beta_k)^2+\gamma_k^2}$$

$$(\alpha_k = \beta_k + i\gamma_k \text{ the roots of } \omega(z)).$$

Analogously

$$\operatorname{Im}\frac{f'(x)}{f(x)} = \delta + \sum_{k=1}^{\infty}\frac{\gamma_k}{(x-\beta_k)^2+\gamma_k^2}.$$

The second terms on the right sides of these equalities are positive, so that

$$\left|\operatorname{Im}\frac{\omega'(x)}{\omega(x)}\right| < \left|\operatorname{Im}\frac{f'(x)}{f(x)}\right|.$$

Since also

$$\operatorname{Re}\frac{\omega'(x)}{\omega(x)} = \operatorname{Re}\frac{f'(x)}{f(x)}, \qquad |\omega(x)| = |f(x)|,$$

therefore

$$|f'(x)| > |\omega'(x)| \qquad (-\infty < x < \infty).$$

The theorem is proved.

Now we shall prove the theorem of N. N. Meĭman, presented in the introduction to this chapter.

THEOREM 13. *If $\omega(z)$ is an $\overline{HB}$-majorant of the entire function $f(z)$, then for all real values of x*

$$|f'(x)| \leqslant |\omega'(x)|.$$

PROOF. Suppose that $\omega(z) \in \overline{HB}$ and that $f(z)$ satisfies conditions (9.00). Then by Lemma 1 the function $\varphi_{\alpha,v}(z) = ve^{i\alpha}\omega(z) + f(z) \in \overline{HB}$ ($v \geqslant 1$), i.e.,

$$\left| \frac{\overline{\varphi_{\alpha,v}(z)}}{\varphi_{\alpha,v}(z)} \right| < 1$$

for Im $z > 0$, and accordingly arg $\varphi_{\alpha,v}(x)$ is a monotonically increasing function. Differentiating the function arg $\varphi_{\alpha,v}(x)$ with respect to x, we obtain

$$\operatorname{Im} \frac{\varphi'_{\alpha,v}(x)}{\varphi_{\alpha,v}(x)} > 0.$$

Without loss of generality we may suppose that $if'(x) > 0$ (it suffices to multiply $f(z)$ by a multiplier of the form $e^{i\gamma}$, which does not change the modulus of $|f'(x)|$). Moreover, $|\omega(x)| \geqslant |f(x)|$, and for $v > 1$ the vector $\varphi_{\alpha,v}(x)$ makes a complete turn as α varies from 0 to 2π. Therefore one may choose α so that $\varphi_{\alpha,v}(x) > 0$. Under this choice we have Im $\{ve^{i\alpha}\omega'(x) + f'(x)\} > 0$, or taking into account the fact that Im $f'(x) = -|f'(x)|$, we find that $|v\omega'(x)| > |f'(x)|$. This is true for any $v > 1$ and accordingly

$$|\omega'(x)| \geqslant |f'(x)|.$$

N. N. Meĭman also proved that the equality here holds only for functions of the form (9.10).

5. Classes of Functions of Several Variables

In this section we extend the fundamental concepts and the theorems of the preceding sections to entire functions of several variables. We shall restrict ourselves to functions of two variables, since the general case differs only in that the formulas are somewhat more complicated to write.

The entire function $\omega(z, u)$ (i.e., a function represented by a power series, converging for all values of the variables) *is said to be a function of class* $\overline{HB}_2$, *if*:

(a) *it does not vanish for* Im $z < 0$, Im $u < 0$,

(b) *for* Im $z < 0$, Im $u < 0$ *the following inequalities are satisfied*:

$$|\omega(z, u)| \geqslant |\omega(\bar{z}, u)|, \quad |\omega(z, u)| \geqslant |\omega(z, \bar{u})|,$$
$$|\omega(z, u)| \geqslant |\omega(\bar{z}, \bar{u})|.$$

Evidently the limit of a uniformly convergent sequence of functions of class $\overline{HB}_2$ is a function of the same class.

The function $\omega(z, u)$ is said to be *an* $\overline{HB}_2$*-majorant* of the entire function $f(z, u)$ if *for* Im $z < 0$, Im $u < 0$

$$|\omega(z, u)| \geqslant |f(z, u)|, \quad |\omega(z, u)| \geqslant |f(\bar{z}, u)|,$$
$$|\omega(z, u)| \geqslant |f(z, \bar{u})|, \quad |\omega(z, u)| \geqslant |f(\bar{z}, \bar{u})|.$$

We shall prove a theorem analogous to Theorem 1.

THEOREM 1′. *Suppose that* $\omega(z, u)$ *and* $f(z, u)$ *are entire functions. In order that the function*

$$\varphi_v(z, u) = f(z, u) - v\omega(z, u)$$

belong to the class $\overline{HB}_2$ *for all* v *taken outside the unit circle* ($|v| \geqslant 1$), *it is necessary and sufficient that* $\omega(z, u)$ *be a* $\overline{HB}_2$*-majorant for* $f(z, u)$.

We first prove the necessity. We have

$$\omega(z, u) = \lim_{v\to\infty} v^{-1}\varphi_v(z, u)$$

so that $\omega(z, u) \in \overline{HB}_2$. Moreover, if $\operatorname{Im} u \leqslant 0$, then the function of z $\varphi_v(z, u) \in \overline{HB}$ in the variable z (we shall write in this case $\varphi_v(z, u) \in \overline{HB}_z$), and accordingly, from Lemma 1

$$|f(z, u)| \leqslant |\omega(z, u)|,$$

$$|f(\bar{z}, u)| \leqslant |\omega(z, u)|.$$

Analogously one proves that

$$|f(z, \bar{u})| \leqslant |\omega(z, u)|. \tag{9.23}$$

Finally, from the inequality

$$|\bar{f}(z, u) - \bar{v}\bar{\omega}(z, u)| \leqslant |f(z, u) - v\omega(z, u)|$$
$$(\operatorname{Im} z \leqslant 0, \ \operatorname{Im} u \leqslant 0)$$

it follows that

$$|\bar{f}(z, u)| \leqslant (1 + 2|v|)|\omega(z, u)|$$

$$(\operatorname{Im} z \leqslant 0, \ \operatorname{Im} u \leqslant 0),$$

and the function

$$\psi(z, u) = \frac{\bar{f}(z, u)}{\omega(z, u)}$$

is bounded in the indicated region. It follows from (9.23) that for real z and for $\operatorname{Im} u < 0$

$$|\psi(z, u)| \leqslant 1.$$

For a fixed u ($\operatorname{Im} u < 0$) we apply the Phragmén-Lindelöf principle for the half-plane $\operatorname{Im} z \leqslant 0$ to the function $\psi(z, u)$ and obtain

$$|\psi(z, u)| \leqslant 1 \qquad (\operatorname{Im} z \leqslant 0, \ \operatorname{Im} u \leqslant 0),$$

or

$$|f(\bar{z}, \bar{u})| \leqslant |\omega(z, u)|.$$

We now prove the sufficiency. If $\omega(z, u)$ is an $\overline{HB}_2$-majorant for $f(z, u)$, then, fixing u ($\operatorname{Im} u < 0$), we see from Lemma 1 that the function $\varphi_v(z, u) \in HB_z$. Accordingly, it has no roots for $\operatorname{Im} z < 0$, $\operatorname{Im} u < 0$, and

$$|\varphi_v(\bar{z}, u)| \leqslant |\varphi_v(z, u)| \qquad (\operatorname{Im} z \leqslant 0, \ \operatorname{Im} u \leqslant 0).$$

Analogously one proves that

$$|\varphi_v(z, \bar{u})| \leqslant |\varphi_v(z, u)| \qquad (\operatorname{Im} z \leqslant 0, \ \operatorname{Im} u \leqslant 0). \tag{9.24}$$

Just as in the proof of Lemma 1, we find that for $\operatorname{Im} z \leqslant 0$, $\operatorname{Im} u \leqslant 0$,

$$\left|\frac{\overline{\varphi}_v(z, u)}{\varphi_v(z, u)}\right| \leqslant \frac{|v|+1}{|v|-1} \qquad (|v| > 1).$$

For a real z and $\operatorname{Im} u < 0$ we have, in view of (9.24),

$$\left|\frac{\overline{\varphi}_v(z, u)}{\varphi_v(z, u)}\right| \leqslant 1$$

and, applying to this function the Phragmén-Lindelöf principle in the half-plane $\operatorname{Im} z \leqslant 0$, we obtain a third inequality

$$|\varphi_v(\bar{z}, \bar{u})| \leqslant |\varphi_v(z, u)|$$

$$(\operatorname{Im} z \leqslant 0, \ \operatorname{Im} u \leqslant 0).$$

The theorem is proved.

Admissible class and $\mathfrak{B}$-operator are defined as in the case of one variable.

THEOREM 5′. *If $\omega(z, u)$ is a T-majorant of the function $f(z, u)$, then $\mathfrak{B}_T[\omega(z, u)]$ is a T-majorant of the function $\mathfrak{B}_T[f(z, u)]$.*

The proof does not differ from that of Theorem 5.

DEFINITION. A polynomial not vanishing for $\operatorname{Im} z < 0$, $\operatorname{Im} u < 0$ is called *an H_2-polynomial in the variables z, u.*

It is easy to see that every H_2-polynomial $P(z, u)$ belongs to the class $\overline{HB}_2$. Indeed, it belongs to $\overline{HB}$ in each of the variables, under the condition that the other variable is fixed in the lower half-plane. Moreover, for fixed u ($\operatorname{Im} u \leqslant 0$) the function

$$\varphi(z, u) = \frac{\overline{P}(z, u)}{P(z, u)}$$

tends to a finite limit as $|z| \to \infty$ and, accordingly, is bounded in the lower half-plane. Since for a fixed real z the polynomial $P(z, u)$ is an H-polynomial in u, therefore $|\varphi(z, u)| \leqslant 1$ for $\operatorname{Im} u \leqslant 0$ and real z. Applying the Phragmén-Lindelöf principle to this function in the half-plane $\operatorname{Im} z \geqslant 0$ for a fixed u ($\operatorname{Im} u \leqslant 0$), we obtain

$$|P(\bar{z}, \bar{u})| = |\overline{P}(z, u)|$$

$$\leqslant |P(z, u)| \ (\operatorname{Im} z \leqslant 0, \ \operatorname{Im} u \leqslant 0).$$

Evidently a product of H-polynomials of different variables, $P_1(z)P_2(u)$, is an H_2-polynomial. Lemma 1 makes it possible to construct nontrivial examples of H_2-polynomials in the following way. Suppose that $Q(z, u)$ is any polynomial of two variables. Suppose that n is the highest degree in which z appears, and that m is the highest degree of u. Then for real z and u the function

$$\left|\frac{Q(z, u)}{(z-i)^n (u-i)^m}\right|$$

is bounded. Choosing a number M larger than or equal to the least upper bound of this function for real z and u, we easily find, using the Phragmén-Lindelöf principle for the half-plane, that the function $M(z-i)^n(u-i)^m$ is an $\overline{HB}_2$-majorant of the polynomial $Q(z, u)$ and that by Lemma 1′,

$$P(z, u) = M(z-i)^n (u-i)^m + Q(z, u)$$

is an H_2-polynomial. Analogously one may construct nontrivial examples of entire functions of arbitrary growth and of class $\overline{HB}_2$, taking as majorants the products of functions of class $\overline{HB}$ and of one variable.

Note that a partial derivative of an H_2-polynomial is an H_2-polynomial. Evidently, the class of functions which are uniform limits of H_2-polynomials is invariant under partial differentiation.

We define as follows the class P_2^* of two variables.

DEFINITION. *The class P_2^* of functions of two variables* is the set of all entire functions $\omega(z, u)$, satisfying the conditions:

1. $\omega(z, u) \in \overline{HB}_2$.
2. *For fixed* $u(\operatorname{Im} u \leqslant 0)$

$$\omega(z, u) \in P_z^*.$$

3. *For fixed* $z(\operatorname{Im} z \leqslant 0)$

$$\omega(z, u) \in P_u^*.$$

It is easy to see that P_2^* *is an admissible class.* Indeed, it follows from Lemma 1′ that if $\omega(z, u)$ is a P_2^*-majorant of the function $f(z, u)$, then

$$\omega(z, u) + f(z, u) \in \overline{HB}_2.$$

Moreover, for a fixed $u(\operatorname{Im} u \leqslant 0)$, $\omega(z, u)$ is a P_2^*-majorant of the function $f(z, u)$, and accordingly

$$\omega(z, u) + f(z, u) \in P_z^*.$$

In the same way

$$\omega(z, u) + f(z, u) \in P_u^* \qquad (\operatorname{Im} z \leqslant 0).$$

We note further that, as one sees from Theorem 4 of Chapter VIII, the limit function of a uniformly convergent sequence of H_2-polynomials belongs to the class P_2^*.

We shall show that the class P_2^* has the same characteristic property as P^*. Indeed, if for positive h and k

$$\omega(z - ih, u - ik) \in \overline{HB}_2,$$

then $\omega(z, u) \in \overline{HB}_2$. For a fixed u ($\operatorname{Im} u \leqslant 0$) $\omega(z - ih, u) \in \overline{HB}_z$, and from the characteristic property of the class P_z^* one obtains $\omega(z, u) \in P_z^*$. Analogously one shows that $\omega(z, u) \in P_u^*$.

Conversely, if $\omega(z, u) \in P_2^*$, then evidently for $h \geqslant 0$ also $\operatorname{Im} z \leqslant 0$ $\omega(z - ih, u) \in P_u^*$. Moreover, in view of the characteristic property of the class P_z^* for $\operatorname{Im} u \leqslant 0$ we have $\omega(z - ih, u) \in P_z^*$. We need still to prove that $\omega(z - ih, u) \in \overline{HB}_2$. This now comes down to the proof of the inequality

$$|\omega(z - ih, u)| \geqslant |\omega(\bar{z} - ih, \bar{u})| \qquad (\operatorname{Im} z \leqslant 0, \operatorname{Im} u \leqslant 0).$$

In order to prove this we note that for an H-polynomial, which means also for a function of class P^*, the following inequality holds:

$$|\omega(z - 2ih)| \geqslant |\omega(z)|$$

for $h \geqslant 0$, $\operatorname{Im} z \leqslant h$. Therefore from $\omega(z, u) \in P_z^*$ (for $\operatorname{Im} u \leqslant 0$) it follows that for $h \geqslant 0$, $\operatorname{Im} z \leqslant h$, $\operatorname{Im} u \leqslant 0$,

$$|\omega(z - 2ih, u)| \geqslant |\omega(z, u)| \geqslant |\omega(\bar{z}, \bar{u})|.$$

Putting $z_1 = z - ih$, we find that for $\operatorname{Im} z_1 \leqslant 0$, $\operatorname{Im} u \leqslant 0$

$$|\omega(z_1 - ih, u)| \geqslant |\omega(\bar{z}_1 - ih, \bar{u})|.$$

Thus, $\omega(z - ih, u) \in P_2^*$. Carrying out the same operation over u, we find that, for $h > 0$, $k > 0$, $\omega(z - ih, u - ik) \in P_2^*$.

We shall prove the following theorem on the representation of functions of class P_2^*.

THEOREM 14. *In order that an entire function $\omega(z, u)$ of two variables belong to the class P_2^*, it is necessary and sufficient that it be representable in the form*

$$\omega(z, u) = e^{-\gamma_1 z^2 - \gamma_2 u^2} \omega_1(z, u), \tag{9.25}$$

where γ_1 and γ_2 are non-negative numbers, and $\omega_1(z, u)$ is an entire function in $\overline{HB}_2$ and of genus one in each of its variables under the condition that the other variable is fixed in the lower half-plane.

PROOF. The sufficiency of this condition for the function $\omega(z, u)$ to belong to the class P_2^* follows immediately from the definitions of the classes P^* and P_2^*. For the proof of necessity we note that if the function $\omega(z, u)$ belongs to the class P_u^* for $\operatorname{Im} z \leqslant 0$, it follows from the representation of a function of class $\overline{HB}$ (Chapter VII, Theorem 6) that the function

$$\varphi_z(u) = \frac{\omega_z'(z, u)}{\omega(z, u)}$$

maps the lower half-plane $\operatorname{Im} u < 0$ onto the upper $\operatorname{Im} u > 0$. Therefore the whole family of functions of u, $\{\varphi_z(u)\}$, is normal for $\operatorname{Im} u < 0$ (z a parameter varying in the lower half-plane).

From the decomposition

$$\varphi_z(u) = -2\gamma(u)z + \nu(u) + A + \sum_{k=1}^{\infty}\left(\frac{1}{z - a_k} + \operatorname{Re}\left(\frac{1}{a_k}\right)\right) \quad (\gamma(u) \geqslant 0)$$

(a_k functions of u) it follows that

$$-\gamma(u) = \lim_{z \to -i\infty} z^{-1}\varphi_z(u).$$

Since the family of functions $\{z^{-1}\varphi_z(u)\}$ is normal, the convergence is uniform, and accordingly $\gamma(u)$ is a holomorphic function in the lower half-plane. Since, on the other hand, $\gamma(u)$ takes on non-negative values in the entire lower half-plane, $\gamma(u)$ is a constant. Thus,

$$\omega(z, u) = e^{-\gamma_1 z^2}\omega_2(z, u),$$

where $\omega_2(z, u)$ is of genus one relative to the variable z for $\operatorname{Im} u \leqslant 0$.

Analogously one separates out the multiplier $e^{-\gamma_2 u^2}$. The theorem is proved.

DEFINITION. The operator $K[f(z)]$, carrying entire functions into entire functions, will be called *continuous* if it carries every sequence of entire functions converging uniformly in every finite region into a sequence converging in the same sense.

THEOREM 15 [9]. *Every continuous $\mathfrak{B}_z^*$-operator over entire functions of one variable z carries functions $\omega(z, u)$ of class P_2^* into functions of the same class.*

PROOF. First of all we verify that the continuous operator $\mathfrak{B}_z^*$ carries a function of class P_2^* into an entire function of two variables. Indeed, by the very definition of the operator $\mathfrak{B}_z^*$ the function

$$\varphi(z, u) = \mathfrak{B}_z^*[\omega(z, u)]$$

belongs to P_z^* for $\operatorname{Im} u \leqslant 0$. Since for $\operatorname{Im} u \leqslant 0$ $\omega(z, u)$ is a P_z^*-majorant of $\omega(z, \bar{u})$ and P_z^* is an admissible class, therefore $\omega(z, u) + \omega(z, \bar{u}) \in P_z^*$. Accordingly,

$$\varphi(z, u) + \varphi(z, \bar{u}) \in P_z^*$$

and therefore $\varphi(z, u)$ is an entire function of z also for $\operatorname{Im} u > 0$. Now we shall show that it is entire in u. We have

$$\frac{\varphi(z, u+h) - \varphi(z, u)}{h} = \mathfrak{B}_z^*\left[\frac{\omega(z, u+h) - \omega(z, u)}{h}\right].$$

In view of the continuity of the operator $\mathfrak{B}_z^*$ it therefore follows that for all u (uniformly in z in a finite region) there exists a limit

$$\varphi_u'(z, u) = \mathfrak{B}_z^*[\omega_u'(z, u)],$$

and accordingly $\varphi(z, u)$ is an entire function of u for all z. Hence it follows that $\varphi(z, u)$ is an entire function of two variables.

Now we shall show that it belongs to P_2^*. Since a $\mathfrak{B}_z^*$-operator carries every function of P_z^* into a function of the same class, therefore for $\operatorname{Im} u \leqslant 0$ $\varphi(z, u) \in P_z^*$, and consequently for $\operatorname{Im} z \leqslant 0$, $\operatorname{Im} u \leqslant 0$,

$$|\varphi(z, u)| \geqslant |\varphi(\bar{z}, u)|.$$

In order to obtain the two other inequalities we observe that since for $\operatorname{Im} u \leqslant 0$ $\omega(z, u)$ is a P_z^*-majorant for $\omega(z, \bar{u})$, and a $\mathfrak{B}_z^*$-operator preserves subordination, therefore

$$|\varphi(z, u)| \geqslant |\varphi(z, \bar{u})|, \quad |\varphi(z, u)| \geqslant |\varphi(\bar{z}, \bar{u})|.$$

Thus $\varphi(z, u) \in \overline{HB}_2$. In view of the characteristic property of the class P_2^* for $k \geqslant 0$ we have $\omega(z, u - ik) \in P_2^*$ and consequently $\varphi(z, u - ik) \in HB_2$.

Making use of the characteristic property of the class P_u^*, we find that for $\operatorname{Im} z \leqslant 0$ we have $\varphi(z, u) \in P_u^*$. Thus conditions 1, 2 and 3 are satisfied and consequently $\varphi(z, u) \in P_2^*$. The theorem is proved.[14]

The apparatus developed for $\mathfrak{B}$-operations makes it now possible to obtain the following generalized theorem of Laguerre-Pólya on functions of several variables.

THEOREM 16. *In order that the entire function $\omega(z, u)$ may be uniformly approximated by a sequence of H_2-polynomials in the variables z and u, it is necessary and sufficient that it belong to the class P_2^*, and accordingly have the representation* (9.25).

PROOF. We have already proved that the limit function of a sequence of H_2-polynomials belongs to P_2^*.

Now suppose that

$$\omega(z, u) = \sum_{k, p=0}^{\infty} a_{kp} z^k u^p$$

is a function of class P_2^*. Applying to it the real $\mathfrak{B}^*$-operators of Jensen, I_n in z and I_m in u (Chapter VIII), we obtain a sequence of polynomials

$$\tilde{\omega}_{n,m}(z, u) = a_{00} + \sum_{k=1}^{n} a_{k0}\left(1 - \frac{1}{n}\right)\cdots\left(1 - \frac{k-1}{n}\right)z^k$$
$$+ \sum_{p=1}^{m} a_{0p}\left(1 - \frac{1}{m}\right)\cdots\left(1 - \frac{p-1}{m}\right)u^p$$
$$+ \sum_{k,p=1}^{n,m} a_{kp}\left(1 - \frac{1}{n}\right)\cdots\left(1 - \frac{k-1}{n}\right)\left(1 - \frac{1}{m}\right)\cdots\left(1 - \frac{p-1}{m}\right) z^k u^p,$$

[14] Evidently the theorem remains true if the class P^* is replaced by the class HB, and the $\mathfrak{B}^*$-operator by a $\mathfrak{B}_{HB}$-operator. In this case the proof of the theorem becomes simpler. The continuity of the $\mathfrak{B}_z^*$-operator is used only in proving that it carries an entire function of two variables into an entire function. We observe that this operator on functions of two variables is also continuous.

which, as $n \to \infty$ and $m \to \infty$, converge to the function $\omega(z, u)$. By Theorem 15 these polynomials are H_2-polynomials.

6. General Form of the Operators $\mathfrak{B}$ and $\mathfrak{B}^*$

Suppose that $K[f(z)]$ is a linear continuous operator defined on some class of entire functions of one variable and mapping these functions into entire functions, and suppose that its region of definition V contains the function e^{-uz} for any fixed u. Evidently,

$$\varphi(z, u) = K[e^{-uz}] \tag{9.26}$$

is an entire function of two variables. Differentiating both sides of the equation n times with respect to u and putting u equal to zero, we obtain

$$K[z^n] = (-1)^n \varphi_u^{(n)}(z, 0). \tag{9.27}$$

Thus if

$$f(z) = \sum_{k=0}^{\infty} a_k z^k$$

is an entire function belonging to V, then in view of the continuity of the operator

$$K[f(z)] = \sum_{k=0}^{\infty} (-1)^k a_k \varphi_u^{(k)}(z, 0),$$

or in other words

$$K[f(z)] = f(-D_u)[\varphi(z, u)]|_{u=0}, \tag{9.28}$$

where D_u is the operator of partial differentiation with respect to u.

Conversely, to every entire function of two variables $\varphi(z, u)$ there corresponds an operator defined by equation (9.28) which is continuous on a certain set of entire functions.

We will impose on the operator a restriction somewhat weaker than continuity. An additive homogeneous operator, defined on some class of entire functions V, will be called *regular* if on each of the functions $f(z)$ of the class V the sequence $\{K[\tilde{f}_n(z)]\}$ converges uniformly in each finite region to the function $K[f(z)]$.

THEOREM 17. *In order that the operator $K[f(z)]$ be a regular $\mathfrak{B}^*$-operator, it is necessary and sufficient that it be representable in the form*

$$K[f(z)] = f(-D_u)[\varphi(z, u)]|_{u=0}, \tag{9.29}$$

where $\varphi(z, u)$ is an entire function of two variables of the class P_2^*, admitting the representation

$$\varphi(z, u) = e^{-\gamma_1 u^2} \omega(z, u), \tag{9.30}$$

in which $\gamma_1 \geqslant 0$, and $\omega(z, u)$ is a function of genus one in each of the variables under the condition that the second variable is fixed in the lower half-plane (i.e., in the representation (9.25) $\gamma_2 = 0$).

PROOF. First we shall prove the necessity.

Suppose that K is a regular $\mathfrak{B}_z^*$ operator. Applying it to the function e^{-uz}, which for $\operatorname{Im} u \leqslant 0$ belongs to P_z^*, we obtain the function

$$\varphi(z, u) = K[e^{-uz}] \in P_z^*.$$

If K were a continuous $\mathfrak{B}_z^*$-operator, then by what has been proved above $\varphi(z, u)$ would be a function in P_z^*. We shall show that this is the situation only in the case of regularity of the operator K. From the regularity of K it follows that for any fixed u the sequence of functions

$$\varphi_n(z, u) = K\left[\left(1 - \frac{uz}{n}\right)^n\right]$$
$$= \varphi_0(z) + \sum_1^n \frac{\left(1 - \frac{1}{n}\right)\cdots\left(1 - \frac{k-1}{n}\right)}{k!} \varphi_k(z) u^k \quad (\varphi_k(z) = K[z^k])$$

converges uniformly to $\varphi(z, u)$. Evidently if $\operatorname{Im} z \leqslant 0$ the function $\varphi_n(z, u)$ is an H-polynomial in u. Without loss of generality we may suppose that $\varphi_0(z)$ is not identically zero. By the remark to Theorem 3 of Chapter VIII the polynomials $\varphi_n(z, u)$ form a normal family. It follows that for any fixed z ($\operatorname{Im} z \leqslant 0$) this sequence converges uniformly, so that $\varphi(z, u) \in P_u^*$ for $\operatorname{Im} z \leqslant 0$.

Finally, e^{-uz} is, for $\operatorname{Im} u \leqslant 0$, a P_z^*-majorant of the function $e^{-\bar{u}z}$, and since subordination is preserved after application of a $\mathfrak{B}^*$-operator, therefore $\varphi(z, u)$ is a majorant of $\varphi(z, \bar{u})$, and thus for $\operatorname{Im} z \leqslant 0$, $\operatorname{Im} u \leqslant 0$

$$|\varphi(z, u)| \geqslant |\varphi(\bar{z}, \bar{u})|.$$

Thus, $\varphi(z, u) \in P_2^*$, and by Theorem 14 it may be represented in the form

$$\varphi(z, u) = e^{-\gamma_1 z^2 - \gamma_2 u^2} \omega_1(z, u),$$

where $\omega_1(z, u) \in \overline{HB}_2$ and is of genus one in each of the variables under the condition that the other variable is fixed in the closed lower half-plane. We shall prove that $\gamma_2 = 0$. To this end we denote by $\psi(z)$ the function

$$\psi(z) = \overline{\varphi}(0, -z) = \sum_0^\infty \frac{\overline{\varphi_k(0)}}{k!}(-z)^k, \tag{9.31}$$

where $\varphi_k(z) = \varphi^{(k)}(z, u)\,|_{u=0} = K[z^k](-1)^k$. Evidently $\psi(z) \in P_z^*$.

Applying the $\mathfrak{B}^*$-operator K to the H-polynomial $\tilde{\psi}_n(\lambda z)$ for $\lambda > 0$, we obtain

$$K[\tilde{\psi}_n(\lambda z)] = \sum_{k=0}^\infty \left(1 - \frac{1}{n}\right)\cdots\left(1 - \frac{k-1}{n}\right)\frac{\overline{\varphi_k}(0)\,\varphi_k(z)}{k!}\lambda^k,$$

or putting $z = 0$ we have

$$K[\tilde{\psi}_n(\lambda z)]\,|_{z=0} = \sum_{k=0}^n \left(1 - \frac{1}{n}\right)\cdots\left(1 - \frac{k-1}{n}\right)\frac{|\varphi_k(0)|^2}{k}\lambda^k. \tag{9.32}$$

In view of the regularity of the operator this sequence must converge for $n \to \infty$ to the quantity

$$K[\psi(\lambda z)]|_{z=0}.$$

We shall show that this is possible only if $\gamma_2 = 0$. Suppose first that $\varphi(0, u)$ is not identically equal to zero. From the connection between the growth of an entire function and the rate of decrease of the coefficients of its power series (Chapter I, § 2) we have

$$\overline{\lim_{k\to\infty}} k^{\frac{1}{2}} \sqrt[k]{\frac{|\varphi_k(0)|}{k!}} = (2\gamma_2 e)^{\frac{1}{2}}$$

and accordingly

$$\overline{\lim_{k\to\infty}} \sqrt[k]{\frac{|\varphi_k(0)|^2}{k!}\lambda^k} = 2\gamma_2\lambda.$$

For any $\gamma_2 > 0$ we may choose $\lambda > 0$ such that $2\gamma_2\lambda > 1$, and then we obtain

$$\lim_{n\to\infty} K[\tilde{\psi}_n(\lambda z)]|_{z=0} = \infty.$$

Thus, in this case it is proved that $\gamma_2 = 0$. If $\varphi(0, u) \equiv 0$, then $\varphi(z, u) = z^p\varphi_1(z, u)$, while $\varphi_1(0, u) \not\equiv 0$. In this case one must take as $\psi(z)$ the function

$$\psi(z) = \bar{\varphi}_1(0, -z) = \frac{1}{p!}\bar{\varphi}_v^{(p)}(v, -z)|_{v=0}.$$

Then we will have

$$\varphi_k(z) = \varphi_u^{(k)}(z, u)|_{u=0} = z^p\theta_k(z)$$

($k = 1, 2, \ldots$), while not all the $\theta_k(0)$ are equal to zero. Substituting $\tilde{\psi}_n(\lambda z)$ into the operator, we obtain

$$K[\tilde{\psi}_n(\lambda z)] = \sum_{k=0}^{n}\left(1 - \frac{1}{n}\right)\cdots\left(1 - \frac{k-1}{n}\right) z^p \frac{\overline{\theta_k(0)}\,\theta_k(z)}{k!}\lambda^k.$$

From the regularity of the operator it follows that the sequence

$$\{z^{-p}K[\tilde{\psi}_n(\lambda z)]|_{z=0}\}$$

has a finite limit. However, this is possible only in case $\gamma_2 = 0$. The first part of the theorem is proved.

Now we shall prove that if $\varphi(z, u)$ is any entire function of the form (9.30), then the operator defined by the equation (9.29) is a regular $\mathfrak{B}^*$-operator.

Suppose that $f(z) \in P^*$ and $\tilde{f}_n(z)$ is the result of the application to $f(z)$ of a Jensen operator. $\tilde{f}_n(z)$ is an H-polynomial and therefore defines a continuous $\mathfrak{B}^*$-operator $\tilde{f}_n(-D_u)$. Defining

$$\psi_n(z, u) = \tilde{f}_n(-D_u)[\varphi(z, u)],$$

we obtain, from Theorem 15, that $\psi_n(z, u) \in P_2^*$. Expanding $\varphi(z, u)$ into series in powers of z, we will have

$$\varphi(z, u) = \vartheta_0(u) + \frac{\vartheta_1(u)}{1!} z + \frac{\vartheta_2(u)}{2!} z^2 + \ldots,$$

where $\vartheta_0(u), \vartheta_1(u), \ldots$ are entire functions of class P^* and of genus no greater than one. Applying the operator $\tilde{f}_n(-D_u)$, we obtain

$$\psi_n(z, u) = \tilde{f}_n(-D)[\vartheta_0(u)] + \frac{z}{1!}\tilde{f}_n(-D)[\vartheta_1(u)] + \frac{z^2}{2!}\tilde{f}_n(-D)[\vartheta_2(u)] + \ldots.$$

From the simple estimates (9.16) and (9.17), obtained in the proof of Theorem 8, it follows that the functions $\tilde{f}_n(-D)[\vartheta_k(u)]$ converge uniformly, as $n \to \infty$, to the functions $f(-D)[\vartheta_k(u)]$. If for a given $u(\operatorname{Im} u \leqslant 0)$ not all of these functions are equal to zero, then by the remark to Theorem 3 of Chapter VIII the sequence $\{\psi_n(z, u)\}$ of functions of class P_z^* forms a normal family, so that they uniformly converge to functions

$$f(-D_u)[\varphi(z, u)] = \sum_0^\infty f(-D_u)\left[\vartheta_k(u)\frac{z^k}{k!}\right],$$

belonging to P_z^* for u fixed and in the closed lower half-plane. Putting $u = 0$, we will have for $f(z) \in P^*$:

$$K[f(z)] = f(-D_u)[\varphi(z, u)]\,|_{u=0} \in P^*, \tag{9.33}$$

while

$$K[f(z)] = \lim_{n\to\infty}[\psi_n(z, u)]\,|_{u=0} = \lim_{n\to\infty} K[\tilde{f}_n(z)],$$

i.e., $K[f(z)]$ is a regular $\mathfrak{B}^*$-operator. The theorem is proved.

We turn to the general form of a $\mathfrak{B}$-operator. We obtain it also under a certain additional requirement related to the continuity of the operator.

We will say that the sequence $\{P_n(z)\}$ of functions of exponential type converge with the weight A $(A > 0)$ to the function of exponential type $P_0(z)(P_n(z) \xrightarrow{A} P_0(z))$, if

$$\lim_{n\to\infty}(\sup_z |P_n(z) - P_0(z)|\, e^{-A|z|}) = 0. \tag{9.34}$$

The operator K over functions of exponential type will be said to be *upper continuous*, if to each positive number A there corresponds a positive number B such that from

$$P_n(z) \xrightarrow{A} P_0(z)$$

it follows that

$$K[P_n(z)] \xrightarrow{B} K[P_0(z)].$$

THEOREM 18. *An upper continuous*[15] *$\mathfrak{B}$-operator $K[f(z)]$ has the form*

$$K[f(z)] = f(-D_u)[\varphi(z, u)]|_{u=0}, \tag{9.35}$$

where the entire function $\varphi(z, u)$ of two variables satisfies the conditions:

(1) *it belongs to the class P_2^* of functions of two variables;*

(2) *it is of exponential type in z for bounded u, i.e.,*

$$\overline{\lim_{|z|\to\infty}} \frac{\ln|\varphi(z, u)|}{|z|} = \sigma_R < \infty \qquad (|u| < R). \tag{9.36}$$

Conversely, if the function $\varphi(z, u)$ satisfies these conditions, then the operator $K[f(z)]$, defined by formula (9.35), *is an upper continuous $\mathfrak{B}$-operator.*

The function $\varphi(z, u)$ is defined here by formula (9.26).

PROOF. First of all we note that a continuous $\mathfrak{B}$-operator $K[f(z)]$ may be extended by continuity to the linear envelope of the class P^*. Indeed, approximating the function $F(z)$ of the class P^* by a sequence of H-polynomials $\bar{F}_n(z)$, we obtain a uniformly convergent sequence $\{K[\bar{F}_n(z)]\}$ of entire functions of class P^*. Extending the operator by continuity, we obtain a $\mathfrak{B}^*$-operator. From Theorem 17 this operator has the form (9.35), while

$$\varphi(z, u) = K[e^{-uz}]$$

and $\varphi(z, u) \in P_2^*$. It remains to prove that the function $\varphi(z, u)$ satisfies condition (2). To this end we use the upper continuity of the operator K.

The set $\{e^{-uz}\}$ of functions of z is bounded in the norm

$$\|f\|_R = \sup_z |f(z)| e^{-R|z|}$$

for $|u| \leqslant R$. We shall show that for some positive B, depending on R, and $|u| \leqslant R$,

$$\sup_z |\varphi(z, u)| e^{-B|z|} \leqslant M. \tag{9.37}$$

Indeed, if this were not the case, then for any $M > 0$ and $B > 0$ there would be a point $u'(|u'| < R)$ at which

$$\sup_z |\varphi(z, u')| e^{-B|z|} > M > 0.$$

Choose a sequence $B_n \to \infty$, $M_n \to \infty$ and a convergent sequence of points u_n such that

$$\sup_z |\varphi(z, u_n)| e^{-B_n|z|} > M_n. \tag{9.38}$$

But for $u_n \to u_0$ $(|u_0| \leqslant R)$

$$\sup_z |e^{-u_n z} - e^{-u_0 z}| e^{-A|z|} \to 0 \qquad (A \geqslant R).$$

[15] It is not clear how essential the requirement of upper continuity is in this theorem, just as with the requirement of regularity in the preceding theorem.

Accordingly, for some $B > 0$

$$\sup_z |\varphi(z, u_n) - \varphi(z, u_0)| e^{-B|z|} \to 0$$

and for any $\varepsilon > 0$, beginning with some n,

$$\sup_z |\varphi(z, u_n)| e^{-B|z|} < \sup_z |\varphi(z, u_0)| e^{-B|z|} + \varepsilon \quad (\varepsilon > 0),$$

which contradicts (9.38). Thus, (9.37) is proved, which evidently is equivalent to condition (2).

To prove the second part of the theorem we suppose that $\varphi(z, u)$ satisfies conditions (1) and (2), and we define an operator over functions $f(z)$ of class P in the following way:

$$K[f(z)] = f(-D_u)[\varphi(z, u)]|_{u=0} = \frac{1}{2\pi i} \int_{(C_R)} \psi(u) \varphi(z, -u)\, du, \tag{9.39}$$

where $\psi(u)$ is the Borel transform of the function $f(z)$, R a number higher than the degree of the function $f(z)$, and C_R a circumference with center at the origin and radius R. By Theorem 17 this operator is a $\mathfrak{B}^*$-operator. Using condition (2), we obtain the estimate

$$|K[f(z)]| < \frac{1}{2\pi} N_f e^{B|z|}, \tag{9.40}$$

where N_f is a constant depending on the function $f(z)$. Accordingly, the operator $K[f(z)]$ carries entire functions of exponential type into functions of the same class, i.e., it is a $\mathfrak{B}$-operator. Further, from the convergence of the sequence $\{f_n(z)\}$ of entire functions of exponential type with the weight A it follows that the sequence $\{\Psi_n(u)\}$ of associated functions converges uniformly outside some circle in the u plane, and it follows from formula (9.39) that the sequence $\{K[f_n(z)]\}$ converges with the weight B. Thus, $K[f(z)]$ is an upper continuous $\mathfrak{B}$-operator. The theorem is proved. It is useful to note the integral representation (9.39) of an upper continuous $\mathfrak{B}$-operator.

From Theorems 17 and 18 it follows that the classes of $\mathfrak{B}$- and $\mathfrak{B}^*$-operators are very wide. Indeed, every entire function of two variables which satisfies very general conditions generates such an operator.

We shall investigate some special types of $\mathfrak{B}$- and $\mathfrak{B}^*$-operators.

(1) Operators commuting with the differentiation operator.

Suppose for definiteness that $K[f(z)]$ is a $\mathfrak{B}^*$-operator commuting with the differentiation operator.

Then, if

$$\varphi(z, u) = K[e^{-uz}],$$

then

$$\varphi'_z(z, u) = K[-ue^{-uz}] = -u\varphi(z, u)$$

and

$$\varphi(z, u) = e^{-uz}\Phi(u). \tag{9.41}$$

By Theorem 17 $\Phi(u)$ is a function of class P^* and of genus one. It is easy to see from (9.41) that on functions of the form e^{-uz} the operator K coincides with the operator $\Phi(-D)$. The operators K and $\Phi(-D)$ coincide also on all polynomials and hence on the entire domain of definition. One obtains a theorem:

THEOREM 19. *A regular $\mathfrak{B}^*$-operator, commuting with the differentiation, operator, has the form*

$$\mathfrak{B}^*[f(z)] = \overline{F}(D)[\overline{f}(z)],$$

where $F(z)$ is a function of class P^* and of genus one.

Analogously one obtains the following theorem.

THEOREM 19′. *An upper continuous $\mathfrak{B}$-operator, commuting with the differentiation operator, has the form*

$$\mathfrak{B}[f(z)] = \overline{F}(D)[f(z)],$$

where $F(z)$ is any function of class P^*.

(2) We consider further operators $K[f(z)]$ commuting with the operator $z(d/dz)$. Putting

$$\varphi_k(z) = K[z^k],$$

we will have

$$z\varphi_k(z) = K[kz^k] = k\varphi_k(z),$$

or

$$\varphi_k(z) = \gamma_k z^k.$$

Applying the operator $K[f(z)]$ to the polynomial

$$Q(z) = \sum_{k=1}^{n} a_k z^k,$$

we obtain

$$K[Q(z)] = \sum_{k=0}^{n} \gamma_k a_k z^k.$$

If K is a $\mathfrak{B}$- or a $\mathfrak{B}^*$-operator, and $Q(z)$ an H-polynomial, then $K[Q(z)]$ is also an H-polynomial. Hence it easily follows that the operator $K[f(z)]$ carries every polynomial with positive roots into a polynomial with positive roots.[16] Accordingly, $\gamma_0, \gamma_1, \gamma_2, \ldots$ is a sequence of multipliers of genus one and all of the same sign. Thus one obtains

[16] In order to prove this, it suffices to consider the polynomial $K[P(e^{i\gamma}z)]$ $(-\pi < \gamma < 0)$, where $P(z)$ is a polynomial with positive roots.

THEOREM 20. *Every* $\mathfrak{B}$- *or* $\mathfrak{B}^*$-*operator, commuting with the operator* $z(d/dz)$, *coincides with the operator* Γ *defined in Chapter* VIII.

Knowledge of the general form of $\mathfrak{B}$- and $\mathfrak{B}^*$-operators makes it possible to construct an infinite set of such operators. In particular, if $\omega(z) \in \Gamma$, then the operator $K[f(z)]$, defined by the equation

$$K[e^{-uz}] = \omega(az + bu) \qquad (a > 0,\ b > 0),$$

is a $\mathfrak{B}$-operator, satisfying the condition

$$a\mathfrak{B}[zf(z)] = b\{\mathfrak{B}[f(z)]\}'.$$

7. Certain Extremal Properties of Entire Functions

In this section we shall prove some theorems related to entire functions of exponential type, least deviating from zero, which are generalizations of corresponding theorems of S. N. Bernšteĭn.[17] Suppose that $\omega(z)$ is a function of class P and exponential type σ_ω. For functions $f(x)$, defined on the real axis, we define the deviation from zero with weight $|\omega(x)|$ as

$$E_\omega[f] = \sup_{-\infty < x < \infty} \frac{|f(x)|}{|\omega(x)|}. \tag{9.42}$$

THEOREM 21. *Of all the real entire functions of exponential type less than or equal to* σ_ω *and of the form*

$$f(x) = a_0 + a_1 x + \ldots + a_{n-1}x^{n-1} + x^n + a_{n+1}x^{n+1} + \ldots \qquad (n > 0), \tag{9.43}$$

the unique function giving the smallest deviation from zero with the weight $|\omega(x)|$ *is the function*

$$f_0(x) = n!\,\mathrm{Re}\frac{\omega(x)}{\omega^{(n)}(0)}. \tag{9.44}$$

PROOF. For the function $f_0(x)$ we have

$$E_\omega[f_0] = \frac{n!}{|\omega^{(n)}(0)|}.$$

Suppose that some entire function $f(x)$ of form (9.43) of exponential type not larger than σ_ω satisfies

$$|f(x)| \leqslant \frac{n!}{|\omega^{(n)}(0)|}|\omega(x)| \qquad (-\infty < x < \infty).$$

Then the function

$$\omega_1(z) = \frac{n!}{\omega^{(n)}(0)}\omega(z)$$

[17] See S. N. Bernšteĭn [1], Chapter III, Theorems I and II of § 1 and Theorems II and III of § 9.

is a majorant for $f(x)$, and since

$$f^{(n)}(0) = \omega_1^{(n)}(0),$$

therefore by Theorem 11 $f(x) = \operatorname{Re} \omega_1(x) = f_0(x)$.

REMARK. If one assumes $n = 0$, i.e., fixes the free term in the expansion

$$f(x) = 1 + a_1 x + a_2 x^2 + \ldots,$$

then evidently the function

$$f_0(x) = \operatorname{Re} \frac{\omega(z)}{\omega(0)}$$

gives the smallest deviation $L = 1/|\omega(0)|$ to the expression (9.42). However, in this case one can not assert the uniqueness of the extremal function.[18]

THEOREM 22. *Suppose that $\omega(z) = s(z) + it(z)$ is a function of class P and of exponential type $\sigma \geqslant 0$ without real roots, not equal to the exponential function $ce^{\pm \sigma z}$ or to a polynomial of degree less than or equal to n. Then among all the real functions of exponential type $\tau \leqslant \sigma$ of the form*

$$f(z) = a_0 + a_1 z + \ldots + a_{n-1} z^{n-1} + a z^n + b z^{n+1} + a_{n+2} z^{n+2} + \ldots \quad (n \geqslant 1), \tag{9.45}$$

where a and b are fixed, the unique function giving the smallest deviation $E_\omega[f]$ is the function

$$f_1(x) = As(x) + Bt(x), \tag{9.46}$$

where A and B are chosen so that $f_1(x)$ has the form (9.45).

PROOF. First of all we shall show that one may choose constants A and B such that the function $f_1(x)$ will have the form (9.45).

This problem reduces to the problem of the solvability of a system of equations

$$\left.\begin{aligned} As^{(n)}(0) + Bt^{(n)}(0) &= n!a, \\ As^{(n+1)}(0) + Bt^{(n+1)}(0) &= (n+1)!\,b, \end{aligned}\right\} \tag{9.47}$$

i.e., to the proof that the determinant

$$\Delta = s^{(n+1)}(0)\,t^{(n)}(0) - s^{(n)}(0)\,t^{(n+1)}(0)$$

is not zero. By the hypothesis of the theorem, $\omega(z)$ has no real roots. From this hypothesis it follows that $\omega^{(n)}(z)$ also has no real roots (see (9.11)), and accordingly $\omega^{(n)}(0) \neq 0$. We have the equality

$$\operatorname{Im} \frac{\omega^{(n+1)}(0)}{\omega^{(n)}(0)} = \frac{\Delta}{|\omega^{(n)}(0)|^2}.$$

[18] For instance, for $\omega(z) = e^{iz}$ all functions $f(x) = \cos \lambda x$ ($|\lambda| \leq 1$) give a deviation equal to $f(0) = 1$.

If Im $[\omega^{(n+1)}(0)/\omega^{(n)}(0)] = 0$, then, as one easily sees from the expansion (9.11) of the logarithmic derivative of the function $\omega^{(n)}(z)$ of class P, $\omega^{(n)}(z)$ is a real function. Since it can not have any real roots, it therefore has no roots at all, so that

$$\omega^{(n)}(z) = ce^{kz}.$$

The equation $k = 0$ is impossible, as in this case $\omega(z) = P_n(z)$. For $k \neq 0$

$$\omega(z) = \frac{c}{k^n} e^{kz} + P_{n-1}(z). \tag{9.48}$$

The case $P_{n-1}(z) \equiv 0$ is excluded by the condition of the theorem. A function $\omega(z)$, represented by equation (9.48), grows on the imaginary axis like a polynomial. Therefore it is of completely regular growth. Its indicator diagram is a segment $[0, k]$ of the real axis, and accordingly it has roots in both half-planes. This contradicts the statement that $\omega(z) \in P$. Thus,

$$\operatorname{Im} \frac{\omega^{(n+1)}(0)}{\omega^{(n)}(0)} \neq 0$$

so that the system (9.47) has a unique solution.

Now suppose that a function of type (9.45) satisfies on the entire axis the inequality

$$\left| \frac{f(x)}{\omega(x)} \right| \leqslant E_\omega [f_1],$$

where $f_1(x)$ is defined by equation (9.46). The function

$$\chi(z) = f(z) - (A - iB)\,\omega(z)$$

belongs, by Lemma 1, to the class P, while

$$\vartheta(x) = R\chi(x) = c_0 + c_1 x + \dots + c_{n-1}x^{n-1} + c_{n+2}x^{n+2} + \dots.$$

If $\vartheta(z) \not\equiv 0$, then all the roots of $\vartheta(z)$ are real. But then all the roots of the Jensen polynomial $\tilde{\vartheta}_{n+2}(z)$ are also real.

On the other hand, by the Laguerre theorem, a polynomial for which two successive coefficients are zero (other than the first two or the last two) must have nonreal roots,[19] and therefore either $c_{n+q} = 0$ $(q = 2, 3, \dots)$, or $c_0 = c_1 = \dots = c_{n-1} = 0$. The first case is impossible, because then

$$I[(A - iB)\,\omega(x)] = I[\chi(x)] = P_{n+1}(x)$$

and the function $\omega(x)$ is a polynomial of degree not larger than $n + 1$. If now $c_0 = c_1 = \dots = c_{n-1} = 0$, then the function $\vartheta(x)$ has at the origin a root of multiplicity not less than three. From the interlacing of the roots of the real and imaginary parts it follows that $(A - iB)\omega(x)$ has a multiple root at the origin. This is impossible, since $\omega(x)$ has no real roots. Thus,

$$f(x) = \operatorname{Re}[(A - iB)\,\omega(x)].$$

The theorem is proved.

---

[19] Chapter VIII, Lemma 3.

APPENDIX I

SOME ADDITIONAL PROBLEMS OF THE GENERAL THEORY

1. The Impossibility of Constructing an Exact Scale of Growth

The scale of growth (of order and type) usually used in the theory of entire functions is very incomplete. Functions with the rates of growth, for example, of e^r and $r^n e^r$ have the same order and type. The rates of growth of functions of infinite order can differ even more widely. The question naturally arises whether it is possible to construct an exact growth scale, so that to every monotonically increasing function there will correspond a positive number in such a way that a function with a greater rate of growth will correspond to a larger number. Here we assume that the rate of growth of $\varphi(r)$ is greater than that of $\psi(r)$ if

$$\lim_{r\to\infty} \frac{\varphi(r)}{\psi(r)} = \infty .$$

The impossibility of finding such a scale is a consequence of the following theorem of Du Bois-Reymond.

THEOREM. *Let*

$$\psi_1(r),\ \psi_2(r),\ \ldots,\ \psi_n(r),\ \ldots$$

be a sequence of positive, monotonically increasing functions such that the rate of growth of any function is greater than that of the preceding function.

There exists a positive, monotonically increasing function $\psi(r)$ with a rate of growth greater than that of any member of this sequence.

In fact, the convergent series

$$\psi(r) = \sum_{n=1}^{\infty} \frac{1}{2^n \psi_n(n)} \psi_n(r)$$

defines a monotonic function, and for any n we have

$$\frac{\psi(r)}{\psi_n(r)} > \frac{1}{2^{n+1}\psi_{n+1}(n+1)} \frac{\psi_{n+1}(r)}{\psi_n(r)} \to \infty \quad \text{for } r\to\infty.$$

The introduction of transfinite numbers does not help, since if we consider the inverse functions, we can state that, for any sequence of positive, monotonically increasing functions in which the rate of growth of any function is smaller than that of the preceding function, there exists a positive, monotonically increasing function of smaller rate of growth than that of any function of the sequence.[1]

2. Convergent and Divergent Types[2]

To give a more precise characterization of the rate of growth of functions of finite order, we introduce the idea of convergent and divergent types. An entire function $f(z)$ of order ρ is said to be of the first or second type, depending on the convergence of the integral

$$\int_{r_0}^{\infty} \frac{\ln M_f(r)}{r^{\rho+1}} dr. \tag{1}$$

THEOREM. *If ρ is not an integer, then the convergence of the integral* (1) *is equivalent to the convergence of the series*

$$\sum_{k=1}^{\infty} \frac{1}{|a_k|^{\rho}}, \tag{2}$$

where a_k $(k = 1, 2, \ldots)$ are the zeros of $f(z)$, excluding a possible zero at the origin.

If the integral (1) converges, then from (1.27) the integral

$$\int_{r_0}^{\infty} \frac{n(r)}{r^{\rho+1}} dr \tag{3}$$

also converges.

We also have

$$\frac{n(r)}{r^{\rho}} \leqslant \rho \int_{r}^{\infty} \frac{n(t)}{t^{\rho+1}} dt,$$

and so, since (3) converges,

$$\lim_{r\to\infty} \frac{n(r)}{r^{\rho}} = 0.$$

Thus

$$\sum_{k=1}^{\infty} \frac{1}{|a_k|^{\rho}} = \int_0^{\infty} \frac{dn(t)}{t^{\rho}} = \frac{n(t)}{t^{\rho}}\Bigg|_0^{\infty} + \rho \int_0^{\infty} \frac{n(t)}{t^{\rho+1}} dr = \rho \int_0^{\infty} \frac{n(t)}{t^{\rho+1}} dt, \tag{4}$$

---

[1] We note that it is also impossible to construct an exact growth scale for the subclass of the monotonic functions consisting of those functions that are the maximum absolute values of entire functions. To prove this, with each monotonic function $\psi(z)$, tending to infinity, we associate the function

$$f(z) = \sum_{n=1}^{\infty} \frac{\psi^2(n+1)}{n^{p_n}} z^{p_n},$$

where the numbers p_n are chosen so that $\lim[\psi(n+1)]^{1/p_n} = 1$. It is clear that $\lim M_f(r)/\psi(r) = \infty$.

[2] See G. Valiron [1].

and the convergence of the series (2) follows from the convergence of the integral (1).

Conversely, (4) shows that the integral (3) converges when the series (2) converges.

Further, it follows from Hadamard's theorem and (1.24) that

$$\ln M_f(r) < kr^\rho\left(\int_0^r \frac{n(t)}{t^{\rho+1}}\,dt + r\int_r^\infty \frac{n(t)}{t^{\rho+2}}\,dt\right) + O(r^\rho).$$

Dividing both sides of this relation by $r^{\rho+1}$, integrating from some lower limit $\alpha > 0$ to ∞, and changing the order of integration in the right hand side, we obtain

$$\int_\alpha^\infty \frac{\ln M_f(r)}{r^{\rho+1}}\,dr < \infty.$$

We note that, when the order ρ is an integer, the preceding reasoning remains valid except for the fact that (1.40) and (1.41) must be used instead of (1.24), and we then obtain the result:

In the case of an integral order ρ, *the convergence of the integral* (1) *is equivalent to the convergence of the series* (2) *and the integral*

$$\int^\infty \frac{|\delta_f(r)|\,dr}{r}. \tag{5}$$

It is not difficult to generalize this theorem to the proximate order $\rho(r)$, by replacing the integral (1) and the series (2) by the integrals

$$\int^\infty \frac{\ln M_f(r)}{r^{\rho(r)+1}}\,dr, \qquad \int^\infty \frac{dn(t)}{t^{\rho(t)}},$$

and the integral (5) by

$$\int^\infty \frac{|\delta_f(r)|}{rL(r)}\,dr.$$

The theorem we have obtained can be used to give a very simple proof of the following theorem due to S. N. Bernšteĭn [6].

THEOREM. *A sum of even entire functions of genus zero is an entire function of genus zero.*

PROOF. Let $f(z)$ and $\varphi(z)$ be even entire functions of genus zero. Then $F(z) = f(\sqrt{z})$ and $\Phi(z) = \varphi(\sqrt{z})$ are entire functions of order not higher than half that of the original functions, and their zeros $\{b_n\}$ and $\{b'_n\}$ are the squares of the zeros of $f(z)$ and $\varphi(z)$.

Since $f(z)$ and $\varphi(z)$ are of genus zero, the series

$$\sum \frac{1}{|b_n|^{\frac{1}{2}}}, \quad \sum \frac{1}{|b_n'|^{\frac{1}{2}}}$$

converge, and thus $F(z)$ and $\Phi(z)$ are of convergent type with $\rho = \frac{1}{2}$.

But if r is large enough, $M_F(r) > 1$, $M_\Phi(r) > 1$, and

$$M_{F+\Phi}(r) \leqslant M_F(r) + M_\Phi(r) \leqslant 2M_F(r)\, M_\Phi(r).$$

It immediately follows that $F(z) + \Phi(z)$ is of convergent type with $\rho = \frac{1}{2}$. Therefore the series

$$\sum \frac{1}{|a_n|^{\frac{1}{2}}},$$

in which the a_n are the zeros of $F(z) + \Phi(z)$, converges,

$$F(z) + \Phi(z) = cz^m \prod_{n=1}^{\infty} \left(1 - \frac{z}{a_n}\right)$$

and the entire function

$$f(z) + \varphi(z) = cz^{2m} \prod_{n=1}^{\infty} \left(1 - \frac{z^2}{a_n}\right)$$

is of genus zero.

We note that the requirement that $f(z)$ and $\varphi(z)$ be even is essential; there are examples of entire functions having the same genus, the sum of which has a genus greater by one than the genus of the original functions. Such an example was given by Lindelöf in [**3**].

3. The Paley-Wiener Theorem

In Fourier-transform theory there is a well-known theorem due to Plancherel:[3]

If $f(x) \in L^2(-\infty < x < \infty)$, *then*

$$\varphi(\lambda) = \text{l.i.m.}\frac{1}{\sqrt{\pi}}\int_{-n}^{n} f(x)\, e^{i\lambda x}\, dx \quad {}^{4} \tag{1}$$

[3] See, for example, Titchmarsh [**2**].

[4] The relation $\varphi(x) = \lim \varphi_n(x)$ denotes that

$$\lim_{n\to\infty} \int_{-\infty}^{\infty} |\varphi(x) - \varphi_n(x)|^2\, dx = 0.$$

In the present case, the convergent sequence is the sequence of functions

$$\varphi_{T_1T_2}(\lambda) = \frac{1}{\sqrt{\pi}} \int_{-T_1}^{T_2} f(x)\, e^{i\lambda x}\, dx \qquad (T_1 \to +\infty \text{ and } T_2 \to +\infty).$$

exists, and

$$f(x) = \text{l.i.m.}\frac{1}{\sqrt{\pi}}\int_{-n}^{n} \varphi(\lambda)\, e^{-i\lambda x}\, d\lambda \tag{2}$$

almost everywhere. Moreover

$$\int_{-\infty}^{\infty} |f(x)|^2\, dx = \int_{-\infty}^{\infty} |\varphi(\lambda)|^2\, d\lambda. \tag{3}$$

We will use this theorem in the proof of the following theorem due to Paley and Wiener [1].[5]

THEOREM. *For a function $f(x)$ $(-\infty < x < \infty)$ to be expressible in the form*

$$f(x) = \int_a^b e^{i\lambda x}\psi(\lambda)\, d\lambda \qquad (\psi(\lambda) \in L^2(a, b)), \tag{4}$$

it is necessary and sufficient that $f(x)$ be square-integrable along the whole real axis, and that it be possible to extend the definition of $f(x)$ to the complex plane as an entire function of exponential type. If the interval (a, b) can not be replaced by a smaller interval, then the segment (ia, ib) of the imaginary axis coincides with the conjugate diagram of $f(z)$.

PROOF. Let

$$f(z) = \int_a^b e^{i\lambda z}\psi(\lambda)\, d\lambda, \tag{4'}$$

where $\psi(\lambda) \in L^2(a, b)$. The function $\psi(\lambda)$ can be approximated in mean square by continuous functions $\psi_n(\lambda)$. Thus

$$\left| f(z) - \int_a^b e^{i\lambda z}\psi_n(\lambda)\, d\lambda \right| \leqslant e^{c\,|\operatorname{Im} z|}\int_a^b |\psi(\lambda) - \psi_n(\lambda)|\, d\lambda$$

$$\leqslant e^{c\,|\operatorname{Im} z|}\sqrt{(b-a)\left(\int_a^b |\psi(\lambda) - \psi_n(\lambda)|^2\, d\lambda\right)}$$

$$(c = \max(|a|, |b|)).$$

On the other hand, every integral of the form

$$f_n(z) = \int_a^b e^{i\lambda z}\psi_n(\lambda)\, d\lambda$$

can be uniformly approximated in any bounded region of the z plane by the finite sums

$$\sum e^{i\lambda_k z}\psi_n(\lambda_k)(\lambda_{k+1} - \lambda_k).$$

[5] The following proof of this theorem was given by M. Plancherel and G. Pólya in [1].

Thus $f(z)$ is the limit of a uniformly convergent sequence of entire functions and is thus an entire function.

Moreover, it follows from (4′) that

$$|f(z)| \leqslant \int_a^b e^{-\lambda \operatorname{Im} z} |\psi(\lambda)| \, d\lambda \leqslant M e^{-a \operatorname{Im} z}$$

for Im $z > 0$, while for Im $z < 0$ we have

$$|f(z)| \leqslant M e^{-b \operatorname{Im} z} \qquad \left(M = \int_a^b |\psi(\lambda)| \, d\lambda\right).$$

It follows from these two equalities that the indicator diagram of $f(z)$ is a segment of the imaginary axis contained in the segment $(-ib, ia)$.

Now let $f(z)$ be an entire function of exponential type and let $f(x) \in L^2(-\infty, \infty)$. We write

$$\varphi(z) = \int_0^\infty e^{-zx} f(x) \, dx \qquad (\operatorname{Re} z > 0).$$

It follows from (3) that

$$\int_{-\infty}^{\infty} |\varphi(\varepsilon_1 + it) - \varphi(\varepsilon_2 + it)|^2 \, dt$$

$$= \frac{1}{\pi} \int_0^\infty [e^{-\varepsilon_1 x} - e^{-\varepsilon_2 x}]^2 |f(x)|^2 \, dx \quad (0 < \varepsilon_1 < \varepsilon_2).$$

For $\varepsilon_1 \to 0$ and $\varepsilon_2 \to 0$, the integral on the right tends to zero and thus, for $\varepsilon \to +0$, the function $\varphi(\varepsilon + it)$ tends to some function $\psi_+(t) \in L^2(-\infty, \infty)$ in the mean square sense. If we define the function $\varphi(z)$ for Re $z < 0$ by

$$\varphi(z) = -\int_0^\infty e^{zx} f(-x) \, dx,$$

we find, as above, that for $\varepsilon < 0$ and $\varepsilon \to 0$ the function $\varphi(\varepsilon + it)$ tends to some function $\psi_-(t) \in L^2(-\infty, \infty)$ in the mean square sense. The function $\varphi(z)$ obviously coincides with the Borel transform of $f(z)$ (Chapter I, § 20).

Taking for the path of integration a rectangle containing in its interior the conjugate diagram $(i\alpha, i\beta)$ of the function $f(z)$, from (1.120) we obtain

$$f(z) = \frac{1}{2\pi i} \int_{\alpha_1}^{\beta_1} e^{(\varepsilon + it) z} \varphi(\varepsilon + it) \, dt - \frac{1}{2\pi i} \int_{\alpha_1}^{\beta_1} e^{(-\varepsilon + it) z} \varphi(-\varepsilon + it) \, dt$$

$$+ \frac{1}{2\pi i} \int_{-\varepsilon}^{\varepsilon} e^{(s + i\alpha_1) z} \varphi(s + i\alpha_1) \, ds - \frac{1}{2\pi i} \int_{-\varepsilon}^{\varepsilon} e^{(s + i\beta_1) z} \varphi(s + i\beta_1) \, ds$$

$$(\varepsilon > 0, \ \alpha_1 < \alpha < \beta < \beta_1).$$

Since $\varphi(z)$ is holomorphic everywhere outside the conjugate diagram, for $\varepsilon \to 0$ the last two integrals here tend to zero, and

$$f(z) = \frac{1}{2\pi} \int_{\alpha_1}^{\beta_1} e^{itz} [\psi_+(t) - \psi_-(t)]\, dt.$$

For points of the imaginary axis outside the conjugate diagram, the limit values $\psi_+(t)$ and $\psi_-(t)$ are equal, and thus

$$f(z) = \frac{1}{2\pi} \int_{\alpha}^{\beta} e^{itz} \psi(t)\, dt \qquad (\psi(t) = \psi_+(t) - \psi_-(t)).$$

The theorem has thus been proved.

The Paley-Wiener theorem has the following generalization.

THEOREM. *Let the indicator diagram I of the entire function $f(z)$ of exponential type be polygonal. Let the rays* $\arg z = \theta_j$ $(j = 1, 2, \ldots, n)$ *be parallel to the normals of the sides of this polygon, and let the functions*

$$f_j(x) = f(xe^{i\theta_j})\, e^{-h(\theta_j)x}$$

belong to the class $L^2(0, \infty)$. *Then*

$$f(z) = \int_{\Gamma} e^{\lambda z} \psi(\lambda)\, d\lambda, \tag{5}$$

where Γ is the boundary of the conjugate diagram, and $\psi(\lambda)$ is a holomorphic function outside $\bar{I}$ such that

$$\int_{\Gamma} |\psi(\lambda)|^2 |d\lambda| < \infty. \tag{6}$$

PROOF. With no loss of generality, we can assume that $\theta_j = 0$ and $h(\theta_j) = 0$. Then the function

$$\varphi_j(\varepsilon + it) = \int_0^{\infty} e^{-(\varepsilon + it)x} f_j(x)\, dx \qquad (\varepsilon > 0)$$

belongs to $L^2(-\infty, \infty)$ and tends to some function $\psi_j(t) \in L^2(-\infty, \infty)$ in the mean square sense when $\varepsilon \to +0$.

The function $f(z)$ can be expressed as an integral taken around the perimeter of a polygon $\bar{I}_\varepsilon$ that contains the conjugate diagram of the function in its interior (Figure 9):

$$f(z) = \frac{1}{2\pi} \int_{(ABCDA)} e^{\lambda z} \varphi(\lambda)\, d\lambda.$$

By Cauchy's theorem, we can replace this integral by the integral around the contour *KHGCDEK*.

By a passage to the limit, as in the proof of the previous theorem, the integration along KH can be replaced by an integration along EG.

It can similarly be shown that, for ε sufficiently small, the difference between the integral along GCM and that along LM can be made as small as we please,

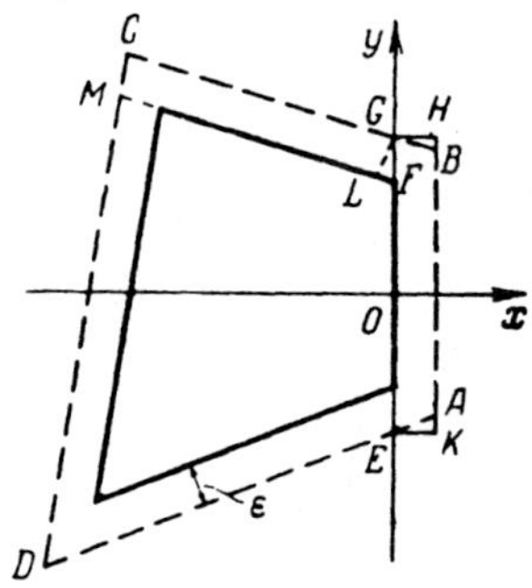

Fig. 9

and, since $\varepsilon > 0$ is arbitrary, the integral along $FGCM$ is equal to that along FM. By a continuation of this process we find that the integral along $ABCDA$ is equal to that about l. Moreover, the limiting values of $\varphi(z)$ on each of the sides are functions of L^2.

The theorem is thus proved.[6]

It follows from the above theorem that every function of exponential type satisfying the condition of the theorem can be expressed in the form

$$f(z) = \sum_{j=1}^{n} e^{\lambda_j z} g_j(-ie^{i\theta_j} z),$$

where the λ_j are the vertices of the conjugate diagram, and the entire functions $g_j(z)$ are of the class $L^2(-\infty, \infty)$ on the real axis. In fact, the integral along any side L_j of the polygon $\lambda_j + te^{i\theta_j}$ $(0 \leqslant t \leqslant d_j)$ can be written

$$f_j(z) = \int_{L_j} e^{\lambda z} \psi_j(\lambda)\, d\lambda = e^{\lambda_j z} e^{i\theta_j} \int_0^{d_j} \exp(te^{i\theta_j} z)\, \psi_j(\lambda_j + te^{i\theta_j})\, dt$$

or

$$f_j(z) = e^{\lambda_j z} g_j(-ize^{i\theta_j}),$$

where

$$g_j(u) = e^{i\theta_j} \int_0^{d_j} e^{iut} \psi_j(\lambda + te^{i\theta_j})\, dt,$$

[6] We note that the converse is also true, i.e., if an entire function of exponential type is expressed by the integral (5) and $\psi(\lambda)$ satisfies (6), then all the $f_j(x)$ $(j = 1, 2, \ldots, n)$ belong to $L^2(0, \infty)$.

i.e., $g_j(u)$ is an entire function of exponential type, belonging to $L^2(-\infty, \infty)$ on the real axis.

Liht [1] extended the theorem proved here to the case when the indicator diagram is a disk. He showed that in this case a necessary and sufficient condition for an entire function of exponential type to be represented in the form (5) is that

$$\int_0^{2\pi}\int_0^{\infty} \sqrt{x}\, e^{-2h(\theta)x}|f(xe^{i\theta})|^2\, dx\, d\theta < \infty.$$

It is clear from this result that the hypotheses of the Paley-Wiener theorem must be modified if it is to be generalized to arbitrary indicator diagrams.

Kacnel'son [1] gave a further generalization of the theorem when the indicator diagram of the entire function of exponential type is the disk $|z| \leqslant R$. He introduced the weight function

$$\rho(r) = r\int_0^{\infty} \frac{e^{-2tr}}{\sqrt{2t + t^2}}\, dt$$

and proved that representability of $f(z)$ in the form (5) is equivalent to

$$\int_0^{2\pi}\int_0^{\infty} |f(re^{i\theta})|^2 e^{-2rR}\rho(rR)dr\, d\theta < \infty,$$

and this in turn is equivalent to having the Borel transform $\psi(\lambda)$ of $f(z)$ belong to the Smirnov class E_2 in the domain $|\lambda| > R$. Finally, with this weight function there is an analog of Parseval's equation,

$$\int_{|\lambda|=R} |\psi(\lambda)|^2\, d\lambda = \int_0^{2\pi}\int_0^{\infty} |f(re^{i\theta})|^2 e^{-2rR}\rho(rR)dr\, d\theta.$$

In the general case of an arbitrary indicator diagram, Kacnel'son showed that a necessary condition for the Borel transform $\psi(\lambda)$ of $f(z)$ to belong to the Smirnov class E_2 in the exterior of the conjugate indicator diagram is that

$$\int_0^{2\pi}\int_0^{\infty} |f(re^{i\theta})|^2 e^{-2h(\theta)r}\, dr\, d\theta \leqslant c_r \int_\Gamma |\psi(\lambda)|^2 |d\lambda|$$

for some constant c_r. If the indicator diagram is a polygon, this condition is both necessary and sufficient.

To study the representability of an entire function of exponential type when the indicator diagram is arbitrary, Kacnel'son [1] introduced the weight function

$$\rho_\Gamma(\zeta) = |\zeta - d_1(\zeta)| + |\zeta - d_2(\zeta)|$$

in the exterior $\mathring{\Gamma}$ of the conjugate diagram, where $a_1(\zeta)$ and $a_2(\zeta)$ are the points of contact of the conjugate diagram with the supporting lines drawn from the point $\zeta \in \mathring{\Gamma}$. For functions holomorphic in $\mathring{\Gamma}$ he defined a scalar product using integration over $\mathring{\Gamma}$ with weight $\rho_\Gamma(\zeta)$, and introduced the corresponding Hilbert space H_r. For entire functions of exponential type with indicator not exceeding $h(\theta)$, he defined the scalar product

$$(f, g) = \int_0^{2\pi}\int_0^{\infty} e^{-2h(\theta)r} f(re^{i\theta})\overline{g(re^{i\theta})}\, dr\, d\theta$$

and introduced the Hilbert space $H_{h(\theta)}$. Kacnel'son showed that the operator

$$f(z) = \frac{1}{2\pi i}\int_c \psi(\lambda)e^{\lambda z}\, d\lambda$$

(C surrounds Γ) maps H_r onto $H_{h(\theta)}$, and $\|f\|_{h(\theta)} = \|\psi\|_\Gamma$.

4. Levitan Polynomials

A very important theorem concerning the approximation by trigonometric polynomials of entire functions of exponential type bounded on the real axis was proved by B. M. Levitan. There are now several proofs of this theorem.[7] The proof we give is due to N. I. Ahiezer and V. A. Marčenko [1].

THEOREM. *Let $f(z)$ be an entire function of exponential type σ and let*

$$|f(x)| \leqslant M \qquad (-\infty < x < \infty). \tag{1}$$

For any $N > 0$, $\eta > 0$ and $\varepsilon > 0$, there is a trigonometric polynomial

$$P_n(x) = \sum_{-(\sigma+\varepsilon) < \frac{k}{n} < \sigma+\varepsilon} a_{k,n} e^{i\frac{k}{n}x} \tag{2}$$

such that

$$|f(x) - P_n(x)| < \eta \tag{3}$$

for $|x| \leqslant N$, and $|P_n(x)| \leqslant M$ for $-\infty < x < \infty$.

PROOF. We consider the function

$$\varphi(x) = f(x)\left(\frac{\sin \varkappa x}{\varkappa x}\right)^2 \qquad (\varkappa > 0), \tag{4}$$

which satisfies the inequality

$$|\varphi(x)| \leqslant \frac{M}{\varkappa^2 x^2} \qquad (-\infty < x < \infty), \tag{5}$$

and construct the series

$$\psi_T(x) = \sum_{n=-\infty}^{\infty} \varphi(x + nT), \tag{6}$$

which converges absolutely on the real axis.

The number $\varkappa > 0$ can be chosen so small that

$$|f(x) - \varphi(x)| < \frac{\varepsilon}{2} \qquad \text{for } |x| \leqslant N. \tag{7}$$

[7] See B. M. Levitan [2], M. G. Kreĭn [5].

If T is chosen so that κT is a multiple of π, we have

$$\psi_T(z) = \frac{\sin^2 \varkappa z}{\varkappa^2} \sum_{n=-\infty}^{\infty} \frac{f(z+nT)}{(z+nT)^2}.$$

It is obvious that

$$|\psi_T(x)| \leqslant M \sin^2 \varkappa x \sum_{n=-\infty}^{\infty} \frac{1}{(\varkappa x + n\pi)^2} = M.$$

To prove this theorem, we need only replace the factor[9] $(\sin \kappa x/\kappa x)^2$ by $(\sin \kappa x/\kappa x)^{2+k}$.

5. Power Series with Coefficients that are the Values of Entire Functions of Exponential Type for Integral Values of Their Arguments

We consider series of the form

$$f(z) = \sum_{n=0}^{\infty} \varphi(n) z^n, \tag{1}$$

where $\varphi(z)$ is an entire function of exponential type.

The distribution of the singularities of a function $f(z)$ expressed in this way is given by the following theorem of G. Pólya.

THEOREM. *If the width of the indicator diagram $h_\varphi(\pi/2) + h_\varphi(-\pi/2)$ of $\varphi(z)$ is less than 2π, then all the singularities of $f(z)$ are in a region G obtained from the conjugate diagram $\bar{I}$ of $\varphi(z)$ by the transformation*

$$w = e^{-z} \tag{2}$$

and $f(\infty) = 0$.

Conversely, if $f(z)$ can be expressed in the neighborhood of the origin by the series

$$f(z) = \sum_{n=0}^{\infty} a_n z^n, \tag{3}$$

all its singularities are in some region G obtained by the transformation (2) of a convex region K with width along the imaginary axis less that 2π, and if $f(\infty) = 0$, then there exists an entire function $\varphi(z)$ of exponential type with its conjugate diagram in K such that $\varphi(n) = a_n$ $(n = 0, 1, 2, \ldots)$.

For the proof of the first part of the theorem, we note that an entire function $\varphi(z)$ of exponential type can be written in the form

$$\varphi(u) = \int_{(C)} e^{u\lambda} \psi(\lambda)\, d\lambda,$$

where C is an arbitrary contour encircling the diagram $\bar{I}$. We may, of course, assume that C is the boundary of a convex region with width on the imaginary axis smaller than 2π.

---

[9] By introducing a new kernel to replace $(\sin \kappa x/\kappa x)^2$, we can generalize Levitan's theorem to the case of functions with rapid growth on the real axis. This was done by V. A. Marčenko in [2]. See also the article [1] by L. I. Ronkin. Ljubarskiĭ [1] has recently obtained a more general result on the approximation of entire functions of exponential type, in various norms, by trigonometric polynomials.

Substituting this expression for $\varphi(u)$ in the series (1), we obtain

$$f(z)=\int_{(C)}\psi(\lambda)\Big(\sum_{n=0}^{\infty}e^{n\lambda}z^n\Big)\,d\lambda.$$

In a sufficiently small circle $|z|<\varepsilon$, the series in the integrand is convergent, and so

$$f(z)=\int_{(C)}\frac{\psi(\lambda)}{1-ze^{\lambda}}\,d\lambda.$$

It follows from this representation of $f(z)$ that it has no singularities outside the one-sheeted region obtained by the transformation (2) from the region bounded by the contour C. In other words, all the singularities of $f(z)$ are in the image of G. From this representation of $f(z)$ it also follows that $f(\infty)=0$.

Conversely, let all the singularities of a function $f(z)$ given in the form (1) lie in some closed region G, where this region is the image of a convex region K with width less than 2π in the direction of the imaginary axis. In this case, the set complementary to G contains the points 0 and ∞.

We have

$$a_n=\frac{1}{2\pi i}\int_{(C)}\frac{f(z)}{z^{n+1}}\,dz \qquad (n=0,\ 1,\ 2,\ \ldots)$$

(C is a circle of small radius with center the origin). Since we also have $f(\infty)=0$, by the fundamental theorem of residue theory we may replace this integral by an integral around any contour L that encircles the region G but does not encircle the origin. If

$$\varphi(u)=\frac{1}{2\pi i}\int_{(L)}\frac{f(z)}{z^{u+1}}\,dz,$$

then

$$\varphi(u)=\frac{1}{2\pi i}\int_{(C)}f(e^{-v})\,e^{uv}\,dv, \tag{4}$$

where C is a contour encircling the region K.

It is evident that $\varphi(n)=a_n$. It also follows from the representation (4) that $\varphi(u)$ is an entire function of exponential type, and that its conjugate diagram is a subset of K. The theorem has been proved.

The following theorem of S. Wiegert is a direct consequence of this theorem.

THEOREM. *In order that a function $f(z)$ given by the series* (3) *have its only singularity at $z=1$ and be zero at infinity, it is necessary and sufficient that the coefficients of the series be the values of an entire function of zero degree, or the values of a polynomial, for integral values of the argument.*

In the latter case, $f(z)$ will be a rational function with its only pole at $z=1$.[10]

[10] The distribution of the singularities of the series (1) has been studied in many works. We cite here Leau [1], Wiegert [1], Lindelöf [2], Carlson [1] and Pólya [1].

In the work of A. O. Gel'fond [3], a substantial improvement in Wiegert's theorem is obtained. A relation is established between the rate of growth of the function $\varphi(z)$ and that of the entire function $G(z)$, where these functions are connected by the relation $f(z) = G(1/(z-1))$.

6. Lower Estimates for Holomorphic Functions

An estimate cruder than Cartan's (given in Chapter I) had already been established by Valiron [**1**]. Ahlfors [**1**] gave a similar lower bound for subharmonic functions in the plane with a finite total mass n. The difference between this inequality and Cartan's is that the sum of the radii of the excluded disks is $5H$ rather than $2H$ as in Cartan's inequality.[11] Landkof [**1**] obtained a similar inequality for subharmonic functions in an arbitrary k-space, with the sum of the radii of the exceptional disks equal to C_kH, where C_k depends only on k. Lower bounds for the moduli of holomorphic functions in a disk are given in Šaginjan [**1**; **2**], Havinson [**1**], and Ušakova [**1–3**]. We also mention lower bounds in $|z| \leqslant 1$ for the modulus of a function that is holomorphic and bounded in $|z| < 1$ (Govorov [**4**], Grišin [**3**], Kudina [**1**]).

In the first group of results there may be infinitely many exceptional disks, all inside $|z| < 1$; in the second group, there are finitely many but they may extend outside the unit disk.

[11] In 1968 Grišin [**2**] obtained the same inequality outside exceptional disks the sum of whose radii is $2H$, as in Cartan's inequality.

Appendix II

THE APPLICATION OF UNIQUENESS THEOREMS TO QUASIANALYTIC CLASSES OF FUNCTIONS

The proofs of some uniqueness theorems in the real domain, for functions expressed by ordinary or generalized trigonometric series, can be obtained by reducing the problems to those of the uniqueness of the definition of holomorphic functions, i.e., by their reduction to problems of the type investigated in Chapter IV. In particular, some theorems concerning quasianalytic classes can be obtained in this way.

1. Functions Defined by Their Values on an Interval

A periodic function is defined by its values on any interval of length equal to its period. But this property is not restricted to periodic functions. We can cite the examples of almost-periodic functions and functions represented by Fourier-Stieltjes integrals, which are also defined by their values on a sufficiently long interval. This property is related to the spectrum of the function.[1]

We consider the class T_Ω of Stepanov almost-periodic functions (S_p),[2] whose Fourier exponents make up some set Ω of real points.

We denote by $n_\Omega(t)$ the number of nonzero points of the set in the interval $(-t, t)$, and write

$$N_\Omega(R) = \int_0^R \frac{n_\Omega(t)}{t}\, dt, \tag{1}$$

[1] M. G. Kreĭn brought this relation between the spectrum of a function and its uniqueness interval to my attention in 1941. The results given here were published by me in 1950 [11]. New results in this direction were recently obtained by J. P. Kahane [1; 2]. These results are connected with the theory of functions "periodic in the mean."

[2] For our purposes it is sufficient to use the following property of S_p-almost-periodic functions, which can be taken as a definition. Every S_p-almost-periodic function $f(x)$ can be approximated by an almost-periodic polynomial $P(x)$ in the sense that

$$\sup_{-\infty < x < \infty} \left\{ \int_x^{x+1} |f(t) - P(t)|^p\, dt \right\}^{\frac{1}{p}} < \varepsilon.$$

when zero is not in the set Ω, and

$$N_\Omega(R) = \int_0^R \frac{n_\Omega(t)}{t}\,dt + \ln R,$$

when zero is in Ω. In particular, we can have $N_\Omega(R) = \infty$. The function $N_\Omega(R)$ characterizes the density of the distribution of points of Ω on the real axis.

THEOREM 1. *If an S_p-function of the class T_Ω*

$$f(t) \sim \sum_{k=-\infty}^{\infty} a_k e^{i\lambda_k t} \qquad (\lambda_k \in \Omega) \tag{2}$$

is zero on some interval of length d and if

$$\varlimsup_{R\to\infty}\left[N_\Omega(R) - \frac{d}{\pi}R - \frac{1}{q}\ln R\right] = -\infty \qquad \left(\frac{1}{p}+\frac{1}{q}=1\right), \tag{3}$$

then $f(x) \equiv 0$ almost everywhere.

Thus a function of T_Ω is determined by its values on an interval of given length if the density of Ω is sufficiently small.

PROOF. We obviously need only consider the case when Ω has a condensation point at infinity only. The Laplace transform of $f(x)$ is

$$\varphi(z) = \int_0^\infty e^{-zt} f(t)\,dt = \sum_{k=-\infty}^{\infty} \frac{a_k}{z - i\lambda_k} \tag{4}$$

and in this case is a meromorphic function with simple poles at the points $i\lambda_k$.

We may assume, with no loss of generality, that $f(t)$ is zero on the interval $(-d/2, d/2)$.[3] We first obtain an estimate of the size of $\varphi(z)$ in the right half-plane. For $m \geqslant 0$ we have

$$\left|\int_m^{m+1} e^{-zt} f(t)\,dt\right| \leqslant \int_m^{m+1} e^{-tx}\,|f(t)|\,dt \leqslant \|f\|_p \left(\int_m^{m+1} e^{-qtx}\,dt\right)^{\frac{1}{q}},$$

where

$$\|f\|_p = \sup_{-\infty < x < \infty}\left[\int_x^{x+1} |f(t)|^p\,dt\right]^{\frac{1}{p}} \quad \left(z = x + iy \text{ and } \frac{1}{p}+\frac{1}{q}=1\right).$$

Hence

$$\left|\int_m^{m+1} e^{-zt} f(t)\,dt\right| \leqslant \|f\|_p \frac{e^{-mx}(1-e^{-qx})^{\frac{1}{q}}}{(qx)^{\frac{1}{q}}}.$$

[3] This follows from the fact that the translation $f(t + t_0)$ does not alter the spectrum of the function.

From this result we infer the convergence of the integral in (4) and obtain

$$|\varphi(z)| \leqslant \|f\|_p \frac{e^{-\frac{d}{2}x}(1-e^{-qx})^{\frac{1}{q}}}{(qx)^{\frac{1}{q}}(1-e^{-x})}$$

for $x > 0$. In the left half-plane we similarly obtain the result

$$|\varphi(z)| \leqslant \|f\|_p \frac{e^{-\frac{d}{2}|x|}(1-e^{-q|x|})^{\frac{1}{q}}}{(q|x|)^{\frac{1}{q}}(1-e^{-|x|})}, \tag{5}$$

which is valid for all values of $z = x + iy$. We note that if $f(t)$ is bounded, in particular, if $f(t)$ is an ordinary almost-periodic function, then this result can be replaced by

$$|\varphi(z)| \leqslant \frac{M}{|x|} e^{-\frac{d}{2}|x|}, \tag{6}$$

where

$$M = \sup_{-\infty < t < \infty} |f(t)|.\text{[3a]}$$

Jensen's formula (Chapter I, § 5) for functions holomorphic in a closed circle $|z| \leqslant R$ can be extended to apply to functions with poles in this circle. If the meromorphic function $\varphi(z)$ is not zero either at infinity or at the origin, then Jensen's formula becomes

$$\int_0^R \frac{n_1(t)}{t}\,dt - \int_0^R \frac{n_2(t)}{t}\,dt = \frac{1}{2\pi}\int_0^{2\pi} \ln|\varphi(Re^{i\theta})|\,d\theta - \ln|\varphi(0)|, \tag{7}$$

where $n_1(t)$ is the number of zeros and $n_2(t)$ the number of poles of $\varphi(z)$ in the circle $|z| \leqslant t$.

We will assume that $\varphi(z) \not\equiv 0$, and will start by assuming that $\varphi(0)$ is neither zero or infinity. Then from (5) we have

$$\int_0^R \frac{n_2(t)}{t}\,dt > \frac{d}{\pi}R + \frac{1}{q}\ln R + O(1) + \int_0^R \frac{n_1(t)}{t}\,dt. \tag{8}$$

[3a] The convergence of the series (4) has to be understood in a generalized sense. We know that there are multipliers

$$\{\alpha_k^\delta\}_{-\infty}^{\infty}, \qquad 0 < \alpha_k^\delta < 1 \qquad (\delta > 0), \qquad \alpha_k^\delta \uparrow 1,$$

such that for a given $\varepsilon > 0$ and sufficiently small $\delta > 0$ we have

$$\|f(t) - \sum_{-n_\delta}^{n_\delta} \alpha_k^\delta a_k e^{i\lambda_k t}\|_p < \varepsilon, \qquad 0 \leqslant p \leqslant \infty.$$

It follows from (5) and (6) that

$$\|\varphi(z) - \sum_{-n_\delta}^{n_\delta} 1 \frac{\alpha_k^\delta a_k}{z - i\lambda_k}\|_p < C_q \frac{\varepsilon}{|x|} e^{-\alpha|x|/2}$$

and $n_\delta \to \infty$ as $\delta \to 0$.

Noting that $\int_0^R (n_1(t)/t)\,dt \geqslant 0$ and $n_\Omega(t) \geqslant n_2(t)$, we obtain

$$N_\Omega(R) > \frac{d}{\pi} R + \frac{1}{q} \ln R + O(1). \tag{8'}$$

If zero is a point of the set Ω and is also a simple pole of $\varphi(z)$, then we again obtain (8′) by applying formula (7) to $\varphi_1(z) = z\varphi(z)$. The case $\varphi(0) = 0$ can be dealt with similarly. But (8′) contradicts the condition (3) of the theorem. Thus if the condition (3) is satisfied we have $\varphi(z) \equiv 0$. Taking the Laplace transform, we find that $f(t) \equiv 0$ almost everywhere.

We note that if $f(t)$ is bounded (in particular, if it is an ordinary almost-periodic function), then we can use (6) instead of (5), and obtain the result that if

$$\varliminf_{R\to\infty} \left[N_\Omega(R) - \frac{d}{\pi} R - \ln R\right] = -\infty \tag{3'}$$

then $f(t) \equiv 0$ almost everywhere. Theorem 1 has thus been extended to the case $p = \infty$.[4]

The estimate (3) is sharp, as is shown by the following example. For any $\delta > 0$, we write

$$f_1(t) = (e^{it} + 1)^{-2\delta}, \qquad f_2(t) = (1 + e^{-it})^{-2\delta}$$

for $-\pi < t < \pi$, and

$$f_i(t + 2\pi) = f_i(t) \qquad (i = 1,\ 2; \quad -\infty < t < \infty).$$

For $2\delta < 1/p$, the function

$$f(t) = e^{i\delta t} f_1(t) - e^{-i\delta t} f_2(t)$$

belongs to the class (S_p) and is zero for $|t| < \pi$. The Fourier exponents of $f(t)$ are $\lambda_n = \pm(n + \delta)$ $(n = 0, 1, 2, \ldots)$. We denote the set of numbers $\pm(n + \delta)$ by Ω, and obtain

$$N_\Omega(R) = 2R + (1 - 2\delta) \ln R + O(1).$$

Thus if we replace the quantity $1/q$ in (3) by a larger number, we obtain a sequence Ω satisfying condition (3), and an S_p-almost-periodic function $f(t)$ not identically equal to zero, with Fourier exponents from Ω, and equal to zero in the interval $|t| < \pi$.

It is of interest to investigate the "boundary functions," i.e., the functions that are zero in the interval $|t| < d/2$ and such that their Fourier exponents satisfy

$$\varliminf_{R\to\infty} \left[N_\Omega(R) - \frac{d}{\pi} R - \sigma \ln R\right] = -\infty$$
$$\left(1 + \frac{1}{q} \geqslant \sigma > \frac{1}{q}\right). \tag{9}$$

---

[4] For $p = 1$, instead of (5) we have

$$|\varphi(z)| \leqslant \|f\|\, e^{-\frac{d}{2}|x|} (1 - e^{-|x|})^{-1},$$

and in this case we must set $q = \infty$ in the relation (3) of Theorem 1.

In this case only one point need be omitted from Ω in order that there exist a function from T_Ω that is zero in the interval $(-d, d)$ but not identically zero. We have the following theorem.

THEOREM 2. *If the set* $\{\lambda_k\}$ *of Fourier exponents of an* S_p*-almost-periodic function* $f(t)$ *satisfies condition* (9), *and if the function itself is zero on the interval* $(-d/2, d/2)$, *then the set* $\{\lambda_k\}$ *is of density* d/π *on the positive and negative rays and*

$$\delta = \lim_{R\to\infty}\left(\sum_{|\lambda_k|<R}\frac{1}{\lambda_k}\right)$$

exists.

Moreover the function $f(t)$ *can be expressed in the form*

$$f(t) \sim \sum_{-\infty}^{\infty}\frac{1}{\psi'(i\lambda_k)}e^{i\lambda_k t}, \tag{10}$$

where

$$\psi(z) = c\lim_{R\to\infty}\prod_{|\lambda_k|<R}\left(1-\frac{z}{i\lambda_k}\right). \tag{10'}$$

PROOF. It follows from the inequality (8) that $n_1(t) = 0$ when condition (9) is satisfied. Thus $\varphi(z)$ has no zeros, and so $\psi(z) = [\varphi(z)]^{-1}$ is an entire function with zeros at the points of Ω. All the points of Ω must be zeros of $\psi(z)$, since it is impossible to remove any points from Ω. The representation

$$\frac{1}{\psi(z)} = \sum_{-\infty}^{\infty}\frac{a_k}{z-i\lambda_k} \tag{4'}$$

(see (4)) shows, by a theorem of M. G. Kreĭn (Chapter V, Theorem 12), that $\psi(z)$ is of exponential type and of completely regular growth, and that its indicator diagram is a segment of the real axis. It thus follows that the set $\{\lambda_k\}$ has a density that is the same on the positive and negative half-axes, and that the limit

$$\delta = \lim_{R\to\infty}\sum_{|\lambda_k|<R}\frac{1}{\lambda_k}$$

exists. From Theorem 11, Chapter V, the function $\psi(z)$ can be represented in the form (10′). The representation (10) of $f(t)$ is obtained from (4′), which yields $a_k = 1/(\psi'(i\lambda_k))$. We show further that the density of Ω is $d/2\pi$ on each of the rays $x > 0$, $x < 0$. It follows directly from (9) that

$$\lim_{R\to\infty}\frac{n_\Omega(R)}{R} \leqslant \frac{d}{\pi}.$$

On the other hand, (5) is a result of the fact that $f(x)$ vanishes on the interval $(-d/2, d/2)$, and thus

$$|\psi(z)| \geqslant c \frac{|z|^{\frac{1}{q}}(1-e^{-|x|})}{(1-e^{-q|x|})^{\frac{1}{q}}} e^{\frac{d}{2}|x|}. \tag{11}$$

By an application of Jensen's theorem, this result easily yields

$$\lim_{R\to\infty} \frac{n_\Omega(R)}{R} \geqslant \frac{d}{\pi}.$$

The set Ω therefore has the density d/π, which means that its density is $d/2\pi$ on each of the rays $x > 0$ and $x < 0$. The theorem is proved.

The following theorem can be considered, in a sense, as the converse of Theorem 2.

THEOREM 3. *If $\psi(z)$ has only the simple zeros $i\lambda_k$ such that* $\inf|\lambda_k - \lambda_n| \geqslant \delta > 0$ *for $k \neq n$, and if, for $\gamma > 1$,*

$$0 < m < |\psi(z)| e^{-\frac{d}{2}|x|}(1+|z|^\gamma)^{-1} < M \tag{11'}$$

outside nonintersecting circles containing the zeros, then the series

$$\sum_{-\infty}^{\infty} \frac{e^{i\lambda_k t}}{\psi'(i\lambda_k)} \tag{12}$$

converges absolutely and uniformly on the whole real axis and defines a function $f(t)$ which is zero on $(-d/2, d/2)$.

PROOF. It follows from (11′) that

$$\left|\frac{\psi(z)}{z-i\lambda_k}\right| > \frac{m}{\delta}(1+|z|^\gamma)$$

on the circle $|z - i\lambda_k| = \delta$, and so an application of the maximum-modulus principle to the function

$$\ln|\psi(z)(z-i\lambda_n)^{-1}|,$$

which is harmonic in the circle $|z - i\lambda_k| \leqslant \delta$, yields

$$|\psi'(i\lambda_n)| > c|\lambda_k|^\gamma,$$

where c is a constant. A consequence of (11′) is that $\psi(z)$ is of completely regular growth. Hence $(n/\lambda_n) \to (d/2\pi)$, and so the series (12) is absolutely convergent. We now show that the representation of the function $f(t)$ as the series (12) yields the value zero in the interval $(-d/2, d/2)$.

Integrating the function $e^{-tz}/\psi(z)$ around the contour Γ (Figure 10), we obtain

$$\sum_{|\lambda_k|<N} \frac{e^{i\lambda_k t}}{\psi'(i\lambda_k)} = \frac{1}{2\pi i}\int_{(\Gamma)} \frac{e^{-tz}}{\psi(z)}\,dz.$$

For $t < +d/2$, the integral along the circular arc tends to zero with increasing N, and hence

$$f(t) = \frac{1}{2\pi i} \int_{\varepsilon - i\infty}^{\varepsilon + i\infty} \frac{e^{-tz}}{\psi(z)}\, dz \qquad \left(t < \frac{d}{2}\right).$$

On the other hand, integrating around Γ_1 (Figure 11) we obtain

$$0 = \frac{1}{2\pi i} \int_{(\Gamma_1)} \frac{e^{-tz}}{\psi(z)}\, dz.$$

For $t > -d/2$, the integral along the circular arc tends to zero, and we have

$$0 = \frac{1}{2\pi i} \int_{\varepsilon - i\infty}^{\varepsilon + i\infty} \frac{e^{-tz}}{\psi(z)}\, dz.$$

Thus $f(t) = 0$ for $-d/2 < t < d/2$. The theorem has thus been proved.

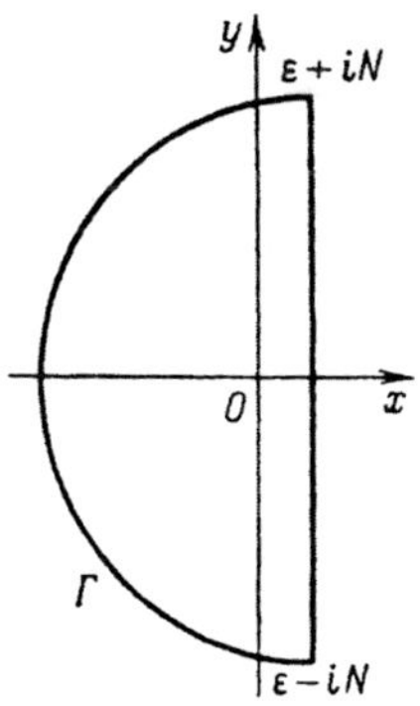

Fig. 10

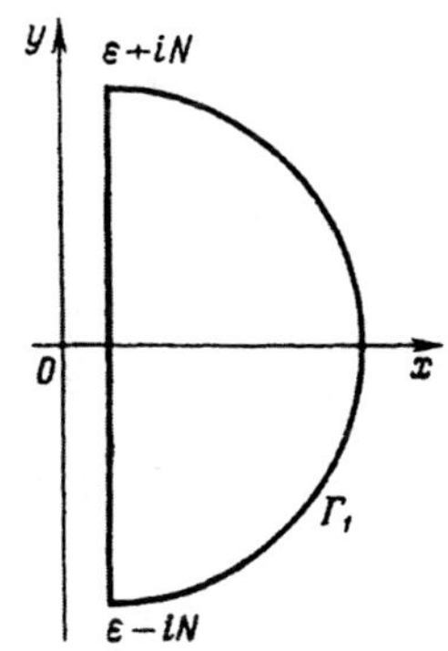

Fig. 11

In the proof of Theorem 1, we discarded the non-negative term

$$\int_0^R \frac{n_1(t)}{t}\, dt$$

from the inequality (8)

$$\int_0^R \frac{n_2(t)}{t}\, dt > \frac{d}{\pi} R + \frac{1}{q} \ln R + \int_0^R \frac{n_1(t)}{t}\, dt + O(1).$$

We now take this term into account. We will consider S_p-almost-periodic functions satisfying the condition $f(-x) = \overline{f(x)}$, i.e., functions $f(x)$ such that

$$f(x) \sim \sum_{-\infty}^{\infty} a_n e^{i\lambda_n x},$$

where the a_n are real.

We denote by λ_0 the Fourier exponent of $f(x)$ nearest to zero, and enumerate the numbers λ_n in order of increasing value $\ldots < \lambda_{-k} < \ldots < \lambda_{-1} < \lambda_0 < \lambda_1 < \ldots < \lambda_k < \ldots$. We use the notation $n_0(t)$ for the number of changes of sign in the sequence of coefficients $\ldots, a_{-k}, a_{-k+1}, \ldots, a_{-1}, a_0, a_1, \ldots, a_k, \ldots$ corresponding to the numbers λ_n of the interval $(-t, t)$, and write

$$N_0(R) = \int^{R} \frac{n_0(t)}{t}\, dt.$$

THEOREM 4. *If an S_p-almost-periodic function $f(x)$ is zero in the interval $(-d/2, d/2)$ but not identically zero, and if $f(-x) = \overline{f(x)}$, then the density of the changes of sign $N_0(R)$ in the Fourier series of this function satisfies the condition*

$$\varliminf_{R\to\infty} \left[N_0(R) - \frac{d}{\pi} R + \frac{1}{p} \ln R \right] > -\infty. \tag{13}$$

PROOF. Let $a_n a_{n+1} > 0$. Then the Laplace transform

$$\varphi(z) = \sum_{-\infty}^{\infty} \frac{a_k}{z - i\lambda_k}$$

of $f(t)$ is purely imaginary on the Oy axis and changes sign in the interval $(i\lambda_n, i\lambda_{n+1})$, and so it takes the value zero in this interval. Thus the number of zeros $n_1(t)$ of $\varphi(z)$ in the circle $|z| \leqslant t$ is not less than the number of changes of sign in the sequence of coefficients $\{a_n\}$ corresponding to $|\lambda_n| \leqslant t$. Thus

$$n_2(t) - n_1(t) \leqslant n_0(t) + 1.$$

If $\varphi(z) \not\equiv 0$, there is no loss of generality in assuming that $\varphi(0) \neq 0$, and from Jensen's formula (7) we obtain

$$N_0(R) + \ln R > -\frac{1}{2\pi} \int_0^{2\pi} \ln |\varphi(Re^{i\theta})|\, d\theta + \ln |\varphi(0)|.$$

It thus follows from (5) that

$$N_0(R) > \frac{d}{\pi} R - \frac{1}{p} \ln R + O(1).$$

This completes the proof of the theorem.

Theorem 4 can also be modified to apply to a function of the form

$$f(x) = \int_{-\infty}^{\infty} e^{itx}\, d\sigma(t), \tag{14}$$

where $\sigma(t)$ is a real function of bounded variation.

An interval on which $\sigma(t)$ is constant will be called an extremal interval if all the values of $\sigma(t)$ in some neighborhood of this interval are either not less than or not greater than the value on the interval.

An extremum point, not in an interval of constancy of $\sigma(t)$, will be considered as a particular case of an extremal interval. If, at a point of discontinuity, both the left and the right limiting values of $\sigma(t)$ are extremal, then we will consider that there are two extrema at this point.

We denote by $n_0(t)$ the number of extremal intervals covered by the whole interval $(-t, t)$, and also use the notation

$$N_0(R) = \int_0^R \frac{n_0(t)}{t}\, dt.$$

THEOREM 5. *If a function $f(x)$ is given by* (14) *and is not identically zero, and if it is zero in the interval* $(-d/2, d/2)$, *then the density of the set of extremal intervals of the function $\sigma(t)$ satisfies the condition*[5]

$$\varliminf_{R\to\infty}\left[N_0(R) - \frac{d}{\pi} R\right] > -\infty.$$

PROOF. The Laplace transform of $f(x)$ is

$$\varphi(z) = \int_{-\infty}^{\infty} \frac{d\sigma(t)}{it - z}. \tag{15}$$

It is not difficult to construct a step function $\sigma_\varepsilon(t)$ such that the extremal intervals of positive length of $\sigma(t)$ are extremal intervals of $\sigma_\varepsilon(t)$, and the extremal points of $\sigma(t)$ coincide with the ends of extremal intervals of $\sigma_\varepsilon(t)$. In addition, this function can be constructed so that

$$|\sigma(t) - \sigma_\varepsilon(t)| < \varepsilon \qquad (-\infty < t < \infty), \tag{16}$$

where ε is an arbitrary given positive number.

The value of $n_0(t)$ for $\sigma(t)$ is not less than the value of $n_{0,\varepsilon}(t)$ for $\sigma_\varepsilon(t)$.

We now define the function

$$\varphi_\varepsilon(z) = \int_{-\infty}^{\infty} \frac{d\sigma_\varepsilon(t)}{it - z} = \sum_{-\infty}^{\infty} \frac{a_k^\varepsilon}{it_k^{(\varepsilon)} - z}, \tag{17}$$

where the $t_k^{(\varepsilon)}$ are the points of discontinuity of $\sigma_\varepsilon(t)$ and the a_k^ε are the jumps of $\sigma_\varepsilon(t)$. We have

$$\varphi(z) - \varphi_\varepsilon(z) = \int_{-\infty}^{\infty} \frac{d[\sigma(t) - \sigma_\varepsilon(t)]}{it - z} = i \int_{-\infty}^{\infty} \frac{\sigma(t) - \sigma_\varepsilon(t)}{(it - z)^2}\, dt,$$

or

$$|\varphi(z) - \varphi_\varepsilon(z)| \leqslant \varepsilon \int_{-\infty}^{\infty} \frac{dt}{|it - z|^2} = \frac{\pi\varepsilon}{|x|} \qquad (z = x + iy). \tag{18}$$

[5] A similar, somewhat stronger result was obtained by M. G. Kreĭn in the theory of continuation of Hermite-positive functions.

Applying Jensen's formula to $\varphi_\varepsilon(z+\delta)$ $(\delta > 0)$, we obtain

$$\int_0^R \frac{n_{1,\varepsilon}(r) - n_{2,\varepsilon}(r)}{r}\, dr = \frac{1}{2\pi}\int_0^{2\pi} \ln|\varphi_\varepsilon(Re^{i\theta}+\delta)|\, d\theta - \ln|\varphi_\varepsilon(\delta)|, \quad (19)$$

where $n_{1,\varepsilon}(r)$ is the number of zeros and $n_{2,\varepsilon}(r)$ the number of poles of $\varphi_\varepsilon(z+\delta)$ in the circle $|z| \leqslant r$.

All the poles of $\varphi_\varepsilon(z)$ are on the imaginary axis, and there is a zero between any two poles corresponding to two successive terms in the sequence $\{a_k^\varepsilon\}$ with the same sign. Thus for $t = \sqrt{r^2 - \delta^2}$, we have

$$n_{2,\varepsilon}(r) - n_{1,\varepsilon}(r) \geqslant -[n_{0,\varepsilon}(t) + 1]$$

and from this result and (19) we obtain

$$\int_0^{\sqrt{R^2-\delta^2}} \frac{n_{0,\varepsilon}(t)+1}{\sqrt{t^2+\delta^2}}\, dt \geqslant \ln|\varphi_\varepsilon(\delta)| - \frac{1}{2\pi}\int_0^{2\pi} \ln|\varphi_\varepsilon(Re^{i\theta}+\delta)|\, d\theta.$$

The inquality

$$\frac{n_0(t)}{t} \geqslant \frac{n_{0,\varepsilon}(t)}{t} > \frac{n_{0,\varepsilon}(t)}{\sqrt{t^2+\delta^2}}$$

thus yields

$$N_0(R) + \ln\frac{2R}{\delta} > -\frac{1}{2\pi}\int_0^{2\pi} \ln|\varphi_\varepsilon(Re^{i\theta}+\delta)|\, d\theta + \ln|\varphi_\varepsilon(\delta)|.$$

We now fix a value of δ so that $\varphi(\delta) \neq 0$, and then, using (18), take the limit for $\varepsilon \to 0$. We obtain

$$N_0(R) + \ln\frac{2R}{\delta} \geqslant \ln|\varphi(\delta)| - \frac{1}{2\pi}\int_0^{2\pi} \ln|\varphi(Re^{i\theta}+\delta)|\, d\theta. \quad (20)$$

Hence, using (6), which is obviously valid in this case, we arrive at the result

$$N_0(R) > \frac{d}{\pi} R + O(1).$$

The theorem has thus been proved. The question naturally arises of the possibility of determining the function $f(z)$ with a given spectrum from its values on an interval $(-d/2, d/2)$, when these values determine the function everywhere. We will limit ourselves to the case when the S_p-almost-periodic function

$$f(t) \sim \sum a_k e^{i\lambda_k t}$$

is holomorphic in the segment $[-d/2, d/2]$, and we will assume that we know its values in some neighborhood of this segment in the complex plane and that the real sequence $\{\lambda_k\}$ is a regular set with density $\Delta = d/\pi$. The function

$$\theta(\lambda) = \prod_{-\infty}^{\infty}\left(1 - \frac{\lambda}{i\lambda_k}\right)$$

is an entire function of exponential type and of completely regular growth, and its indicator diagram is the segment $[-d/2, d/2]$ of the real axis. We use this function to construct a biorthogonal system for the sequence of functions $\{e^{i\lambda_k t}\}$.

The functions

$$\varphi_k(z) = \int_0^\infty \frac{\theta(\lambda) e^{-\lambda z}}{\theta'(i\lambda_k)(\lambda - i\lambda_k)} d\lambda \qquad (\pm k = 0, 1, 2, \ldots)$$

are holomorphic everywhere outside the segment $[-d/2, d/2]$, and if C is a contour encircling this segment, then

$$\frac{1}{2\pi i} \int_{(C)} \varphi_k(z) e^{i\lambda_p z} dz = \begin{cases} 1 & \text{for} \quad p = k, \\ 0 & \text{for} \quad p \neq k. \end{cases}$$

It immediately follows that

$$a_k = \int_{(C)} \varphi_k(z) f(z) dz \qquad (\pm k = 0, 1, \ldots).$$

The contour (C) must be chosen so that it contains the segment $[-d/2, d/2]$ and is contained in that neighborhood of this segment where the function $f(t)$ is given. Thus from the values of $f(z)$ in a neighborhood of $[-d/2, d/2]$ we have obtained its Fourier series, and we have thus determined the function itself.

We note that if the sequence $\{\lambda_k\}$ is such that the function

$$\frac{\theta(\lambda)}{1 + |\lambda|}$$

is in L^2, then we must obtain the biorthogonal system $\{\varphi_k(t)\}$ from the formulas

$$\varphi_k(t) = \text{l.i.m} \int_{-\infty}^{\infty} \frac{\theta(\lambda) e^{-\lambda t}}{\theta'(i\lambda_k)(\lambda - i\lambda_k)} d\lambda \qquad (\pm k = 0, 1, 2, \ldots).$$

If the function is given on the segment $[-d/2, d/2]$ and belongs to $L^2(-d/2, d/2)$, then

$$a_k = \frac{1}{2\pi} \text{l.i.m} \int_{-\frac{d}{2}}^{\frac{d}{2}} f(t) \varphi_k(t) dt \qquad (\pm k = 0, 1, 2, \ldots).$$

2. Quasianalytic Classes of Almost-Periodic Functions

In this section, we will also consider S_p-almost-periodic functions having spectra $\{\lambda_n\}$ that belong to some set Ω without any finite limit points. If we also have

$$\lim_{R \to \infty} \frac{N_\Omega(R)}{R} = 0,$$

then we infer from Theorem 1 that if a function of this class is zero on any arbitrarily small interval, then it is zero almost everywhere.[6] In other words, if two functions of this class take the same values on any arbitrarily small interval, they are equal almost everywhere, i.e., this class is [*I*]-quasianalytic.

By taking a very "sparse" set Ω, we can obtain classes of functions T_Ω in which nonzero functions can not be zero on any interval and also can not tend to zero rapidly when the independent variable approaches any finite value.

In [**1**], Mandelbrojt introduced the special concept of classes of quasianalytic functions for periodic functions. For integrable functions, he introduced the idea of a mean zero of exponential order ρ, defining this order by

$$\rho = \overline{\lim_{\alpha \to +0}} \frac{\ln(-\ln \theta(\alpha))}{-\ln \alpha}, \tag{21}$$

where

$$\theta(\alpha) = \int_{t_0}^{t_0+\alpha} |f(t)|\, dt. \tag{21'}$$

Mandelbrojt's fundamental *Theorem C* is the following: *If a function $f(t)$, expressible in the form*

$$f(t) \sim \sum_{i=1}^{\infty} (a_{n_i} \cos n_i t + b_{n_i} \sin n_i t), \tag{22}$$

has a zero of exponential order ρ at some point, if the numbers n_i form a sequence with convergence index $\sigma < 1$, and if

$$\rho > \frac{\sigma}{1-\sigma},$$

then $f(t) \equiv 0$ almost everywhere.

A simple proof of this theorem, based on the application of a uniqueness theorem from the theory of entire functions, was given by the author in collaboration with M. S. Livšic in [**1**]. This new method introduced the possibility of a significant generalization and strengthening of the theorem.

In the article [**12**] by the author, this strengthening of the theorem concerning quasianalyticity was obtained for the class of almost-periodic functions.[7] We will describe this result in a somewhat simpler but more complete fashion than in [**12**].

We use the same method here as was used in the investigation of functions defined by their values on an interval.

---

[6] N. Levinson [**1**].

[7] The results of the author and M. S. Livšic were improved in a paper by I. Hirschman and J. Jenkins [**1**]. A further development of this method was obtained in recent articles by J. P. Kahane [**1**; **2**].

As before, we will characterize the sequence $\Lambda = \{\lambda_n\}$ ($\pm n = 0, 1, 2, \ldots$) by the function

$$N_\Lambda(R) = \int_0^R \frac{n_\Lambda(t)}{t}\,dt. \tag{23}$$

Let $\psi(\alpha)$ be a continuous, nondecreasing, non-negative function, and let $\psi(0) = 0$. We will say that $f(t)$ has a right mean zero of order not less than $\psi(\alpha)$ at t_0 if

$$\int_{t_0}^{t_0+\alpha} |f(t)|\,dt \leqslant \psi(\alpha) \qquad (0 < \alpha < \alpha_0). \tag{24}$$

A left zero is defined similarly, and also a two-sided zero.

For the function $\psi(\alpha)$ we can take, for example, (1) $\psi(\alpha) = |\alpha|^n$ ($n = 1, 2, \ldots$), (2) $\psi(\alpha) = e^{-|1/\alpha|^p}$, (3) $\psi(\alpha) \equiv 0$ for $0 < \alpha < \alpha_0$.

Using the function $\psi(\alpha)$, whose rate of decrease determines the order of the mean zero of the function $f(t)$, we form the monotonically decreasing function $\alpha = \eta(r)$ $(0 < r < \infty)$ defined as the inverse of the monotonic function

$$r = -\frac{\ln \psi(\alpha)}{\alpha}. \tag{25}$$

A higher multiplicity of the zero t_0 of $f(t)$ obviously leads to a more slowly decreasing function $\eta(r)$.

THEOREM 1′. *Let the Fourier exponents of the almost-periodic function*

$$f(t) \sim \sum_{-\infty}^{\infty} a_k e^{i\mu_k t} \tag{26}$$

form the sequence $\Lambda = \{\lambda_k\}$.

If this function has a one-sided mean zero of order $\psi(\alpha)$ *at some point* t_0 *and if*

$$\varliminf_{R\to\infty} [\pi N_\Lambda(R) - R\eta(R) - \ln R] = -\infty, \tag{27}$$

then $f(t) \equiv 0$.

Thus if the sequence Λ is "not too dense," the function $f(t)$ can not have a mean zero of high order unless it is identically zero.

PROOF OF THE THEOREM. It can be assumed, without loss of generality, that $f(t)$ has a right zero of order not greater than $\psi(\alpha)$ at $t_0 = 0$. We consider the Laplace transform of $f(t)$, which is

$$\varphi(z) = \int_0^\infty e^{-zt} f(t)\,dt = \sum_{-\infty}^{\infty} \frac{a_n}{z - i\lambda_n} \tag{28}$$

and estimate the size of the function $\varphi(z)$ in the right half-plane.

In order to do this, we break up the integral in (28) into the sum of two integrals:

$$\varphi(z) = \int_0^\alpha e^{-zt} f(t)\,dt + \int_\alpha^\infty e^{-zt} f(t)\,dt.$$

For Re $z > 0$, we have

$$\left|\int_0^\alpha e^{-zt} f(t)\,dt\right| < \int_0^\alpha |f(t)|\,dt \leqslant \psi(\alpha) \text{ and } \left|\int_\alpha^\infty e^{-zt} f(t)\,dt\right| < \frac{M}{x} e^{-\alpha x}$$

$$\left(z = x + iy, \quad M = \sup_{0 < t < \infty} |f(t)|\right)$$

and so

$$|\varphi(z)| < \psi(\alpha) + \frac{M}{x} e^{-\alpha x}.$$

Setting $\alpha = \eta(x)$, we have $\psi(\alpha) = e^{-\alpha x} = e^{-x\eta(x)}$ (see (25)) and

$$|\varphi(z)| < \left(1 + \frac{M}{x}\right) e^{-x\eta(x)} \qquad (x > 0). \tag{29}$$

A direct estimate of the size of the Laplace integral in the left half-plane yields

$$|\varphi(z)| \leqslant \frac{M}{|x|} \quad \text{for} \quad x < 0. \tag{30}$$

If $\varphi(z)$ is not identically zero, then $\varphi(z) = z^m \varphi_1(z)$, where $m \geqslant -1$ is an integer and $\varphi_1(z)$ is a meromorphic function with $\varphi_1(0)$ neither zero nor infinite.

When we apply Jensen's formula to $\varphi_1(z)$, we obtain

$$N_\Lambda(R) - \int_0^R \frac{n_2(t)}{t}\,dt = \ln|\varphi_1(0)| - \frac{1}{2\pi}\int_0^{2\pi} \ln|\varphi_1(Re^{i\theta})|\,d\theta,$$

and from (29) and (30)

$$N_\Lambda(R) > \frac{1}{2\pi}\int_{-\frac{\pi}{2}}^{\frac{\pi}{2}} [R\cos\theta\,\eta(R\cos\theta) + \ln(R\cos\theta)]\,d\theta$$

$$+ \frac{1}{2\pi}\int_{\frac{\pi}{2}}^{\frac{3}{2}\pi} \ln|R\cos\theta|\,d\theta + O(1).$$

Since $\eta(R\cos\theta) > \eta(R)$, we have

$$N_\Lambda(R) > \frac{1}{\pi} R\eta(R) + \ln R + O(1).$$

It thus follows that $\varphi(z) \equiv 0$ if condition (27) is to be satisfied. The inverse Laplace transformation yields $f(t) = 0$, and the theorem is proved.

This theorem can be considered as a theorem on quasianalytic classes, assuming that a class of functions is defined by the sequence Λ in which the Fourier exponents of the functions lie. The proximity of two functions at a point t_0 is given by the mean zero of the difference of the functions at the point. The theorem means that if two functions of a class are too close together at any point, then these functions are identical.

If, in an ε-half-neighborhood of t_0, the function $f(t)$ is differentiable and

$$|f^{(n)}(t)| < m_n \qquad (t_0 \leqslant t \leqslant t_0 + \varepsilon,\ n = 1, 2, \ldots)$$

and if $f^{(n)}(t_0 + 0) = 0$ for $n = 0, 1, \ldots$, then Taylor's theorem yields

$$|f(t)| < \frac{m_n}{n!}(t - t_0)^n \qquad (t_0 \leqslant t \leqslant t_0 + \varepsilon)$$

and we have

$$\int_{t_0}^{t_0+\alpha} |f(t)|\,dt < \frac{m_n}{(n+1)!}\alpha^{n+1} \qquad (0 < \alpha < \varepsilon).$$

If

$$P(\alpha) = \min_n \frac{m_n}{(n+1)!}\alpha^{n+1},$$

then we can use $P(\alpha)$ for the function $\psi(\alpha)$. In this case, the function $f(t)$ will have a mean zero of order not less than $\psi(\alpha)$ at t_0.

Here Theorem 1′ implies that if the sequence $\{m_n\}$ does not increase too rapidly and the "density" $N_\Lambda(r)$ of the sequence $\Lambda = \{\lambda_n\}$ is not too high, then the infinitely differentiable, almost-periodic functions with Fourier exponents from Λ, satisfying the conditions

$$|f^{(n)}(x)| \leqslant m_n k^n \qquad (-\infty < x < \infty) \tag{31}$$

for any $k > 0$, will form the class of quasianalytic functions (Δ).

We also note that if a zero of a function is of exponential order ρ (see (21)) and if the convergence exponent of the sequence Λ is σ, then we have $N_\Lambda(R) < \eta R^{\sigma+\varepsilon}$ ($\varepsilon > 0$) asymptotically, and on some sequence of points $r_n \to \infty$ we will have $\eta(r_n) > r_n^{-1/(\rho+1)-\varepsilon}$.

Thus for $\sigma < \rho/(\rho + 1)$ condition (27) will be satisfied, and we obtain the following corollary.

COROLLARY. *If the exponential order ρ of a zero of the almost-periodic function* (26) *satisfies the condition*

$$\rho > \frac{\sigma}{1-\sigma},$$

where σ is the convergence exponent of the sequence Λ, then $f(t) \equiv 0$.

This corollary provides the direct transfer of S. Mandelbrojt's theorem to the case of almost-periodic functions.

REMARK 1. By a slight variation of the proof, we can extend this result to S_p-almost-periodic functions. The inequalities (29) and (30) must be altered somewhat, and the condition (27) must be replaced by

$$\varliminf_{R\to\infty}\left[\pi N_\Lambda(R) - R\eta(R) - \frac{1}{q}\ln R\right] = -\infty.$$

REMARK 2. If the point x_0 is a two-sided zero of order $\theta(\alpha)$, i.e.,

$$\int_{x_0}^{x_0+\alpha} |f(x)|\,dx \leqslant \theta(\alpha), \qquad \int_{x_0-\alpha}^{x_0} |f(x)|\,dx \leqslant \theta(\alpha),$$

then for $x \neq 0$ we have

$$|\varphi(z)| < \left(1 + \frac{M}{|x|}\right) e^{-|x|\,\eta(|x|)},$$

and the condition (27) is replaced by the condition

$$\varliminf_{R\to\infty}\left[N_\Omega(R) - \frac{2}{\pi} R\eta(R)\right] = -\infty.$$

We will show that the result given in the corollary is exact in the following sense.

Let the sequence Λ have a density satisfying the condition $0 < \sigma < 1$, let $\inf(\lambda_{k+1} - \lambda_k) = \delta > 0$, and let $-\lambda_k$ belong to the set Λ when λ_k is in this set. Then there is an almost-periodic function $f(t)$ with spectrum in Λ which has a zero of order $\rho = \sigma/(1-\sigma)$ at the origin.

To find this function, we set

$$\Delta(z) = z\prod_{n=1}^{\infty}\left(1 + \frac{z^2}{\lambda_n^2}\right).$$

From Theorem (25) in Chapter I, we have the asymptotic relation

$$\ln|\Delta(re^{i\theta})| \approx \frac{\pi r^\sigma}{\sin\frac{\pi\sigma}{2}}\cos\sigma\left(|\theta| - \frac{\pi}{2}\right), \tag{32}$$

valid in the whole plane with the possible exception of some E^0-set of values of r.

We consider, for $t > 0$, the inverse Laplace transform

$$f(t) = \frac{1}{2\pi i}\int_{-i\infty+\varepsilon}^{i\infty+\varepsilon} e^{tz}\frac{dz}{\Delta(z)},$$

and use Cauchy's theorem to obtain

$$f(t)=\frac{1}{2\pi i}\int\limits_{C(\varepsilon,r)} e^{zt}\frac{dz}{\Delta(z)}+\frac{1}{\pi}\int\limits_{-i\infty+\varepsilon}^{-ir+\varepsilon}+\int\limits_{ir+\varepsilon}^{i\infty+\varepsilon} e^{zt}\frac{dz}{\Delta(z)}\quad(r_1^2=r^2-\varepsilon^2), \tag{33}$$

where $C(\varepsilon, r)$ is the circular arc $|z| = r$, $\operatorname{Re} z \geqslant \varepsilon > 0$.

Using (32) in (33), we obtain

$$|f(t)| < Me^{rt-ar^{\sigma}}$$

($a = \cot(\sigma/2)$ and M is a constant).

Setting $r = (t/a\sigma)^{1/(1-\sigma)}$, we find that

$$|f(t)| < M\exp\left(-c_{\delta}t^{\frac{\sigma}{1-\sigma}}\right),$$

for some $c_\delta > 0$, and thus $f(t)$ has a right zero of order not less than $\rho = \sigma/(1-\sigma)$.

We now show that $f(t)$ is an almost-periodic function. For $t > 0$, the function $f(t)$ is the limit for $r \to \infty$ of the function

$$f_r(t)=\frac{1}{2\pi i}\int\limits_{\varepsilon-ir}^{\varepsilon+ir} e^{tz}\frac{dz}{\Delta(z)}.$$

An application of the theory of residues transforms this integral into

$$f_r(t)=\sum_{|\lambda_n|<r}\frac{1}{\Delta'(i\lambda_n)}e^{i\lambda_n t}+\int\limits_{C'(\varepsilon,r)} e^{zt}\frac{dz}{\Delta(z)},$$

where $C'(\varepsilon, r)$ is the arc of the circle $|z| = r$ with $\operatorname{Re} z \leqslant \varepsilon$. The last integral, as can easily be seen from (32), tends uniformly to zero on the ray $t \geqslant 0$ for $r \to \infty$ and $r \bar{\in} E^0$. Thus $f(t)$ is the uniform limit of the almost-periodic polynomials

$$\sum_{|\lambda_n|<r}\frac{1}{\Delta(i\lambda_n)}e^{i\lambda_n t}.$$

A similar method can be used for $t < 0$. Thus $f(t)$ is an almost-periodic function, whose spectrum is in Ω and which has a two-sided zero of order $\rho = \sigma/(1-\sigma)$ at the origin.

APPENDIX III

COMPLETENESS AND LINEAR INDEPENDENCE ON A FINITE INTERVAL OF SYSTEMS OF FUNCTIONS $\{e^{i\lambda_k x}\}$

1. Completeness Theorems

In Chapter IV, we studied the completeness in the complex plane of systems of functions $\{e^{i\lambda_k x}\}$ with $\{\lambda_k\}$ a sequence of complex numbers. Here we consider the completeness of $\{e^{i\lambda_k x}\}$ on some interval of the real axis. The most detailed results in this direction were obtained by Paley and Wiener in **[1]** and by N. Levinson **[1]**.

In what follows we will assume, with no loss of generality, that the system $\{e^{i\lambda_k x}\}$ is considered on the interval $(-\pi, \pi)$.

THEOREM 1 (N. LEVINSON). *For the system $\{e^{i\lambda_k x}\}$, in which the λ_k $(k = 1, 2, \ldots)$ are complex, to be incomplete in $L^p(-\pi, \pi)$ $(p \geqslant 1)$, it is necessary and sufficient that there exist an entire function $f(\lambda)$, zero for all λ_k $(k = 1, 2, \ldots)$, that can be expressed in the form*

$$f(\lambda) = \int_{-\pi}^{\pi} e^{i\lambda x}\psi(x)\,dx, \tag{1}$$

where $\psi(x) \in L^q(-\pi, \pi)$ $(1/p + 1/q = 1)$ (for $p = 1$, $\psi(x)$ is a bounded, measurable function).

PROOF. If the system $\{e^{i\lambda_k x}\}$ is not complete in $L^p(-\pi, \pi)$, then it follows from Riesz's theorem that there is a linear functional operating in the $L^p(-\pi, \pi)$ space, not identically zero, that is zero for all the functions of the system $\{e^{i\lambda_k x}\}$.

It follows from the general form of a functional in $L^p(-\pi, \pi)$ that there is a function $\psi(x)$ of $L^q(-\pi, \pi)$, not equivalent to zero, that is orthogonal to all functions of the system $\{e^{i\lambda_k x}\}$, i.e., such that

$$\int_{-\pi}^{\pi} e^{i\lambda_k x}\psi(x)\,dx = 0 \qquad (k = 1, 2, 3, \ldots).$$

Thus $f(\lambda)$ is zero at all the points λ_k $(k = 1, 2, \ldots)$.

REMARK 1. From the Paley-Wiener theorem (Appendix I, § 3) for $p = 2$, the fact that $f(\lambda)$ can be expressed in the form (1) is equivalent to the fact that $f(\lambda)$ is an entire function of exponential type $\sigma \leqslant \pi$ and belongs to $L^p(-\infty, \infty)$ on the real axis.

REMARK 2. If we assume that the system $\{e^{i\lambda_k x}\}$ is complete in $C(-\pi, \pi)$, i.e., that any function continuous on $[-\pi, \pi]$ can be uniformly approximated by a linear combination of functions of this system, then the incompleteness of the system is equivalent to the existence of an entire function $f(\lambda)$, equal to zero for all λ_k ($k = 1, 2, \ldots$), and expressible in the form

$$f(\lambda) = \int_{-\pi}^{\pi} e^{i\lambda x}\, d\sigma(x), \tag{2}$$

in which $\sigma(x)$ is a function of bounded variation on the interval $(-\pi, \pi)$.

This result follows directly from the general form of the linear functional in $C(-\pi, \pi)$.

REMARK 3. If μ is a zero of multiplicity $m > 1$ of a function $f(\lambda)$ of the form (1) or (2), then differentiating (1) (or (2)) with respect to λ we see that $\psi(x)$ (the differential $d\sigma(x)$) is orthogonal to all the functions

$$e^{i\mu x},\ xe^{i\mu x},\ \ldots,\ x^{m-1}e^{i\mu x}.$$

Conversely, the orthogonality of the function $\psi(x)$ (the differential $d\sigma(x)$) to all functions of this form implies that μ is a zero of multiplicity m of $f(\lambda)$.

In the sequel theorems will be given concerning the completeness of systems of the form $\{e^{i\lambda_k x}\}$, but the proofs can be immediately generalized to apply to wider classes of functions consisting of groups of the form

$$e^{i\lambda_k x},\ xe^{i\lambda_k x},\ \ldots,\ x^{m_k-1}e^{i\lambda_k x}.$$

Here each number λ_k is to be counted m_k times.

THEOREM 2 (N. LEVINSON [1]). *If the entire function $f(\lambda)$ can be expressed in the form* (1) *with $\psi(\lambda) \in L^q(-\pi, \pi)$* (*or in the form* (2)), *if λ_0 is a zero of $f(\lambda)$, and if μ is any complex number, then*

$$f_1(\lambda) = \frac{\lambda - \mu}{\lambda - \lambda_0} f(\lambda)$$

can also be expressed in the form (1) (*in the form* (2)).

PROOF. We prove the theorem for functions expressible in the form (2). We set

$$d\tau(x) = d\sigma(x) + i(\mu - \lambda_0)\, e^{-i\lambda_0 x} \int_{-\pi}^{x} e^{i\lambda_0 y}\, d\sigma(y)\, dx.$$

Multiplying both sides here by $e^{i\lambda x}$ and integrating from $-\pi$ to π, we obtain

$$\int_{-\pi}^{\pi} e^{i\lambda x}\, d\tau(x) = f(\lambda) + i(\mu - \lambda_0) \int_{-\pi}^{\pi} e^{i(\lambda-\lambda_0)x} \left(\int_{-\pi}^{x} e^{i\lambda_0 y}\, d\sigma(y) \right) dx.$$

Changing the order of integration and using the fact that

$$\int_{-\pi}^{\pi} e^{i\lambda_0 x}\, d\sigma(x) = 0,$$

we find that

$$\int_{-\pi}^{\pi} e^{i\lambda x}\, d\tau(x) = \frac{\lambda - \mu}{\lambda - \lambda_0} f(\lambda),$$

i.e., that

$$f_1(\lambda) = \int_{-\pi}^{\pi} e^{i\lambda x}\, d\tau(x).$$

The theorem has thus been proved. This theorem and Remark 3 lead to the following obvious corollary.

COROLLARY. *If the system* $\{e^{i\lambda_k x}\}$ *is complete in* $L^p(-\pi, \pi)$ *or in* $C(-\pi, \pi)$, *and if an arbitrary number n of functions are removed from this system and replaced by n other functions* $e^{i\mu_j x}$ $(j = 1, 2, \ldots, n)$ *where* $\mu_1, \mu_2, \ldots, \mu_n$ *are arbitrary complex numbers*[1] *not equal to any* λ_k, *then the new system will be complete in the same sense as the original system.*

This corollary enables us to assume, without loss of generality, that the set λ_k does not contain the number zero, i.e., that the system $\{e^{i\lambda_k x}\}$ does not contain the function identically equal to unity.

As usual, $n_\lambda(t)$ denotes the number of points λ_k in the circle $|z| \leqslant t$ and we write

$$N_\lambda(R) = \int_0^R \frac{n_\lambda(t)}{t}\, dt.$$

The function $N_\lambda(R)$ characterizes the density of the set $\{\lambda_k\}$.

THEOREM 3 (N. LEVINSON [1]). *If the sequence* $\{\lambda_k\}$ *of complex numbers is such that*

$$\overline{\lim_{R\to\infty}} \left\{ N_\lambda(R) - 2R + \frac{1}{p} \ln R \right\} > -\infty, \tag{3}$$

then the system $\{e^{i\lambda_k x}\}$ *is complete in* $L^p(-\pi, \pi)$ $(p > 1)$.

If (3) *is satisfied for* $p = \infty$, *then* $\{e^{i\lambda_k x}\}$ *is complete in* $C(-\pi, \pi)$.

PROOF. If the system is not complete in $L^p(-\pi, \pi)$, then there is an entire function $f(\lambda)$, equal to zero at all the points λ_k, that can be expressed in the form

$$f(\lambda) = \int_{-\pi}^{\pi} e^{i\lambda x} \psi(x)\, dx \qquad \left(\psi(x) \in L^q(-\pi, \pi),\ \frac{1}{p} + \frac{1}{q} = 1\right).$$

[1] We could choose $\mu_j = \lambda_k$ $(j = 1, 2, \ldots, j_k)$, but then we would have to consider the system consisting of the system $\{e^{i\mu_j x}\}$ and the functions $e^{i\mu_j x}, x^2 e^{i\mu_j x}, \ldots, x^{j_k} e^{i\mu_j x}$.

From this representation of the function, we obtain the inequality

$$|f(\lambda)| \leqslant \int_{-(\pi-\varepsilon)}^{\pi-\varepsilon} e^{|vx|}\,|\psi(x)|\,dx + \Big(\int_{-\pi}^{-\pi+\varepsilon} + \int_{\pi-\varepsilon}^{\pi}\Big) e^{|vx|}\,|\psi(x)|\,dx$$

$$(\lambda = u + iv)$$

and, using the Hölder inequality, we have

$$|f(\lambda)| \leqslant e^{(\pi-\varepsilon)|v|}\,|pv|^{-\frac{1}{p}} \Big(\int_{-\pi+\varepsilon}^{\pi-\varepsilon} |\psi(x)|^q\,dx\Big)^{\frac{1}{q}}$$

$$+ e^{\pi|v|}\,|pv|^{-\frac{1}{p}} \Big[\Big(\int_{-\pi}^{-\pi+\varepsilon} + \int_{\pi-\varepsilon}^{\pi}\Big)|\psi(x)|^q\,dx\Big]^{\frac{1}{q}},$$

and finally

$$|f(\lambda)| < Ae^{\pi|v|}\,|v|^{-\frac{1}{p}}\,(e^{-\varepsilon|v|} + \delta), \tag{4}$$

where A is a constant and

$$\delta = \Big[\Big(\int_{-\pi}^{-\pi+\varepsilon} + \int_{\pi-\varepsilon}^{\pi}\Big)|\psi(x)|^q\,dx\Big]^{\frac{1}{q}}.$$

It is obvious that $\delta \to 0$ when $\varepsilon \to 0$. (This is also true when $p = 1$.)

When we apply Jensen's theorem to $f(\lambda)$ and use (4), we find that

$$N_f(R) \leqslant 2R - \frac{1}{p}\ln R + \frac{1}{2\pi}\int_0^{2\pi} \ln|e^{-\varepsilon|v|} + \delta|\,d\theta + O(1). \tag{5}$$

Since all the points λ_k are zeros of $f(\lambda)$, we have $N_f(R) \geqslant N_\lambda(R)$. In addition, if $\varepsilon > 0$ is small enough and $R > 0$ large enough, the third term on the right in (5) can be made less than any arbitrarily large negative number, and so

$$\lim_{R\to\infty}\Big\{N_\lambda(R) - 2R + \frac{1}{p}\ln R\Big\} = -\infty.$$

Hence the theorem on closure in $L^p(-\pi, \pi)$ follows. The theorem for $C(-\pi, \pi)$ can be obtained somewhat more simply.

COROLLARY. *If*

$$|\lambda_n| \leqslant |n| + \frac{1}{2p} \qquad (\pm n = 0,\ 1,\ 2,\ \ldots),$$

then the system $\{e^{i\lambda_n x}\}$ *is complete*[2] *in* $L^p(-\pi, \pi)$.

This result is easily obtained from Theorem 3. We omit the proof.

Many different criteria for completeness follow from Theorem 1. To derive these criteria, we note first of all that $f(\lambda)$, expressed in the form (1) or (2), is a

---

[2] This theorem was proved by N. Levinson [1].

function of completely regular growth and belongs to the class A (Chapter V, Theorem 11). The set of its zeros μ_j ($j = 1, 2, \ldots$) must therefore satisfy the following conditions:

1. *The set* $\{\mu_j\}$ *has an angular density with exponent* $\rho = 1$, *and* $\Delta(\theta)$ *is a piecewise-constant function with two different jumps, not exceeding unity, at* 0 *and* π.

2. $\sum_{k=0}^{\infty}\left|\operatorname{Im}\frac{1}{\mu_k}\right| < \infty.$

3. *The limit*

$$\lim_{R\to\infty}\sum_{|\mu_k|<R}\mu_k^{-1}$$

exists.

We call such a set "*smooth.*" For example, the set of even real numbers, having a density, is smooth. A set $\{\lambda_k\}$ will be called N-smooth if the magnitude of the jumps of $\Delta(\theta)$ is equal to $N \leqslant 1$.

The following theorem is obvious.

THEOREM 4. *If the set of points* $\{\lambda_k\}$ *can not be included in any N-smooth set, then the system* $\{e^{i\lambda_k x}\}$ *is complete in* $C(-N\pi, N\pi)$.

Various sufficient conditions for completeness can be obtained from this result.

$$\lim_{\xi\to 1}\overline{\lim_{r\to\infty}}\frac{n_1(\xi r)-n_1(r)}{(\xi-1)r} > 1 \tag{6}$$

where $n_1(r)$ *is the number of the points* λ_k *inside the sector* $|z| \leqslant r$, $\arg z \leqslant \gamma < \pi$, *then the system* $\{e^{i\lambda_k x}\}$ *is complete in* $C(-\pi, \pi)$.

In fact, for every smooth sequence $\{\mu_k\}$, from condition 1 we have

$$\lim_{\xi\to 1}\overline{\lim_{r\to\infty}}\frac{m(\xi r)-m(r)}{(\xi-1)r} \leqslant 1, \tag{7}$$

where $m(r)$ is the number of points μ_k in the sector $|z| \leqslant r$, $|\arg z| \leqslant \gamma < \pi$.

It is evident that condition (7) also holds for any partial sequence taken from $\{\mu_k\}$ and so when condition (6) is satisfied the sequence $\{\lambda_k\}$ can not be included in a smooth set.

(b) *If* $|\arg\lambda_k - \pi/2| \leqslant \tau < \pi/2$ *and*

$$\sum_{k=1}^{\infty}\frac{1}{|\lambda_k|} = \infty, \tag{8}$$

then the system $\{e^{i\lambda_k x}\}$ *is complete in* $C(-L, L)$ *for any* $L > 0$.

In fact, any subsequence of an N-smooth sequence must satisfy condition (2). But

$$\sum_{k=1}^{\infty}\left|\operatorname{Im}\frac{1}{\lambda_k}\right| \geqslant \frac{1}{\cos\tau}\sum_{k=1}^{\infty}\frac{1}{|\lambda_k|} = \infty.$$

[8] The left-hand side of this inequality is called by Pólya the maximal density.

Consequently the sequence $\{\lambda_k\}$ is not included in a smooth sequence, and the system $\{e^{i\lambda_k x}\}$ is complete. For purely imaginary λ_k this theorem coincides with the Müntz theorem concerning the completeness of systems of the form x^{λ_k} on any segment $a \leqslant x \leqslant b$, $b > a > 0$.

(c) *If $\{\lambda_k\}$ is the set of zeros of an entire function $\varphi(\lambda)$ of exponential type, the width of whose indicator-diagram in the direction of the Oy axis is not less than $2\pi(h_\varphi(\pi/2) + h_\varphi(-\pi/2) \geqslant 2\pi)$, and if on any straight line parallel to the real axis* (Im $\lambda = a$) *the inequality*

$$|\varphi(u+ia)| > \frac{m}{(1+|u|)^{\frac{1}{p}}} \qquad (-\infty < u < \infty, \quad p \geqslant 2) \tag{9}$$

is satisfied for some $m > 0$, then the system $\{e^{i\lambda_k x}\}$ is complete in $L^p(-\pi, \pi)$.

PROOF. If the system of functions $\{e^{i\lambda_k x}\}$ is not complete, then there is an entire function $f(\lambda)$, equal to zero at the points λ_k $(k = 1, 2, \ldots)$, that can be expressed in the form (1). It follows from this representation that $f(\lambda)$ is of exponential type not greater than π and is of completely regular growth.

The quotient

$$\chi(\lambda) = \frac{f(\lambda)}{\varphi(\lambda)}$$

is a function of exponential type, satisfying on the straight line $v = a$ ($v =$ Im λ) the inequality

$$|\chi(u+ia)| < M(1+|u|)^{\frac{1}{p}}, \tag{10}$$

in which M is a constant. From Theorem 11, Chapter V, the function $\chi(\lambda)$ is of completely regular growth and its indicator diagram is a *segment I_χ of the imaginary axis.* The indicator diagram of $f(\lambda)$ is the sum of the indicator diagrams of the functions $\varphi(\lambda)$ and $\chi(\lambda)$ (Chapter III, § 3, Theorem 4). Since the width of the indicator diagram of $\varphi(\lambda)$ is not less than 2π and the type of $f(\lambda)$ does not exceed π, therefore the length of the segment I_χ is zero and $\chi(\lambda)$ is of exponential type zero. If $\chi(\lambda)$ has the zero μ, then from (10) the function

$$\chi(\lambda)(\lambda-\mu)^{-1}$$

is bounded on the straight line parallel to the real axis and is thus constant. Since $\chi(\lambda)(\lambda - \mu)^{-1}$ tends to zero on this line, we have $\chi(\lambda) \equiv 0$, and so $f(\lambda) \equiv 0$. If $\chi(\lambda)$ has no zeros, then it is constant and

$$f(\lambda) = c\varphi(\lambda),$$

where c is a constant. From (9) we obtain

$$|f(u+ia)| > \frac{cm}{(1+|u|)^{\frac{1}{p}}} \qquad (-\infty < u < \infty),$$

and so

$$\int_{-\infty}^{\infty} |f(u+ia)|^p \, du = \infty.$$

On the other hand, for $p \geqslant 2$ the representation (1) of $f(\lambda)$ yields[4]

$$\int_{-\infty}^{\infty} |f(u+ia)|^p \, du \leqslant \left(\int_{-\pi}^{\pi} e^{aqv} |\psi(v)|^2 \, dv \right)^{p-1} < \infty.$$

Thus a function $f(\lambda)$ with the properties described in Theorem 1 does not exist, and so the system $\{e^{i\lambda_k x}\}$ is complete in $L^p(-\pi, \pi)$.

A similar result also holds for the completeness of the system $\{e^{i\lambda_k x}\}$ in $C(-\pi, \pi)$. We need only replace condition (9) by the condition

$$|\varphi(u+ia)| \geqslant m > 0, \qquad \overline{\lim_{|u| \to \infty}} |\varphi(u+ia)| = \infty$$
$$(-\infty < u < \infty)$$

in (c). In fact, a repetition of the preceding reasoning yields

$$f(\lambda) = c\varphi(\lambda),$$

and so the condition $\overline{\lim}_{|u| \to \infty} |(u \pm ia)| = \infty$ is incompatible with the representation (2) of $f(\lambda)$. Thus, from Remark 2 following Theorem 1, the system $\{e^{i\lambda_k x}\}$ is complete in $C(-\pi, \pi)$.

If the condition $h_\varphi(-\pi/2) + h_\varphi(\pi/2) \geqslant 2\pi$ in (c) is replaced by a somewhat stronger condition, then condition (9) becomes superfluous. More precisely:

(d) *If $\{\lambda_k\}$ is the set of zeros of an entire function $\varphi(\lambda)$ of exponential type and*

$$\underline{\lim_{|v| \to \infty}} |\varphi(iv)| e^{-\pi|v|} > 0, \tag{11}$$

then the system $\{e^{i\lambda_k x}\}$ is complete in $L^p(-\pi, \pi)$ $(p \geqslant 1)$.

Proof. If the system $\{e^{i\lambda_k x}\}$ is not complete in $L^p(-\pi, \pi)$, then there is an entire function $f(\lambda)$, zero at all the points λ_k $(k = 1, 2, \ldots)$ and expressible in the form (1), which consequently satisfies the inequality (4). From this inequality and condition (11), it follows that the entire function

$$\chi(\lambda) = \frac{f(\lambda)}{\varphi(\lambda)}$$

of exponential type is bounded on the imaginary axis, and from Corollary 2 of Theorem 6 (§ 4, Chapter V) it is of completely regular growth and its indicator diagram is a segment I_χ of the real axis. The indicator diagram of $f(\lambda)$ is the sum of the indicator diagrams of $\varphi(\lambda)$ and $\chi(\lambda)$. Therefore the length of I_χ is not greater than the width of the indicator diagram of $f(\lambda)$ in the direction of

[4] See Zygmund [1], p. 311.

the real axis and is thus zero. Thus $\chi(\lambda)$ is of zero type, and since it is bounded on the imaginary axis it must be constant. But then $f(\lambda) = c\varphi(\lambda)$ and

$$\overline{\lim_{|v|\to\infty}} |f(iv)| e^{-\pi|v|} > 0,$$

which contradicts (4). Hence the completeness of the system $\{e^{i\lambda_k x}\}$ in $L^p(-\pi, \pi)$ follows.

2. Minimality (Strong Linear Independence) of the System $\{e^{i\lambda_k x}\}$

Of particular interest are complete systems that do not contain superfluous functions, i.e., systems such that one member can not be approximated by linear combinations of the others. In other words, it is a question of systems $\{e^{i\lambda_k x}\}$ for which it follows from the inequality

$$\lim_{n\to\infty} \left\| \sum_{k=1}^{n} a_k^{(n)} e^{i\lambda_k x} \right\| = 0$$

that

$$\lim_{n\to\infty} a_k^{(n)} = 0.$$

Here the $L^p(-\pi, \pi)$ norm will be used. A system of functions with this property (not necessarily complete) will be called *minimal*.

THEOREM 5 (PALEY AND WIENER [1]). *For a system of functions $\{e^{i\lambda_k x}\}$ to be minimal, it is necessary and sufficient that there be an entire function $\varphi(\lambda)$ of exponential type $p \leqslant \pi$, equal to zero at all the points $\{\lambda_k\}$, such that*

$$\frac{|\varphi(\lambda)|}{\sqrt{1+\lambda^2}} \in L^2(-\pi, \pi).^{4a} \tag{12}$$

PROOF. If such a function $\varphi(\lambda)$ exists, then a system of functions $G_k(\lambda) = s_k!\varphi(\lambda)/\varphi^{(s_k)}(\lambda_k)(\lambda - \lambda_k)^{s_k}$ can be constructed, where s_k is the multiplicity of the zero λ_k, each function belonging to the class $L^2(-\pi, \pi)$. Setting

$$h_k(x) = \frac{1}{2\pi} \int_{-\infty}^{\infty} G_k(\lambda) e^{-i\lambda x} d\lambda,$$

we obtain a system $\{h_k(x)\}$ of functions of $L^p(-p, p)$ which are zero outside the interval $(-p, p)$. These functions form a biorthogonal system, i.e.,

$$\int_{-p}^{p} h_k(x) e^{i\lambda_j x} dx = \begin{cases} 0 & \text{for } k \neq j, \\ 1 & \text{for } k = j. \end{cases}$$

Now let the sequence of sums

$$\sum_{k=1}^{n} a_j^{(n)} e^{i\lambda_j x}$$

[4a] Similar conditions can be given for systems

$$\{e^{i\lambda_k x}, xe^{i\lambda_k x}, \ldots, x^{s_k - 1}e^{i\lambda_k x}\}.$$

tend to zero in $L^2(-\pi, \pi)$. If we multiply these sums by $h_k(x)$ and integrate from $-\pi$ to π, we find that $\lim_{n\to\infty} a_j^{(n)} = 0$.

Conversely, if the system $\{e^{i\lambda_k x}\}$ is minimal, then the function $e^{i\lambda_1 x}$ is at a positive distance from the linear span of the remaining functions, and thus there is a functional, equal to zero for all these remaining functions, and equal to one for $e^{i\lambda_1 x}$.

By the general Riesz theorem this functional is specified by some function $h(x) \in L^2(-\pi, \pi)$. We have

$$\int_{-\pi}^{\pi} e^{i\lambda_k x} h(x)\, dx = \begin{cases} 0 & k \neq 1, \\ 1 & k = 1. \end{cases}$$

The entire function

$$\varphi(\lambda) = (\lambda - \lambda_1) \int_{-\pi}^{\pi} e^{i\lambda x} h(x)\, dx$$

is zero for all the λ_k $(k = 1, 2, \ldots)$ and $\varphi(\lambda)(\lambda - \lambda_1)^{-1} \in L^2(-\infty, \infty)$. The theorem is proved.

THEOREM 6. *For a system $\{e^{i\lambda_k x}\}$ to be complete and minimal in $L^2(-\pi, \pi)$, it is necessary and sufficient that the following conditions hold:*

1° *the limit*

$$\varphi(\lambda) = \lim_{R\to\infty} \prod_{|\lambda_k| < R} \left(1 - \frac{\lambda}{\lambda_k}\right)$$

exists;

2° *$\varphi(\lambda)$ is an entire function of exponential type with indicator diagram of width 2π;*

3° $$|\varphi(\lambda)|(1 + |\lambda|)^{-1} \in L^2(-\infty, \infty);$$

4° *if $\chi(\lambda)$ is any entire function of exponential type zero, then*

$$\varphi(\lambda)\chi(\lambda) \overline{\in} L^2(-\infty, \infty).$$

PROOF. We first prove the necessity of the conditions 1°-4°. From Theorem 5, it follows from the minimality of the system $\{e^{i\lambda_k x}\}$ that there is an entire function $\varphi(\lambda)$ of exponential type $p \leqslant \pi$, equal to zero at all the points $\{\lambda_k\}$ and satisfying condition (12).

Without loss of generality, we may assume that $h_\varphi(\pi/2) = h_\varphi(-\pi/2)$. If the degree p of $\varphi(\lambda)$ is less than π, then the function

$$\varphi(\lambda) \frac{\sin(\pi - p)\lambda}{\lambda} \in L^2(-\infty, \infty)$$

and has degree π. Hence, from Remark 1 to Theorem 1, if we adjoin to the system $\{e^{i\lambda_k x}\}$ all the functions of the form $e^{i\mu x}$ with μ any nonzero zero of

$\sin(\pi - p)\lambda$, the resulting system is complete. Since the system $\{e^{i\lambda_k x}\}$ is complete, this is impossible, and thus $p = \pi$, i.e., the width of the indicator diagram of $\varphi(\lambda)$ is 2π.

Condition 3° follows from Theorem 5 and the minimality of the system $\{e^{i\lambda_k x}\}$.

The function $\varphi(\lambda)$ has no zeros different from λ_k. If this were not true, further exponential functions could be adjoined to the system $\{e^{i\lambda_k x}\}$ without destroying the minimality of the system. But this contradicts the completeness of the system $\{e^{i\lambda_k x}\}$.

From Theorem 11 of Chapter V, it follows that when condition 3° is satisfied for a function $\varphi(\lambda)$ of exponential type, this function is of completely regular growth, belongs to the class A, has an indicator diagram that is a segment of the imaginary axis, and can be expressed in the form

$$\varphi(\lambda) = ce^{il\lambda} \lim_{R\to\infty} \prod_{|\lambda_k|<R} \left(1 - \frac{\lambda}{\lambda_k}\right).$$

With no loss of generality, it can be assumed that $l = 0$, since if $\varphi(\lambda)$ satisfies conditions 1°-4°, then $e^{-il\lambda}\varphi(\lambda)$ satisfies the same conditions.

It only remains to deal with condition 4°. If for any function $\chi(\lambda)$ of genus zero we have

$$\varphi(\lambda)\chi(\lambda) \in L^2(-\infty, \infty),$$

then we obtain a system that is not complete in $L^2(-\pi, \pi)$ when we adjoin the exponentials $\{e^{i\mu_k x}\}$ to the system $\{e^{i\lambda_k x}\}$ which is originally complete in $L^2(-\pi, \pi)$. Here the μ_k are the zeros of $\chi(\lambda)$. But this is not possible. Such a function $\chi(\lambda)$ therefore does not exist. The necessity of conditions 1°-4° has thus been proved.

For the proof of sufficiency, we note that by Theorem 5 the minimality follows from conditions 1°-3°. Moreover, it follows from these conditions that $\varphi(\lambda)$ is of completely regular growth and that its indicator diagram is a segment of the imaginary axis.

If the system $\{e^{i\lambda_k x}\}$ is not complete in $L^2(-\infty, \infty)$, then, from Remark 1 following Theorem 1, it follows that there is an entire function $f(\lambda)$ of exponential type p not greater than π which belongs to the class $L^2(-\infty, \infty)$ and which is zero at all the points $\{\lambda_k\}$. With no loss of generality it may be assumed that its indicator diagram is symmetric about the real axis.

The quotient

$$\chi(\lambda) = \frac{f(\lambda)}{\varphi(\lambda)} \tag{13}$$

is an entire function of genus zero, and $\varphi(\lambda)\chi(\lambda) \in L^2(-\infty, \infty)$. But this contradicts condition 4°. The system $\{e^{i\lambda_k x}\}$ is therefore complete and minimal in $L^2(-\infty, \infty)$. The theorem has thus been proved.

From Theorem 6, which is a very general theorem, various conditions of a more special type can be obtained for the completeness and minimality of a system $\{e^{i\lambda_k x}\}$.

It will be shown in Appendix VI (Theorem 7) that *if*

$$\lambda_k = k + \psi(k) \quad (\pm k = 0, 1, 2, \ldots), \tag{14}$$

where $\psi(z)e^{-i\pi z}$ is an entire function of finite degree $\sigma \leqslant \pi$ and has a bounded absolute value on the real axis, then the entire function of exponential type

$$\varphi(\lambda) = \lim_{R\to\infty} \prod_{|\lambda_k|<R} \left(1 - \frac{\lambda}{\lambda_k}\right)$$

satisfies the inequality

$$0 < m_h < |\varphi(x \pm ih)| < M_h \quad (-\infty < x < \infty), \tag{15}$$

where $h >$ sup $|\psi(k)|$ and M_h and m_h are positive numbers.

It easily follows from this theorem that in this case the system $\{e^{i\lambda_k x}\}$ is complete and minimal in $L^2(-\pi, \pi)$.

Conditions 1°, 2°, and 3° are obviously satisfied, and we need only verify condition 4°. Let $\chi(\lambda)$ be an entire function of zero type, and let

$$\varphi(u)\chi(u) \in L^2(-\infty, \infty).$$

It follows from the Paley-Wiener representation of an entire function of exponential type belonging to $L^2(-\infty, \infty)$ that

$$\varphi(u \pm ih)\chi(u \pm ih) \in L^2(-\infty, \infty).$$

If the left inequality in (15) is used, then it follows that $\chi(u \pm ih) \in L^2(-\infty, \infty)$, and so $\chi(u) \in L^2(-\infty, \infty)$.

The function

$$\psi(u) = \int_0^u \chi^2(t)\, dt$$

is of exponential type zero and is bounded on the real axis. Thus, from the inequality of S. N. Bernšteĭn, it follows that $\psi'(u) \equiv 0$, i.e., $\chi(\lambda) \equiv 0$. Condition 4° is thus satisfied and the system $\{e^{i\lambda_k x}\}$ is complete and minimal.

It can also be shown that $\{e^{i\lambda_k x}\}$ is complete and minimal in $L^2(-\pi, \pi)$ if $\psi(z)$ satisfies the same conditions as in (14) and $|\lambda_k - k - \psi(k)| < \frac{1}{4}$. We do not prove this result here.

It is interesting to compare some of the results obtained in this appendix with similar results from Appendix II. From Theorem 5, Jensen's formula, and the inequality (4), it is easy to obtain the following result:

If the system $\{e^{i\lambda_k x}\}$ is minimal in the interval $(-\pi, \pi)$, then

$$\lim_{R\to\infty} \left[N_\Lambda(R) - 2R - \frac{1}{q}\ln R\right] = -\infty. \tag{16}$$

In Theorem 1 of Appendix II it was established that *if*

$$\underline{\lim}_{R\to\infty}\left[N_\Lambda(R)-2R-\frac{1}{q}\ln R\right]=-\infty \tag{17}$$

and if the sequence of functions

$$P_n(x)=\sum_{k=1}^{p_n} a_k^{(n)}e^{i\lambda_k x}$$

converges in S^p to a function equal to zero on the segment $(-\pi,\pi)$, then $\lim_{n\to\infty} a_k^{(n)}=0$.

The condition (16) is *necessary* if the L^p convergence of $P_n(x)$ to zero in the interval $(-\pi,\pi)$ is to imply $\lim_{n\to\infty} a_k^{(n)}=0$.

The condition (17) is *sufficient* that the S^p convergence of the sequence $P_n(x)$ to a function equal to zero in $(-\pi,\pi)$ should imply $\lim_{n\to\infty} a_k^{(n)}=0$. We note further that from Theorem 3 the condition

$$\overline{\lim_{R\to\infty}}\left[N_\Lambda(R)-2R-\frac{1}{q}\ln R+\ln R\right]>-\infty$$

is a sufficient condition for the completeness of the system $\{e^{i\lambda_k x}\}$ in $L^p(-\pi,\pi)$.

APPENDIX IV

APPLICATION OF UNIQUENESS THEOREMS TO SOME PROBLEMS IN THE THEORY OF DIFFERENTIAL EQUATIONS

1. Completeness of a System of Solutions of a Differential Equation of Second Order in Regions in the Complex Plane

In the investigation of the boundary-value problem

$$\left.\begin{aligned} y'' - q(x)\,y + \lambda^2 y = 0, \\ y'(0) - hy(0) = 0, \quad y'(\pi) + Hy(\pi) = 0, \end{aligned}\right\} \tag{1}$$

in the theory of differential equations, where $q(x)$ is a real function integrable in $(0, \pi)$, an essential role is played by the completeness of the system of solutions $\varphi(x, \lambda_k)$ of this problem in the interval $(0, \pi)$.

We will assume that in the interval $(0, \pi)$ the function $q(x)$ takes values that are the same as those of some entire function $q(z)$ in the same interval. Then the general solution of equation (1) is an entire function of z, and so the solutions $\alpha(z, \lambda_k)$ of the boundary-value problem are also entire functions of z.

We propose to investigate the completeness of the system of solutions $\alpha(z, \lambda_k)$ of this boundary-value problem in a region of the complex plane (in the simplest case, when $y(0) = y(\pi) = 0$). We will also investigate the completeness of the system of solutions $e(z, \lambda_k)$ of equation (1) for $\lambda = \lambda_k$ with the initial conditions $e(0, \lambda_k) = 1$, $e'(0, \lambda_k) = i\lambda_k$, where λ_k is a sequence of complex numbers. For $q(x) = 0$, these conditions determine solutions of the form $e^{i\lambda_k x}$. The results given here are closely related to the theorems concerning completeness given in § 7 of Chapter IV.

For this investigation, we will use "transformation operators" which convert the function $e^{i\lambda x}$ into a solution $\varphi(x, \lambda)$ of the differential equation (1).

These operators were introduced by Delsarte in **[1]** and were subsequently used in several applications in the work of B. M. Levitan [3] and A. Ja. Povzner **[1]**. The latter gave a construction of these operators using Riemann functions. A detailed study of these operators and their systematic application in the theory of differential equations was carried out by V. A. Marčenko **[1]**.

To obtain the transformation operators, we follow A. Ja. Povzner and consider the partial differential equation

$$u''_{yy} - q(y)\,u = u''_{xx}. \tag{2}$$

The solution of this equation for the initial conditions

$$u\big|_{y=0} = \varphi(x), \qquad u'_y\big|_{y=0} = \psi(x) \tag{3}$$

and $y > 0$ is given by[1]

$$u(x, y) = \frac{\varphi(x+y) + \varphi(x-y)}{2} + \frac{1}{2}\int_{x-y}^{x+y} [\psi(\xi)\, v(\xi, 0, x, y) - \varphi(\xi)\, v'_\eta(\xi, 0, x, y)]\, d\xi, \tag{4}$$

where $v(\xi, \eta, x, y)$ is the Riemann function of equation (2), i.e., the function that is the solution of the equation

$$u''_{\eta\eta} - u''_{\xi\xi} - q(\eta)\, u = 0$$

inside the triangle $\eta > 0$, $\xi + \eta < x + y$ and $\xi - \eta > x - y$ for any fixed x, y $(y > 0)$, and equal to unity for $\xi + \eta = x + y$ and for $\xi - \eta = x - y$.

If $q(x)$ is integrable on $(0, \pi)$, then the Riemann function $v(\xi, \eta, x, y)$ and its partial derivatives are continuous. Starting from (4), we construct the operator that transforms a solution of the equation $y'' + \lambda^2 y = 0$ into a solution of equation (1).

Let $\alpha_h(x, \lambda)$ be the solution of the equation

$$\alpha'' - q(x)\,\alpha + \lambda^2\alpha = 0 \tag{5}$$

satisfying the initial conditions

$$\alpha(0, \lambda) = 1, \qquad \alpha'(0, \lambda) = h, \tag{6}$$

where h is an arbitrary complex number. The function $u(x, y, \lambda) = \alpha_h(y, \lambda) \cos \lambda x$ is the solution of (2) for the initial conditions

$$\varphi(x) = \cos \lambda x, \quad \psi(x) = h \cos \lambda x.$$

Substituting these results in (4) and setting $x = 0$, we obtain for the function $\alpha_h(y, \lambda)$ the representation

$$\alpha_h(y, \lambda) = \cos \lambda y + \frac{1}{2}\int_{-y}^{y} [v(\xi, 0, 0, y)\, h - v'_\eta(\xi, 0, 0, y)] \cos \lambda\xi\, d\xi.$$

This equation can obviously be transformed into

$$\alpha_h(x, \lambda) = \cos \lambda x + \int_0^x K_h(x, t) \cos \lambda t\, dt. \tag{7}$$

[1] See, for example, A. N. Tihonov and A. A. Samarskiĭ, *Equations of mathematical physics*, Gostehizdat, 1953, pp. 127–130.

The Volterra operator corresponding to

$$K[f]=f(x)+\int_0^x K_h(x, t)f(t)\,dt$$

will be called the transformation operator. This operator transforms the function $\cos\lambda x$ *into a solution of equation* (5) *for the initial conditions* (6).

Denoting the solution of equation (5) for $h=0$ by $c(x, \lambda)$, from (7) we obtain

$$c(x, \lambda)=\cos\lambda x-\frac{1}{2}\int_{-x}^{x} v'_\eta(\xi, 0, 0, x)\cos\lambda\xi\,d\xi. \tag{8}$$

If $s(x, \lambda)$ is the solution of (5) for the initial conditions $s(0, \lambda)=0$ and $s'(0, \lambda)=\lambda$, we obtain for $s(y, \lambda)\sin\lambda x$ the expression

$$s(y, \lambda)\sin\lambda x=+\frac{\lambda}{2}\int_{x-y}^{x+y} v(\xi, 0, x, y)\sin\lambda\xi\,d\xi.$$

Differentiating both sides of this relation with respect to x and setting $x=0$, we obtain

$$s(y, \lambda)=\sin\lambda y+\frac{1}{2}\int_{-y}^{y} v'_x(\xi, 0, 0, y)\sin\lambda\xi\,d\xi. \tag{9}$$

We note that the solution $e(x, \lambda)$ of (5) that satisfies the conditions $e(0, \lambda)=1$, $e'(0, \lambda)=i\lambda$ can be written $e(x, \lambda)=c(x, \lambda)+is(x, \lambda)$, and then deduce from (8) and (9) that

$$e(x, \lambda)=e^{i\lambda x}+\int_{-x}^{x}[K_1(x, \xi)e^{i\lambda\xi}+K_2(x, \xi)e^{-i\lambda\xi}]\,d\xi,$$

where $K_1(x, \xi)=v'_x(\xi, 0, 0, x)-v'_\eta(\xi, 0, 0, x)$ and $K_2(x, \xi)=-v'_x(\xi, 0, 0, x)-v'_\eta(\xi, 0, 0, x)$. Finally, writing $K(x, \xi)=K_1(x, \xi)+K_2(x, -\xi)$, we obtain

$$e(x, \lambda)=e^{i\lambda x}+\int_{-x}^{x} K(x, \xi)e^{i\lambda\xi}\,d\xi. \tag{10}$$

It is obvious from (10) that

$$c(x, \lambda)=\cos\lambda x+\int_{-x}^{x} K(x, \xi)\cos\lambda\xi\,d\xi$$

and

$$s(x, \lambda)=\sin\lambda x+\int_{-x}^{x} K(x, \xi)\sin\lambda\xi\,d\xi.$$

We have thus arrived at

THEOREM 1.[2] *If the function $q(x)$ in (1) is integrable on the interval $(0, \pi)$, then there exists a kernel $K(x, \xi)$, continuous for $0 < x, \xi < \pi$, for which the Volterra operator*

$$A[\varphi(x)] = \varphi(x) + \int_{-\infty}^{\infty} K(x, \xi)\varphi(\xi)\,d\xi \tag{11}$$

transforms every solution $y(x)$ of

$$y'' + \lambda^2 y = 0$$

into a solution of (5) satisfying the same initial conditions as $y(x)$ for $x = 0$.

The Volterra operator (11) will be called a *transformation operator* (the same as the operator (7)). For our purposes, the following fact is fundamental. If the values of $q(x)$ in (5) are the values of an entire function $q(z)$ in the interval $(0, \pi)$, then *the kernel $K(x, \xi)$ of the operator $A[\varphi(x)]$ is an entire function of both variables.*[2a]

To prove this, it is sufficient to show that the Riemann function $v(\xi, \eta, x, y)$ of equation (2) is an entire function of all four variables.

The change of variables $t = x + y$, $s = x - y$ reduces equation (2) to

$$\frac{\partial^2 u}{\partial t\,\partial s} = -\frac{1}{4} q\left(\frac{t-s}{2}\right) u,$$

and the Riemann function $v(\xi, \eta, x, y)$ is obtained if we make the substitution $t = \xi + \eta$, $s = \xi - \eta$, $\tau = x + y$, $\sigma = x - y$ in the solution of the integral equation

$$u = 1 - \frac{1}{4}\int_{\tau}^{t}\int_{\sigma}^{s} q\left(\frac{t-s}{2}\right) u(t, s)\,dt\,ds. \tag{12}$$

Solving this Volterra equation by the iterative method, we obtain

$$u = 1 + u_1 + u_2 + \dots + u_n + \dots, \tag{13}$$

where

$$u_n(\sigma, \tau, s, t) = -\frac{1}{4}\int_{\tau}^{t}\int_{\sigma}^{s} q\left(\frac{t-s}{2}\right) u_{n-1}(\sigma, \tau, s, t)\,ds\,dt.$$

If $|\frac{1}{4}q((t-s)/2)| < M$ in the region $|t - \tau| \leqslant a$, $|s - \sigma| \leqslant b$, then

$$|u_n(\sigma, \tau, s, t)| < M\frac{|s-\sigma|^n |t-\tau|^n}{n!^2}. \tag{14}$$

If $q(z)$ is an entire function, then $u_n(\sigma, \tau, s, t)$ will obviously be an entire function of all its variables, and thus from (14) we see that the series (13) converges to an entire function of σ, τ, s, and t. In this case the kernel $K(z, \xi)$ is therefore an entire function of z and ξ.

---

[2] V. A. Marčenko [1].

[2a] Leont'ev [8] showed that if $q(z)$ is not entire, but only holomorphic in a convex set G, the kernel $K(z, \zeta)$ of the transformation operator is holomorphic in the domain $z \in G$, $\zeta \in G$.

The resolvent of $K(x, \xi)$, which is

$$R(x, \xi, -1) = K(x, \xi) - K_2(x, \xi) + \dots + (-1)^{n+1} K_n(x, \xi) + \dots,$$

also takes values which are the same as those of an entire function $R(z, \zeta)$ in the real domain.

Let $\varphi(z)$ be holomorphic in a region D containing the origin. Then

$$f(z) = \varphi(z) + \int_0^z K(z, \zeta)\varphi(\zeta)\,d\zeta \tag{15}$$

is also holomorphic in D, and so is

$$\psi(z) = f(z) + \int_0^z R(z, \zeta) f(\zeta)\,d\zeta. \tag{16}$$

For z real, we can integrate along the real axis and obtain $\psi(x) = \varphi(x)$. From the equality on the real axis we infer equality in the whole region D.

We have thus derived the following theorem.

THEOREM 2. *The operators $I + K$ and $I + R$ transform every function which is holomorphic in a region D containing the origin into a function holomorphic in the same region, and conversely.*

We will assume here that $q(z)$ is an even entire function. It follows from (1) and the initial conditions that, for any λ, the function $c(z, \lambda)$ is even while $s(z, \lambda)$ is odd. Thus the operator $A[\varphi]$ transforms an even function into an even function and an odd function into an odd function.

THEOREM 3. *If $q(z)$ is an entire function and $\{\lambda_n\}$ is a sequence of complex numbers such that*

$$\lim_{n\to\infty} \frac{n}{\lambda_n} = 1,$$

then every odd function that is holomorphic in the simply connected region D, where D contains the origin and lies in the strip $|\mathrm{Re}\, z| < \pi$, can be uniformly approximated by linear combinations of the functions $s(z, \lambda_n)$ in every interior region D' ($\bar{D}' \subset D$).

PROOF. We transform $f(z)$ with the operator $I + R$. From Theorem 21, Chapter IV, the resulting function can be uniformly approximated by linear combinations of the functions $e^{i\lambda_k z}$ in D' so that

$$\Big|\psi(z) - \sum_0^{k_n} a_k e^{i\lambda_k z}\Big| < \varepsilon \qquad (z \in D')$$

and, since (z) is odd ($\psi(z) = \frac{1}{2}[\psi(z) - \psi(-z)]$), this reduces to

$$\Big|\psi(z) - i\sum_0^{k_n} a_k \sin\lambda_k z\Big| < \varepsilon \qquad (z \in D').$$

Applying the inverse operator $I + K$, we obtain

$$\left| f(z) - t\sum_{0}^{k_n} a_k s(z, \lambda_k) \right| < \varepsilon (1 + P_{D'}) \qquad (z \in D'),$$

where $P_{D'}$ is a constant depending on the region D'. The theorem is proved.

LEMMA. *Every function $f(z)$, holomorphic in a simply-connected region D not containing any pair of points located symmetrically about the origin, can be uniformly approximated by an odd entire function in every bounded, interior region D' ($\bar{D}' \subset D$).*

PROOF. The transformation $u = z^2$ yields a one-sheeted transformation of D into some region G not containing the origin, and it transforms D' into G' ($\bar{G}' \subset G$). The function

$$F(u) = \frac{f(\sqrt{u})}{\sqrt{u}},$$

holomorphic in G, can be uniformly approximated in G' by a polynomial such that

$$|F(u) - P(u)| < \varepsilon$$

or, after the transformation to the z plane,

$$|f(z) - zP(z^2)| < \varepsilon |z|.$$

The proof of the lemma is complete.

THEOREM 4. *The set of functions $\{s(z, \lambda_n)\}$, where*

$$\lim_{n \to \infty} \frac{n}{\lambda_n} = 1,$$

is complete in $0 < \operatorname{Re} z < \pi$.

PROOF. An arbitrary function $f(z)$, holomorphic in some region G contained in the strip $0 < \operatorname{Re} z < \pi$, can be approximated with any desired accuracy by an odd entire function $\chi(z)$, and this latter function can be approximated by a linear combination of the functions $s(z, \lambda_n)$ in a region containing the region G and the origin. The theorem is proved.

COROLLARY. *If $q(z)$ in (1) is an entire function, then the solutions of the boundary-value problem for $y(0) = y(\pi) = 0$ form a complete system in the strip $0 < \operatorname{Re} z < \pi$.*[3]

[3] It is known that $\lambda_n - n \to 0$. This follows immediately, if it is recalled that the eigenvalues λ_n are the roots of the equation $\sin \pi\lambda + \int_0^\pi K(\pi, \zeta) \sin \lambda\zeta \, d\zeta = 0$, and Rouché's theorem is used. If the reasoning used here is generalized somewhat, the completeness of the solutions can also be established for more general boundary conditions.

2. The Inverse Problem for the Sturm-Liouville Equation

We consider the solution of the differential equation

$$y'' - q(x)\,y + \lambda^2 y = 0 \qquad \left(0 < x < 1,\ \int_0^1 |q(x)|\,dx < \infty\right) \tag{1}$$

with the boundary conditions

$$y'(0) - hy(0) = 0, \quad y'(1) + h_1 y(1) = 0. \tag{1'}$$

Here h and h_1 are complex numbers and $q(x)$ is a complex-valued function that is integrable in (0, 1).

The set of values of λ for which this problem has a nontrivial solution make up what is called *the spectrum of the operator* $L[y] = y'' - q(x)y$ for the boundary conditions $\{h; h_1\}$. The spectrum of the operator (1) for the boundary conditions $\{h; h_1\}$ will be denoted by $S_q(h; h_1)$.

The problem called the inverse Sturm-Liouville problem is that of *finding the operator* $L[y]$ from a knowledge of two of its spectra corresponding to different boundary conditions. The problem of the unique determination of an operator from the spectra for different boundary values was first posed in a particular case by V. A. Ambarzumian in [1]. A detailed study of inverse boundary problems was made by G. Borg in [1]. In his work on the unique determination of an operator from its boundary-value spectra, G. Borg showed that, if $q_1(x)$, $q_2(x)$, h, h_1 and h_2 are real, then $q_1(x) = q_2(x)$ almost everywhere in $0 \leqslant x \leqslant 1$ whenever $S_{q_1}(h; h_1) = S_{q_2}(h; h_1)$ and $S_{q_1}(h; h_2) = S_{q_2}(h; h_2)$.

V. A. Marčenko showed in [1] that not only $q(x)$ but also the numbers h, h_1 and h_2 are uniquely determined by the spectra $S_q(h; h_1)$ and $S_q(h; h_2)$. He also extended the uniqueness theorem to the case of operators on a semi-infinite interval.[4] Some of Marčenko's reasoning leads to a uniqueness theorem for complex[5] $q(x)$, h, h_1 and h_2. The problem of finding an operator from two of its spectra was solved recently in articles by M. G. Kreĭn [6; 7], by I. M. Gel'fand and B. M. Levitan [2], and by V. A. Marčenko [1].

We will prove here a uniqueness theorem for non-self-adjoint operators (on a finite interval), mainly following an article by T. M. Karasev [1], but with

---

[4] M. G. Kreĭn in [6] gave another generalization of the uniqueness theorem that shows that the non-negative function $\rho(x)$ is uniquely determined by two spectra $S_\rho(h; h_1)$ and $S_\rho(h; h_2)$ of the equation

$$y'' + \lambda\rho(x)\,y = 0.$$

We note that equation (1) can be reduced to this form by a change of variables, but here the function $\rho(x)$ obtained is not only positive but has an absolutely continuous derivative. The Kreĭn result is therefore stronger than the Borg theorem.

[5] See [1], pp. 385–394. V. A. Marčenko indicates a method for constructing an operator and boundary conditions from two spectra, and this yields a new proof of the uniqueness theorem. In this proof, it is not necessary that $q(x)$ and the numbers h, h_1 and h_2 be real.

some simplifications based on the application of elementary uniqueness theorems from the theory of entire functions.

We first of all note that the condition at the origin determines the solution of equation (1) up to a constant factor, and so there can not be two linearly independent eigenfunctions corresponding to a single point of the spectrum.

We now define the multiplicity of a point λ of the spectrum. We consider the equations

$$\left.\begin{aligned} &L[y_1]+\lambda^2 y_1=0,\\ &L[y_2]+\lambda^2 y_2=y_1,\\ &\cdots\cdots\cdots\quad (L[y]=y''-q(x)y).\\ &\cdots\cdots\cdots\\ &L[y_m]+\lambda^2 y_m=y_{m-1}\end{aligned}\right\} \tag{2}$$

We will define the multiplicity of an eigenvalue λ to be the largest value n of the number m for which the system (2) has a solution, all of whose components $y_1, y_2, \ldots, y_m$ satisfy the boundary conditions (1′).[6] A point λ of a spectrum with $n = 1$ is called a simple point.

In the sequel, we will use the transformation operators already discussed in the preceding section. The main theorem concerning these operators is that, to every differential equation of the form (1) with the boundary condition $y'(0) - hy(0) = 0$, there corresponds an operator

$$V[f]=f(x)+\int^{x} K_h(x,\ t)\,f(t)\,dt \tag{3}$$

such that the solution $y(x, \lambda)$ of equation (1), for any value of λ and for the boundary condition $y'(0) - hy(0) = 0$, can be expressed in the form

$$y(x,\ \lambda)=\cos\lambda x+\int_0^{x} K_h(x,\ t)\cos\lambda t\,dt. \tag{4}$$

The kernel $K_h(x, t)$ is continuous in the square $0 \leqslant x \leqslant 1$, $0 \leqslant t \leqslant 1$.

Substituting this expression for $y(x, \lambda)$ in the second boundary condition $y'(1) + h_1 y(1) = 0$, we obtain the characteristic equation

$$-\lambda\sin\lambda+a\cos\lambda+\int_0^1 Q(t)\cos\lambda t\,dt=0, \tag{5}$$

in which $a = K_h(1, 1) + h_1$ and $Q(t) = K_h'(1, t) + h_1 K_h(1, t)$.

---

[6] It is easily seen that the multiplicity n of a point of a spectrum coincides with the dimension of the corresponding invariant subspace.

LEMMA. *The multiplicity of a point of the spectrum* λ_0 *is the same as that of the corresponding root of the characteristic equation* (5).

PROOF. We write

$$\varphi(\lambda) = -\lambda \sin \lambda + a \cos \lambda + \int_0^1 Q(t) \cos \lambda t \, dt \tag{5'}$$

and assume that

$$\varphi(\lambda_0) = \varphi'(\lambda_0) = \ldots = \varphi^{(n-1)}(\lambda_0) = 0 \text{ and } \varphi^{(n)}(\lambda_0) \neq 0. \tag{6}$$

Let $y(x, \lambda_0)$ be the eigenfunction. Differentiating equation (1) $n - 1$ times with respect to λ^2, we find that the functions

$$z_1(x) = y(x, \lambda_0), \quad z_2(x) = -\left.\frac{\partial y}{\partial \lambda^2}\right|_{\lambda=\lambda_0},$$
$$\ldots, z_n(x) = \frac{(-1)^{n-1}}{n-1} \left.\frac{\partial^{n-1} y}{\partial (\lambda^2)^{n-1}}\right|_{\lambda=\lambda_0}$$

satisfy the equations (2). It is evident that all these functions satisfy the boundary condition at zero, and also, from the relation (6), the condition at $x = 1$.

The function $z_{n+1}(x) = \frac{(-1)^n}{n} \left.\frac{\partial^n y}{\partial(\lambda^2)^n}\right|_{\lambda=\lambda_0}$ satisfies the equation

$$L[z_{n+1}] + \lambda^2 z_{n+1} = z_n \tag{7}$$

and the boundary condition at zero. In view of (6), it does not satisfy the boundary condition at the point 1.

It is easy to see that any solution $(y_1, y_2, \ldots, y_n, y_{n+1})$ of the system (2) for $m = n + 1$, all the components of which satisfy the boundary condition at zero, has the form

$$y_1 = c_1 z_1, \quad y_2 = c_1 z_2 + c_2 z_1,$$
$$\ldots, y_{n+1} = c_1 z_{n+1} + z_2 z_n + \ldots + c_{n+1} z_1,$$

and so no solution of the system (2) for $m = n + 1$ can simultaneously satisfy the conditions at $x = 0$ and $x = 1$.

The system (2) thus has a solution for $m = n$ but has no solution for $m = n + 1$. The lemma is proved.

From the expression (5′) for $\varphi(\lambda)$, it follows that this function satisfies the asymptotic relation

$$\varphi(\lambda) = -\lambda \sin \lambda \left[1 + O\left(\frac{1}{|\lambda|}\right)\right] \tag{8}$$

outside certain exceptional circles, and so, using Rouché's theorem, we find that the number of multiple zeros of $\varphi(\lambda)$ is finite.[7]

[7] The relation (8) also yields the simple asymptotic formula $\lambda_n = \pi n + \varphi(n)$ for the eigenvalues, where $\varphi(n)$ is a complex-valued function tending to zero for $n \to \infty$.

We have thus proved that the number of multiple points in the spectrum is finite, and *the multiplicity of each of these points is also finite*. It remains to note that in the self-adjoint case, i.e., for real $q(x)$, h and h_1, the spectrum is made up of simple points. In fact, for the second equation in (2) to have a solution satisfying the condition (1′), the right-hand side of this equation must be orthogonal to $y_1(x)$, i.e., the relation $\int_0^1 y_1^2(x)\,dx = 0$ must hold. But for real $q(x)$, h and h_1, the eigenfunction $y_1(x)$ is real (up to a constant factor), and this equality can not hold. This means that a spectrum with multiple points can not occur.

After these general remarks from the theory of differential equations, we now attack the inverse problems. Here two spectra will be said to coincide only if their elements and the multiplicity of each element coincide.

THEOREM 1. *The function* $q(x)$ $(0 \leqslant x \leqslant 1)$ *and the numbers* h, h_1 *and* h_2 *are uniquely determined by two spectra of the operator* (1) *corresponding to two different sets of boundary conditions* $\{h; h_1\}$ *and* $\{h; h_2\}$.

We first prove that the spectrum uniquely determines the characteristic equation. Let the equations

$$\left.\begin{aligned} -\lambda\sin\lambda + a_1\cos\lambda + \int_0^1 Q_1(t)\cos\lambda t\,dt = 0,\\ -\lambda\sin\lambda + a_2\cos\lambda + \int_0^1 Q_2(t)\cos\lambda t\,dt = 0 \end{aligned}\right\} \tag{9}$$

be equivalent. The left-hand sides of these equations are entire functions of exponential type with the same set of zeros, and so they can only differ by an exponential factor. It follows from the asymptotic formula (8) that this factor is unity, and so

$$(a_2 - a_1)\cos\lambda + \int_0^1 [Q_2(t) - Q_1(t)]\cos\lambda t\,dt = 0.$$

Since the integral in this equation tends to zero if λ is real and tends to infinity, we have

$$a_2 = a_1 \quad\text{and}\quad \int_0^1 [Q_2(t) - Q_1(t)]\cos\lambda t\,dt = 0$$

and so $Q_2(t) \equiv Q_1(t)$.

We now show that, for given h, two spectra for the same operator (1) determine h_1 and h_2 uniquely. Actually, the spectra uniquely determine the numbers

$$a_1 = K(1,\ 1) + h_1, \quad Q_1(1) = K_h'(1,\ 1) + h_1K_h(1,\ 1),$$

$$a_2 = K(1,\ 1) + h_2, \quad Q_2(1) = K_h'(1,\ 1) + h_2K_h(1,\ 1),$$

and so also the numbers

$$h_1 = \frac{Q_1(1) - Q_2(1)}{a_2 - a_1} + a_1 \text{ and } h_2 = \frac{Q_1(1) - Q_2(1)}{a_2 - a_1} + a_2.$$

Now let the spectra $\{\lambda_k\}$ and $\{\mu_k\}$ of the operator

$$y'' - q_1(x)y + \lambda^2 y = 0 \tag{10}$$

for the boundary conditions $\{h; h_1\}$ and $\{h; h_2\}$ coincide with the spectrum of

$$y'' - q_2(x)y + \lambda^2 y = 0 \tag{11}$$

for the boundary conditions $\{g; g_1\}$ and $\{g; g_2\}$. Then, from what has been proved, $g_1 = h_1$ and $g_2 = h_2$.

Let $y(x, \lambda)$ be the solution of equation (10) for the initial conditions $y(0) = 1$, $y'(0) = h$ and let $z(x, \lambda)$ be the solution of (11) for the initial conditions $z(0) = 1$, $z'(0) = g$.

We substitute $y(x, \lambda)$ in (10), $z(x, \lambda)$ in (11), multiply the first identity by $z(x, \lambda)$ and the second by $y(x, \lambda)$, subtract the second from the first, and integrate the resulting relation from zero to one. We obtain

$$[y'(x, \lambda)z(x, \lambda) - y(x, \lambda)z'(x, \lambda)]\Big|_0^1 + \int_0^1 [q_1(x) - q_2(x)]\, y(x, \lambda)\, z(x, \lambda)\, dx = 0.$$

If λ is one of the points of the spectra λ_k or μ_k, then from the relations $h_1 = g_1$, $h_2 = g_2$ the first term in this identity is equal to $g - h$. Thus the entire function

$$\psi(\lambda) = g - h + \int_0^1 [q_1(x) - q_2(x)]\, y(x, \lambda)\, z(x, \lambda)\, dx \tag{12}$$

is zero for all the λ_k and μ_k, while the multiplicity of any zero of $\psi(\lambda)$ is not less than the multiplicity of the corresponding point of the spectrum. We will show that the rate of growth of $\psi(\lambda)$ is insufficient for it to have the indicated zeros without being identically zero. In fact, the functions $y(x, \lambda)$ and $z(x, \lambda)$ can be expressed in the form

$$y(x, \lambda) = \cos \lambda x + \int_0^x K_h(x, t) \cos \lambda t\, dt, \tag{13}$$

$$z(x, \lambda) = \cos \lambda x + \int_0^x L_g(x, t) \cos \lambda t\, dt, \tag{14}$$

and so it easily follows that $\psi(\lambda)$ satisfies

$$|\psi(\lambda)| < M e^{2|\mathrm{Im}\, \lambda|} \tag{15}$$

in the whole complex plane for some constant M.

From previous results, we know that the spectra $\{\lambda_k\}$ and $\{\mu_k\}$ coincide with the set of zeros of functions $\varphi_1(\lambda)$ and $\varphi_2(\lambda)$ of the form (5') satisfying the asymptotic relation (8). Thus $\psi(\lambda)$ is zero for all the zeros of the function

$$\varphi_1(\lambda)\varphi_2(\lambda) = \lambda^2 \sin^2\lambda\left[1 + O\left(\frac{1}{|\lambda|}\right)\right], \tag{16}$$

where the asymptotic relation (16) holds outside exceptional nonintersecting circles with centers at the zeros of this function.

From (15) and (16), we find that the entire function

$$\Theta(\lambda) = \frac{\psi(\lambda)}{\varphi_1(\lambda)\varphi_2(\lambda)}$$

satisfies, outside the exceptional circles, the relation

$$\Theta(\lambda) = O\left(\frac{1}{|\lambda|^2}\right)$$

and so the maximum-modulus principle yields[8] $\Theta(\lambda) \equiv 0$.

We thus have the identity

$$g - h + \int_0^1 [q_1(x) - q_2(x)]\, y(x, \lambda)\, z(x, \lambda)\, dx \equiv 0. \tag{17}$$

From this it is not difficult to prove that $g = h$ and $q_1(x) = q_2(x)$ almost everywhere. From the representations (13) and (14), it follows that

$$y(x, \lambda) z(x, \lambda) = \cos^2 \lambda x + \int_0^x K_h(x, t) \cos \lambda t \cos \lambda x\, dt + \int_0^x L_g(x, t) \cos \lambda t \cos \lambda x\, dt$$
$$+ \int_0^x \int_0^x K_h(x, s)\, L_g(x, t) \cos \lambda s \cos \lambda t\, dt\, ds.$$

Extending the range of the functions $K_h(x, s)$ and $L_g(x, t)$ by specifying that they are even, i.e., so that $K_h(x, -t) = K_h(x, t)$ and $L_g(x, -t) = L_g(x, t)$, we can write

$$y(x, \lambda) z(x, \lambda) = \frac{1}{2} + \cos 2\lambda x + \frac{1}{2} \int_{-x}^{x} [K_h(x, t) + L_g(x, t)] \cos \lambda (x - t)\, dt$$
$$+ \frac{1}{4} \int_{-x}^{x} \int_{-x}^{x} K_h(x, s) L_g(x, t) \cos \lambda (t - s)\, dt\, ds \quad (x \geqslant 0).$$

---

[8] If it is given that $\psi(\lambda)$ is zero, not for all the zeros of the product $\varphi_1(\lambda)\varphi_2(\lambda)$, but for all these zeros except one, i.e., the spectra do not correspond exactly but are the same except for one point, then $\psi(\lambda)(\lambda - \lambda_0)$ divides into $\varphi_1(\lambda)\varphi_2(\lambda)$ and we again obtain $\psi(\lambda) \equiv 0$. This fact shows that some strengthening of the theorem is possible.

By changing variables, we obtain

$$y(x, \lambda)\, z(x, \lambda) = \frac{1}{2} + \frac{1}{2}\cos 2\lambda x + \frac{1}{2}\int_{-x}^{x} \tilde{K}(x, \tau)\cos 2\lambda\tau\, d\tau, \tag{18}$$

where

$$\tilde{K}(x, \tau) = \begin{cases} 2\,[K_h(x, x - 2\tau) + L_g(x, x - 2\tau)] \\ \quad + \displaystyle\int_{-x+2\tau}^{x} K_h(x, s)\, L_g(x, s - 2\tau)\, ds & \text{for } \tau > 0, \\ \displaystyle\int_{-x}^{x} K_h(x, s)\, L_g(x, s - 2\tau)\, ds & \text{for } \tau < 0. \end{cases}$$

After the change of notation $q_1(x) - q_2(x) = r(x)$ $[r(-x) = r(x)]$ and the substitution in (17) of the expression in (18) for $y(x, \lambda)z(x, \lambda)$, we obtain

$$g - h + \frac{1}{2}\int_0^1 r(x)\, dx$$

$$+ \frac{1}{2}\int_{-1}^{1}\left[\frac{1}{2} r(x) + \int_x^1 \tilde{K}(y, x)\, r(y)\, dy\right]\cos 2\lambda x\, dx = 0.$$

The last term on the left tends to zero when $\lambda \to \infty$ along the real axis. Therefore

$$g - h + \frac{1}{2}\int_0^1 r(x)\, dx = 0 \tag{19}$$

and

$$\frac{1}{2} r(x) + \int_x^1 \tilde{K}(y, x)\, r(y)\, dy = 0.$$

By substituting $r(1 - x) = \varphi(x)$ and $y = 1 - t$ in this equation, we obtain

$$\varphi(x) + 2\int_0^x K(1 - t,\ 1 - x)\, \varphi(t)\, dt = 0.$$

This is a homogeneous Volterra equation, and so $\varphi(x) \equiv 0$. Hence $r(x) \equiv 0$ and $h = g$. The theorem has thus been proved.

APPENDIX V

REPRESENTATION OF A POSITIVE ENTIRE FUNCTION OF EXPONENTIAL TYPE AS THE SQUARE OF THE ABSOLUTE VALUE OF AN ENTIRE FUNCTION

The starting point of the results described in this appendix is the following Fejér-Riesz theorem:

A non-negative trigonometric polynomial

$$T(x)=\sum_{-n}^{n} c_k e^{ikx} \qquad (T(x)\geqslant 0,\ -\infty<x<\infty)$$

can be expressed in the form

$$T(x)=|S(x)|^2 \qquad (-\infty<x<\infty),$$

where

$$S(x)=\sum_{0}^{n} b_k e^{ikx}.$$

The trigonometric polynomial $S(x)$ *can be chosen so that all its zeros are in the closed upper half-plane.*

We note that, in this representation, the trigonometric polynomial $S(x)$ can be replaced by the polynomial $e^{-inx/2}S(x)$, where the degree of the latter is half the degree of $T(x)$.

In [5], M. G. Kreĭn generalized this theorem to bounded entire functions of exponential type, showing that an entire function of exponential type k, bounded and non-negative on the real axis, can be expressed as the square of the absolute value of a function of exponential type $k/2$ with its zeros in the closed upper half-plane. Kreĭn's proof was based on the approximation of a bounded, entire function of exponential type by Levitan trigonometric polynomials (Appendix I, § 4).

N. I. Ahiezer, in [1], further generalized this theorem, giving the following criterion for the expressibility of an entire function of exponential type as the square of the absolute value of another entire function.

THEOREM 1. *For an entire function* $f(z)$ *of exponential type to have the representation*

$$f(x)=|\varphi(x)|^2, \tag{1}$$

where $\varphi(z)$ is an entire function of exponential type $k/2$ with zeros in one of the half-planes $\operatorname{Im} z \geqslant 0$ *or* $\operatorname{Im} z \leqslant 0$, *it is necessary and sufficient that $f(z)$ belong to the class A and be non-negative on the real axis.*

PROOF. The necessity is obvious. The fact that $f(x)$ must be non-negative follows directly from (1). Furthermore, the entire function $\varphi(z)$ of exponential type with its zeros in a half-plane must be of class A (Chapter V), and so $f(z)$ also belongs to A, since its zeros are those of $\varphi(z)$ and their conjugates.

We now prove the sufficiency of the conditions. A function $f(z)$ of exponential type can be expressed in the form

$$f(z) = z^{2p} e^{2az+2b} \prod_{1}^{\infty} \left(1 - \frac{z}{a_k}\right) e^{\frac{z}{a_k}}. \tag{2}$$

The zeros of $f(z)$ on the real axis are of even multiplicity, since $f(x) \geqslant 0$. We denote the zeros of $f(z)$ in the half-plane $\operatorname{Im} z \geqslant 0$ by a'_k. In this enumeration of the zeros, each zero in the open half-plane $\operatorname{Im} z > 0$ is repeated a number of times equal to its multiplicity, while each of the zeros on the real axis is repeated a number of times equal to one half its multiplicity. We write

$$\varphi(z) = z^p e^{az+b+i\gamma z} \prod_{1}^{\infty} \left(1 - \frac{z}{a'_k}\right) e^{\frac{z}{a'_k}} \quad \left(\gamma = -\sum_{1}^{\infty} \operatorname{Im} \frac{1}{a'_k}\right) \tag{3}$$

and

$$\bar{\varphi}(z) = z^p e^{az+b-i\gamma z} \prod_{1}^{\infty} \left(1 - \frac{z}{\bar{a}'_k}\right) e^{\frac{z}{\bar{a}'_k}}. \tag{4}$$

It is evident that

$$f(z) = \varphi(z)\bar{\varphi}(z). \tag{5}$$

We will show that $\varphi(z)$ is of exponential type. From Lindelöf's theorem (Chapter I, Theorem 15) and the fact that $f(z)$ is of exponential type, we have

$$\left| \sum_{|a_k|<r} \frac{1}{a_k} \right| < M \qquad (M \text{ a constant}). \tag{6}$$

Since

$$\sum_{1}^{\infty} \left| \operatorname{Im} \frac{1}{a_k} \right| < \infty, \tag{7}$$

it follows from (6) that

$$\left| \sum_{|a_k| \leqslant r} \operatorname{Re} \frac{1}{a_k} \right| < M_1 \qquad (M_1 \text{ a constant}).$$

From the definition of the a'_k, we have

$$\left| \sum_{|a'_k| \leqslant r} \operatorname{Re} \frac{1}{a'_k} \right| = \frac{1}{2} \left| \sum_{|a_k| \leqslant r} \operatorname{Re} \frac{1}{a_k} \right| < \frac{1}{2} M_1,$$

and using (7) we obtain

$$\Big| \sum_{|a'_k| \leqslant r} \frac{1}{a'_k} \Big| < M_2 \qquad (M_2 \text{ a constant}). \tag{8}$$

We also have

$$\overline{\lim_{r\to\infty}} \frac{n_\varphi(r)}{r} = \frac{1}{2} \overline{\lim_{r\to\infty}} \frac{n_f(r)}{r} < \infty. \tag{9}$$

From (8) and (9) and Lindelöf's theorem, it follows that $\varphi(z)$ is of exponential type. It remains to prove that the exponential type of $\varphi(z)$ is $k/2$. The function $f(z)$ can be expressed in the form

$$f(z) = \overline{\varphi}^2(z)\chi(z)$$

where

$$\chi(z) = \prod_1^\infty \left(1 - \frac{z}{a'_k}\right)\left(1 - \frac{z}{\overline{a}'_k}\right)^{-1}.$$

Lemma 5, Chapter V, yields the result that $\chi(z)$ has completely regular growth in the upper half-plane and has an indicator equal to zero. From Theorem 5, Chapter III, concerning the addition of indicators for products of holomorphic functions when one of these functions is of completely regular growth, we have

$$h_f(\theta) = 2h_{\overline{\varphi}}(\theta) = 2h_\varphi(\theta) \qquad (0 < \theta < \pi).$$

Similar reasoning applies in the lower half-plane. Thus

$$h_f(\theta) = 2h_\varphi(\theta) \qquad (0 \leqslant \theta < 2\pi),$$

and the degree of $\varphi(z)$ is one half that of $f(z)$. The theorem is proved.

Another generalization of the Fejér-Riesz theorem can be obtained by considering, instead of finite trigonometric sums, infinite series of the type

$$f(x) = \sum_{-\infty}^{\infty} a_k e^{i\lambda_k x} \qquad \left(|\lambda_k| \leqslant \Delta,\ \sum_1^\infty |a_k| < \infty\right). \tag{10}$$

Every such series obviously represents an entire function of exponential type, the conjugate diagram of which coincides with the smallest segment of the imaginary axis containing all the points $i\lambda_k$. It follows from the preceding theorem, of course, that if $f(x) > 0$, then it can be expressed as the square of the absolute value of an entire function $\varphi(z)$ with degree one half that of $f(x)$ and with zeros in one half-plane, but it still remains to prove that $\varphi(x)$ is also a function of the form (10).

THEOREM 2. *Let $f(x)$ be of the form* (10) *and of exponential type Δ, and let $f(x) \geqslant m > 0$. Then there is a function $\varphi(x)$ of the form* (10), *with exponential type $\Delta/2$ and with zeros in a half-plane, such that*[1]

$$f(x) = |\varphi(x)|^2. \tag{11}$$

[1] See B. Ja. Levin [13].

PROOF. From Theorem 1, we know that a function $\bar{\varphi}(z)$ of the form (4) is an entire function of exponential type $\Delta/2$, satisfying (11), with zeros in one half-plane $\operatorname{Im} z > 0$. It is only necessary to prove that it can be expanded in an absolutely convergent series of the form (10). The basis of the proof is the following theorem proved by Calderón and Pitt.[2] *If $\psi(x)$ can be expressed by the absolutely convergent series*

$$\psi(x) = \sum_{-\infty}^{\infty} c_k e^{i\lambda_k x}, \tag{12}$$

in which the λ_k are real, and if the closure of the set of values of $\psi(x)$ lies in the region in which the function $\Phi(z)$ is holomorphic, then $\varphi(x) = \Phi[\varphi(x)]$ can also be expressed as an absolutely convergent series of the form (12).

In particular, it follows from this theorem that

$$\frac{1}{2}\ln f(x) = \sum_{-\infty}^{\infty} b_k e^{i\mu_k x} \qquad \left(b_{-k} = \bar{b}_k,\ \mu_{-k} = -\mu_k,\ \sum_{-\infty}^{\infty} |b_k| < \infty\right).$$

The function

$$\zeta(z) = 2\sum_{k=0}^{\infty} b_k e^{i\mu_k z} \qquad (\mu_k \geqslant 0)$$

is holomorphic and bounded in the upper half-plane and

$$\operatorname{Re}\zeta(x) = \frac{1}{2}\ln f(x).$$

It thus follows from (11) that $\operatorname{Re}\zeta(x)$ and $\ln|\bar{\varphi}(x)|$ take the same values. We will show that $\bar{\varphi}(z)$ and

$$\vartheta(z) = e^{\zeta(z)}$$

are essentially the same function. For this we note that $\bar{\varphi}(z)$ is of exponential type and bounded on the real axis. Therefore $\bar{\varphi}(z)$ is of completely regular growth (Chapter V, Theorem 11) and

$$h_{\bar{\varphi}}(\theta) = \frac{\Delta}{2}|\sin\theta|.$$

We know that $\bar{\varphi}(z)$ has no zeros in the upper half-plane, and so the indicator of the function

$$\bar{\varphi}^{-1}(z)e^{-i\frac{\Delta}{2}z}$$

is zero in the upper half-plane.

The function $\vartheta(z)$ is bounded in the upper half-plane, and the indicator of the function

$$\xi(z) = \vartheta(z)\bar{\varphi}^{-1}(z)e^{-i\frac{\Delta}{2}z}$$

is therefore not positive in the upper half-plane.

---

[2] See Gel'fand, Raĭkov and Šilov [1], §29, Corollary 2 to Theorem 2.

Moreover, on the real axis we have

$$|\xi(x)| = 1,$$

and consequently, from the Phragmén-Lindelöf theorem (Chapter I, Theorem 22),

$$|\xi(z)| \leqslant 1 \qquad (\operatorname{Im} z \geqslant 0).$$

The function $\xi(z)$ can be continued by symmetry to the lower half-plane, giving an entire function without zeros. But the function $-i \ln \xi(z)$ satisfies all the conditions of Čebotarev's lemma (Chapter V, Lemma 1).

Therefore

$$\xi(z) = e^{i\alpha z + i\beta} \qquad (\alpha \geqslant 0, \ \operatorname{Im} \beta = 0)$$

and

$$\bar{\varphi}(z) = e^{\zeta(z) - i\left(\frac{\Delta}{2} + \alpha\right) z - i\beta}.$$

Again applying Wiener's theorem, we obtain

$$\bar{\varphi}(x) = \sum_{-\infty}^{\infty} d_k e^{i\varkappa_k x} \qquad \left(\sum_{-\infty}^{\infty} |d_k| < \infty\right).$$

We now show that $|\varkappa_k| \leqslant \Delta/2$. It is not difficult to see that the points $i\varkappa_k$ are singularities of the function associated with $\bar{\varphi}(z)$ in the Borel sense.

But the conjugate diagram of $\bar{\varphi}(z)$ is the segment $[-i\,\Delta/2, i\,\Delta/2]$ of the imaginary axis, and so $|\varkappa_k| \leqslant \Delta/2$. The theorem is proved.

It should be noted that this theorem can not be applied to products of almost-periodic functions with bounded spectra. For if it is not required that the Fourier series of the entire, almost-periodic function $f(x)$ of exponential type be absolutely convergent, then a positive, almost-periodic function $f(x)$ with bounded spectrum can be constructed for which the corresponding function $\varphi(z)$ is not almost periodic.[3]

If it is required that the Fourier series of $f(x)$ be absolutely convergent, then as was shown in Theorem 2 the function $\varphi(x)$ is not only almost-periodic but has an absolutely convergent Fourier series.

It is interesting to note that there are *finite* positive sums of the form (12) that can not be expressed as the squares of the absolute values of *finite sums*. It can be proved, in particular, that for $c > 2$ and irrational λ, the almost-periodic polynomial

$$P(x) = C + \cos x + \cos \lambda x$$

can not be expressed as a product of almost-periodic polynomials.[4]

A theorem similar to Theorem 2 is obtained when, instead of entire functions of exponential type expanded in trigonometric series, we consider functions expressed as Fourier integrals.

[3] An example of such a function is given in Appendix VI.

[4] We, of course, exclude the trivial polynomials $e^{i\mu x}$.

THEOREM 3. *If $f(x)$ is expressible in the form*

$$f(x) = \nu + \int_{-k}^{k} e^{i\lambda x}\psi(\lambda)\,d\lambda \quad \left(\int_{-k}^{k} |\psi(\lambda)|\,d\lambda < \infty,\ \nu \text{ a constant}\right)$$

and if $f(x) \geqslant q > 0$ for $-\infty < x < \infty$, then

$$f(x) = |\varphi(x)|^2,$$

where $\varphi(x)$ can be expressed in the form

$$\varphi(x) = \nu_1 + \int_{-\frac{k}{2}}^{\frac{k}{2}} e^{i\lambda x}\chi(\lambda)\,d\lambda \qquad \left(\int_{-\frac{k}{2}}^{\frac{k}{2}} |\chi(\lambda)|\,d\lambda < \infty\right).$$

This theorem is proved by repeating, step by step, the proof of Theorem 2, but using the following theorem due to Wiener and Pitt.[4a]

If

$$f(x) = \nu + \int_{-\infty}^{\infty} e^{i\lambda x}\psi(\lambda)\,d\lambda,$$

and $\Phi(z)$ is holomorphic in a region containing the closure of the set of values of $f(x)$ $(-\infty < x < \infty)$, then the function $\Phi[f(x)]$ can also be expressed in the form

$$\Phi[f(x)] = \nu_1 + \int_{-\infty}^{\infty} e^{i\lambda x}\psi_1(\lambda)\,d\lambda$$

$$\left(\int_{-\infty}^{\infty} |\psi_1(\lambda)|\,d\lambda < \infty,\ \nu_1 = \Phi(\nu)\right).$$

A more general result is obtained if a theorem due to Wiener and Pitt **[1]** concerning functions expressible as Fourier-Stieltjes integrals is used. We will only give the result without proof.[5]

THEOREM 4. *Let*

$$f(x) = \int_{-k}^{k} e^{i\lambda x}\,d\sigma(\lambda),$$

where $\sigma(\lambda)$ is a function of bounded variation in $[-k, k]$.

Let $\sigma(\lambda) = \sigma_a(\lambda) + \sigma_s(\lambda) + \sigma_d(\lambda)$, where $\sigma_a(\lambda)$ is the absolutely continuous part of the function $\sigma(\lambda)$, $\sigma_s(\lambda)$ is its function of singularities, and $\sigma_d(\lambda)$ is its jump function.

[4a] See Gel'fand, Raĭkov and Šilov **[1]**, §17, Corollary 2 to Theorem 1.

[5] See B. Ja. Levin **[13]**.

Then if $f(x) \geqslant m > 0$ *and*

$$\int_{-k}^{k} |d\sigma_s(\lambda)| < \left| \int_{-k}^{k} e^{i\lambda x}\, d\sigma_d(\lambda) \right|,$$

we have

$$f(x) = |\varphi(x)|^2,$$

where

$$\varphi(x) = \int_{-\frac{k}{2}}^{\frac{k}{2}} e^{i\lambda x}\, d\tau(\lambda),$$

with $\tau(\lambda)$ *of bounded variation on* $[-k/2, k/2]$.

Moreover, if $\sigma_s(\lambda) = 0$, *then* $\tau_s(\lambda) = 0$, *and if* $\sigma_a(\lambda) = 0$, *then* $\tau_a(\lambda) = 0$.

APPENDIX VI

ALMOST-PERIODIC FUNCTIONS WITH BOUNDED SPECTRA

In § 2, Chapter VI, we gave some results concerning the distribution of the zeros of almost-periodic functions with bounded spectra. In 1948, M. G. Kreĭn and the author carried out a much more complete investigation of the zeros of these functions and obtained a precise characterization for the set of zeros of functions of the class $[\Delta]$. These results were published without proofs in **[1]**.

Here for the first time we give these results with proofs, basing our discussion on the contents of § 2, Chapter VI. We will assume (with no loss of generality) that a function of the class $[\Delta]$ is of degree Δ in the upper and lower half-planes.

1. An Expression for the Real Part of a Zero of a Function of the Class $[\Delta]$

It was proved in Chapter VI (Theorem 3) that all the zeros of a function of $[\Delta]$ are in some strip $|y| < h$. We denote by $n(t)$, for $t > 0$, the number of zeros of a function $f(z)$ of $[\Delta]$ in the rectangle $0 < \operatorname{Re} z \leqslant t$, $|y| \leqslant h$, while $n(t)$ for $t < 0$ will be minus the number of zeros in the rectangle $t < \operatorname{Re} z \leqslant 0$, $|y| < h$.

THEOREM 1. *For functions of* $[\Delta]$, *we have*

$$n(t) = \frac{\Delta}{\pi} t + \omega(t),$$

where $\omega(t)$ *is a bounded function, almost-periodic in the sense of V. V. Stepanov*[1] (for any $p > 0$), *with the numerical module of this function included in the numerical module of* $f(z)$.

PROOF. We start from (6.16), § 2, Chapter VI:

$$2\pi n(t) = 2\Delta t - \varphi_H(t) + \varphi_{-H}(t) + \operatorname{Re} \int_{-H}^{H} \frac{f'(t+is)}{f(t+is)} ds + c, \tag{1}$$

where

$$c = \operatorname{Re} \int_{-H}^{H} \frac{f'(s)}{f(s)} ds$$

[1] See B. M. Levitan [1].

and $\varphi_H(t)$ and $\varphi_{-H}(t)$ are almost-periodic functions. If the sequence $\{h_n\}$ is such that $\{f(t + h_n)\}$ tends uniformly to $f(t)$ on the whole axis, then $\varphi_{\pm H}(t + h_n) \rightrightarrows \varphi_{\pm H}(t)$ (Chapter VI, proof of Lemma 4). From a well-known theorem of Favard,[2] it thus follows that the modules of the functions $\varphi_{\pm H}(t)$ belong to the module of $f(t)$.

In § 2, Chapter VI, the boundedness of the function

$$\omega_1(t) = \operatorname{Re} \int_{-H}^{H} \frac{f'(t+is)}{f(t+is)}\, ds$$

was established.

We now show that this function is S_p-almost-periodic with a module belonging to that of $f(t)$.

Writing

$$M = \sup |f(x+iy)|$$

in the strip $-\infty < x < \infty$, $|y| \leqslant H + \delta$, we will have

$$|f'(x+iy)| = \left| \frac{1}{2\pi} \int_{|\zeta|=\delta} \frac{f(x+iy+\zeta)}{\zeta^2}\, d\zeta \right| \leqslant \frac{M}{\delta} \quad (|y| \leqslant H).$$

Similarly, for $|y| \leqslant H$, we have

$$|f'(x+iy+\tau) - f'(x+iy)| < \frac{1}{\delta} \sup_{\substack{-\infty < x < \infty \\ |y| \leqslant H+\delta}} |f(x+iy+\tau) - f(x+iy)|. \tag{2}$$

The intervals

$$(\alpha_n - \delta,\ \alpha_n + \delta) \qquad (\pm n = 0,\ 1, \ldots),$$

where α_n is the real part of the zero a_n of $f(z)$ and δ is positive, will be called "exceptional" intervals.

It follows from Lemma 2, Chapter VI, that the length of the intersection of the exceptional intervals with any interval of unit length does not exceed $2(k + 1)\delta$, where k is the number occurring in the statement of the lemma.

Let τ be an arbitrary real number and $(x, x + 1)$ an arbitrary unit interval. Let $E_{\delta,\tau}$ be the set of $\tau \in (x, x + 1)$ for which either t or $t + 1$ belongs to an exceptional interval. Obviously

$$\operatorname{mes} E_{\delta,\tau} \leqslant 4(k+1)\delta.$$

If $t \bar{\in} E_{\delta,\tau}$, then from Lemma 1, Chapter VI, we have $|f(t + iy)| \geqslant m > 0$ and $|f(t + \tau + iy)| \geqslant m > 0$, and from (2) we obtain

$$\left| \frac{f'(t+\tau+iy)}{f(t+\tau+iy)} - \frac{f'(t+iy)}{f(t+iy)} \right| \leqslant \frac{2M}{\delta m^2} \sup_{\substack{-\infty < t < \infty \\ |y| < H+\delta}} |f(t+\tau+iy) - f(t+iy)|, \tag{3}$$

[2] See B. M. Levitan [1].

where $M = \sup|f(t + iy)|$ for $-\infty < t < \infty$ and $|y| \leqslant H + \delta$. We write $\sup|\omega_1(t)| = M_1$, and using (3) we obtain

$$\|\omega_1(t+\tau)-\omega_1(t)\|_p^p = \sup_{-\infty<x<\infty}\int_x^{x+1}|\omega_1(t+\tau)-\omega_1(t)|^p\,dt$$

$$\leqslant \frac{AMH}{m^2\delta}\sup_{\substack{-\infty<t<\infty\\|y|<H+\delta}}|f(t+\tau+iy)-f(t+iy)|+4(k+1)\delta M_1.$$

We thus have proved that with an arbitrary $\varepsilon > 0$ is associated a number $\eta > 0$ such that any η-translation of $f(t)$ is an S_p-translation of the function $\omega_1(t)$. Consequently $\omega_1(t)$ is an (S_p, ε)-almost-periodic function with a module belonging to the module of $f(t)$. It remains to remark that

$$\omega(t) = \frac{1}{2\pi}[\omega_1(t)+\varphi_H(t)-\varphi_{-H}(t)+c].$$

The theorem has thus been proved. The conversion of the formula to one for $n(t)$ yields the theorem.

THEOREM 2. *Let $\{a_k\}_{-\infty}^{\infty}$ be the sequence of all the zeros of the function $f(z)$ of the class $[\Delta]$, enumerated in the order of increasing real parts. Then*

$$\operatorname{Re} a_k = \frac{\pi}{\Delta}k+\varphi(k), \tag{4}$$

where $\varphi(x)$ is a Bohr almost-periodic function such that its smallest numerical module, containing all the products formed by multiplying Δ/π with every Fourier exponent of $\varphi(x)$, is contained in the module of $f(z)$.

PROOF. The function $u = n(t)$ is piecewise constant, nondecreasing, continuous from the right and takes only integral values. Its inverse $n_{-1}(u)$ can be determined, and this inverse is continuous from the right, nondecreasing, and constant between successive integral values of its argument (Figure 12).

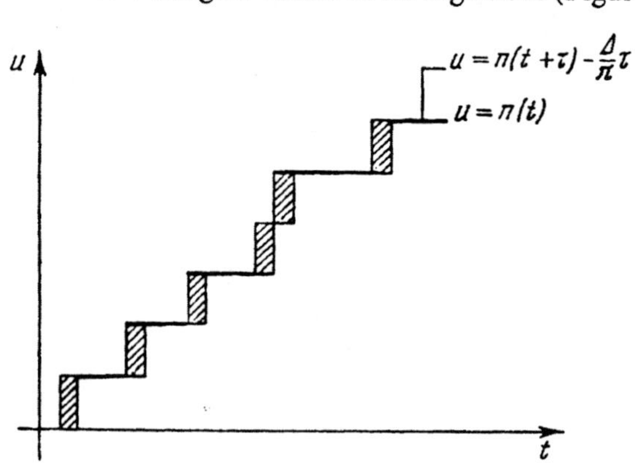

Fig. 12

We form the function

$$v^{(\tau)}(t) = n(t+\tau) - \frac{\Delta}{\pi}\tau.$$

Solving this relation for t, we obtain the inverse function

$$v^{(\tau)}_{-1}(u) = n_{-1}\left(u + \frac{\Delta}{\pi}\tau\right) - \tau.$$

We choose τ so that it is an ε-translation of the S_1-almost-continuous function $\omega(t)$ and such that $q = \Delta\tau/\pi$ is an integer.[3] Then $v^{(\tau)}(t) - n(t)$ is an integer. On the other hand, Theorem 1 yields

$$\int_x^{x+1} |v^{(\tau)}(t) - n(t)|\,dt = \int_x^{x+1} |\omega(t+\tau) - \omega(t)|\,dt < \varepsilon.$$

This integral is equal to the sum of the areas of certain rectangles (see Figure 12) the heights of which are integers, and so the bases of these rectangles are not greater than ε.

We consider t as a function of u, obtaining

$$|v^{(\tau)}_{-1}(u) - n_{-1}(u)| < \varepsilon \qquad (-\infty < u < \infty).$$

Introducing the function $\varphi_1(u) = n_{-1}(u) - (\pi/\Delta)u$, we have

$$\left|\varphi_1\left(u + \frac{\Delta}{\pi}\tau\right) - \varphi_1(u)\right| = \left|n_{-1}\left(u + \frac{\Delta}{\pi}\tau\right) - \tau - n_{-1}(u)\right| < \varepsilon. \tag{5}$$

We now consider the expression for Re a_k. From the definition of $n(t)$ we have

$$n(\operatorname{Re} a_k) = k,$$

or for the inverse function

$$\operatorname{Re} a_k = n_{-1}(k) = \frac{\pi}{\Delta}k + \varphi_1(k) \qquad (\pm k = 0, 1, 2, \ldots).$$

In view of (5), to every positive ε there corresponds a relatively dense set of integers $\nu = \Delta\tau/\pi$ such that

$$|\varphi_1(k+\nu) - \varphi_1(k)| < \varepsilon \qquad (\pm k = 0, 1, 2, \ldots).$$

Interpolating with a linear function $\varphi_1(k)$ between successive integral points, we obtain a continuous, almost-periodic function $\varphi(x)$, and

$$\operatorname{Re} a_k = \frac{\pi}{\Delta}k + \varphi(k).$$

We further note that for any $\varepsilon > 0$ we can find a $\delta > 0$ such that if τ is a δ-translation of $\omega(t)$, then $\Delta\tau/\pi$ is an ε-translation of $\varphi(x)$. Hence the numerical module of the function $\varphi(x)$ multiplied by Δ/π belongs to the numerical module of $\omega(t)$. The theorem has thus been proved.

---

[3] It is not difficult to show that there is a relatively dense set of such ε-translations, and that this set determines the module of the function up to terms that are multiples of 2Δ (see B. M. Levitan [1]).

COROLLARY. *Since all the zeros of a function of the class* $[\Delta]$ *lie in some strip* $|y| \leqslant H$, *it follows from the representation* (4) *that*

$$a_k = \frac{\pi}{\Delta} k + \psi(k), \tag{6}$$

where $\psi(k)$ *is a bounded function.*

It is interesting to note that there are functions of the class $[\Delta]$ for which $\operatorname{Im} \psi(k)$ is not an almost-periodic function. The function

$$g(z) = \sin z \sin \lambda (z - i)$$

with λ irrational is an example. In fact, the function $\psi_1(k) = \operatorname{Im} a_k$ in this case takes only the two values 0 and 1. For $\varepsilon < 1$, every ε-translation of $\psi_1(k)$ must be simply a period of this function.

Let the integer τ be the period of $\psi_1(k)$. Then all the zeros a_k with subscripts $k = n\tau + r$ $(\pm n = 0, 1, 2, \ldots)$ are zeros of only one of the functions $\sin z$ or $\sin \lambda(z - i)$. Thus the set of points a_k can be divided up into τ classes $(r = 0, 1, \ldots, \tau - 1)$, each of which consists of the points of the form $m\pi$ or of points of the form $m\pi/\lambda + i$ $(\pm m = 0, 1, 2, \ldots)$.

Let p_1 be the number of classes of the first type and p_2 the number of the second type $(p_1 + p_2 = \tau)$. Let T_n be a number such that $n\tau = n(T_n)$. Then it is obvious that

$$T_n = np_1\pi + \varepsilon_1(T_n) = np_2 \frac{\pi}{\lambda} + \varepsilon_2(T_n),$$

where $0 \leqslant \varepsilon_1(T_n) < \pi$, $0 \leqslant \varepsilon_2(T_n) < \pi/\lambda$. Dividing this equality by n and taking the limit for $n \to \infty$, we obtain

$$p_1 = \frac{p_2}{\lambda},$$

i.e., λ is rational and the assumption is contradicted. Thus $\psi_1(k)$ can not be almost-periodic.

The following theorem is of interest in connection with this example.

THEOREM 3. *If the sequence of zeros of an almost-periodic function of the class* $[\Delta]$ *satisfies the condition*

$$\operatorname{Re}(a_{k+1} - a_k) \geqslant \delta > 0 \qquad (\pm k = 0, 1, 2, \ldots),$$

then $a_k = (\Delta/\pi)k + \psi(k)$, *where* $\psi(x)$ *is almost-periodic.*

PROOF. We denote the rectangle $x = \pm\delta/2$, $y = \pm(H + \delta)$ by I, and consider the integral

$$\varphi_1(k) = \frac{1}{2\pi i} \int_{(I)} \frac{f'\left(z + \frac{\pi}{\Delta} k + \varphi(k)\right)}{f\left(z + \frac{\pi}{\Delta} k + \varphi(k)\right)} z \, dz, \tag{7}$$

in which $\varphi(k)$ is the almost-periodic function equal to $\operatorname{Re} a_k - (\pi/\Delta)k$. It is evident that the rectangle (I) contains one simple zero of the function $f(z + (\pi/\Delta)k + \varphi(k))$, and so $\varphi_1(k) = i \operatorname{Im} a_k$.

From Lemma 1 we have $|f(z + (\pi/\Delta)k + \varphi(k))| \geqslant m > 0$ on the contour of integration (I). Moreover, since the numerical module of $\varphi(k)$ multiplied by Δ/π belongs to the numerical module of $f(z)$, to every number $\sigma > 0$ we can associate a number $\eta \geqslant 0$ such that every η-translation of $f(z)$ multiplied by Δ/π will be a σ-translation of $\varphi(k)$.

Let $\varepsilon \geqslant 0$. We choose $\sigma > 0$ and $\eta > 0$ so small that for any η-translation $\tau = (\pi/\Delta)q$ (q an integer) of the function $f(z)$ the inequality

$$\left| \frac{f'\left(z+\tau+\frac{\pi}{\Delta}k+\varphi(k+q)\right)}{f\left(z+\tau+\frac{\pi}{\Delta}k+\varphi(k+q)\right)} - \frac{f'\left(z+\frac{\pi}{\Delta}k+\varphi(k)\right)}{f\left(z+\frac{\pi}{\Delta}k+\varphi(k)\right)} \right| < \frac{\varepsilon}{2(H+2\delta)^2}$$

holds. From the integral representation of $\varphi_1(k)$ we obtain

$$|\varphi_1(k+q) - \varphi_1(k)| < \varepsilon.$$

Thus $\varphi_1(k)$ is an almost-periodic function, and so $\psi(k) = \varphi(k) + \varphi_1(k)$ is also almost-periodic. The theorem has been proved.

We note that here also we can find an $\eta > 0$ for any $\varepsilon > 0$ such that an arbitrary η-translation of $f(z)$ multiplied by Δ/π is an ε-translation of $\varphi_1(k)$. Hence the module of $\psi(k)$ multiplied by Δ/π is contained in the module of $f(z)$.

2. A Characterization of the Set of Zeros of an Almost-Periodic Function of the Class [Δ]

DEFINITION 1. A number τ is called an *ε-translation of the point set* $\{a\}$ if there is a one-to-one correspondence between the points of the set $\{a + \tau\}$ and the points of $\{a\}$ such that the distance between corresponding points is less than ε.

DEFINITION 2. A set $\{a\}$ of points is called *an almost-periodic point set* if to every $\varepsilon > 0$ there corresponds a relatively dense set of its ε-translations.

LEMMA 1. *The set of zeros of an almost-periodic function of the class* [Δ] *is an almost-periodic point set.*

PROOF. Let the number ε $(0 < \varepsilon < \frac{1}{2})$ be given. We choose $\delta > 0$ small enough so that $2k\delta < \varepsilon$, where k is not smaller than the number of zeros in the rectangle $t \leqslant x \leqslant t + 1$, $|y| \leqslant H$. Then the diameter of every region covered by the exceptional circles of radius δ with centers at the zeros a_k of $f(z)$ does not exceed ε.

From Lemma 1, Chapter VI, there is a number $m_\delta > 0$ such that

$$|f(z)| \geqslant m_\delta > 0 \tag{8}$$

outside such regions.

Let τ be an $(m_\delta/2)$-translation of $f(z)$ in the strip $|y| \leqslant H + \delta$. Then from (8) the function $f(z - \tau)$ has no zeros outside the exceptional regions. Thus from Rouché's theorem the functions $f(z)$ and $f(z - \tau)$ have the same number of zeros inside each of these regions.

Having established a one-to-one correspondence between the set $\{a_k\}$ of zeros of $f(z)$ and the set $\{a_k + \tau\}$ of zeros of $f(z - \tau)$ inside every such region, we find that τ is an ε-translation of the set of zeros of $f(z)$. Therefore each $(m_\delta/2)$-translation of $f(z)$ in the strip $|y| \leqslant H + \delta$ is an ε-displacement of the set $\{a_k\}$.

We note that if a_k can be expressed in the form

$$a_k = ck + \psi(k), \tag{6}$$

where c is a constant and $\psi(x)$ is almost-periodic, then the set $\{a_k\}$ is an almost-periodic point set. The example in § 1 shows that such a representation does not exist for every almost-periodic point set.

LEMMA 2. *Let $\psi(k)$ be a bounded function, let the points*

$$a_k = ck + \psi(k)$$

form an almost-periodic point set,[4] *and let*

$$f(z) = P.V. \prod_{-\infty}^{\infty}\left(1 - \frac{z}{a_k}\right). \tag{9}$$

Then

$$\chi_y(x) = \frac{f'(x+iy)}{f(x+iy)}$$

is an almost-periodic function of x for $|y| \geqslant H =$ sup (Im $\psi(k)$).

PROOF. It follows from the representation (9) of $f(z)$ that

$$\chi_y(x) = P.V. \sum_{-\infty}^{\infty} \frac{1}{z - a_k}.$$

If a_{n_k} denotes the zero corresponding to a_k after an ε-translation τ, so that

$$|a_{n_k} + \tau - a_k| < \varepsilon,$$

then

$$|\chi_y(x+\tau) - \chi_y(x)| \leqslant \sum_{-\infty}^{\infty} \frac{|a_k - (a_{n_k} + \tau)|}{|z - a_k|\,|z - a_{n_k} - \tau|} \tag{10}$$

$$< M^2_\delta \varepsilon \sum_{-\infty}^{\infty} \frac{1}{|z - ck|^2}.$$

[4] We may assume, with no loss of generality, that the set $\{a_k\}$ does not contain 0.

Here M_δ is a constant such that, for $|y| \geqslant H + \delta$ ($\delta > 0$),

$$\left|\frac{z-ck}{z-a_k}\right| < M_\delta \quad \text{and} \quad \left|\frac{z-ck}{z-a_{n_k}-\tau}\right| < M_\delta .$$

The sum on the right in (10) is periodic and therefore bounded. The lemma has been proved.

LEMMA 3. *If*

$$a_k = ck + \psi(k) \qquad (\pm k = 0, 1, 2, \ldots),$$

where $\psi(k)$ *is a bounded function and*

$$f(z) = P.V \prod_{-\infty}^{\infty}\left(1 - \frac{z}{a_k}\right), \tag{9}$$

then in order that the function

$$\theta(x+ia) = \ln\left\{f(x+ia)e^{i\frac{\pi}{c}x}\right\} \qquad (a > 0)$$

be bounded for all sufficiently large, fixed values of a, *it is necessary and sufficient that there exist a constant* M *such that for any* $h \geqslant 0$ *and arbitrary integral* τ

$$\left|\sum_{-\infty}^{\infty}[\psi(k+\tau) - \psi(k)]\frac{k}{k^2+h^2}\right| < M, \tag{11}$$

where M *is independent of* τ.

PROOF. From (9) we obtain

$$\theta(x+c\tau+ia) - \theta(x+ia) = \ln\frac{f(x+c\tau+ia)}{f(x+ia)}e^{i\pi\tau}$$

$$= \lim_{N\to\infty}\left\{\ln\prod_{-N}^{N}\frac{ck+\psi(k)-c\tau-x-ia}{ck+\psi(k)-x-ia} + i\pi\tau\right\}$$

$$= \lim_{N\to\infty}\ln\prod_{-N}^{N-\tau}\frac{ck+\psi(k+\tau)-x-ia}{ck+\psi(k)-x-ia}$$

$$+ \lim_{N\to\infty}\left\{\ln\prod_{q=1}^{\tau}\frac{c(-N-q)+\psi(-N-q+\tau)-x-ia}{c(N-\tau+q)+\psi(N-\tau+q)-x-ia} + i\pi\tau\right\}.$$

Since every factor in the product in the second term tends to $e^{-i\pi}$, the logarithm in this term tends to $-i\pi\tau$ and so the whole term tends to zero. Hence,

$$\theta(x+c\tau+ia) - \theta(x+ia)$$

$$= P.V.\sum_{-\infty}^{\infty}\ln\left[1 + \frac{\psi(k+\tau)-\psi(k)}{ck+\psi(k)-x-ia}\right]. \tag{12}$$

The boundedness of $\theta(x+ia)$ in $-\infty < x < \infty$ is obviously equivalent to the uniform boundedness of $\theta(x+c\tau+ia) - \theta(x+ia)$ for $0 \leqslant x \leqslant c$ and for all integral values of τ. We will show that the uniform boundedness of the right-hand side of (12) is equivalent to the condition (11). We first estimate the size of

$$\left|\frac{\psi(k+\tau)-\psi(k)}{ck+\psi(k)-x-ia}\right|.$$

We write $M = \sup_k |\psi(k)|$ and

$$l_a = \sup_{-\infty<t<\infty} \frac{|t-ia|}{|t-ia|-M} = \frac{a}{a-M}$$

for $a > M$. We then have

$$\left|\frac{\psi(k+\tau)-\psi(k)}{ck+\psi(k)-x-ia}\right| \leqslant \frac{2Ml_a}{|ck-x-ia|},$$

and for sufficiently large a this expression can be made less than $\frac{1}{2}$. Using the inequality $|\ln(1-\alpha)-\alpha| < |\alpha|^2$, which is valid for $|\alpha| < \frac{1}{2}$, we find that the uniform boundedness of $\theta(x + c\tau + ia) - \theta(x + ia)$ for $0 \leqslant x \leqslant c$ and for all integral τ is equivalent to the boundedness of

$$\sum_{-\infty}^{\infty} \frac{\psi(k+\tau)-\psi(k)}{ck+\psi(k)-x-ia} \tag{13}$$

for all integral τ. Actually this equivalence can be obtained directly by noting that

$$\sum_{-\infty}^{\infty}\left|\frac{\psi(k+\tau)-\psi(k)}{ck+\psi(k)-x-ia}\right|^2 < (2Ml_a)^2 \sum_{-\infty}^{\infty} \frac{1}{|ck-x-ia|^2} < k_a$$

$$(0 \leqslant x \leqslant c),$$

where k_a is a constant. From a comparison of the expression (13) with

$$\sum_{-\infty}^{\infty} \frac{\psi(k+\tau)-\psi(k)}{ck-ia}, \tag{14}$$

we obtain

$$\left|\sum_{-\infty}^{\infty} \frac{\psi(k+\tau)-\psi(k)}{ck+\psi(k)-x-ia} - \sum_{-\infty}^{\infty} \frac{\psi(k+\tau)-\psi(k)}{ck-ia}\right|$$

$$\leqslant 2M \sum_{-\infty}^{\infty} \frac{|\psi(k)-x|}{|ck+\psi(k)-x-ia|\,|ck-ia|} \leqslant K' \sum_{-\infty}^{\infty} \frac{1}{|ck-ia|^2} < K'',$$

where K' and K'' are constants. Finally, a comparison of (14) with the left side of (11), and the use of the notation $h = a/c$, yields

$$\left|\sum_{-\infty}^{\infty} \frac{\psi(k+\tau)-\psi(k)}{k-ih} - \sum_{-\infty}^{\infty} \frac{\psi(k+\tau)-\psi(k)}{k^2+h^2}\,k\right| < 2M \sum_{-\infty}^{\infty} \frac{h}{k^2+h^2}$$

Therefore the boundedness of $\theta(x + ia)$ for all sufficiently large, fixed values of $a > 0$ is equivalent to condition (11). The lemma is proved.

It is useful to note further that if $\theta(x + ia)$ is bounded for any value of a ($a > \sup_k \psi(k)$), then it is bounded for all larger values of a.[4a]

We can now give the following complete characterization of the set of zeros of an almost-periodic function belonging to the class $[\Delta]$.

THEOREM 4. *For $f(z)$ to be an almost-periodic function of the class $[\Delta]$, it is necessary and sufficient that the following conditions hold*:

(a) *$f(z)$ can be expressed in the form*

$$f(z) = c \lim_{N \to \infty} \prod_{-N}^{N} \left(1 - \frac{z}{a_k}\right) \qquad (c \text{ a constant}); \tag{15}$$

(b) *the zeros a_k of $f(z)$ form an almost-periodic point set and can be given by the formula*

$$a_k = \frac{\pi}{\Delta} k + \psi(k),$$

where $\psi(k)$ is a complex-valued, bounded function;

(c) *for any integral τ we have the inequality $|\mathfrak{L}_\tau[\psi]| < M$, where*

$$\mathfrak{L}_\tau[\psi] = \sum_{-\infty}^{\infty} [\psi(k+\tau) - \psi(k)] \frac{k}{k^2 + h^2}$$

and M is independent of τ.

PROOF. NECESSITY. If $f(z)$ belongs to the class $[\Delta]$, then from Lemma 1 the set $\{a_k\}$ of its zeros is an almost-periodic point set. Moreover, from the corollary to Theorem 2 we have

$$a_k = \frac{\pi}{\Delta} k + \psi(k), \tag{6}$$

where $\psi(k)$ is a bounded function. Thus condition (b) is satisfied.

It follows from (7) that the product

$$\varphi(z) = \lim_{N \to \infty} \prod_{-N}^{N} \left(1 - \frac{z}{a_k}\right)$$

converges and that $\varphi(z)$ is of completely regular growth, while the indicator diagram of $\varphi(z)$ is the segment $[-i\Delta, i\Delta]$. The quotient

$$f(z) : \varphi(z)$$

[4a] By a slight modification of the argument used to prove Lemma 3, one can prove that if $a_k = ck + \psi(k)$, where $\psi(k)$ is bounded, then the condition $|\mathfrak{L}_\tau[\mathrm{Re}\,\psi(k)]| \leqslant L < \infty$ is necessary and sufficient for an entire function $f(z)$ of the form (9) to satisfy

$$0 < m_a \leqslant |f(x + ia)| \leqslant M_a < \infty,$$

where m_a and M_a are constants and $|a| > \sup \mathrm{Im}\,\psi(k)$. Heĭfic [1] has remarked that the condition $|\mathfrak{L}_\tau[\mathrm{Re}\,\psi(k)]| \leqslant c\nu \ln(1 + |\tau|) + o(1)$ $(\nu > 0)$ is necessary and sufficient for the inequalities $m_a(1 + |x|)^{-\nu} \leqslant |f(x + ia)| \leqslant M_a(1 + |x|)^{\nu}$.
Finally, Heĭfic has shown that the condition

$$\sum_{\tau = -\infty}^{\infty} \exp\left\{\frac{p}{c} \mathfrak{L}_\tau[\mathrm{Re}\,\psi]\right\} < \infty$$

is necessary and sufficient for $f(x)$ to belong to $L^p(-\infty, \infty)$ $(p > 0)$.

is an entire function of exponential type with no zeros, and its indicator diagram consists of a single point, the origin. Thus the quotient $f(z): \varphi(z)$ is constant, and so condition (a) is satisfied.

Further, if $f(z)$ is of the class $[\Delta]$, then for $|a| > H$ $(\delta > 0)$ the function $\ln |f(x + ia)|$ is bounded. Moreover, $\arg f(x + ia) = -\Delta x + \chi(x)$ where $\chi(x)$ is almost-periodic (Chapter VI). Thus for $|a| > H$ the function

$$\theta(x + ia) = \ln f(x + ia)\, e^{i\Delta x}$$

is bounded, and so from Lemma 3 it follows that condition (c) holds.

SUFFICIENCY. Let $f(z)$ be expressed as in (15) with $a_k = (\pi/\Delta)k + \psi(k)$. Then $f(z)$ is an entire function of exponential type Δ. It also follows from Lemma 2 that

$$\frac{f'(x + iy)}{f(x + iy)} + i\Delta$$

is an almost-periodic function of x for $|y| > \sup_{-\infty<k<\infty} |\operatorname{Im} \psi(k)|$.

The integral of this function with respect to the real variable from zero to x is $\theta(x + ia) - \theta(ia)$, and from condition (c) and Lemma 5 this integral is bounded. From a well-known theorem due to Bohr,[5] if an indefinite integral of an almost-periodic function is bounded, then it is also an almost-periodic function. Thus

$$\ln f(x + ia)\, e^{i\Delta x}$$

is almost-periodic for all sufficiently large $|a|$. The same result holds for $f(x + ia)$.

From the Phragmén-Lindelöf theorem for a strip, $f(z)$ is bounded in any strip parallel to the real axis, and since it is almost-periodic on any straight line, it is almost-periodic in the whole plane. Its whole spectrum is in the segment $[-i\Delta, i\Delta]$. Since all its zeros are in a strip, from Corollary 2, Theorem 2, § 2, Chapter VI, it follows that $f(z)$ belongs to the class $[\Delta]$. The theorem has thus been proved.

REMARK. If the zeros a_k can be given by the formula

$$a_k = \frac{\pi}{\Delta} k + \psi(k),$$

where $\psi(k)$ is almost-periodic, then for any $\varepsilon > 0$ there is a $\delta > 0$ such that every δ-translation of $\psi(k)$ becomes, after multiplication by π/Δ, an ε-translation of $f(z)$. Hence the module of $f(z)$ belongs to the module obtained when the smallest module $\mathfrak{M}(\psi, \pi)$ containing the number π and the module of $\psi(k)$ is multiplied by Δ/π.

---

[5] See B. M. Levitan [1].

When this result is compared with Theorem 2, we find that when the conditions of this theorem are fulfilled the module of $f(z)$ is the same as the module $\mathfrak{M}(\psi, \pi)$ multiplied by Δ/π.

We now prove that condition (c) is independent of the remaining conditions of Theorem 4.

In addition to this, we prove that there is an almost-periodic function $\psi(k)$ for which the set of functionals $L_\tau[\psi]$ ($\pm\tau = 0, 1, 2, \ldots$) is not bounded.

The set of all almost-periodic functions forms a complete, normed space with norm

$$\|\psi\| = \sup_{-\infty < k < \infty} |\psi(k)|.$$

The functionals $L_\tau[\psi]$ are linear in this space. By a well-known theorem, a set of linear functionals, bounded for every element of the space in which they are defined, is bounded in norm, i.e., if for all almost-periodic functions we have

$$|L_\tau[\psi]| \leqslant M_\psi \qquad (\pm\tau = 0, 1, 2, \ldots),$$

then there is a number M such that

$$|L_\tau[\psi]| < M\|\psi\| \qquad (\pm\tau = 0, 1, 2, \ldots). \tag{16}$$

The functional $L_\tau[\psi]$ can be written in the form

$$L_\tau[\psi] = \sum_{-\infty}^{\infty} \psi(k)\left[\frac{k-\tau}{(k-\tau)^2+h^2} - \frac{k}{k^2+h^2}\right].$$

The kernel $(k - \tau)/[(k - \tau)^2 + h^2] - k/(k^2 + h^2)$ changes sign at two points k_1 and k_2, where $k_2 \to \tau$ and $k_1 \to 0$ for $\tau \to \infty$. For fixed τ, we set $\psi(k) = -1$ ($k_1 < k < k_2$). We also choose the number $T > 0$ so that $k_2 - k_1 + T$ will be an integer and so that

$$\left(\sum_{-\infty}^{-T+k} + \sum_{T+k_2}^{\infty}\right)\left|\frac{k-\tau}{(k-\tau)^2+h^2} - \frac{k}{k^2+h^2}\right| < 1,$$

and define $\psi(k) = 1$ on the partial interval $k_2 < k \leqslant T + k_2$. We then continue $\psi(k)$ to the whole axis by periodicity, using the period $\omega = k_2 - k_1 + T$.

We then have

$$L_\tau[\psi] > \sum_{-\infty}^{\infty}\left|\frac{k-\tau}{(k-\tau)^2+h^2} - \frac{k}{k^2+h^2}\right| - 1 = 2\ln\tau + O(1),$$

which contradicts (16). Thus the set of linear functionals $L_\tau[\psi]$ is not bounded, and so there is an almost-periodic function $\Phi(k)$ for which the family of functionals is unbounded.[6] Condition (c) is difficult to verify, and so it is useful to have the following sufficient condition that it be satisfied.

[6] Such a function can in fact be constructed.

THEOREM 5. *If the Fourier series of an almost-periodic function $\psi(k)$ is absolutely convergent, then* $|L_\tau[\psi]| < M$, *i.e., condition* (c) *holds.*

PROOF. We write

$$\vartheta(\lambda) = L_\tau[e^{i\lambda x}] = 2i(e^{i\lambda\tau} - 1)\sum_{k=1}^{\infty}\frac{k\sin k\lambda}{k^2+h^2}$$

or, otherwise,

$$\vartheta(\lambda) = \begin{cases} \dfrac{\pi \operatorname{sh} h(\pi-\lambda)}{\operatorname{sh}\pi h}(e^{i\lambda\tau}-1) & (0<\lambda\leqslant\pi), \\ -\dfrac{\pi \operatorname{sh} h(\pi+\lambda)}{\operatorname{sh}\pi h}(e^{i\lambda\tau}-1) & (-\pi\leqslant\lambda<0). \end{cases} \tag{17}$$

Thus $\vartheta(\lambda + 2\pi) = \vartheta(\lambda)$.[6a]

Since $\psi_n(t)$ tends uniformly to $\psi(t)$, then for fixed τ we have

$$L_\tau[\psi_n] \to L_\tau[\psi].$$

Hence if

$$\psi(k) = \sum_{-\infty}^{\infty} c_j e^{i\lambda_j k} \qquad \left(\sum_{-\infty}^{\infty}|c_j| < \infty\right),$$

then

$$L_\tau[\psi] = \sum_{-\infty}^{\infty} c_j \vartheta(\lambda_j)$$

or

$$|L_\tau[\psi]| \leqslant 2\left(\sum_{-\infty}^{\infty}|c_j|\right)\frac{\pi \operatorname{sh} h\pi}{\operatorname{sh} h}.$$

The theorem has thus been proved.

Condition (c) can be transformed into another form. To perform this transformation, we prove an interpolation theorem that is interesting in itself.

THEOREM 6. *In order that there exist an entire function of exponential type not exceeding π, bounded on the real axis, and taking the values $(-1)^k\psi(k)$ for all integral values k of the argument where $\psi(k)$ is a bounded function, it is necessary and sufficient that for all integral values of τ we have*

$$|L_\tau[\psi]| < M, \tag{11}$$

where M is a constant not depending on τ.

PROOF. We form the interpolation series

$$\vartheta(z) = \frac{\sin \pi z}{\pi}\left\{\sum_{-\infty}^{\infty}{}' \psi(k)\left[\frac{1}{z-k}+\frac{1}{k}\right]+\frac{\psi(0)}{z}\right\}.$$

This series converges uniformly in every finite region to an entire function of exponential type not greater than π. We have the relation $\vartheta(k) = (-1)^k\psi(k)$.

[6a] Note that in this derivation we do not assume that τ is an integer.

We will prove that when condition (11) holds the value of $|\vartheta(x)|$ $(-\infty < x < \infty)$ is bounded. Let $z = x \pm ih$, $|x| \leqslant 1$ and let τ be an integer. Then

$$\vartheta(x \pm ih + \tau) - \vartheta(x \pm ih) = \frac{\sin \pi (x \pm ih)}{\pi} \sum_{-\infty}^{\infty} \frac{\psi(k-\tau) - \psi(k)}{x - k \pm ih}.$$

If condition (11) is satisfied, the absolute value of the sum on the right is bounded for all τ, and so

$$|\vartheta(x \pm ih + \tau) - \vartheta(x \pm ih)| < M_1 \quad (|x| \leqslant 1),$$

where M_1 is a constant. Thus $\vartheta(x \pm ih)$ is bounded for $-\infty < x < \infty$, and from the Phragmén-Lindelöf theorem $\vartheta(x)$ is bounded on the real axis.

Conversely, let it be given that there exists an entire function $\vartheta(z)$ of exponential type not greater than π, bounded on the axis, and such that $\vartheta(k) = (-1)^k \psi(k)$. We will prove that

$$|L_\tau[\psi]| < M,$$

where M is independent of τ. We consider the function

$$\gamma(z) = \frac{\sin \pi z}{\pi} \left\{ \frac{\psi(0)}{z} + \sum_{-\infty}^{\infty}{}' \psi(k) \left[\frac{1}{z-k} + \frac{1}{k} \right] \right\}.$$

This is an entire function of exponential type, with an indicator diagram in the segment $[-i\pi, i\pi]$. The function

$$\chi(z) = \frac{\vartheta(z) - \gamma(z)}{\sin \pi z}$$

is entire and of exponential type, and its indicator diagram consists of the origin only. We will obtain information on the size of this function on the imaginary axis. From the definition of $\gamma(z)$, we have

$$|\gamma(iy)| \leqslant \frac{M}{\pi} |\operatorname{sh} \pi y| \left\{ \frac{1}{|y|} + 2 \sum_{1}^{\infty} \frac{|y|}{k\,|k - iy|} \right\}$$

$$\leqslant \frac{M}{\pi} |\operatorname{sh} \pi y| \left\{ \frac{1}{|y|} + \frac{2\pi}{\sqrt{6}} |y| \left(\sum_{1}^{\infty} \frac{1}{k^2 + y^2} \right)^{\frac{1}{2}} \right\}$$

$$< \frac{M}{\pi} |\operatorname{sh} \pi y| \left\{ \frac{1}{|y|} + \frac{\pi \sqrt{\pi |y|}}{\sqrt{3}} \right\}.$$

Moreover, the boundedness of $\vartheta(z)$ on the real axis yields

$$|\vartheta(z)| < Ne^{\pi |y|}.$$

It follows from the last two inequalities that $\chi(iy) = O(\sqrt{|y|}\,)$, and since $\chi(z)$ is of type zero, we have $\chi(z) = \text{const}$.

We thus have

$$\vartheta(z) = \frac{\sin \pi z}{\pi} \left\{ c + \frac{\psi(0)}{z} + \sum_{-\infty}^{\infty}{}' \psi(k) \left[\frac{1}{z-k} - \frac{1}{k} \right] \right\}.$$

For τ even, we thus have

$$\vartheta(ih+\tau)-\vartheta(ih)=\frac{\operatorname{sh}\pi h}{i\pi}\sum_{-\infty}^{\infty}\psi(k)\left[\frac{1}{ih-k+\tau}-\frac{1}{ih-k}\right]$$

$$=\frac{\operatorname{sh}\pi h}{i\pi}\sum_{-\infty}^{\infty}\frac{\psi(k+\tau)-\psi(k)}{ih-k},$$

and from the boundedness of $\vartheta(ih+x)$ we obtain

$$\left|\sum_{-\infty}^{\infty}[\psi(k+\tau)-\psi(k)]\frac{k}{k^2+h^2}\right|\leqslant\frac{\pi}{\operatorname{sh}\pi h}|\vartheta(ih+\tau)-\vartheta(ih)|$$

$$+2M\sum_{-\infty}^{\infty}\frac{h}{k^2+h^2}<CM,$$

where C is a constant. When τ is odd, we have

$$L_\tau[\psi]=\sum_{-\infty}^{\infty}[\psi(k+\tau-1)-\psi(k)]\frac{k}{k^2+h^2}$$

$$+\sum_{-\infty}^{\infty}\psi(k+\tau-1)\left[\frac{k-1}{(k-1)^2+h^2}-\frac{k}{k^2+h^2}\right],$$

and so

$$|L_\tau[\psi]|<M_1 \qquad (M_1 \text{ a constant }).$$

The theorem has been proved.

When we compare this theorem with Lemma 3, we obtain the following theorem.

THEOREM 7. *Let* $a_k=ck+\psi(k)$ $(\pm k=0,1,2,\ldots)$, *where* $\psi(k)$ *is bounded. For the function*

$$f(z)=P.V.\prod_{-\infty}^{\infty}\left(1-\frac{z}{a_k}\right)$$

to satisfy the inequality

$$|\ln[f(x+ia)e^{i\pi x/c}]|\leqslant M_a{}^{6b}$$

on every straight line $y=a$ $(|a|>\sup_k\psi(k))$, *where* M_a *is a constant, it is necessary and sufficient that there exist an entire function of exponential type not exceeding* π, *bounded on the real axis, that takes the values* $(-1)^k\psi(k)$ *for integral* k.

Theorem 7 can be used as follows in the reformulation of Theorem 4.

[6b] It can be shown that this condition is equivalent to having the entire function $f(z)$ be of sine type; that is,

$$0<m_a\leqslant|f(x+iy)|e^{-\pi|y|}\leqslant M_a<\infty;\qquad |y|>a.$$

THEOREM 8. *For $f(z)$ to be a function of the class $[\Delta]$, it is necessary and sufficient that the following conditions be satisfied:*

(a) *$f(z)$ can be expressed in the form*

$$f(z)=\lim_{N\to\infty} c\prod_{-N}^{N}\left(1-\frac{z}{a_k}\right);$$

(b) *the zeros a_k form an almost-periodic point set;*

(c) *the zeros a_k can be given by the formula*

$$a_k=\frac{\pi}{\Delta}k+\psi(k),$$

where $\psi(z)e^{-i\pi z}$ is an entire function of exponential type $H \leqslant \pi$, bounded on the real axis.

3. The Relation Between the Fourier Series of $\psi(x)$ and $f(x)$

THEOREM 9. *Let $\psi(x)$ be any almost-periodic function with absolutely convergent Fourier series, and let*

$$a_k=\frac{\pi}{\Delta}k+\psi(k) \qquad (\pm k=0,\ 1,\ 2,\ \ldots;\ \Delta>0).$$

Then $f(z)$, defined by (15), *is a function of the class $[\Delta]$ with an absolutely convergent Fourier series.*

PROOF. We start from the relation (12). Setting $x = 0$ and replacing τ by x/c in this relation, we obtain

$$\ln\frac{f(x+ia)\,e^{i\frac{\pi}{c}x}}{f(ia)}=P.V.\sum_{k=-\infty}^{\infty}\ln\left[1+\frac{\psi\left(k+\frac{x}{c}\right)-\psi(k)}{ck+\psi(k)-ia}\right]\left(c=\frac{\pi}{\Delta}\right),$$

for $x = cn$ ($\pm n = 0, 1, 2, \ldots$). For sufficiently large $|a|$,

$$\ln\frac{f(x+ia)\,e^{i\Delta x}}{f(ia)}=P.\,V.\sum_{k=-\infty}^{\infty}\frac{\psi\left(k+\frac{x}{c}\right)-\psi(k)}{ck+\psi(k)-ia}$$

$$+\sum_{n=2}^{\infty}\frac{(-1)^{n-1}}{n}\sum_{k=-\infty}^{\infty}\left(\frac{\psi\left(k+\frac{x}{c}\right)-\psi(k)}{ck+\psi(k)-ia}\right)^n, \tag{18}$$

for all real values of x. The hypothesis of the theorem yields

$$\psi(x)=\sum_{-\infty}^{\infty}c_je^{i\lambda_jx} \qquad \left(\sum_{-\infty}^{\infty}|c_j|<\infty\right). \tag{19}$$

If we use $||\psi|| = \sum_{-\infty}^{\infty}|c_j|$ for the norm in the ring (W) of functions of the form (19), then for any fixed k we have

$$\left\|\left[\psi\left(k+\frac{x}{c}\right)-\psi(k)\right]^n\right\|\leqslant(2\,\|\psi\|)^n. \tag{20}$$

In addition to this, for $n \geqslant 2$, we have

$$\sum_{-\infty}^{\infty} \frac{1}{(ck+\psi(k)-ia)^n} = O\left(\frac{1}{|a|^{n-1}}\right).$$

Thus for $|a| > 2\,||\psi||$ the second series in (18) converges in norm, and so it converges to an almost-periodic function with an absolutely convergent Fourier series.

The first sum in (18) can be expressed in the form

$$P.\ V. \sum_{-\infty}^{\infty} \frac{\psi\left(k+\frac{x}{c}\right)-\psi(k)}{ck+\psi(k)-ia} = P.V. \sum_{-\infty}^{\infty}\left[\psi\left(k+\frac{x}{c}\right)-\psi(k)\right]$$

$$\times \frac{ck}{c^2k^2+a^2} + ia\sum_{-\infty}^{\infty} \frac{\psi\left(k+\frac{x}{c}\right)-\psi(k)}{c^2k^2+a^2}$$

$$-\sum_{-\infty}^{\infty}\left[\psi\left(k+\frac{x}{c}\right)-\psi(k)\right]\frac{\psi(k)}{(ck+\psi(k)-ia)(ck-ia)}.$$

The convergence in norm of the second and third series follows from (20). From (17) we infer that the first series is also convergent in norm. Therefore the function

$$\theta(x) = P.V. \sum_{-\infty}^{\infty} \ln\left[1 + \frac{\psi\left(k+\frac{x}{c}\right) - \psi(k)}{ck + \psi(k) - ia}\right] \qquad (c = \pi/a)$$

is almost periodic with an absolutely convergent Fourier series, and all the frequencies in this series belong to the numerical module $c^{-1}\mathfrak{M}_\psi$ or $(\Delta/\pi)\mathfrak{M}_\psi$. The same statement applies to the function

$$\chi(x) = e^{\theta(x)} = \sum_{-\infty}^{\infty} c_k e^{i\mu_k x};$$

that is, $\sum_{-\infty}^{\infty}|c_k| < \infty$ and $\mu_k \in (\Delta/\pi)\mathfrak{M}_\psi$. If we consider the function $\chi(x)$ only at points of the form $x = cn$ $(\pm n = 0, 1, 2, \ldots)$, the frequencies μ_k can be replaced by numbers that are congruent to μ_k modulo $2\pi/c = 2\Delta$ and lie in the interval $[0, 2\Delta]$. We obtain an expansion

$$\chi_1(x) = \sum^{\infty} d_k e^{i\gamma_k x},$$

where $\sum_{-\infty}^{\infty}|d_k| < \infty$ and $0 \leqslant \gamma_k \leqslant 2\Delta$. In addition, $\chi_1(x)$ coincides with $\chi(x)$ at the points cn, and (because of (18)) with the function

$$f^{-1}(ia)f(x+ia)e^{i\Delta x}.$$

The difference $f(x+ia) - f(ia)\chi_1(x)e^{-i\Delta x}$ is clearly an entire function of exponential type $\leqslant \Delta$, bounded on the real axis, and vanishing at the points cn. Hence

$$f(x+ia) \equiv f(ia)\chi_1(x)e^{-i\Delta x} + c\sin\Delta x,$$

where c is a constant. Consequently $f(x)$ is represented by the absolutely convergent Fourier series

$$f(x) = \sum_{-\infty}^{\infty} c_k e^{i\lambda_k x},$$

where the exponents λ_k belong to the smallest numerical module that contains $(\Delta/\pi)\mathfrak{M}_\psi$ and the number Δ.

The example given at the end of § 1 shows that the converse of this theorem does not hold. In view of this fact, the following theorem is of interest.

THEOREM 10. *If the Fourier series of a function $f(z)$ of the class $[\Delta]$ converges absolutely, and if the sequence $\{a_k\}_{-\infty}^{\infty}$ of its zeros satisfies the condition*

$$\left|\operatorname{Re} a_k - \frac{\pi}{\Delta} k - s\right| < \frac{\pi}{4\Delta},$$

where s is any real number, then $a_k = (\pi/\Delta)k + \psi(k)$, where $\psi(k)$ is an almost-periodic function with absolutely convergent Fourier series.

PROOF. We prove this theorem by using the general theory of rings, as described by I. M. Gel'fand in [1]. We consider the ring of almost-periodic functions $\varphi(k)$ defined on the set of integers $\pm k = 0, 1, 2, \ldots$ with absolutely convergent Fourier series, and take for the norm the sum of the absolute values of the Fourier-series coefficients.[7] The set of values of the functions $\varphi(k)$ in all the maximal ideals of the ring coincides with the closure of the set of values of the $\varphi(k)$ for all integral points.

The functions $f(z + (\pi/\Delta)k + s)$ and $f'(z + (\pi/\Delta)k + s)$ are obviously elements of this ring for all complex values of z. It is also easy to see that these functions are continuous in the norm of the ring relative to z.

We denote by I the boundary contour of the rectangle $x = \pm\pi/2\Delta$, $y = \pm H + \delta$ $(\delta > 0)$. For $z \in I$ we have $|f(z + (\pi/\Delta)k + s)| \geqslant m > 0$ for all integral k, and so $f(z + (\pi/\Delta)k + s)$ is not in any maximal ideal of the ring. Hence, for $z \in I$, the function

$$\frac{f'\left(z + \frac{\pi}{\Delta} k + s\right)}{f\left(z + \frac{\pi}{\Delta} k + s\right)},$$

belongs to the ring and is continuous in norm. Since the operations of summation and taking a limit do not lead out of the ring, the function

$$\psi(k) = \frac{1}{2\pi i} \int_{(I)} \frac{f'\left(z + \frac{\pi}{\Delta} k + s\right)}{f\left(z + \frac{\pi}{\Delta} k + s\right)} z\, dz - s$$

belongs to the ring. In other words $\psi(k)$ can be expanded in an absolutely convergent Fourier series.

The following theorem follows easily from the last two theorems.

[7] To avoid ambiguity in the expansion of such a function in a Fourier series, we may assume that its spectrum is in the half interval $(-\pi, \pi]$.

THEOREM 11. *For the sequence* $\{a_k\}_{-\infty}^{\infty}$ *of zeros of a function* $f(z)$ *of the class* $[\Delta]$ *to be expressible in the form* $a_k = \frac{\pi}{\Delta} k + \psi(k)$, *with* $\psi(k)$ *almost-periodic and expressible as an absolutely convergent Fourier series, it is necessary and sufficient that, for any positive integer* p *exceeding some* p_0, *all divisors of* $f(z)$ *of the form*

$$f_{p,r}(z) = P.V.\prod_{-\infty}^{\infty}\left(1 - \frac{z}{a_{pk+r}}\right) \qquad (r = 0, 1, 2, \ldots, p-1)$$

can be expanded in absolutely convergent Fourier series.

PROOF. If $\psi(k)$ can be expanded in an absolutely convergent Fourier series, then so can $\psi(pk + r)$, and from Theorem 9 so can $f_{p,r}(z)$. Conversely, let all the functions $f_{p,r}(z)$ be expansible in absolutely convergent Fourier series for all p greater than some p_0. If p is chosen so that the inequality $|\psi(k)| < \pi p/4\Delta$ is satisfied, then

$$\left|\operatorname{Re} a_{pk+r} - \frac{\pi p}{\Delta} k - \frac{\pi r}{\Delta}\right| < \frac{\pi p}{4\Delta}.$$

From Theorem 10, taken together with this equality and the absolute convergence of the Fourier series of $f_{p,r}(z)$, it follows that a_{pk+r} can be expressed in the form

$$a_{pk+r} = \frac{\pi p}{\Delta} k + \psi_{p,r}(k) \qquad (r = 0, 1, 2, \ldots, p-1),$$

where $\psi_{p,r}(k)$ is an almost-periodic function, expansible in an absolutely convergent Fourier series. Setting

$$Q_j(z) = \left(\frac{\prod\limits_{r \neq j} \sin\frac{\pi}{p}(z - r)}{\prod\limits_{r \neq j} \sin\frac{\pi}{p}(j - r)}\right)^2$$

and defining $\psi(k)$ by the absolutely convergent series

$$\psi(k) = \sum_{j=0}^{p-1} Q_j(k)\left[\psi_{p,j}\left(\frac{k-j}{p}\right) - \frac{\pi}{\Delta}\right],$$

we obtain the representation $a_k = \frac{\pi}{\Delta} k + \psi(k)$. The theorem has thus been proved. We have shown that if the zeros of a function of class $[\Delta]$ can be expressed in the form (6) and if $\psi(k)$ can be expanded in an absolutely convergent series, then the corresponding function $f(z)$ can be expended in an absolutely convergent series and $f(z)$ has an infinite set of divisors of the same type.[8]

[8] V. P. Potapov in [1], using results obtained by M. G. Kreĭn and the author in [1], showed that every almost-periodic polynomial is infinitely divisible in the ring of entire, almost-periodic functions of the class $[\Delta]$ with absolutely convergent Fourier series, and gave a complete classification of all such divisors. We mention that, as was shown in Appendix V, almost-periodic polynomials are not necessarily divisible in the ring of almost-periodic polynomials.

In Appendix V we considered a special factorization of an almost-periodic function $f(x)$ with a bounded spectrum. We showed that, if $f(z)$ can be expanded in an absolutely convergent Fourier series and $f(x) \geqslant 0$ $(-\infty < x < \infty)$, then

$$f(x) = |\varphi(x)|^2,$$

where $\varphi(z)$ is almost-periodic, has a spectrum half the length of that of $f(x)$, and has an absolutely convergent Fourier series, and the zeros of $\varphi(z)$ are in one half-plane.

We can immediately show that the requirement in this theorem that the Fourier series of $f(x)$ be absolutely convergent is essential. In fact, we can construct an example of an almost-periodic function of class $[\Delta]$, expressible in the form

$$f(z) = P.V.\prod_{-\infty}^{\infty}\left(1 - \frac{z}{a_k}\right)\left(1 - \frac{z}{\bar{a}_k}\right) \qquad (\operatorname{Im} a_k > 0)$$

and having as a divisor the function

$$\varphi(z) = P.V.\prod_{-\infty}^{\infty}\left(1 - \frac{z}{a_k}\right),$$

which is not almost-periodic. To obtain this function we set

$$a_k = k + i\psi(k),$$

where $\psi(k)$ is a positive, almost-periodic function not satisfying condition (c) of Theorem 4. From this theorem, it follows that $\varphi(z)$ is not almost-periodic.

We number all the zeros of $f(z)$ so that the zeros a_k have even numbers and the $\bar{a}_k$ odd numbers and form the sum

$$L_\tau[\psi] = \sum_{-\infty}^{\infty}[\psi(p+\tau) - \psi(p)]\frac{p}{p^2+h^2}.$$

For even p, the quantity $\psi(p + \tau) - \psi(p)$ differs in sign from $\psi(p + 1 + \tau) - \psi(p + 1)$ and so

$$L_\tau[\psi] = \sum_{-\infty}^{\infty}[\psi(2k+\tau) - \psi(2k)]\left[\frac{2k}{(2k)^2+h^2} - \frac{2k+1}{(2k+1)^2+h^2}\right].$$

Hence

$$|L_\tau[\psi]| < c_h \sup_k |\psi(k)|,$$

where c_h is a constant independent of τ and ψ.

Thus $f(z)$ is an almost-periodic function of the class $[\Delta]$, is positive on the real axis, and can be written as $f(z) = \varphi(z)\bar{\varphi}(z)$ with $\varphi(z)$ not almost-periodic.

MISCELLANEOUS THEOREMS AND PROBLEMS

1. *Let $f(z)$ be a holomorphic function of proximate order $\rho(r)$ inside the angle $|\arg z| \leqslant \alpha$ $(0 < \alpha \leqslant \pi/2\rho)$ and let $h_f(\pm\alpha) = h_f(0)\cos\rho\alpha$. Then*

$$h_f(\theta) = h_f(0)\cos\rho\theta$$

for all values of θ satisfying $(-\alpha \leqslant \theta \leqslant \alpha)$.

PROOF. Writing $\varphi(z) = f(z^{1/\rho})$, we obtain a function of order unity, holomorphic in the right half-plane, such that the relation

$$h_f(\pm\alpha_1) = h_\varphi(0)\cos\alpha_1 \qquad (\alpha_1 = \rho\alpha)$$

is satisfied.

We find that the indicator diagram of $\varphi(z)$ lies inside the angle formed by the rays $\arg[z - h_f(0)] = (\pi/2 + \alpha_1)$, and the point $h_f(0)$ belongs to the indicator diagram. The above assertion follows directly from this.

2. *Let $f(z)$ be an entire function of order $\rho(r)$ $(\rho < 1)$, let all the zeros of $f(z)$ be on the negative real axis, and let*

$$\ln f(r) \approx \frac{\pi}{\sin\pi\rho} r^{\rho(r)}.$$

Then

$$n(r) \approx r^{\rho(r)} \qquad \text{(Titchmarsh)}.$$

PROOF. It follows from the equality

$$|\ln f(z)| = \left|\int_0^\infty \ln\left(1+\frac{z}{t}\right) dn(t)\right|$$
$$= \left| n(t)\ln\left(1+\frac{z}{t}\right)\Big|_0^\infty + \int \frac{zn(t)}{t(t+z)}\, dt \right|$$

and the inequality $n(r) < cr^{\rho(r)}$ that

$$|\ln f(z)| \leqslant cr \int_0^\infty \frac{t^{\rho(t)-1}}{|t+z|}\, dt,$$

and that for $|\arg z| \leqslant \pi - \delta$ $(\delta > 0)$ we have

$$|\ln f(z)| < c_1 r^{\rho(r)} \int_0^\infty \frac{t^{\rho(t)-1}\, dt}{|t - e^{i\delta}|} < c_\delta r^{\rho(r)},$$

where c_δ is a constant. Let $\varphi(z)$ be an entire function with zeros on the negative axis such that $n_\varphi(r) \approx r^{\rho(r)}$. Then (Chapter I, Theorem 25)

$$\ln \varphi(z) \approx J(\pi - \theta) r^{\rho(r)} \qquad (z = re^{i\theta}).$$

The function

$$\psi(z) = \frac{\ln f(z)}{\ln \varphi(z)}$$

is holomorphic and bounded in the angle $|\arg z| \leqslant \pi - \delta$, and it tends to unity on the negative real axis. It thus follows that $\psi(re^{i\theta}) \to 1$ on any ray in this angle. Thus $f(z)$ has completely regular growth and so $n(r) \approx r^{\rho(r)}$.

3. *If $f(z)$ is holomorphic and of order ρ inside and on the boundary of the angle $|\theta| \leqslant \alpha$, $\theta = \arg z$, $\pi/2\rho < \alpha < \pi/\rho$, and if $h_f(\pm\alpha) \leqslant h_f(0) \cos \rho\alpha$, then $h_f(\theta) = h_f(0) \cos \rho\theta$ for $|\theta| \leqslant \alpha$ and $f(z)$ is of completely regular growth in this angle.*

In fact, again writing $\varphi(z) = f(z^{1/\rho})$ (see § 1), we find that $\varphi(z)$ is a holomorphic function of order one inside the angle $|\arg z| \leqslant \alpha_1$ ($\alpha_1 = \rho\alpha$, $\pi/2 < \alpha_1 < \pi$) and $h_\varphi(\pm\alpha_1) \leqslant h_\varphi(0) \cos \alpha_1$. On the other hand, the fundamental relation for the indicator (Chapter I, § 16) yields $h_\varphi(\alpha_1) + h_\varphi(-\alpha_1) \geqslant 2h_\varphi(0) \cos \alpha_1$, and so $h_\varphi(\pm\alpha_1) = h_\varphi(0) \cos \alpha_1$. The three supporting lines $x = h_\varphi(0)$, $x \cos \alpha_1 + y \sin \alpha_1 - h_\varphi(\alpha_1) = 0$, and $x \cos \alpha_1 - y \sin \alpha_1 - h(-\alpha_1) = 0$ intersect at $z = h_\varphi(0)$, and thus the whole indicator diagram of $\varphi(z)$ consists of this point. Hence $h_\varphi(0) = h_\varphi(0) \cos \rho\theta$ for $|\theta| \leqslant \alpha_1$.

Considering now the function $f(z)$, we obtain $h_f(\theta) = h_f(0) \cos \rho\theta(|\theta| \leqslant \alpha)$. Since the angle corresponding to $|\arg z| \leqslant \alpha$ is greater than π/ρ, the function $f(z)$ is of completely regular growth by a theorem of M. Cartwright (Chapter IV, § 2).

4. *If $f(z)$ is an entire function of order ρ with $\frac{1}{2} < \rho < 1$ and if $h_f(\pi) \leqslant h_f(0) \cos \pi\rho$, then $f(z)$ is of completely regular growth, and any angle containing the negative real axis contains almost all the zeros of this function (i.e., all except perhaps a set of zero density), and*

$$\lim_{r\to\infty} \frac{n(r)}{r^{\rho(r)}} = \frac{\sin \pi\rho}{\pi} h_f(0).$$

PROOF. For $\alpha = \pi$ we have $\pi/2\rho < \alpha < \pi/\rho$ and so $f(z)$ is of completely regular growth for $|\theta| < \pi$ and $h_f(\theta) = h_f(0) \cos \rho\theta$ (see § 3). Since the set of rays of completely regular growth is closed (Chapter III, § 1), therefore $f(z)$ is of completely regular growth. In the angle in which the indicator is trigonometric, the density of the set of zeros is zero. Therefore the density of the set of zeros of $f(z)$ inside any angle not containing the negative real axis is zero. The function $\Delta(\psi)$ is discontinuous only for $\psi = \pi$, and the amount of the jump is equal to the density Δ of the set of zeros of $f(z)$. From (2.04), Chapter II, we have $h_f(\theta) = \Delta\pi/\sin \pi\rho$, and so

$$\Delta = \frac{\sin \pi\rho}{\pi} h_f(0).$$

5. If $\rho < \frac{1}{2}$, $h_f(\pi) < h_f(0)\cos\pi\rho$, and the function $f(z)$ is of completely regular growth along any ray $\arg z = \theta_0 \neq \pi$, then $f(z)$ is of completely regular growth in the whole plane, $\Delta_f(\psi)$ has only one discontinuity at π, and the density of the zeros is

$$\Delta = \frac{\sin\pi\rho}{\pi} h_f(0). \qquad (*)$$

Conversely, if the density of the set of zeros exists and $\Delta(\psi)$ has only one discontinuity at π, then the density can be obtained from (*) and $h_f(\pi) = h_f(0)\cos\pi\rho$.

The second assertion follows from Theorem 25, Chapter I, the first from Theorem 6, Chapter III.

6. If an entire function $f(z)$ is of completely regular growth and of integral proximate order $\rho(r)$, then

$$\delta_f = \frac{1}{\pi}\int_0^{2\pi} h(\theta)\, e^{i\rho\theta}\, d\theta.$$

For the proof we use the formula

$$h(\theta) = \int_{-2\pi}^{0} \varphi \sin\rho\varphi\, d\Delta(\varphi+\theta) + \delta_f \cos\rho\theta$$

(Chapter II, § 1) and the identity

$$\int_{-2\pi}^{0} e^{i\rho\varphi}\, d\Delta(\varphi) = 0.$$

7. *Every trigonometrically convex function $h(\theta)$ with exponent ρ can be expressed in the form*

$$h(\theta) = h(\theta_0)\cos\rho(\theta-\theta_0) + \frac{h'(\theta_0)}{\rho}\sin\rho(\theta-\theta_0) + 2\pi\int_{\theta_0}^{\theta} \sin\rho(\varphi-\theta)\, ds(\varphi), (*)$$

where

$$s(\varphi) = \frac{1}{2\pi\rho}\left[h'(\varphi) + \rho^2\int_0^{\varphi} h(\varphi)\, d\varphi\right] \qquad (\text{ A. Pfluger }).$$

In fact if $s(\varphi)$ is a differentiable function, then $h(\theta)$ is a solution of the equation

$$h'' + \rho^2 h = 2\pi\rho s'.$$

Writing the solution of this equation in terms of a Cauchy function (in § 16, Chapter I, Green's functions with periodic boundary conditions were used) we obtain (*). The general result can be obtained by approximating $s(\varphi)$ by differentiable functions.

If we now differentiate (*) with respect to θ and write $Q(\theta) = h(\theta) - (i/\rho)h'(\theta)$, we also obtain

$$Q(\theta)\, e^{-i\rho\theta} = Q(\theta_0)\, e^{-i\rho\theta_0} - 2\pi i \int_{\theta_0}^{\theta} e^{-i\rho\varphi}\, ds(\varphi).$$

8. *Let $f(z)$ and $\varphi(z)$ be entire functions of exponential type τ_f and τ_φ, where $h_f(-\pi/2) = h_f(\pi/2)$, $h_\varphi(-\pi/2 + \alpha) = h_\varphi(\pi/2 + \alpha)$, and the integrals*

$$\int_{-\infty}^{\infty} \frac{\ln|f(x)|}{1+x^2}\, dx, \qquad \int_{-\infty}^{\infty} \frac{\ln|\varphi(xe^{i\alpha})|}{1+x^2}\, dx$$

exist. Then the exponential type of the function $f(z)\,\varphi(z)$ is

$$\tau = (\tau_f^2 + \tau_\varphi^2 + 2\,|\cos\alpha\,|\,\tau_f\tau_\varphi)^{\frac{1}{2}} \quad \text{(S. N. Bernšteĭn).}$$

In this case the indicator diagrams of $f(z)$ and $\varphi(z)$ are segments. Both functions are of completely regular growth. When they are multiplied, their indicator diagrams are added. Thus the indicator diagram of the function $f(z)\,\varphi(z)$ is a parallelogram and τ is one half the longer diagonal of this parallelogram.

9. *If $|f(x)| = O(e^{+ax})$ $(x \to \pm\infty)$ and $f(z)$ is an entire function of exponential type a, then $f(z) = ce^{az}$* (Titchmarsh).

In fact

$$\varphi(z) = e^{-az} f(z)$$

is an entire function of exponential type, bounded on the real axis. Its indicator diagram is a segment of the imaginary axis or the point 0.

The first of these alternatives is impossible, since when functions are multiplied their indicator diagrams are added, and the indicator diagram of $f(z)$ is a vertical segment passing through the point a. Thus the exponential type of $f(z)$ would have to be greater than a. Therefore $\varphi(z)$ is of exponential type zero, and since it is bounded on the real axis, we have $\varphi(z) \equiv \text{const}$.

10. *Let $F(z)$ be an entire function of exponential type and let*

$$|F(x)| \leqslant M \qquad (-\infty < x < \infty). \tag{*}$$

Let $\Omega(z)$ be an entire function of exponential type, let all the zeros of $\Omega(z)$ be zeros of $F(z)$, let $h_\Omega(0) = h_\Omega(\pi)$, and let

$$\varliminf_{y \to \pm\infty} |\Omega(iy)|\, e^{-\sigma|y|} y^m > 0. \tag{**}$$

Then the function

$$\varphi(z) = \frac{F(z)}{\Omega(z)}$$

is a polynomial of degree not greater than m (N. I. Ahiezer).

PROOF. The function $\varphi(z)$ is entire and of exponential type. Since $\varphi(z)$ is of completely regular growth (Chapter V, Theorem 11), the indicators are connected by the relation

$$h_\varphi(\theta) = h_F(\theta) - h_\Omega(\theta).$$

We have

$$h_\varphi(0) + h_\varphi(\pi) = h_F(0) + h_F(\pi) - [h_\Omega(0) + h_\Omega(\pi)].$$

The indicator diagram of $F(z)$ is a segment of the imaginary axis, i.e., we have $h_F(0) + h_F(\pi) = 0$, $h_\Omega(0) + h_\Omega(\pi) \geqslant 0$ because of the properties of indicators (Chapter I, § 16); consequently

$$h_\varphi(0) + h_\varphi(\pi) \leqslant 0$$

or

$$h_\varphi(0) = h_\varphi(\pi) = 0.$$

On the other hand it follows from (**) that

$$h_\varphi\left(\pm\frac{\pi}{2}\right) \leqslant 0.$$

Hence the indicator diagram of $\varphi(z)$ consists only of the origin, i.e., it is of zero degree.

It follows from (*) that

$$|F(z)| \leqslant Me^{\sigma|y|},$$

and using (**) we obtain

$$|\varphi(iy)| \leqslant c|y|^m \qquad (c \text{ constant }).$$

The entire function of exponential type zero

$$\psi(z) = \left[\varphi(z) - \sum_{k=0}^{m-1} \frac{\varphi^{(k)}(0)}{k!} z^k\right] z^{-m}$$

is bounded on the imaginary axis, and so is a constant. Hence

$$\varphi(z) = \sum_{k=0}^{m} \frac{\varphi^{(k)}(0)}{k!} z^k.$$

11. *Let d_f be the length of the shortest interval containing all the points for which the function of a real variable $f(x)$ differs from zero. Then if $f(x)$ and $\varphi(x)$ are such that $d_f < \infty$ and $d_\varphi < \infty$, and if*

$$\psi(x) = \int_{-\infty}^{\infty} f(x-t)\varphi(t)\,dt,$$

it follows that $d_\psi = d_f + d_\varphi$ (Titchmarsh).

PROOF. It follows from the Paley-Wiener theorem that the indicator diagram for the entire function

$$\chi_f(\lambda) = \int_{-\infty}^{\infty} e^{i\lambda x} f(x)\,dx$$

is a segment of the imaginary axis of length d_f, and that the indicator diagram of

$$\chi_\varphi(\lambda) = \int_{-\infty}^{\infty} e^{i\lambda x} \varphi(x)\,dx$$

is a segment of the imaginary axis of length d_φ. The functions $\chi_f(\lambda)$ and $\chi_\varphi(\lambda)$ are of completely regular growth. Therefore the indicator diagram of

$$\chi_\psi(\lambda) = \chi_f(\lambda)\chi_\varphi(\lambda)$$

is the sum of those for $\chi_f(\lambda)$ and $\chi_\varphi(\lambda)$, and it is thus a segment of the imaginary axis of length $d_f + d_\varphi$. It follows from the representation

$$\chi_\psi(\lambda) = \int_{-\infty}^{\infty} e^{i\lambda x} \psi(x)\,dx$$

and the Paley-Wiener theorem that the shortest interval outside of which $\psi(x) = 0$ coincides with the conjugate diagram of $\chi_\psi(\lambda)$. Thus $d_\psi = d_f + d_\varphi$.

12. *Let $f(x)$ and $\varphi(x)$ belong to the class $L(0, \gamma)$ and let*

$$\int_0^x f(t)\varphi(x-t)\,dt = 0$$

in $(0, \gamma)$. Then $f(x) = 0$ almost everywhere in $(0, \alpha)$, $\varphi(x) = 0$ almost everywhere in $(0, \beta)$, and $\alpha + \beta = \gamma$ (Titchmarsh).

PROOF. The Laplace transforms

$$F(z) = \int_0^\gamma f(t)e^{-tz}\,dt, \qquad \Phi(z) = \int_0^\gamma \varphi(t)e^{-tz}\,dt$$

of the functions (assumed to be zero outside the range of integration) are entire functions of exponential type, bounded on the imaginary axis. They are of completely regular growth, and their indicators in the right half-plane are

$$h_F(\theta) = -\alpha\cos\theta, \qquad h_\Phi(\theta) = -\beta\cos\theta.$$

The indicator of the entire function of exponential type

$$F(z)\Phi(z) = \int_\gamma^{2\gamma} e^{-xz}\left\{\int_0^x f(t)\varphi(x-t)\,dt\right\}dx$$

thus has the form

$$h_{F\Phi}(\theta) = -(\alpha+\beta)\cos\theta \leqslant -\gamma\cos\theta.$$

Hence $\alpha + \beta \geqslant \gamma$. It also follows from the "delay theorem" that $f(x) = 0$ almost everywhere in $(0, \alpha)$ and $\varphi(x) = 0$ almost everywhere in $(0, \beta)$.

APPENDIX VIII

FURTHER DEVELOPMENT OF THE THEORY OF FUNCTIONS OF COMPLETELY REGULAR GROWTH

We present a brief survey of results from the further development of the theory of functions of completely regular growth. Some other results have been given in footnotes added to the relevant parts of the book; we do not repeat these here.

1. Subharmonic functions

The theory of functions of completely regular growth was generalized by Azarin [**4**–**7**] to subharmonic functions on the n-dimensional space $\mathbf{R}^n$. The order and type of subharmonic functions are defined in the natural way:

$$\rho = \overline{\lim_{|x|\to\infty}} \frac{\ln u(x)}{\ln|x|}; \qquad \sigma = \overline{\lim_{|x|\to\infty}} \frac{u(x)}{|x|^\rho},$$

where x denotes an n-vector $(x_1, \ldots, x_n)$ and $|x| = (x_1^2 + \cdots + x_n^2)^{1/2}$. The indicator $h(\vec{e})$ of a subharmonic function (where $\vec{e}$ is a unit vector), relative to a proximate order $\rho(r)$, is defined by the equation

$$h(\vec{e}) = \overline{\lim_{r\to\infty}} \frac{u(r\vec{e})}{r^{\rho(r)}}.$$

A C^0-set is defined in the n-dimensional case in a way similar to that used in the plane case, as a set of hyperspheres such that

$$\lim_{R\to\infty} \frac{1}{R^{n-1}} \sum^{R} r_j^{n-1} = 0.$$

The function $u(r\vec{e})$ is said to be of completely regular growth if the limit

$$\lim_{r\to\infty} \frac{u(r\vec{e})}{r^{\rho(r)}} = h(\vec{e})$$

exists uniformly in $\vec{e}$ under the hypothesis that the end of the vector $r\vec{e}$ lies outside a C^0-set. The role of the zeros is played by the masses $\mu(E)$ that appear in the Riesz representation of $u(r\vec{e})$ in an arbitrary region. Angular density is defined as a function of domains U on the unit sphere by the equation

$$\Delta(U) = \lim_{R\to\infty} \frac{\mu(K_R)}{R^{\rho(R)+n-2}},$$

where $K_R = \{r\vec{e}: \vec{e} \in U, 0 \leqslant r \leqslant R\}$. When ρ is nonintegral the masses are said to be regularly distributed if the limit exists for all domains U for which $\Delta(\partial U) = 0$.[1] For integral ρ we require additional hypotheses similar to condition (B′) of Chapter II.

Azarin showed that a subharmonic function is of completely regular growth if and only if the corresponding masses are regularly distributed. Moreover, using the notion of the regularized indicator

$$\tilde{h}(\vec{e}) = \lim_{\delta\to 0} \sup_{|\vec{e}_1 - \vec{e}|<\delta} h(\vec{e}_1),$$

he obtained the representation

$$\tilde{h}(\vec{e}) = \int_{|\vec{\zeta}|=1} \Phi(\vec{e}, \vec{\zeta}\,)\, dm_{\vec{\zeta}} + Y(\vec{e}), \tag{1}$$

where $Y(\vec{e})$ is a spherical function (which is absent for nonintegral ρ),

$$m_{\vec{\zeta}}(u) = \frac{1}{(n-2)(\rho+n-2)\operatorname{mes}\{|\vec{\zeta}|=1\}} \Delta(U),$$

and $\Phi(\vec{e}, \vec{\zeta})$ is a fundamental solution of

$$\mathfrak{L}\Phi + \rho(\rho+n-2)\Phi = 0,$$

on the unit sphere, where $\mathfrak{L}$ is the spherical part of the Laplacian. Azarin also showed that the "regularized" indicator $\tilde{h}(\vec{e})$ of an arbitrary subharmonic function of proximate order $\rho(r)$ can be represented in the form (1), where $m_{\vec{\zeta}}(u)$ is a mass distribution on the unit sphere. Consequently the representation (1) is a generalization of the representations (1.80) and (1.82) of the indicator of an entire function. For regularly distributed masses this formula yields a generalization of (2.04) and (2.08).

A very useful tool for investigating the possible behavior of entire functions is the following theorem of Azarin [**1**] on the asymptotic approximation of a given subharmonic function of finite order by the logarithm of the modulus of an entire function.

To every function $u(z)$ that is subharmonic in the plane $|z| < \infty$ and of proximate order $\rho(r)$, there corresponds an entire function $f(z)$ such that

$$|u(z) - \ln|f(z)|| = o(r^{\rho(r)})$$

---

[1]This condition is to be understood in the sense that zero is the lower bound of the measures of the measurable subsets of the unit sphere that contain ∂U.

outside some C^0-set. In particular, as Azarin showed, it is easy to construct a function $u(z)$ that is subharmonic in the plane and for which the set of rays of completely regular growth is an arbitrary closed set. The approximation theorem implies the existence of a corresponding entire function. This is a converse of the corollary at the end of Chapter III, §1.

2. Functions of completely regular growth in an angle

Govorov [**1**; **2**] investigated completely regular growth for functions holomorphic in an angle. We may suppose without loss of generality that the angle is the half plane $\Pi = (0 \leqslant \arg z \leqslant \pi)$, since the general case is reducible to this special case. It is necessary to define the order of a function that is holomorphic in Π. Titchmarsh defined the order of a function $f(z)$ that is holomorphic in Π to be the greatest lower bound of numbers ν for which

$$\overline{\lim_{r\to\infty}} \frac{\ln|f(re^{i\theta})|}{r^\nu} = 0 \tag{2}$$

uniformly in θ, $0 < \theta < \pi$. Govorov introduced a definition of order that is more convenient for applications. Suppose that we have the asymptotic inequality

$$\sup_{0<\theta<\pi} |f(re^{i\theta})| < \exp r^\mu$$

for some $\mu > 0$; he then defines the order of $f(z)$ to be the greatest lower bound of numbers ν for which (2) is satisfied for $0 < \theta < \pi$. Here uniformity in θ is not required. Govorov established the nontrivial fact that these definitions are equivalent. We note that the order introduced in this way describes not only the growth of $f(z)$ in the half plane Π, but also the rapidity of its decrease. For entire functions this order coincides with the order defined in terms of $M_f(r)$.

Just as for entire functions, the indicator is defined by

$$h(\theta) = \overline{\lim_{r\to\infty}} \frac{\ln|f(re^{i\theta})|}{r^\rho}.$$

One can also define the indicator with respect to a proximate order. The whole investigation depends on the following representation of holomorphic functions of finite order $\rho \geqslant 0$ in the half plane $\operatorname{Im} z > 0$:

$$\begin{aligned} f(z) = \exp\left[i(a_0 + a_1 z + \cdots + a_q z^q)\right] \\ \times \exp \frac{1}{\pi i}\int_{-\infty}^{\infty}\left(\frac{tz+1}{t^2+1}\right)^{q+1}\frac{d\sigma(t)}{t-z} \times \prod_{|z_k|<1} \frac{z-z_k}{z-\bar z_k} \\ \times \prod_{|z_k|>1} \frac{1-z/z_k}{1-z/\bar z_k} \exp 2i\left(z \operatorname{Im}\frac{1}{z_k} + \frac{z^2}{2}\operatorname{Im}\frac{1}{z_k^2} + \cdots + \frac{z^q}{q}\operatorname{Im}\frac{1}{z_k^q}\right), \end{aligned} \tag{3}$$

where $q = [\rho]$, a_k are real constants, z_k are the zeros of $f(z)$ in Π, and

$$\sigma(t) = \lim_{y\to+0} \int_0^t \ln|f(x+iy)|\,dx,$$

which is clearly of locally bounded variation. This representation plays the same role in the theory of functions of finite order in a half plane as Hadamard's theorem in the theory of entire functions.

Govorov obtained a necessary and sufficient condition for a function $f(z)$ of finite order and normal type in a half plane to be a function of completely regular growth. The theorems involve both the distribution of the zeros of the function and its behavior on the boundary. More precisely, they involve both the masses in the half plane $\operatorname{Im} z > 0$ that are involved in the representation of the subharmonic function $\ln|f(z)|$ as a potential, and also masses on the boundary.

In place of the function $n(r, \vartheta, \theta)$, Govorov introduces the function

$$c(r, \vartheta, \theta) = \sum_{\substack{1<r_k<r \\ \vartheta<\theta_k<\theta}} \sin\theta_k,$$

where $r_k e^{i\theta}$ are the zeros of $f(z)$. To take account of the boundary behavior he considers the function

$$\tau(t) = \int_0^t \frac{d\sigma(x)}{1+|x|}. \tag{4}$$

Finally, he introduces the function

$$a(r,\theta) = \begin{cases} c(r, 0, \theta) - \tau(r), & 0 \leqslant \theta < \pi, \\ c(r, 0, \pi) - \tau(r) + \tau(-r), & \theta = \pi. \end{cases} \tag{5}$$

We say that the set of zeros of a function $f(z)$ of order ρ has an argument-boundary density if the limit

$$\lambda(\theta) = \lim_{r\to\infty} \frac{a(r,\theta)}{r^\rho}$$

exists for all θ, $0 \leqslant \theta \leqslant \pi$, except for an at most countable set that does not contain 0 or π.

Govorov showed that, *for nonintegral ρ, completely regular growth in the open half plane is equivalent to the existence of an argument-boundary density. A necessary and sufficient condition for $f(z)$ to be of completely regular growth in the closed half plane is that the argument-boundary density exists, is continuous at 0 and π, and satisfies the following condition:*

The limit

$$\lim_{r\to\pm\infty} \frac{1}{r^\rho} \int_0^r \frac{d\varphi(x)}{1+|x|} \tag{6}$$

exists, where $\varphi(x)$ *is the singular part of* $\sigma(x)$*; that is,*

$$\varphi(x) = \sigma(x) - \int_0^x \ln|f(t)|\, dt;$$

$$\lim_{r\to\infty} \frac{\tau(r)}{r^\rho} = \frac{1}{\rho} \overline{\lim_{r\to\infty}} \frac{\ln|f(r)|}{r^\rho};$$

$$\lim_{r\to\infty} \frac{\tau(-r)}{r^\rho} = \frac{1}{\rho} \overline{\lim_{r\to\infty}} \frac{\ln|f(-r)|}{r^\rho}.$$

Here the indicator function is given by the formula

$$h_f(\vartheta) = \frac{\pi}{\sin \pi\rho} \int_0^\pi g(\psi, \vartheta) \mathrm{cl}\, \lambda(\psi), \tag{6'}$$

where $\lambda(\psi)$ is the argument-boundary density and

$$g(\psi, \vartheta) = \begin{cases} \dfrac{1}{\sin\psi}\left[\cos\rho(|\vartheta - \psi| - \bar{x}) - \cos\rho(\vartheta + \psi - \pi)\right], & 0 < \psi < \pi, \\ 2\rho \sin\rho(\vartheta - \pi), \quad \psi = 0, \\ -2\rho \sin\rho\vartheta, \quad \psi = \pi. \end{cases}$$

Govorov also obtained a necessary and sufficient condition for the function $f(z)$ to be of completely regular growth when the order is an integer. For the open half plane $0 < \theta < \pi$ the condition is that the argument-boundary density exists and that the limit

$$\sigma = \lim_{r\to\infty} \left[\frac{1}{\rho} \sum_{1<r_n<r} \frac{\sin\rho\theta_n}{r_n^\rho} - \frac{1}{2\pi} \int_{1<|x|<r} \frac{\ln|f(x)|}{x^{\rho+1}}\, dx - \frac{1}{2\pi} \int_{1<|x|<r} \frac{d\varphi(x)}{x^{\rho+1}} \right] \tag{7}$$

exists and is finite. The latter condition is similar to condition (B) in Chapter II, §1.

In this case the indicator is given by the formula

$$\begin{aligned} h_f(\theta) = (2\sigma - a_\rho)\sin\rho\theta &+ 2\cos\rho\theta \int_0^\pi \psi \frac{\sin\rho\psi}{\sin\psi}\, d\lambda(\psi) \\ &+ 2\int_0^\theta \frac{\psi}{\sin\psi} \sin\rho(\theta - \psi)\, d\lambda(\psi) \\ &+ 2\int_\theta^\pi \frac{\psi - \pi}{\sin\psi} \sin\rho(\theta - \psi)\, d\lambda(\psi), \end{aligned} \tag{8}$$

where $\lambda(\psi)$ is the argument-boundary density and a_ρ is the constant that appears in the representation of $f(z)$.

Govorov also found a necessary and sufficient condition for completely regular growth in the closed half plane for functions of integral order, and gave a representation for the indicator.

3. Regularity of the mass distribution and exceptional sets

In Chapter II we considered entire functions with a regular set of zeros; that is, a regularly distributed set satisfying a supplementary condition of "thinness", namely Condition (C) or (C′) of Chapter II, §1. For this case we defined exceptional sets. For example, under Condition (C) the asymptotic equations (2.03) or (2.05) are satisfied outside the disks $|z - a_n| \leqslant d|a_n|^{1-\rho(|a_n|)/2}$.

In order to define an exceptional set in the general case of functions of completely regular growth, Kolomiĭceva [1] sharpened Lemma 7 of Chapter II in the following way:

Let $\{a_n\}$ be an infinite set with a limit point at infinity, and let

$$F_\delta(z) = \Pi(1 - z/a_n), \tag{9}$$

where the product extends over the factors in which $|a_n - z| \leqslant \delta r$; here $|z| = r$ and $0 < \delta < \frac{1}{2}$. Then for $z \neq a_n$

$$\left| |\ln|F_\delta(z)|| - \int_0^{\delta r} (n_z(t)/t)\, dt \right| < \ln(1 + 1/\delta) n_z(\delta r). \tag{10}$$

In addition, let an Ω-set be a set of points z in the plane such that for every $\varepsilon > 0$ there exist $\delta > 0$ and $r_{\varepsilon,\delta} > 0$ such that

$$\int_0^{hr} \frac{n_z(t)}{t}\, dt < \varepsilon r^{\rho(r)} \tag{11}$$

for $h < \delta$ and $|z| > r_{\varepsilon,\delta}$. Then *a necessary and sufficient condition for the asymptotic equality $\ln|f(z)| \approx H(\theta)r^{\rho(r)}$ on an infinite set is that the set is an Ω-set.*

Returning to the "thinness" conditions (C) and (C′), we may remark that the exceptional set can be more restricted than in Theorem 5 of Chapter II. Suppose, for example, that (C) holds. Then the asymptotic equation (2.05′) is satisfied on the boundary of each C_R-disk $|z - a_n| = d|a_n|^{1-\rho(|a_n|)/2}$, and

$$\ln|f(z)| = \ln\left|\frac{f(z)}{z - a_n}\right| + \ln|z - a_n|.$$

Applying the maximum and minimum principles to the first term on the right-hand side of this equation, and using (2.05′), we find that the asymptotic equation

$$\ln|f(re^{i\theta})| \approx H(\theta)r^{\rho(r)} + \ln|z - a_n|$$

holds throughout the C_R-disk.

If we delete the disks $|z - a_n| \leqslant \varphi(|a_n|)$, where $-\ln \varphi(r) = o(r^{\rho(r)})$, the asymptotic equation (2.05) is valid in the rest of the plane. Consequently "uniformity" of the distribution of the zeros (or of masses in the case of subharmonic functions) leads to narrowing the exceptional set outside which the asymptotic equations (2.03) and (2.05) hold.

We also note the following result of Azarin [**11**]: *Let* C_μ^0 *denote a set of disks* $|z - z_\nu| < r_\nu$ *satisfying the condition*

$$\lim_{R\to\infty} R^{-\mu}\left\{\sum_{|z_\nu|<R} r_\nu^{-\mu}\right\} = 0$$

and let C_0^0 *denote a set that can be covered by sets* C_μ^0 *for all* $\mu > 0$. *Then if the asymptotic formula* (2.03) *or* (2.05) *holds outside a* C_2^0 *set, it also holds outside a* C_0^0 *set.*

Grišin [**2–4**] studied the following general problem: let $u(z)$ be subharmonic in the whole complex plane (or in $\Pi_+ = \{z\colon \operatorname{Im} z > 0\}$), and have formal order $\rho(r)$; that is, for some constant M

$$\sup_{|z|<r} u(z) \leqslant Mr^{\rho(r)}.$$

Corresponding to this $u(z)$ we construct the function

$$u_h(z) = r^{-\rho(r)}[u(z + hz) + u(z)]. \tag{12}$$

A set C is called "exceptional" if $u_h(z) \rightrightarrows 0$ outside this set as $h \to 0$. This means that to every $\varepsilon > 0$ there corresponds a $\delta(\varepsilon)$ such that $|u_h(z)| < \varepsilon$ for $|h| < \delta(\varepsilon)$, $z \overline{\in} C$ and $z + zh \overline{\in} C$.

In this case if the masses $\mu(E)$ in the Riesz representation of the subharmonic function $u(z)$ are regularly distributed, the asymptotic formulas (2.03) and (2.05), which hold outside some C^0 set, are easily seen to hold outside an exceptional set C.

The regularity of growth of the subharmonic function is evidently greater, the narrower the excluded set is. Grišin distinguishes classes a), b), c) and d) of subharmonic functions with the formal proximate order $\rho(r)$, according as the exceptional set C can be covered by a collection $\tilde{C}$ of disks respectively satisfying the conditions:[2]

a) $\tilde{C}$ has arbitrarily small upper linear density, $\rho^*(\tilde{C}) < \varepsilon$.

b) $\tilde{C}$ has linear density zero.

c) $\tilde{C}$ subtends a finite angle at the origin; that is,

$$\sum_1^\infty \frac{r_j}{|z_j|} < \infty,$$

where r_j is the radius of the excluded disk of index j, and z_j is the distance of its center from the origin.

---

[2]We state Grišin's results for functions that are subharmonic in the whole plane.

d) $\tilde{C}$ has a finite sum of radii, $\Sigma_1^\infty r_j < \infty$.

Evidently each class is contained in its predecessor.

If the set of zeros of an entire function $f(z)$ is an R-set for the proximate order $\rho(r)$, then, as we have just seen, the subharmonic function $u(z) = \ln|f(z)|$ belongs to class d). In Hayman's theorem[3] (as sharpened by Azarin) the subharmonic function in the half plane $\operatorname{Im} z > 0$ evidently belongs to class c).

The following functions are used to describe the regularity of a mass distribution:

1. The density function $\Phi(\alpha)$ is defined by

$$\Phi(\alpha) = \overline{\lim_{r\to\infty}} \sup_{\theta} \frac{\mu(C(z, \alpha r))}{r^{\rho(r)}}, \qquad z = re^{i\theta}, \qquad 0 \leqslant \alpha \leqslant 1, \tag{13}$$

where $C(z, \alpha r) = \{\zeta\colon |\zeta - z| < \alpha r\}$. If $\Phi(\alpha)$ tends slowly to zero as $\alpha \to 0$, there is a sequence of points $z_n \to \infty$ in whose neighborhoods most of the mass is concentrated; that is, in this case the mass distribution is not regular.

2. In the following approximation the regularity of the mass distribution is described by the function

$$\varepsilon_B(r, \alpha) = \sup_{\theta} \frac{\mu(C(z, \alpha r))}{r^{\rho(r)}} - B(\alpha), \tag{14}$$

where $B(\alpha)$ is an arbitrary nondecreasing function which is continuous on the right and satisfies the condition $\Phi(\alpha + 0) \leqslant B(\alpha - 0)$. The function $\varepsilon_B^+(r, \alpha)$, as it turns out, tends to zero as $r \to \infty$, uniformly with respect to α ($0 \leqslant \alpha \leqslant 1$). Hence the function

$$\varepsilon(r) = \sup_{t > r} \sup_{0 < \alpha \leqslant 1} \varepsilon_B^+(t, \alpha) \tag{15}$$

is monotonic and tends to zero as $r \to \infty$ (we omit the subscript B to simplify the typography).

We can now state the fundamental theorem:

THEOREM. *Let $u(z)$ be subharmonic in the whole plane and of normal order $\rho(r)$. Then*:

1. *$u(z)$ belongs to class* a).

2. *If the density function of the Riesz masses satisfies*

$$\int_0 \ln(1/\alpha)\, d\Phi(\alpha) < \infty, \qquad \Phi(0) = 0, \tag{16}$$

then $u(z)$ belongs to class (b).

If (16) *is satisfied and in addition the integral*

---

[3]See footnote 11 in Chapter V, §3.

$$\int^{\infty} e^{-\varepsilon/\varepsilon(r/2)} \frac{d\mu[C(0, r)]}{r^{\rho(r)}} < \infty, \tag{17}$$

converges for every $\varepsilon > 0$, *then* $u(z)$ *belongs to class* c).

4. *If* (16) *is satisfied and*

$$\int^{\infty} re^{-\varepsilon/\varepsilon(r/2)} \frac{d\mu[C(0, r)]}{r^{\rho(r)}} < \infty, \tag{18}$$

then $u(z)$ *belongs to class* d).

In particular, (17) is satisfied if $\varepsilon(r) = o(1/\ln \ln r)$, and condition (18) is satisfied if $\varepsilon(r) = o(1/\ln r)$. Statement 1 is a generalization to subharmonic functions of Theorem 6 of Chapter II, and also contains the corresponding result of Krasičkov [**1**; **3**] for entire functions.

Grišin extended these results to subharmonic functions in the half plane $\operatorname{Im} z > 0$, with the stated formal order. He also investigated subharmonic functions of order zero in the half plane $\operatorname{Im} z > 0$. This case is essentially different from the case $\rho > 0$.

4. The second term of an asymptotic formula

Let the zeros of the canonical product $f(z)$ be positive, let $n(t)$ be the counting function of the zeros, and let

$$n(t) = \Delta t^{\rho} + \Delta_1 t^{\rho_1} + \varphi(t), \tag{19}$$

where $\rho > \rho_1 > [\rho]$, $\Delta > 0$, and $\varphi(t) = o(t^{\rho_1})$ as $t \to +\infty$. Then it is easy to see that for arbitrary $\delta > 0$ we have the asymptotic equation

$$\ln f(re^{i\theta}) = \frac{\pi \Delta r^{\rho}}{\sin \pi\rho} e^{i\rho(\theta-\pi)} + \frac{\pi \Delta_1 r^{\rho_1}}{\sin \pi\rho_1} e^{i\rho_1(\theta-\pi)} + o(r^{\rho_1}), \tag{20}$$

as $r \to \infty$, uniformly for $\theta \in [\delta, 2\pi - \delta]$. It is natural to ask whether (20) holds uniformly in the plane except on a "small" set.

Logvinenko [**2**] gave a positive answer to this question. Moreover, if we assume even somewhat less about $\varphi(t)$, namely that

$$\int_T^{2T} |\varphi(t)|^q \, dt = o(T^{\rho_1 q+1}), \qquad T \to \infty, \tag{21}$$

for some $q \geqslant 1$, we have the asymptotic equation

$$\ln f(re^{i\theta}) = \frac{\pi \Delta r^{\rho}}{\sin \pi\rho} e^{i\rho(\theta-\pi)} + \frac{\pi \Delta_1 r^{\rho_1}}{\sin \pi\rho_1} e^{i\rho_1(\theta-\pi)} + \psi(re^{i\theta}), \tag{22}$$

where $\psi(re^{i\theta}) = o(r^{\rho_1})$ as $r \to \infty$, uniformly for $\theta \in [0, 2\pi]$, provided that $z = re^{i\theta}$ is outside an exceptional set $E = E_f$. This exceptional set is the union of a countable collection of rectangles $x_n' < \operatorname{Re} z < x_n''$, $|\operatorname{Im} z| \leqslant y_n$ $(n = 1, 2, \ldots)$ that satisfy

a) $x'_n > 0, n = 1, 2, \ldots,$

b) $\Sigma_{x'_n \leqslant R}(x''_n - x'_n) = o(R),$

c) $\sup_{x'_n \leqslant R}|y_n| = o(R).$

Moreover, if (21) holds with $q > 1$, we have the following asymptotic estimate for $\psi(re^{i\theta})$:

$$\sup_\theta \int_T^{2T} |\psi(re^{i\theta})|^q \, dr = o(T^{\rho_1 q+1}), \qquad T \to \infty. \tag{23}$$

Finally, there is a converse theorem. If an entire function $f(z)$ satisfies the asymptotic equation (22) in the complex plane except on the small set E_f, and the zeros of this function lie on the positive real axis, then the counting function $n(t)$ has the form (19) with $\varphi(t) = o(t^{\rho_1})$.

In the theorem of Valiron and Titchmarsh (see Appendix VII, assertion 2) if all the zeros of an entire function of order ρ lie on the positive real axis and

$$\ln f(-r) = \frac{\pi\Delta}{\sin \pi\rho} r^\rho + o(r^\rho), \tag{24}$$

then $n(t) = \Delta t^\rho + o(t^\rho)$.

However, a two-term asymptotic formula

$$\ln f(-r) = \frac{\pi\Delta}{\sin \pi\rho} r^\rho + \frac{\pi\Delta_1}{\sin \pi\rho_1} r^{\rho_1} + o(r^{\rho_1}) \tag{25}$$

for $f(z)$ with only positive zeros does not imply the two-term formula (19). A number of papers have been devoted to the behavior of $n(t)$ when (25) or similar formulas are assumed. Tjan **[1]** and Anderson **[1]** showed that $n(t)$ satisfies the sharp asymptotic formula

$$n(t) = \Delta t^\rho + o(t^\rho/\ln t), \qquad t \to \infty.$$

Logvinenko **[3]** showed that even under the more restrictive condition that (20) is satisfied uniformly in the angle $0 < \delta \leqslant \theta \leqslant 2\pi - \delta$, the asymptotic equation (19) may fail.[4]

The situation is different if we require that (22) is satisfied on both rays $\theta = 0, \pi$. Logvinenko **[3]** proved the following theorem:

Let $f(z)$ be an entire function of nonintegral order with only positive zeros, let (22) hold for $\theta = 0, \pi$, where $\rho > \rho_1 > [\rho]$, and let $\psi_1(x)$ satisfy

$$\int_{T<|x|<2T} |\psi(x)|^q \, dx = o(T^{\rho_1 q+1}), \qquad T \to \infty,$$

on the real axis.

---

[4]For this circle of ideas, see also Subhankulov and Tolbaev **[1]**, Tolbaev **[1; 2]**, Jurov **[1]**, and Chapter VII of the book **[1]** by Subhankulov.

Then $n(t)$ has the form (19), *and the remainder term $\varphi(t) = o(t^{\rho_1})$ as $t \to \infty$ except on a set of relative measure zero. If also $q > 1$, then $\varphi(t)$ satisfies* (21).

Logvinenko also proved theorems for two-term asymptotic formulas that are similar to the theorems of Chapter II of this book. One of these is as follows.

THEOREM. *Let $f(z)$ be an entire function of order ρ and let the distribution function $n(t, \theta_1, \theta_2)$ of the zeros of $f(z)$ satisfy*

$$n(t, \theta_1, \theta_2) = \Delta(\theta_1, \theta_2)t^{\rho} + \Delta_1(\theta_1, \theta_2)t^{\rho_1} + \varphi(t, \theta_1, \theta_2),$$

where $1 > \rho > \rho_1 > 2\rho/3$ and $\varphi(t, \theta_1, \theta_2) = o(t^{3\rho_1 - 2\rho})$, $t \to \infty$, uniformly in θ_1 and θ_2, $0 \leqslant \theta_1 \leqslant \theta_2 \leqslant 2$. Then the following asymptotic equation holds:

$$\ln|f(re^{i\theta})| = H(\theta)r^{\rho} + H_1(\theta)r^{\rho_1} + \psi(re^{i\theta}).$$

Here

$$H(\theta) = \frac{\pi}{\sin \pi\rho} \int_{\theta - 2\pi}^{\theta} \cos \rho(\theta - \omega - \pi)\, d\Delta(\omega),$$

$$H_1(\theta) = \frac{\pi}{\sin \pi\rho_1} \int_{\theta - 2\pi}^{\theta} \cos \rho_1(\theta - \omega - \pi)\, d\Delta_1(\omega),$$

where $\Delta(\omega) = \Delta(0, \omega)$; $\Delta_1(\omega) = \Delta_1(0, \omega)$ for $\omega \in [0; 2\pi]$ and $\Delta_1(\omega + 2\pi) = \Delta_1(\omega) + \Delta_1(2\pi)$. We have $|\psi(re^{i\theta})| = o(r^{\rho_1})$ if r does not belong to a certain exceptional set $E = E_f$ of relative measure zero.[5]

5. Derivatives and primitives of entire functions of completely regular growth

Gol′dberg and Ostrovskiĭ [2] investigated the derivatives and primitives of entire functions of completely regular growth. They proved that *the primitive of an entire function of completely regular growth is again an entire function of completely regular growth,*[6] and described the set of values φ for which the ray $\arg z = \varphi$ is not a ray of completely regular growth for the derivative. Azarin [8] showed that this description is exhaustive. In order to present it, we introduce some notation.

---

[5]Entire functions whose zeros are on a single ray have been investigated by many authors. For a summary and references see Boas [4]. We note the following result of Ostrovskiĭ [4]: *Let $f(z)$ be an entire function whose zeros are on the ray $\arg z = \alpha$. If*

$$\overline{\lim_{r\to\infty}}\, r^{-p} \ln M_f(r) = \infty, \qquad \underline{\lim_{r\to\infty}}\, r^{-p-1/2} \ln M_f(r) = 0, \qquad (*)$$

where p is an integer, then $f(z)$ cannot be bounded on that ray.

Condition $(*)$ is satisfied, in particular, for $p < \rho < p + 1/2$. When $p = 0$ this implies Wiman's theorem (Chapter I, §18, Theorem 30).

[6]This result was obtained independently by Gruman [1], but part of his proof is inconclusive.

Let $h(\varphi)$ be the indicator of a function $f(z)$ that is of completely regular growth with respect to a proximate order $\rho(r)$. Let S be the interior of the set of values φ for which $h(\varphi) = 0$ (when $\rho \leqslant \frac{1}{2}$ this set is empty). We remove from S those of its component intervals (α, β) for which at least one of the conditions

$$1)\ \beta - \alpha > \pi/\rho, \qquad 2)\ h'_-(\alpha) = 0, \qquad 3)\ h'_+(\beta) = 0,$$

is satisfied, and let S_1 be the set of remaining intervals. Let $I(f')$ be the set of values φ for which $\arg z = \varphi$ is not a ray of completely regular growth for $f'(z)$; we recall that this set is open, by Theorem 1 of Chapter III, §1. This open set can be described as follows. We have $I(f') \subset S_1$, and if $I(f')$ intersects a component interval (α, β) of S_1 for which $\beta - \alpha = \pi/\rho$ then $(\alpha, \beta) \subset I(f')$.

In particular, it follows that the derivative of an entire function of order $\rho \leqslant \frac{1}{2}$ and of completely regular growth is also an entire function of completely regular growth. When $\rho > \frac{1}{2}$ it can fail to be of completely regular growth only on the rays for which $h(\varphi) = 0$.

Gol′dberg and Korenkov [**1**; **2**] investigated the asymptotic behavior of the logarithmic derivative of an entire function of completely regular growth. Their sharp asymptotic formula for the logarithmic derivative holds outside a collection of disks $C(r_j, z_j)$ of μ-density zero with $1 < \mu < 2$; that is, a set of disks such that

$$\lim_{R\to\infty} \left\{ R^{-\mu} \sum_{|z_j| < R} r_j^{\mu} \right\} = 0.$$

They established the following results.

1. *Let $f(z)$ be an entire function of nonintegral order of completely regular growth with respect to a proximate order $\rho(r) \to \rho$, let $\Delta(\psi)$ be the angular density function of its zeros, and let $F(z) = f'(z)/f(z)$. Then for every μ, $1 < \mu < 2$, there is a collection K of disks, of μ-density zero, such that, when $re^{i\varphi} \in \mathbf{C} \setminus K$,*

$$\begin{aligned} F(re^{i\varphi}) = {} & -\pi\rho/\sin\pi\rho\; r^{\rho(r)-1} \\ & \times \int_0^{2\pi} \exp\{-i\psi + i(\rho - 1)(|\varphi - \psi| - \pi)\operatorname{sgn}(\varphi - \psi)\}\, d\Delta(\psi) \\ & + o(r^{\rho(r)-1}) \end{aligned}$$

for $0 \leqslant \varphi < 2\pi$.

2. *If the order* $\rho = [\rho]$ *then*

$$F(re^{i\varphi}) = \rho r^{\rho(r)-1}\exp(i(\rho - 1)\varphi) \times \left\{\delta + i\int_0^{2\pi}[\varphi - \psi - \pi \operatorname{sgn}(\varphi - \psi)]\exp(-i\rho\psi)\, d\Delta\psi\right\} + o(r^{\rho(r)-1})$$

outside the exceptional set.

For example, for the Weierstrass ζ-function this yields

$$\zeta(z) = r\left\{2\bar{\eta}_1 e^{i\varphi} - 2i\pi\frac{\sin\varphi}{\lambda\sin\alpha}\right\} + o(r)$$

outside a set of disks of μ-density zero. Here α is the angle in the period parallelogram, and $\eta_1 = \zeta(\frac{1}{2})$.

We note that a set of μ-density zero can be rather large. For example, the domain bounded by the parabola $x > 0$, $|y| < x$, has μ-density zero for every $\mu > 1$. Hence it is important to have examples that show that when $\mu = 1$ the asymptotic formula does not necessarily hold. Such an example is, as is easily seen, the function $f(z) = \cos\sqrt{z}$.

For integral order the existence of an angular density is not a sufficient condition for completely regular growth. However, even in this case there are asymptotic formulas, but for orders different from those for which the angular density exists. For this kind of order see §6.

6. Some complements to the theory of entire functions of integral order

If all the zeros of an entire function of nonintegral order ρ are on the positive real axis and

$$n(r) = \Delta r^{\rho(r)}, \tag{26}$$

then, by Theorem 25 of Chapter I, when ρ is nonintegral we have the asymptotic equation

$$\ln|f(re^{i\theta})| \approx \frac{\pi}{\sin\pi\rho} e^{i\rho(\theta-\pi)} r^{\rho(r)} \qquad (0 < \theta < 2\pi);$$

that is, the functions $n(r)$ and $\ln M_f(r)$ have the same proximate order.

In the case of integral order, if a canonical product has its zeros on the positive real axis and satisfies (26), its logarithm grows more rapidly than $r^{\rho(r)}$. Gol'dberg [3][7] gave sharp formulas and applied them to the theory of functions of completely regular growth.

For integral ρ there are two cases:

$$\alpha)\ \sum_1^\infty a_n^{-\rho} = \infty, \qquad \beta)\ \sum_1^\infty a_n^{-\rho} < \infty.$$

[7]See Gol'dberg and Ostrovskiĭ [1], Chapter II, §5.

In case α) we have

$$\ln f(re^{i\theta}) = \ln f_r(re^{i\theta}) + \frac{r^\rho e^{i\rho\theta}}{\rho} \sum_{a_n<r} a_n^{-\rho},$$

and by Lemma 9 of Chapter I

$$\ln f(re^{i\theta}) \approx -\Delta\left[\frac{1}{\rho} - i(\theta - \pi)\right] e^{i\rho\theta} \cdot r^{\rho(r)} + \frac{r^\rho e^{i\rho\theta}}{\rho} \int_0^r \frac{dn(t)}{t^\rho}.$$

The last term, which we denote by $\mathcal{G}(r, \theta)$, can be transformed by integration by parts, and we obtain

$$\mathcal{G}(r, \theta) \approx n(r)\frac{e^{i\rho\theta}}{\rho} + e^{i\rho\theta} r^\rho \int_1^r \frac{dt}{t^{\rho+1-\rho(t)}}.$$

It is easily verified (by L'Hospital's rule) that

$$L(r) = \int_1^r \frac{dt}{t^{\rho+1-\rho(r)}}$$

is a function of slow growth and therefore

$$r^\rho \int_1^r t^{\rho(t)-\rho-1}\, dt = r^{\rho_1(r)},$$

where $\rho_1(r)$ is a new proximate order. It is easily seen that $r^{\rho_1(r)-\rho(t)} \to \infty$. Consequently $\mathcal{G}(r, \theta) \approx \Delta e^{i\rho\theta} r^{\rho_1(r)}$ and in addition

$$\ln f(re^{i\theta}) \approx \Delta e^{i\rho\theta} r^{\rho_1(r)}. \tag{27}$$

In case β) it is convenient to write the canonical product in the form

$$f(z) = e^{-sz^\rho} \prod_{n=1}^{\infty} G(z/a_n; \rho),$$

where $S = \rho^{-1}\Sigma_1^\infty a_n^{-\rho}$. This can also be written in the form

$$f(z) = f_r(z)\exp\left[-\frac{z^\rho}{\rho} \sum_{a_n>r} a_n^{-\rho}\right],$$

where

$$f_r(z) = \prod_{a_n \leqslant r} G(z/a_n; \rho - 1) \prod_{a_n>r} G(z/a_n; \rho).$$

By Lemma 9 of Chapter I we have

$$\ln f_r(re^{i\theta}) \approx -\Delta[1/\rho - i(\theta - \pi)] e^{i\rho\theta} r^{\rho(r)}. \tag{28}$$

The logarithm of the second factor,

$$\mathcal{G}_1(r, \theta) = -e^{i\rho\theta} \frac{r^\rho}{\rho} \int_r^\infty \frac{dn(t)}{t^\rho},$$

can be transformed as in the previous case, and we obtain

$$\mathcal{G}_1(r, \theta) = -e^{i\rho\theta}\frac{n(r)}{\rho} - r^{\rho}\int_r^{\infty}\frac{dt}{t^{\rho+1-\rho(t)}}.$$

It is easily verified that in this case $\mathcal{G}_1(r, \theta) \approx -\Delta e^{i\rho\theta} r^{\rho_1(r)}$ and $r^{\rho_1(r)-\rho(r)} \to \infty$. Thus we have the following proposition:

If $f(z)$ is a canonical product with integral exponent of convergence, if its zeros are positive, and if it satisfies (27), *then*

$$\ln f(re^{i\theta}) \approx \pm\Delta e^{i\rho\theta} r^{\rho_1(r)}, \tag{29}$$

where

$$L_1(r) = r^{\rho_1(r)-\rho} = \begin{cases} \displaystyle\int_1^r \frac{dt}{t^{\rho+1-\rho(t)}} & \text{when } \displaystyle\int_1^{\infty}\frac{dt}{t^{\rho+1-\rho(t)}} = \infty, \\ \displaystyle\int_r^{\infty} \frac{dt}{t^{\rho+1-\rho(t)}} & \text{when } \displaystyle\int_1^{\infty}\frac{dt}{t^{\rho+1-\rho(t)}} < \infty. \end{cases} \tag{30}$$

In (29) the sign is $+$ if the integral $\int^{\infty} t^{-\rho-1+\rho(t)}\,dt$ diverges and $-$ if it converges. As in Chapter II we can deduce from (29) that when ρ is integral and an angular density $\Delta(\theta)$ exists, the canonical product satisfies the asymptotic equation

$$\ln f(re^{i\theta}) \approx \pm r^{\rho_1(r)}\int_{\theta-2\pi}^{\theta} e^{i\rho(\theta-\psi)}\,d\Delta(\psi)$$

or

$$\ln f(re^{i\theta}) \approx C_{\Delta} e^{i\rho\theta} r^{\rho_1(r)}, \tag{31}$$

where $C_{\Delta} = \pm\int_0^{2\pi} e^{-i\rho\psi}\,d\Delta(\psi)$. Of course, this formula is of interest only when C_{Δ} is different from zero.

As was shown in Chapter II, the condition $C_{\Delta} = 0$ is necessary for an entire function of integral order with angular density $\Delta(\psi)$ to be of completely regular growth for the proximate order $\rho(r)$. Lunc **[1]** has shown that if the angular density $\Delta(\psi)$ exists and $C_{\Delta} = 0$, one can augment the given set $\{a_k\}$ of zeros by a set $\{b_k\}$ of density zero in such a way that the new canonical product satisfies condition (B′) of Chapter II, §1; that is, it is of completely regular growth with respect to the proximate order $\rho(r)$.[8]

7. Upper and lower indicators and their application to the distribution of the zeros with respect to their arguments

In order to study the connection between the growth of an entire function in various directions and the distribution of its zeros with respect to their

---

[8]The case of functions of zero order is exceptional. Grišin **[4]** has investigated the theory of functions of order zero that are of completely regular growth in an angle.

arguments as well as their moduli, it is of course not essential to assume the existence of an angular density for the set of zeros. We can, for example, describe the distribution of the zeros in terms of the upper and lower angular densities

$$\overline{\Delta}(\psi) = \varlimsup_{r\to\infty} \frac{n(r, \psi)}{r^{\rho(r)}}, \qquad \underline{\Delta}(\psi) = \varliminf_{r\to\infty} \frac{n(r, \psi)}{r^{\rho(r)}}. \tag{32}$$

One can describe the growth of an entire function by means of the lower indicator as well as by the indicator. This notion was introduced in a special case by Leont'ev **[9]** and in the general case by Gol'dberg **[4]**. The lower indicator $\underline{h}_f(\theta)$ is defined as follows: remove an arbitrary C^0-set from the plane, form

$$\underline{h}_f(\theta, C^0) = \varliminf_{\substack{r\to\infty \\ re^{i\theta}\overline{\in} C^0}} \frac{\ln|f(re^{i\theta})|}{r^{\rho(r)}},$$

and then take the supremum over all C^0-sets, i.e.

$$\underline{h}_f(\theta) = \sup_{C^0} \underline{h}_f(\theta, C^0). \tag{33}$$

It is easily seen that if $f(z)$ has no zeros inside an angle $\alpha < \theta < \beta$, then for these values of θ the lower indicator $\underline{h}_f(\theta)$ is simply equal to the corresponding lower limit (without preliminary exclusion of a C^0-set).

The ordinary indicator is sometimes called the upper indicator. Clearly the equation $\underline{h}_f(\theta) = h_f(\theta)$ means that $f(z)$ is of completely regular growth. A number of papers have investigated the problem of estimating the upper and lower indicators of an entire function of finite order in terms of the upper and lower angular densities of its zeros.

If $\rho(r)$ denotes the proximate order in terms of which we have defined the upper and lower densities of the set of zeros, it is natural, in the light of the theorems of §6 of this appendix, to define the indicators in terms of the same proximate order $\rho(r)$ when the order is nonintegral, but in terms of the proximate order $\rho_1(r)$ defined by (30) when the order is an integer. To begin with, we can investigate entire functions whose zeros are all positive, and estimate the upper and lower indicators in this case in terms of the upper and lower densities $\overline{\Delta}$ and $\underline{\Delta}$ of the set of zeros on the positive real axis.

Theorem 12 of Chapter V can be considered as a result of this kind. We have only to replace z^2 by z in the representation of the entire function to obtain an entire function $F(z)$ of order $\rho = \frac{1}{2}$ with positive zeros, and the set of zeros has $\overline{\Delta} = 1$ and $\underline{\Delta} = 0$.

Theorem 12 of Chapter V provides a sharp estimate for the indicator of such a function. In fact, an extremal function is constructed for which the estimated value is obtained for each $\theta \in [0, 2\pi]$. Andraško [1] obtained a similar result for entire functions with positive zeros and arbitrary order ρ, $0 < \rho < 1$. Then Gol'dberg [1] obtained a similar result for functions of arbitrary positive order. Govorov [3] found, for $0 < \rho < 1$, sharp estimates for the upper and lower indicators of an entire function with zeros on the positive real axis and given upper and lower densities $\bar{\Delta}$ and $\underline{\Delta}$. Finally Gol'dberg [4] obtained such estimates for arbitrary order.[9] In the general case the estimates are rather complicated. The simplest are for integral ρ:

$$\frac{\bar{\Delta}+\underline{\Delta}}{2}\cos\rho\theta - \frac{\bar{\Delta}-\underline{\Delta}}{2}|\cos\rho\theta| \leqslant \underline{h}(\theta) \leqslant h(\theta)$$

$$\leqslant \frac{\bar{\Delta}+\underline{\Delta}}{2}\cos\rho\theta + \frac{\bar{\Delta}-\underline{\Delta}}{2}|\cos\rho\theta|, \quad (34)$$

where equality can occur at any of the $\leqslant$ signs. Naturally in attempting to obtain similar results in the general case, one does not require that the zeros of the entire function are all on one ray. The results have been worked out by Gol'dberg in a series of papers [2–5].

Let the set $X \subset [0, 2\pi]$ and let $n(r, x)$ be the number of zeros of $f(z)$ in the set $\{|z| = r; \arg z \in X\}$. Put

$$\bar{\Delta}(X) = \overline{\lim_{r\to\infty}} \frac{n(r, x)}{r^{\rho(r)}}, \qquad \underline{\Delta}(X) = \underline{\lim_{r\to\infty}} \frac{n(r, x)}{r^{\rho(r)}}.$$

By analogy with the results of Chapter II we might suppose that we could make the transition from the case where the zeros are on a single ray to the general case by using an interpolation process on the upper and lower densities of the set of zeros. However, as is easily seen, the functions $\bar{\Delta}(X)$ and $\underline{\Delta}(X)$ are, in general, not additive. We may, for example, take a set of zeros such that $\bar{\Delta}(X) \equiv 1$ and $\underline{\Delta}(X) \equiv 0$ for all intervals $X \subset [0, 2\pi]$. Thus the problem of estimating indicators of entire functions leads to an integral of a new kind, corresponding to two nonadditive measures. Such an integral was introduced and investigated in detail by Gol'dberg [2, 4]. Let A be a collection of Borel sets on the half-open interval $[0, 2\pi)$, and let $\mu_1(a)$ and $\mu_2(a)$ be two nonnegative functions (measures) defined for all subsets a of A, with $\mu_2(a) \geqslant \mu_1(a)$. We may suppose without loss of generality that $\mu_1(X)$ and $\mu_2(X)$ are defined for all subsets X of $[0, 2\pi)$ by

$$\mu_j(X) = \inf \mu_j(a), \qquad j = 1, 2, \qquad \text{for } a \in A.$$

---

[9]Other contributions to the same circle of ideas have been made by Leont'ev [9], Krasičkov [2], Ostrovskiĭ [4], Kondratjuk [1], [2], Baillette [1], Kondratjuk and Fridman [1, 2], and Azarin [9].

Let N denote the class of finitely additive measures $\nu(X)$ that satisfy $\mu_1(X) \leqslant \nu(X) \leqslant \mu_2(X)$ for all subsets X of $[0, 2\pi)$. It is easily seen that this class is not empty, provided that $\mu_1(a)$ and $\mu_2(a)$ satisfy the subadditivity conditions

$$\mu_1(a_1 \cup a_2) \geqslant \mu_1(a_1) + \mu_1(a_2), \qquad \mu_2(a_1 \cup a_2) = \mu_2(a_1) + \mu_2(a_2);$$

the lower and upper densities of the set of zeros certainly satisfy these conditions.

We define the integral of $f(t)$ with respect to the measures $\mu_1(X)$ and $\mu_2(X)$ by

$$\int_0^{2\pi} f(t)\,d[\mu_1(t); \mu_2(t)] = \max_{\nu(x)\in N}\left[\int_0^{2\pi} f(t)\,d\nu(t)\right], \tag{35}$$

where the integral on the right is taken in the Radon sense, and the existence of the maximum can be established. We also define the "lower integral"[10]

$$\underline{\int_0^{2\pi}} f(t)\,d[\mu_1(t); \mu_2(t)] = \min_{\nu(x)\in N}\left[\int_0^{2\pi} f(t)\,d\nu(t)\right]$$
$$= -\int_0^{2\pi} [-f(t)]\,d[\mu_1(t); \mu_2(t)].$$

Gol'dberg's estimates were established by the use of this integral. We quote some of them.

Let $\mu_1(X)$ and $\mu_2(X)$ be measures on $[0, 2\pi)$. Let $F(\mu_1, \mu_2, \rho(r))$ be the class of entire functions of order $\rho(r)$ when ρ is nonintegral, and of order $\rho_1(r)$ when ρ is an integer, for which

$$\mu_1(X) \leqslant \underline{\Delta}(X) \leqslant \overline{\Delta}(X) \leqslant \mu_2(X). \tag{36}$$

If ρ is an integer, we have

$$-\int_0^{2\pi} [-\cos\rho(\theta - \psi)]\,d[\mu_1(\psi); \mu_2(\psi)]$$
$$\leqslant h(\theta) \leqslant \int_0^{2\pi} \cos\rho(\theta - \psi)\,d[\mu_1(\psi); \mu_2(\psi)] \tag{37}$$

for all $f \in F(\mu_1, \mu_2, \rho(r))$.

---

[10] We note that the measure $d\nu(t)$ for which the maximum of the Radon integral on the right of (35) is attained depends on the integrand $f(t)$. In his papers, Gol'dberg gave a more general definition of the integral with respect to a nonadditive measure. He also gave a different but constructive definition, rather complicated to state but adequate for his investigation. The equivalence of the two definitions was proved by Levin, Macaev and Ostrovskiĭ when $\mu_1(x) = 0$. This proof is reproduced by Gol'dberg ([2], Theorem 2.10). He also [4] proved the equivalence for $\mu_1(x) \neq 0$.

These inequalities are sharp in the sense that the class $F(\mu_1, \mu_2, \rho(r))$ contains functions for which equality holds either on the left or on the right. Moreover, there is an $f \in F(\mu_1, \mu_2, \rho(r))$ such that the right-hand inequality is satisfied simultaneously for all arguments.

If ρ is nonintegral, we have the inequality

$$h(\theta) \geqslant -\pi \int_0^{2\pi} \left[-\frac{\cos \rho(|\theta - \psi| - \pi)}{\sin \pi\rho} \right] d[\mu_1(\psi); \mu_2(\psi)] \qquad (0 \leqslant \theta < 2\pi).$$

When $\mu_1(X) \equiv 0$ the indicator of every function in the class $F(0, \mu_2, \rho(r))$ satisfies the inequality

$$h(\theta) \leqslant \int_0^{\infty} t^{\rho - p - 1} dt \int_0^{2\pi} K(t, \theta - \psi) d\{0; \mu_2(\psi)\} \qquad (p = [\rho]), \quad (38)$$

where $K(t, \theta)$ $(0 \leqslant t \leqslant \infty; 0 \leqslant \theta < 2\pi)$ is the kernel equal to

$$K(t, \theta) = \frac{\cos p\theta - t \cos(p + 1)\theta}{1 - t \cos \theta + t^2}$$

at the points (t, θ) at which both inequalities

$$|G(e^{i\theta}/t, p)| > 1 \qquad \text{and} \qquad \frac{d}{dt}|G(e^{i\theta}/t, p)| < 0$$

are satisfied, and equal to zero at the remaining points. On the other hand, one can construct an entire function $f(z) \in F(0, \mu_2, \rho(r))$ such that

$$h_f(\theta) \geqslant \int_0^{2\pi} d\{0, \mu_2(\psi)\} \int_0^{\infty} t^{\rho - p - 1} K(t, \theta - \psi)\, dt \qquad (39)$$

for all θ. The integrals on the right-hand sides of (38) and (39) differ only in the order of integration, but they are not necessarily equal, and hence (38) is not always sharp, as can be shown by examples. However, if $\mu_2(X)$ is an additive interval function, we can replace $d\{0, \mu_2(\psi)\}$ in both integrals by $d\mu_2(\psi)$ and take the corresponding integrals as Lebesgue-Stieltjes integrals. The repeated integrals are then equal, and (38) becomes sharp. In particular, this will be true if $\mu_2(X)$ is taken to be the "maximum density" of the set of zeros. Let us consider this case in more detail. Pólya [**1**] introduced the concept of maximum density for sets on the positive real axis:

$$\overline{\overline{\Delta}} = \lim_{k \downarrow 1} \overline{\lim_{r \to \infty}} \frac{n(kr) - n(r)}{(kr)^{\rho(kr)} - r^{\rho(r)}}.$$

This concept can be generalized as follows to sets more generally distributed in the complex plane.

We define the maximum angular density of a plane set to be

$$\overline{\overline{\Delta}}(\psi) = \lim_{\max|\psi_{j+1} - \psi_j| \to 0} \left\{ \sum_{j=0}^{n-1} \lim_{k \downarrow 1} \overline{\lim_{r \to \infty}} \frac{n(kr, \psi_j, \psi_{j+1}) - n(r, \psi_j, \psi_{j+1})}{(kr)^{\rho(kr)} - r^{\rho(r)}} \right\},$$

where $0 = \psi_0 < \psi_1 < \cdots < \psi_n = \psi$. Evidently $\bar{\bar{\Delta}}(\psi) \geqslant \bar{\Delta}(\psi)$ and $\bar{\bar{\Delta}}(\psi)$ is an additive interval function.[11]

We mention two other cases in which sharp upper estimates can be obtained for the indicator:

1) $\mu_1(X) = 0$ and $\mu_2(X) = \Delta = \text{const}$; that is, for the class $F(\mu_1, \mu_2, \rho(r))$ given a global (not angular) density for the set of zeros;

2) $\mu_1(X)$ and $\mu_2(X)$ are additive interval functions and satisfy some additional conditions. In the special case $\mu_1(X) \equiv \mu_2(X)$ we obtain (2.04) of Chapter II.

We also note that inequalities of the same kind as Gol'dberg's for entire functions were obtained by Faĭnberg [**1, 2**] for functions holomorphic in a half-plane; instead of ordinary angular density he used the argument-boundary density (see §2 of this appendix).

8. Miscellaneous results from the theory of functions of completely regular growth

Azarin [**10**] established a connection between regularity of the growth of an entire function and regularity of growth of its coefficients:

$$C_k(r) = \frac{1}{2\pi}\int_{-\pi}^{\pi} \ln|f(re^{i\psi})|e^{-ik\psi}\,d\psi,$$

$k = 0, \pm 1, \pm 2, \ldots$, by expanding $\ln|f(re^{i\vartheta})|$ in a Fourier series. His result is as follows:

A necessary and sufficient condition for $f(z)$ to be of completely regular growth with respect to the proximate order $\rho(r)$ is that the limits

$$\lim_{r\to\infty} \frac{C_k(r)}{r^{\rho(r)}} = C_k; \qquad k = 0, \pm 1, \pm 2, \ldots, \tag{40}$$

exist. In this case the indicator of $f(z)$ is represented by the series

$$h(\vartheta) = \sum^{\infty} C_k e^{ik\vartheta}.$$

It is important that the conditions (40) are independent.

Kondratjuk [**3**] extended this result to the more general orders introduced and investigated by Rubel and Taylor [**1**]. Let $\lambda(r)\uparrow\infty$ and let $\lambda(2r) < M\lambda(r)$ for all $r > 0$. An entire function $f(z)$ is said to be of completely regular growth with respect to $\lambda(r)$ if

---

[11]We have the following fact. *Among all sets containing the given set and having angular densities, there is a set* $\mathfrak{M}$ *such that* $\Delta_{\mathfrak{M}}(\psi) \equiv \bar{\bar{\Delta}}(\psi)$.

Sharp estimates for the indicators of entire functions with given maximum and minimum densities were obtained by Kondratjuk [**1**].

$$\lim_{r\to\infty} \frac{1}{\lambda(r)} \left\{ \int_\alpha^\beta \ln|f(re^{i\vartheta})|\, d\vartheta \right\}$$

exists for all intervals[12] $0 \leqslant \alpha < \beta \leqslant 2\pi$. As indicated in the footnote to Theorem 2 in Chapter III, §2, for $\lambda(r) = r^{\rho(r)}$ this definition is equivalent to the one given in the text. In this more general case the theorem can be stated just like Azarin's theorem.

The theory of functions of completely regular growth can be generalized by considering the growth of the function on logarithmic spirals instead of on rays. Correspondingly, the angular density of the set of zeros can be replaced by the density of the set of zeros between two spirals.

Balašov [**1, 2**] considered more general curves

$$\mathfrak{L}_\varphi^\gamma(a, b) = \left\{t;\, t = |t|e^{i[\gamma(|t|)+\varphi]},\, a \leqslant |t| \leqslant b,\, -\infty < \varphi < \infty\right\}$$

for which the limit

$$\lim_{x\to+\infty} x\gamma'(x) = c$$

exists. He calls these curves of regular rotation. When $\gamma(x) = c \ln x$ they are logarithmic spirals. Introduce the notation

$$D_r^\gamma(\alpha, \beta) = \left\{z;\, z \in \mathfrak{L}_\vartheta^\gamma(a, x),\, \alpha \leqslant \vartheta \leqslant \beta,\, |z| \leqslant r\right\},$$
$$D_\infty^\gamma(\alpha, \beta) = D^\gamma(\alpha, \beta),$$

and the function

$$M_f^\gamma(r, \alpha, \beta) = \max_{\alpha\leqslant\vartheta\leqslant\beta} \left\{f(re^{i(\gamma(r)+\vartheta)})\right\}.$$

The order and type are defined in the natural way, and the indicator is now[13]

$$h_f^\gamma(\vartheta) = \overline{\lim_{z\in\mathfrak{L}_\vartheta^\gamma}} \left\{r^{-\rho(r)}\ln|f(z)|\right\}.$$

The angular density of the set of zeros of an entire function is defined by

$$\Delta^\gamma(\alpha, \beta) = \lim_{r\to\infty} \frac{n^\gamma(r;\, \alpha, \beta)}{r^{\rho(r)}},$$

where $n^\gamma(r;\, \alpha, \beta)$ is the number of zeros of $f(z)$ in $D_r^\gamma(\alpha, \beta)$ and

$$\mathfrak{S}_f^\gamma = \lim_{r\to\infty} r^{\rho\cdot\rho(r)} e^{i\rho\gamma(r)} \left[c + \frac{1}{\rho} \sum_{|a_k|<r} a_k^{-\rho}\right].$$

---

[12]We also note the papers of Černoljas [**1, 2**], who considers orders more general than proximate orders. He assumes that $\overline{\lim}_{r\to\infty} \rho(r) = \rho < \infty$, $\underline{\lim}_{r\to\infty} \rho(r) = \lambda > 0$ and $r\rho'(r)\ln r \to 0$. He carries over the theory of functions of completely regular growth to such "oscillatory" orders, which are somewhat more general than Boutroux orders.

[13]This indicator was studied by Macintyre [1] in the case when $\mathfrak{L}_\vartheta^\gamma$ is a logarithmic spiral.

The set of zeros is called γ-regularly distributed for a nonintegral ρ if the angular density Δ^{γ} exists for all α and β except perhaps for a countable set; and for integral ρ, if the limit δ_f^{γ} exists.

A function $f(z)$ is said to be of γ-completely regular growth if, after exclusion of a C^0-set, the limit

$$h_f^{\gamma}(\vartheta) = \lim_{z \in \mathfrak{L}_{\vartheta},\, z \overline{\in} C^0} \{r^{-\rho(r)} \ln |f(z)|\}$$

exists. With these definitions, we have the theorem: *For an entire function with proximate order $\rho(r) \to \rho$ to be of γ-completely regular growth, it is necessary and sufficient that the set of its zeros is γ-regularly distributed.*

There are also asymptotic formulas of the same kind as (2.03), (2.04) and (2.06).

LIST OF IMPORTANT IDEAS AND THEOREMS

Proximate order. A function $\rho(r)$ $(r > 0)$ is called a *proximate order* if

(1) $\lim_{r\to\infty} \rho(r) = \rho$ exists,

(2) $\lim_{r\to\infty} r\rho'(r) \ln r = 0$.

If

$$\overline{\lim_{r\to\infty}} \frac{\ln M_f(r)}{r^{\rho(r)}} \neq 0, \infty$$

for an entire function $f(z)$, then $\rho(r)$ is called the proximate order of $f(z)$. Every entire function of finite order has a proximate order.

A slowly increasing function $L(r)$ is a positive function for which

$$\lim_{r\to\infty} \frac{L(kr)}{L(r)} = 1$$

uniformly on any segment $0 < a \leqslant k \leqslant b < \infty$.

If $\rho(r)$ satisfies conditions (1) and (2), then

$$r^{\rho(r)} = r^{\rho} L(r) \qquad \text{(p. 32)}.$$

A function designated by $L^*(r)$ is one that can be expressed in the form

$$\ln L^*(r) = \vartheta_2(\ln r) - \vartheta_1(\ln r),$$

where the $\vartheta_i(r)$ are concave functions satisfying the conditions

$$\lim_{x\to+\infty} \vartheta_i(x) = \infty, \qquad \lim_{x\to+\infty} \frac{\vartheta_i(x)}{x} = 0, \qquad \lim_{x\to+\infty} \frac{\vartheta_i''(x)}{\vartheta_i'(x)} = 0.$$

Every function $L^*(r)$ is slowly increasing.

A function $\rho^*(r)$, defined by the relation

$$r^{\rho^*(r)} = r^{\rho} L^*(r),$$

is called a *strong proximate order* (p. 41).

Every entire function of finite order has a strong proximate order (pp. 35–41).

Density. Let $\mathfrak{N}$ be any point set in the plane, let $n(r)$ be the number of points of $\mathfrak{N}$ in the circle $|z| < r$ and $n(r, \vartheta, \theta)$ be the number of points of $\mathfrak{N}$ in the sector $|z| < r$, $\vartheta < \arg z \leqslant \theta$, and let $\rho(r)$ be a proximate order.

The density and the upper density (a similar definition yields the *lower density*) of the set $\mathfrak{N}$ relative to the exponent $\rho(r)$ are defined to be

$$\Delta = \lim_{r\to\infty} \frac{n(r)}{r^{\rho(r)}} \quad \text{(density)},$$

$$\bar{\Delta} = \overline{\lim_{r\to\infty}} \frac{n(r)}{r^{\rho(r)}} \quad \text{(upper density)}.$$

The set $\mathfrak{N}$ has an *angular density*, if

$$\Delta(\vartheta, \theta) = \lim_{r\to\infty} \frac{n(r, \vartheta, \theta)}{r^{\rho(r)}}$$

exists for all ϑ and θ with the possible exception of an enumerable set. In this case the relation

$$\Delta(\theta) - \Delta(\vartheta) = \Delta(\vartheta, \theta)$$

defines a nondecreasing function $\Delta(\theta)$ with an arbitrary additive constant (pp. 89–90).

The *linear density* $\rho^*(C)$ of a set C of circles C_j in the complex plane is defined to be

$$\rho^*(C) = \lim_{r\to\infty} \frac{1}{r} \sum\nolimits^r r_j,$$

where r_j is the radius of C_j and Σ^r denotes the sum over all the circles with centers in the circle $|z| < r$. If $\rho^*(C)$ does not exist, we introduce the *upper* linear density

$$\bar{\rho}^* = \overline{\lim_{r\to\infty}} \frac{1}{r} \sum\nolimits^r r_j$$

of the set and the *lower* linear density $\underline{\rho}^*(C)$. A set of circles with zero linear density is called a *C^0-set* (p. 90).

The *relative measure* of a set E of positive numbers is the limit

$$m^*(E) = \lim_{r\to\infty} \frac{\operatorname{mes}(E^r)}{r},$$

where E^r is the intersection of E with the interval $(0, r)$.

The *upper and lower relative measures* are defined similarly:

$$\overline{m}^*(E) = \overline{\lim_{r\to\infty}} \frac{\operatorname{mes}(E^r)}{r}, \qquad \underline{m}^*(E) = \underline{\lim_{r\to\infty}} \frac{\operatorname{mes}(E^r)}{r}.$$

The symbol E^0 is used for a set of zero relative measure (p. 96).

Sets of points $\mathfrak{N}$, regularly distributed for the exponent $\rho(r)$.

I. ρ not an integer.

A set $\mathfrak{N}$ is said to be *regularly distributed for the exponent* $\rho(r)$ if it has an angular density for this exponent (p. 90).

II. ρ an integer.

A set $\mathfrak{N}$ is said to be *regularly distributed for the exponent* $\rho(r)$ if it has an angular density for the exponent $\rho(r)$, and if the limit

$$\delta = \lim_{r\to\infty} \frac{1}{L(r)}\left\{c + \frac{1}{\rho}\sum_{|a_n|<r} \frac{1}{a_n^\rho}\right\} \qquad (a_n \in \mathfrak{N})$$

exists (is finite) for any value of the constant c (p. 92).

The zeros of an entire function are regularly distributed if the set of all these zeros is regularly distributed for the exponent $\rho(r)$, where $\rho(r)$ is the proximate order of this function.

***R*-sets.** A regularly distributed set $\{a_n\}$ is called an R-set if

(C) there is a number $d > 0$ such that the circles of radius

$$r_n = d\,|a_n|^{1-\frac{1}{2}\rho(|a_n|)}$$

with centers at a_n are nonintersecting, or if

(C′) all the points a_n lie in a finite number of nonoverlapping sectors $\varphi_j' < \arg z < \varphi_j''$ $(j = 1, 2, \ldots, m)$, and

$$|a_{n+1,j}| - |a_{n,j}| > d\,|a_{n,j}|^{1-\rho(|a_{nj}|)} \qquad \text{(p. 95)}$$

for any $d > 0$, where all the points a_{nj} in each of the sectors are numbered in order of increasing absolute value.

The indicator of an entire function. *The indicator of an entire function $f(z)$ of exponential type is*

$$h(\theta) = h_f(\theta) = \overline{\lim_{r\to\infty}} \frac{\ln|f(re^{i\theta})|}{r}.$$

The indicator of an entire function $f(z)$ of finite order ρ and of normal type is

$$h(\theta) = h_f(\theta) = \overline{\lim_{r\to\infty}} \frac{\ln|f(re^{i\theta})|}{r^\rho} \qquad \text{(p. 51).}$$

The generalized indicator of an entire function $f(z)$ of proximate order $\rho(r)$ is

$$h(\theta) = h_f(\theta) = \overline{\lim_{r\to\infty}} \frac{\ln|f(re^{i\theta})|}{r^{\rho(r)}} \qquad \text{(p. 70).}$$

The indicators of functions that are holomorphic in some angle $\alpha \leqslant \arg z \leqslant \beta$ are defined similarly. The characteristic property of an indicator (and of a generalized indicator) is its trigonometric convexity.

A trigonometrically convex function is a function satisfying the inequality

$$h(\theta_1)\sin\rho(\theta_2-\theta_3) + h(\theta_2)\sin\rho(\theta_3-\theta_1) + h(\theta_3)\sin\rho(\theta_1-\theta_2) \leqslant 0$$

$$\text{for } \theta_3-\theta_1 < \frac{\pi}{\rho},\ \theta_1<\theta_2<\theta_3 \qquad \text{(p. 53).}$$

We will use the notation $s(\theta) = h'(\theta) + \rho^2 \int^\theta h(\varphi)\,d\varphi$, $s(\theta, \vartheta) = s(\theta) - s(\vartheta)$.

For a function $h(\theta)$ to be trigonometrically convex in some interval, it is necessary and sufficient that $s(\theta)$ be nondecreasing in this interval.

When ρ is not an integer we have

$$h(\theta) = \frac{1}{2\rho \sin \pi\rho} \int_{\theta-2\pi}^{\theta} \cos \rho (\theta - \psi - \pi) \, ds(\psi);$$

when ρ is an integer

$$h(\theta) = \frac{1}{2\pi\rho} \int_{\theta-2\pi}^{\theta} (\theta-\psi) \sin \rho (\theta - \psi) \, ds(\psi) + A \cos \rho\theta + A \sin \rho\theta \quad \text{(p. 60)}.$$

An indicator of the form

$$a \cos \rho\theta + b \sin \rho\theta$$

is called *sinusoidal.*

THEOREM. *For any strong proximate order $\rho^*(r)$ and any trigonometrically convex, periodic function $h(\theta)$ with period 2π, there is an entire function with proximate order $\rho^*(r)$ and indicator $h(\theta)$* (p. 94).

When ρ is not an integer, a similar assertion is valid for any arbitrary proximate order $\rho(r)$. A strong proximate order can also be replaced by a proximate order for integral ρ if $h(\theta) \geqslant 0$ $(0 \leqslant \theta < 2\pi)$.

Supporting functions. The supporting function of a convex region G is the function

$$k(\theta) = \sup_{z \in G} (x \cos \theta + y \sin \theta),$$

while the straight line l_θ with the equation

$$x \cos \theta + y \sin \theta - k(\theta) = 0$$

is called the supporting line.

The relations

$$k(\theta + 2\pi) = k(\theta),$$
$$k(\theta_1) \sin(\theta_2 - \theta_3) + k(\theta_2) \sin(\theta_3 - \theta_1) + k(\theta_3) \sin(\theta_1 - \theta_2) \leqslant 0,$$
$$\theta_1 \leqslant \theta_2 \leqslant \theta_3, \quad \theta_3 - \theta_2 < \pi, \quad \theta_2 - \theta_1 < \pi \quad \text{(p. 76)}$$

are characteristic of a supporting function.

Let the region G contain the origin, and let s be the arc length of the boundary of G measured counterclockwise from some fixed point to the supporting point. Then θ is a nondecreasing function of s.

At angular points of the boundary $\theta(s)$ is discontinuous, while for straight-line segments of the boundary $\theta(s)$ is constant. The arc length along the boundary between the points of the support lines l_α and l_β is

$$s(\beta + 0) - s(\alpha - 0) = k'_+(\beta) - k'_-(\alpha) + \int_\alpha^\beta k(\theta) \, d\theta.$$

The radius of curvature

$$R = \frac{ds}{d\theta}$$

exists for all values of θ with the possible exception of a set of zero measure, and

$$R = k''(\theta) + k(\theta).$$

We also have

$$k(\theta) = \int_0^\theta \sin(\theta - \psi)\, ds(\psi),$$

while, when the origin is displaced to any arbitrary point,

$$k(\theta) = A\cos(\theta - \theta_0) + \int_0^\theta \sin(\theta - \psi)\, ds(\psi).$$

There is also another representation:

$$k(\theta) = B\cos(\theta - \theta_0) + \frac{1}{2\pi}\int_{\theta-2\pi}^{\theta} (\theta - \psi)\sin(\theta - \psi)\, ds(\psi) \quad \text{(p. 81)}.$$

The indicator diagram. The indicator for any entire function of exponential type is the supporting function of a bounded, convex region I which is called the *indicator diagram* of the function.

The conjugate diagram of an entire function of exponential type

$$f(z) = \sum_0^\infty \frac{a_n}{n!} z^n$$

is the smallest region containing all the singularities of

$$\varphi(z) = \sum_0^\infty \frac{a_n}{z^{n+1}}.$$

The function $\varphi(z)$ is called the Borel transform of $f(z)$.

THEOREM. *The conjugate diagram of an entire function of exponential type is the mirror reflection of its indicator diagram in the real axis.*

Functions of completely regular growth. A function $F(z)$, holomorphic in the sector $\alpha < \arg z < \beta$, is said to be of *completely regular growth* on some set of rays $R_{\mathfrak{M}}$ ($\mathfrak{M}$ a set of values of θ) if a set $E^0_{\mathfrak{M}}$ of zero relative measure exists such that for $r \,\overline{\in}\, E^0_{\mathfrak{M}}$ and $r \to \infty$ the function

$$h_{F,r}(\theta) = \frac{\ln|F(re^{i\theta})|}{r^{\rho(r)}}$$

tends uniformly to $h_F(\theta)$ for $\theta \in \mathfrak{M}$.

If a function $F(z)$, holomorphic and of proximate order $\rho(r)$ inside the angle (α, β), is of completely regular growth on some set of rays $R_{\mathfrak{M}}$ in the angle defined by ($\alpha < \theta_1 \leqslant \theta \leqslant \theta_2 < \beta$), then it is of completely regular growth on the set of rays $R_{\overline{\mathfrak{M}}}$.

An entire function $f(z)$ of proximate order $\rho(r)$ is called an *entire function of completely regular growth* if there is a set E^0 of zero relative measure such that, when $r \to \infty$ with $r \,\overline{\in}\, E^0$, the function

$$h_{F,r}(\theta) = \frac{\ln |f(re^{i\theta})|}{r^{\rho(r)}}$$

tends uniformly to $h_F(\theta)$. The expression $\lim^*_{r\to\infty} f(r)$ will denote the limit of $f(r)$, if it exists, when $r \to \infty$ without taking the values in some E^0-set (p. 139).

Let

$$J_F^{r}(\theta) = \int_0^r \frac{\ln |F(te^{i\theta})|}{t}\, dt \qquad \text{(p. 143)}.$$

If $F(z)$ is holomorphic and of proximate order $\rho(r)$ in the interior of some angle (α, β), and if it is of completely regular growth on some ray $\arg z = \theta$ in this angle, then

$$\lim_{r\to\infty} r^{-\rho(r)} J_F^r(\theta) = \rho^{-1} h_F(\theta).$$

Conversely, if this equality holds on all the rays in some angle, then $F(z)$ is of completely regular growth in this angle.

THEOREM 1. *Let $\rho(r)$ be a proximate order of growth and let $\rho = \lim_{r\to\infty} \rho(r)$ be nonintegral. Let the set $\{a_k\}$ have the angular density $\Delta(\theta)$ for the exponent $\rho(r)$. Then if we consider only values of $z = re^{i\theta}$ which do not lie in a certain set C of circles of zero density (a C^0-set), the entire function*

$$f(z) = z^m e^{P(z)} \prod_1^\infty G\left(\frac{z}{a_k}; p\right),$$

with

$$P(z) = c_0 + c_1 z + \ldots + c_p z^p \qquad (p = [\rho]),$$

satisfies, uniformly in θ, the relation

$$\lim_{r\to\infty} \frac{\ln |f(re^{i\theta})|}{r^{\rho(r)}} = H_f(\theta),$$

where

$$H_f(\theta) = \frac{\pi}{\sin \pi\rho} \int_{\theta-2\pi}^{\theta} \cos \rho(\theta - \psi - \pi)\, d\Delta(\psi) \qquad \text{(p. 91)}.$$

THEOREM 2. *If ρ is an integer, the conclusion of Theorem 2 remains true with*

$$H_f(\theta) = \int_{\theta-2\pi}^{\theta} (\psi - \theta) \sin \rho(\psi - \theta)\, d\Delta(\psi) + \tau \cos \rho(\theta - \theta_0)$$

if we also impose condition II *in the definition of regularly distributed sets with* $\delta = \tau e^{i\theta_0}$ *and with* $c = c_p$ (p. 91).

If $f(z)$ has regularly distributed zeros, then

$$h_f(\theta) = H_f(\theta).$$

For $\rho = 1$, the angular density of the zeros of $f(z)$ in the angle (ϑ, θ) is $(2\pi)^{-1}$ times the arc length of the indicator diagram between the supporting points for the supporting lines l_ϑ and l_θ, these lines being perpendicular to the rays corresponding to the two angles (p. 93).

The converse theorems are also true. We have the following

COROLLARY. *For an entire function $f(z)$ of proximate order $\rho(r)$ to be of completely regular growth, it is necessary and sufficient that the zeros of this function be regularly distributed for the exponent $\rho(r)$* (p. 158).

COROLLARY. *An entire function of exponential type and completely regular growth can be expressed in the form*

$$f(z) = z^m e^{az+b} \lim_{R \to \infty} \prod_{|a_k| < R} \left(1 - \frac{z}{a_k}\right) \quad \text{(p. 158)}$$

Entire functions of completely regular growth are those entire functions, with a given proximate order and a given indicator, *which have the greatest possible lower density of zeros*

$$\lim_{r \to \infty} \frac{n_f(r)}{r^{\rho(r)}} = \frac{\rho}{2\pi} \int_0^{2\pi} h_f(\theta)\, d\theta.$$

For all other entire functions, the left hand side is smaller than the right hand side (p. 173).

In §§ 2 and 3, Chapter IV, there are corresponding uniqueness theorems for functions holomorphic in some angle.

The indicator of the product of two functions. If $f(z)$ and $\varphi(z)$ are holomorphic and have the same proximate order in the angle (α, β) and if one of them is of completely regular growth, then the indicator of the product of these functions is equal to the sum of indicators of the separate functions (p. 159).

COROLLARY 1. *If $F(z)$ is holomorphic and of proximate order $\rho(r)$ in the angle (α, β), and if the entire function $\varphi(z)$ is such that*

$$\lim_{r \to \infty} \frac{\ln M_\varphi(r)}{r^{\rho(r)}} = 0,$$

then the indicator of the product of these functions in the angle (α, β) is the same as the indicator of $F(z)$ (p. 160).

The canonical function of a regularly distributed set. Let $\mathfrak{A}$ denote the set $\{a_k\}$, regularly distributed for the proximate order $\rho(r)$. For nonintegral values of ρ we write

$$f(z, \mathfrak{A}) = \prod_{k=1}^{\infty} G\left(\frac{z}{a_k}; p\right) \qquad (p = [\rho]),$$

and for integral ρ

$$f(z, \mathfrak{A}) = e^{c_A z^\rho} \prod_{k=1}^{\infty} G\left(\frac{z}{a_k}; \rho\right).$$

Here c_A is the constant in the expression for δ. If this constant is nonnegative, we write $c_A = 0$.

The function $H_f(\theta)$ (see the discussion of functions of completely regular growth) for the set $\mathfrak{A}$ is called the indicator of $\mathfrak{A}$, and is denoted by $H_{\mathfrak{A}}(\theta)$ (p. 92).

The Lagrange interpolation formula. Let the interpolation points $\mathfrak{A} = \{a_n\}$ be regularly distributed for the proximate order $\rho(r)$, let $f(z, \mathfrak{A})$ be the canonical function, and let $H_{\mathfrak{A}}(\theta)$ be the indicator.

For brevity, we set $f(z) = f(z, \mathfrak{A})$.

The series

$$\sum_{n=1}^{\infty} \frac{f(z)\, b_n}{(z - a_n) f'(a_n)} \tag{*}$$

is called the *Lagrange interpolation series.*

If $\mathfrak{A}$ is an R-set, and if for any $\varepsilon > 0$, $N > 0$, and $n > N$ we have

$$\ln |b_n| < [H_{\mathfrak{A}}(\psi_n) - \varepsilon]\, |a_n|^{\rho(|a_n|)} \qquad (\psi_n = \arg a_n),$$

then the Lagrange series (*) converges uniformly in any bounded region to an entire function with indicator smaller than $H_{\mathfrak{A}}(\theta)$ (p. 195).

The function $F(z)$ defined by the series (*) is not the only function satisfying the conditions of the interpolation problem and the inequality $h_F(\theta) \leqslant H_{\mathfrak{A}}(\theta)$. The general solution is

$$\Phi(z) = F(z) + \omega(z) f(z),$$

where $\omega(z)$ is an entire function of minimal type for the proximate order $\rho(r)$.

With the additional condition

$$\lim_{|z| \to \infty} \frac{|\Phi(z)|}{|f(z)|} = 0$$

outside the exceptional circles, we have $\omega(z) \equiv 0$ and $\Phi(z)$ is uniquely defined (p. 198).

If the points a_n form an R-set $\mathfrak{A}$ with exponent $\rho(r)$, then any entire function $F(z)$ satisfying

$$h_F(\theta) < H_{\mathfrak{A}}(\theta)$$

can be expressed in the form

$$F(z) = \sum_{n=1}^{\infty} \frac{F(a_n) f(z)}{f'(a_n)(z - a_n)},$$

the series converging in any bounded region.

If the requirement that $\{a_n\}$ be an R-set is discarded, and it is only assumed that $\mathfrak{A}$ is a regularly distributed set, then

$$F(z) = \lim_{\substack{R \to \infty \\ R \in E^0}} \sum_{|a_n| < R} \frac{F(a_n) f(z)}{f'(a_n)(z - a_n)} \qquad \text{(p. 198)}.$$

If $\mathfrak{A}$ is an R-set with exponent $\rho(r)$, then an entire function solving the interpolation problem $b_n = F(a_n)$ and satisfying the condition

$$h_F(\theta) \leqslant H_{\mathfrak{A}}(\theta)$$

will exist if and only if

$$\overline{\lim_{n \to \infty}} \frac{\ln |b_n|}{H_{\mathfrak{A}}(\psi_n) |a_n|^{\rho(|a_n|)}} \leqslant 1.$$

The spectra of entire functions that can be approximated by exponential polynomials. For any bounded, convex region I, the expression

$$\|F\|_I = \sup_{\substack{r > 0, \\ 0 \leqslant \theta < 2\pi}} |F(re^{i\theta})| e^{-h(\theta) r}$$

can be used as a norm for entire functions, where $h(\theta)$ is the supporting function of the region I, symmetric relative to the real axis. The complete space E_I of entire functions is defined correspondingly (p. 288). If an entire function $F(z)$ can be approximated in the norm of B_I by exponential polynomials

$$P_n(z) = \sum_{k=1}^{n} a_n(\lambda_k) e^{\lambda_k z}$$

with exponents from the boundary of I, then *the spectrum of the entire function* $F(z)$ *in* B_I is the name given to the set of points λ on the boundary of I for which the quantity

$$a(\lambda) = \lim_{n \to \infty} a_n(\lambda)$$

is not zero.

For other equivalent definitions of the spectrum, see Chapter VI, pp. 290 and 291.

THEOREM. *If the smallest convex region containing the whole spectrum of* $F(z)$ *coincides with its conjugate diagram, then* $F(z)$ *is of completely regular growth.*

In particular, if the numbers λ_n are on the boundary of a convex, bounded region, then the entire function

$$f(z) = \sum_{n=1}^{\infty} a_n e^{\lambda_n z}, \quad \sum_{n=1}^{\infty} |a_n| < \infty$$

of exponential type is of completely regular growth.

Functions of the class [Δ]. A function of this class is almost-periodic with its whole spectrum in an interval of length 2Δ, the ends of the interval belonging to the spectrum. Such a function is an entire function of exponential type, the conjugate diagram of which coincides with the smallest interval containing its whole spectrum. For an almost-periodic, entire function of exponential type to be in the class [Δ], it is necessary and sufficient that all its zeros be in a strip $|y| \leqslant H$ parallel to the real axis (p. 270). The formula

$$n(t) = \frac{\Delta}{\pi} t + \omega(t)$$

holds for the zeros of a function of the class [Δ], where $n(t)$ is the number of zeros of the function in the rectangle $0 < x \leqslant t$, $|y| \leqslant H$, and $\omega(t)$ is bounded (p. 271).

Functions of the class A. A function of this class satisfies the condition

$$\sum_{n=1}^{\infty} \left| \operatorname{Im} \frac{1}{a_n} \right| < \infty,$$

where the a_n are all the zeros of this function.

Function of the class A in the half-plane $\operatorname{Im} z > 0$ (or $\operatorname{Im} z < 0$) are defined similarly (p. 223).

An entire function of exponential type belongs to the class A if the integral

$$\int_0^R \frac{\ln |F(x) F(-x)|}{1+x^2} dx$$

is bounded for all $R > 0$ (pp. 225 and 228).

(a) For an entire function $f(z)$ of exponential type to be of completely regular growth and to belong to the class A, it is necessary and sufficient that

(1) the integral

$$\int_0^{\infty} \frac{\ln |f(x) f(-x)|}{1+x^2} dx$$

exists and

(2) $h_f(0) + h_f(\pi) = 0$.

(b) The conditions (1) and (2) are *sufficient* to ensure that a function $F(z)$, holomorphic and of exponential type in the half-plane $\operatorname{Im} z \geqslant 0$ ($\operatorname{Im} z \leqslant 0$), belong to the class A and be of completely regular growth.

(c) Another sufficient condition for an entire function (or a function holomorphic in a half-plane) to belong to the class A and to be of completely regular growth is the existence of either of the integrals

$$\int_{-\infty}^{\infty} \frac{\ln |F(x)|}{1+x^2}\, dx \quad \text{or} \quad \int_{-\infty}^{\infty} \frac{\ln_+ |F(x)|}{1+x^2}\, dx.$$

THEOREM. *If an entire function $f(z)$ belongs to the class A, and if*

$$\frac{1}{f(z)} = R(z) + \sum_{k=1}^{\infty} \frac{1}{f'(a_n)} \left[\frac{1}{z-a_n} + \frac{1}{a_n} + \frac{z}{a_n^2} + \dots + \frac{z^{p-1}}{a_n^p} \right],$$

where $R(z)$ is a polynomial and

$$\sum_{n=1}^{\infty} \frac{1}{|a_n|^{p+1} |f'(a_n)|} < \infty,$$

then

(1) *$f(z)$ is of exponential type,*

(2) $$\int_{-\infty}^{\infty} \frac{\ln_+ |f(x)|}{1+x^2}\, dx < \infty.$$

It follows from (1) and (2) that $f(z)$ is of completely regular growth and that its indicator diagram is a segment of the imaginary axis.

The class HB. An entire function $\omega(z)$ belongs to the class HB ($\overline{HB}$) if

(1) it has no zeros in the closed (open) lower half-plane,

(2) for $\operatorname{Im} z < 0$ it satisfies the inequality

$$|\omega(z)| \geqslant |\omega(\bar{z})|.$$

For an entire function $\omega(z)$ to be in the class $\overline{HB}$, it is necessary and sufficient that it be in the class A and be expressible in the form

$$\omega(z) = z^m e^{u(z)+i(\nu z+\delta)} \prod_{k=1}^{\infty} \left(1 - \frac{z}{a_k}\right) e^{RP_k\left(\frac{z}{a_k}\right)},$$

where $\nu \geqslant 0$, $u(z)$ is a real entire function, and

$$R \sum (a_k + ib_k) z^k = \sum a_k z^k.$$

THEOREM. *Let*

$$\omega(z) = P(z) + iQ(z),$$

where

$$P(z) = Ae^{u(z)} (z-a_0) \prod_{-\infty}^{\infty}{}' \left(1 - \frac{z}{a_n}\right) e^{P_n\left(\frac{z}{a_n}\right)}$$

and

$$Q(z) = Be^{v(z)}(z - b_0)\prod_{-\infty}^{\infty}{}'\left(1 - \frac{z}{b_n}\right)e^{P_n\left(\frac{z}{b_n}\right)}$$

are real entire functions. For $\omega(z)$ to belong to the class HB, it is necessary and sufficient that the following conditions be satisfied:

(1) *the zeros of $P(z)$ and $Q(z)$ are all real, simple and interlacing;*

(2) *the entire functions $u(z)$ and $v(z)$ and the Weierstrass exponents $P_n(z/a_n)$ and $P_n(z/b_n)$ satisfy the relation*

$$u(z) - v(z) + \sum_{n=-\infty}^{\infty}\left[P_n\left(\frac{z}{a_n}\right) - P_n\left(\frac{z}{b_n}\right)\right] \equiv 0;$$

(3) *the constants A and B have the same sign.*

When the first two conditions are satisfied, the last condition is equivalent to the inequality

$$Q'(x)P(x) - Q(x)P'(x) > 0 \qquad (-\infty < x < \infty) \text{ (p. 313).}$$

A real pair of entire functions is the name given to two real, entire functions $P(z)$ and $Q(z)$ if they have no common zeros and if any linear combination $\mu P(z) + \nu Q(z)$ with real μ and ν has no complex zeros. If $P(z)$ and $Q(z)$ are permitted to have common zeros, then they are called a *generalized real pair.*

THEOREM. *For an entire function $\omega(z) = P(z) + iQ(z)$ to belong to the class HB ($\overline{HB}$), it is necessary and sufficient that $P(z)$ and $Q(z)$ be a real pair (a generalized real pair) and that $Q'(x_0)P(x_0) - P'(x_0)Q(x_0) > 0$ for some real point x_0.*

The class *P*. An entire function $\omega(z)$ of exponential type is said to be a *function of the class P* if

(1) it has no zeros in the open lower half-plane,

(2) $h_\omega\left(\frac{\pi}{2}\right) \leqslant h_\omega\left(-\frac{\pi}{2}\right)$ (p. 320).

The quantity $d_\omega = h_\omega(-\pi/2) - h_\omega(\pi/2)$ is called the *defect* of a function of the class *P* (p. 320).

The class *P* consists of entire functions of exponential type of the class $\overline{HB}$ (p. 320).

THEOREM. *For an entire function of exponential type*

$$\omega(z) = P(z) + iQ(z)$$

to belong to the class P, it is necessary and sufficient that

(1) *the zeros of $P(z)$ and $Q(z)$ be real and interlacing,*

(2) *the indicator diagrams of $P(z)$ and $Q(z)$ coincide, and*

(3) $Q'(x)P(x) - Q(x)P'(x) > 0$ (pp. 321–322).

The class P^*. This is the class of all entire functions in $\overline{HB}$ expressible in the form

$$\omega(z) = e^{-\gamma z^2} \omega_1(z),$$

where the genus of $\omega_1(z)$ is not larger than one and $\gamma \geqslant 0$.

THEOREM. *For an entire function $\omega(z)$ to be of the class P^*, it is necessary and sufficient that it be the uniform limit of a sequence of polynomials, where these polynomials have no zeros in the lower half-plane* (p. 334).

An *N-polynomial* is a polynomial having no zeros in the lower half-plane (p. 331).

Majorants. A function $\omega(z)$ of $\overline{HB}$ is said to be a majorant of $f(z)$ if

$$|\omega(z)| \geqslant |f(z)|, \qquad |\omega(z)| \geqslant |f(\bar{z})|$$

for $\operatorname{Im} z < 0$.

Majorants of the class P or P^* are called *P-majorants* or *P^*-majorants*, respectively.

The above definition of a P-majorant is equivalent to the following: A function $\omega(z)$ of the class P is a *P-majorant* for the entire function $f(z)$ of exponential type if

(1) the exponential type of $\omega(z)$ is not lower than that of $f(z)$,

(2) $|f(x)| \leqslant |\omega(x)|$ $(-\infty < x < \infty)$ (p. 362).

Admissible classes. A subclass T of the class $\overline{HB}$ is called an *admissible class* if, when $\omega(z) \in T$ and $f(z)$ is subordinate to $\omega(z)$, it follows that $\omega(z) + f(z) \in T$. The classes $\overline{HB}$, P and P^* are admissible classes (pp. 350 and 355).

A $\mathfrak{B}_T$-operator is a homogeneous, additive operator, defined on the linear span of some T-class of functions and mapping this class into itself.

A $\mathfrak{B}_P$-operator is denoted simply by $\mathfrak{B}$, while a $\mathfrak{B}_{P^*}$-operator is denoted by $\mathfrak{B}^*$. A $\mathfrak{B}_T$-operator is said to be continuous if it maps every sequence of functions of a class T, uniformly convergent in every finite region, into a sequence of functions from T convergent in the same sense.

THEOREM. *If T is an admissible class, then a $\mathfrak{B}_T$-operator preserves subordination. More precisely: If $\omega(z) \in T$ and $f(z)$ is subordinate to (majorized by) $\omega(z)$, then $\mathfrak{B}_T[f(z)]$ is an entire function subordinate to the function $\mathfrak{B}[\omega(z)]$* (p. 356).

THEOREM. *The operation of differentiation is a $\mathfrak{B}$- and a $\mathfrak{B}^*$-operator* (p. 356).

THEOREM. *Let a subclass Q of the class $\overline{HB}$ satisfy the conditions*

(1) *from $\omega(z) \in Q$ it follows that $e^{\tau z}\omega(z) \in Q$ for any real τ,*

(2) *from $\omega(z) \in Q$ it follows that $\omega'(z) \in Q$ (i.e., differentiation is a $\mathfrak{B}_Q$-operator).*

Then $Q \in P^*$. In other words, P^* is the largest subclass of $\overline{HB}$ satisfying the conditions (1) and (2) (p. 353).

THEOREM. *If $\omega(z)$ is a P^*-majorant of the function $f(z)$, then $\omega^{(k)}(z)$ is a P^*-majorant of $f^{(k)}(z)$ $(k = 1, 2, \ldots)$, and if $\omega(z)$ is not a polynomial and $|\omega^{(k)}(z_0)| = |f^{(k)}(z_0)|$ for some point z_0* ($\operatorname{Im} z_0 \leqslant 0$)*, then $f(z) = c_1\omega(z) + c_2\bar{\omega}(z)$* ($|c_1| + |c_2| = 1$) (p. 356).

THEOREM. *If $\omega(z) \in P$, if the exponential type of $\omega(z)$ is not lower than that of $f(z)$, and if $|f(x)| \leqslant |\omega(x)|$ $(-\infty < x < \infty)$, then $|f^{(k)}(x)| \leqslant |\omega^{(k)}(x)|$ $(-\infty < x < \infty)$. If $\omega(x)$ is not a polynomial, then if $|f^{(k)}(x_0)| = |\omega^{(k)}(x_0)|$ for some point x_0 of the real axis, it follows that $f(x) = c_1\omega(x) + c_2\bar{\omega}(x)$* ($|c_1| + |c_2| = 1$).

For $\omega(x) = Me^{i\sigma x}$, we obtain a theorem due to S. N. Bernšteĭn: If $f(x)$ is an entire function of exponential type not exceeding σ and $|f(x)| \leqslant M$ $(-\infty < x < \infty)$, then $|f'(x)| \leqslant M\sigma$, with equality only for $f(x) = c_1e^{i\sigma x} + c_2e^{-i\sigma x}$.

The class P is the largest class of majorants in the following sense.

If $\omega(z) \,\overline{\in}\, P$ and $\bar{\omega}(z) \,\overline{\in}\, P$, then there is an entire function $f(z)$, of the same proximate order and the same type as $\omega(z)$, satisfying the inequality $|f(x)| \leqslant |\omega(x)|$ $(-\infty < x < \infty)$, and $|f'(x_0)| \geqslant |\omega'(x_0)|$ for some real point x_0 (p. 366).

A sequence of multipliers of the first kind is a sequence of numbers $\gamma_0, \gamma_1, \gamma_2, \ldots, \gamma_n, \ldots$ such that, if all the zeros of the polynomial

$$P(z) = \sum_{k=0}^{n} a_k z^k$$

are real, then so are all the zeros of the polynomial

$$\sum_{k=0}^{n} \gamma_k a_k z^k \quad \text{(p. 341).}$$

The numbers γ_k $(k = 0, 1, 2, \ldots)$ either all have the same sign or have alternating signs (p. 342).

THEOREM. *For the sequence of numbers $\gamma_0, \gamma_1, \ldots, \gamma_n, \ldots$ to be a sequence of multipliers of the first kind, it is necessary and sufficient that the series*

$$\sum_{n=0}^{\infty} \frac{\gamma_n}{n!} z^n \qquad (0! = 1)$$

converge to an entire function $\varphi(z)$, expressible in the form

$$\varphi(z) = e^{\sigma z} \prod_{k=1}^{\infty} (1 + \varkappa_k z),$$

$$\textit{where } \sigma \geqslant 0 \textit{ and } \varkappa_k > 0 \ (k = 1, 2, 3, \ldots) \textit{ and } \sum_{k=1}^{\infty} \varkappa_k < \infty \quad \text{(p. 346).}$$

Γ-operators. These are operators that transform any entire function

$$f(z) = \sum_{k=0}^{\infty} a_k z^k$$

into an entire function

$$\Gamma[f(z)] = \sum_{k=0}^{\infty} \gamma_k a_k z^n,$$

where γ_k ($k = 0, 1, 2, \ldots$) is a sequence of multipliers of the first kind all with the same sign (pp. 340–344).

THEOREM. *Γ-operators give the general form of $\mathfrak{B}$- and $\mathfrak{B}^*$-operators permutable with the operator $z(d/dz)$.*

A Jensen operator transforms any entire function

$$f(z) = \sum_{k=0}^{\infty} a_k z^k$$

into the polynomial

$$\tilde{f}_n(z) = a_0 + \sum_{k=1}^{n} \left(1 - \frac{1}{n}\right)\left(1 - \frac{2}{n}\right) \cdots \left(1 - \frac{k-1}{n}\right) a_k z^k.$$

Obviously $\tilde{f}_n(z) \rightrightarrows f(z)$ for $n \to \infty$. A Jensen operator is a Γ-operator (p. 343).

The class HB_2 ($\overline{HB_2}$) consists of the entire functions $\omega(z, u)$ satisfying the conditions:

(1) $\omega(z, u)$ is not zero for $\operatorname{Im} z \leqslant 0$ and $\operatorname{Im} u \leqslant 0$ ($\operatorname{Im} z < 0$, $\operatorname{Im} u < 0$);
(2) for $\operatorname{Im} z < 0$, $\operatorname{Im} u < 0$ we have

$$|\omega(z, u)| \geqslant |\omega(\bar{z}, u)|, \ |\omega(z, u)| \geqslant |w(z, \bar{u})|, \ |\omega(z, u)| \geqslant |\omega(\bar{z}, \bar{u})|.$$

The class P_2 consists of the entire functions of exponential type belonging to the class HB_2 (p. 366).

The class P_2^* consists of the functions of HB_2 of the form

$$\omega(z, u) = e^{-\gamma_1 z^2 - \gamma_2 u^2}\omega_1(z, u),$$

where $\omega_1(z, u)$ is of genus one for each of the variables when the other is fixed in the lower half-plane and $\gamma_1 \geqslant 0$, $\gamma_2 \geqslant 0$.

An H_2-polynomial is a polynomial of two variables $P(z, u)$ which is not zero for $\operatorname{Im} z < 0$, $\operatorname{Im} u < 0$. Such polynomials belong to HB_2.

THEOREM. *For an entire function $\omega(z, u)$ to be in the class P_2^*, it is necessary and sufficient that it be the uniform limit of a sequence of H_2-polynomials.*

Majorants for functions of more than one variable.[1] A function $\omega(z, u)$ of the class $\overline{HB_2}$ is said to be a *majorant* of the entire function $f(z, u)$ if, for $\operatorname{Im} z < 0$ and $\operatorname{Im} u < 0$,

$$|\omega(z, u)| \geqslant |f(z, u)|, \qquad |\omega(z, u)| \geqslant |f(\bar{z}, u)|,$$
$$|\omega(z, u)| \geqslant |f(z, \bar{u})|, \qquad |\omega(z, u)| \geqslant |f(\bar{z}, \bar{u})|.$$

[1] For simplicity we limit our discussion to functions of two variables.

Admissible classes. A subclass T_2 of HB_2 is called an *admissible class* if, when $\omega(z, u) \in T_2$ and $f(z, u)$ is subordinate to $\omega(z, u)$, it follows that $\omega(z, u) + f(z, u) \in T_2$. The classes $\overline{HB}_2$, P_2, and P_2^* are admissible classes.

$\mathfrak{B}_{T_2}$-operators are defined in the same way as for functions of one variable. If T_2 is an admissible class, then a $\mathfrak{B}_{T_2}$-operator preserves subordination.

THEOREM. *Every continuous $\mathfrak{B}^*$-operator for functions of the single variable z is also a $\mathfrak{B}^*$-operator for functions of the two variables z and u.*

The general form of the operators $\mathfrak{B}$ and $\mathfrak{B}^*$. An operator $K[f]$ applicable to entire functions is called regular if it satisfies the condition $K[\tilde{f}_n] \to K[f]$ for $n \to \infty$ uniformly in every finite region, where $\tilde{f}_n(z)$ is the function obtained by applying the Jensen operator to $f(z)$ (p. 373).

Upper continuous operators. An operator $K[f]$ applicable to entire functions of exponential type is called *upper continuous* if for every $A > 0$ there is a number $B > 0$ such that, if

$$\sup |f_n(z) - f_0(z)| e^{-A|z|} \to 0,$$

then

$$\sup |K[f_n(z)] - K[f_0(z)]| e^{-B|z|} \to 0 \quad \text{(p. 376)}.$$

THEOREM. *A regular $\mathfrak{B}^*$-operator can be expressed in the form*

$$\mathfrak{B}^*[f(z)] = f(-D_u)[\varphi(z, u)]|_{u=0};$$

here $f(-D_u)[\varphi(u)] = \sum_{n=0}^{\infty} (-1)^n a_n \varphi^{(n)}(u)$ *for* $f(z) = \sum_{n=0}^{\infty} a_n z^n$,

and $\varphi(z, u)$ is of the class P_2^* and of genus one in u for $\operatorname{Im} z < 0$ (p. 373).

THEOREM. *An upper continuous $\mathfrak{B}$-operator can be written in the form*

$$\mathfrak{B}[f(z)] = f(-D_u)[\varphi(z, u)]|_{u=0},$$

where $\varphi(z, u)$ belongs to P_2^* and is of exponential type in z for $\operatorname{Im} u \leqslant 0$ (p. 377).

THEOREM. *A regular $\mathfrak{B}^*$-operator permutable with the operation of differentiation can be expressed in the form*

$$\mathfrak{B}[f(z)] = \overline{F}(D)[f(z)],$$

where $F(z)$ is a function of the class P^* and of genus one (p. 379).

BIBLIOGRAPHY

AGMON, S.

[1] *Functions of exponential type in an angle and singularities of Taylor series*, Trans. Amer. Math. Soc. **70** (1951), 492–508.

AGRANOVIČ, P. Z.

[1] *On functions of several variables of completely regular growth*, Teor. Funkciĭ, Funkcional. Anal. i Priložen. Vyp. 30 (1978), 3–13. (Russian)

AHIEZER, N. I.

[1] *On the theory of entire functions of finite degree*, Dokl. Akad. Nauk SSSR **63** (1948), 475–478. (Russian)

[2] *On some properties of integral transcendental functions of exponential type*, Izv. Akad. Nauk SSSR Ser. Mat. **10** (1946), 411–428. (Russian; English summary)

[3] *On the polynomials of B. M. Levitan*, C. R. (Dokl.) Acad. Sci. URSS **54** (1946), 3–5.

[4] *On entire transcendental functions of finite degree having a majorant on a sequence of real points*, Izv. Akad. Nauk SSSR Ser. Mat. **16** (1952), 353–364. (Russian)

[5] *Lectures on the theory of approximation*, OGIZ, Moscow, 1947; English transl., Ungar, New York, 1956.

AHIEZER, N. I., KREĬN, M. G.

[1] *Some questions in the theory of moments*, Gos. Naučno.-Tehn. Izdat. Ukrain., Kharkov, 1938; English transl., Amer. Math. Soc., Providence, R. I., 1962.

AHIEZER, N. I., LEVIN, B. JA.

[1] *On interpolation of entire transcendental functions of finite degree*, Har'kov. Gos. Univ. Učen. Zap. 40 = Zap. Mat. Otd. Fiz.-Mat. Fak. i Har'kov. Mat. Obšč. (4) **23** (1952), 5–26. (Russian)

AHIEZER, N. I., MARČENKO, V. A.

[1] *On some questions of approximations of continuous functions on the whole real axis*. II, Har'kov. Gos. Univ. Učen. Zap. 29 = Zap. Mat. Otd. Fiz.-Mat. Fak. i Har'kov. Mat. Obšč. (4) **21** (1949), 5–9. (Russian)

AHLFORS, L.

[1] *Ein Satz von Henri Cartan und seine Anwendung auf die Theorie der meromorphen Funktionen*, Soc. Sci. Fenn. Comment. Phys.-Math. **5** (1931), no. 16.

AHLFORS, L., HEINS, M.

[1] *Questions of regularity connected with the Phragmén-Lindelöf principle*, Ann. of Math. (2) **50** (1949), 341–346.

ÅLANDER, M.

[1] *Sur le déplacement des zéros des fonctions entières par leur dérivation*, Thèse, Uppsala, 1914.

AMBARZUMIAN, V. [AMBARCUMJAN, V. A.]

[1] *Über eine Frage der Eigenwerttheorie*, Z. Phys. **53** (1929), 690–695.

ANDERSON, J. M.

[1] *Integral functions and Tauberian theorems*, Duke Math. J. **32** (1965), 597–606.

ANDRAŠKO, M. I.

[1] *The extremal indicator of an entire function of order less than one with positive zeros*, Dopovīdī Akad. Nauk Ukraïn. RSR **1960**, 869–872. (Ukrainian)

AZARIN, V. S.

[1] *On rays of completely regular growth of an entire function*, Mat. Sb. **79 (121)** (1969), 463–476; English transl. in Math. USSR Sb. **8** (1969).

[2] *A characteristic property of functions of completely regular growth in the interior of an angle*, Teor. Funkciĭ, Funkcional. Anal. i Priložen. Vyp. 2 (1966), 55–66. (Russian)

[3] *Generalization of a theorem of Hayman on subharmonic functions in an n-dimensional cone*, Mat. Sb. **66 (108)** (1965), 248–264; English transl. in Amer. Math. Soc. Transl. (2) **80** (1969).

[4] *Functions of regular growth that are subharmonic in the whole space*, Har′kov. Gos. Univ. Učen. Zap. **120** = Zap. Meh. Mat. Fak. i Har′kov. Mat. Obšč. (4) **28** (1961), 128–148. (Russian)

[5] *On the indicator of a function subharmonic in a multidimensional space*, Mat. Sb. **58 (100)** (1962), 87–94. (Russian)

[6] *On the indicator of a function subharmonic in a multidimensional space*, Dokl. Akad. Nauk SSSR **139** (1961), 1033–1036; English transl. in Soviet Math. Dokl. **2** (1961).

[7] *Subharmonic functions of completely regular growth in multidimensional space*, Dokl. Akad. Nauk SSSR **146** (1962), 743–746; English transl. in Soviet Math. Dokl. **3** (1962).

[8] *Regularity of growth of entire functions*, Dokl. Akad. Nauk SSSR **200** (1971), 511–512; English transl. in Soviet Math. Dokl. **12** (1971).

[9] *Extremal problems on entire functions*, Teor. Funkciĭ, Funkcional. Anal. i Priložen. Vyp. 18 (1973), 18–50. (Russian)

[10] *The regularity of the growth of the Fourier coefficients of the logarithm of the modulus of an entire function*, Teor. Funkciĭ, Funkcional. Anal. i Priložen. Vyp. 27 (1977), 9–22. (Russian)

[11] *Asymptotic behavior of subharmonic and entire functions*, Dokl. Akad. Nauk SSSR **229** (1976), 1289–1291; English transl. in Soviet Math. Dokl. **17** (1976).

[12] *Example of an entire function with given indicator and lower indicator*, Mat. Sb. **89 (131)** (1972), 541–557; English transl. in Math. USSR Sb. **18** (1972).

BAILLETTE, A.

[1] *Sur la co-indicatrice des produits canoniques*, Ann. Inst. Fourier (Grenoble) **17** (1967), fasc. 1, 109–118.

BALAŠOV, S. K.

[1] *On entire functions of finite order with zeros on curves of regular rotation*, Izv. Akad. Nauk SSSR Ser. Mat. **37** (1973), 603–629; English transl. in Math. USSR Izv. **7** (1973).

[2] *On entire functions of completely regular growth along curves of regular rotation*, Izv. Akad. Nauk SSSR Ser. Mat. **40** (1976), 338–354; English transl. in Math. USSR Izv. **10** (1976).

BERNSTEIN, S. N.

[1] *Extremal properties of polynomials*, ONTI, Moscow, 1937 (rev. Russian transl. of [**6**]).

[2] *The extension of properties of trigonometric polynomials to entire functions of finite degree*, Izv. Akad. Nauk SSSR Ser. Mat. **12** (1948), 421–444. (Russian)

[3] *On entire functions of finite degree of several complex variables*, Dokl. Akad. Nauk SSSR **60** (1948), 949–952. (Russian)

[4] *On majorants of finite or quasi-finite growth*, Dokl. Akad. Nauk SSSR **65** (1949), 117–120. (Russian)

[5] *On additive majorants of finite growth*, Dokl. Akad. Nauk SSSR **66** (1949), 545–548. (Russian)

[6] *Leçons sur les propriétés extrémales et la meilleure approximation des fonctions analytiques d'une variable réelle*, Gauthier-Villars, Paris, 1926; reprint, S. N. Bernstein and Ch.-J. de la Vallée Poussin, *L'approximation*, Chelsea, New York, 1970.

[7] *Collected works*. Vol. I: *The constructive theory of functions* [1905–1930], Izdat. Akad. Nauk SSSR, Moscow, 1952. (Russian)

[8] *Sur les propriétés extrémales des polynomes et des fonctions entières sur l'axe réel*, C. R. Acad. Sci. Paris **176** (1923), 1782–1785.

BERNSTEIN, V.

[1] *Sur les propriétés caractéristiques des indicatrices de croissance*, C. R. Acad. Sci. Paris **202** (1936), 108–110.

[2] *Sopra una proposizione relativa alla crescenza delle funzioni olomorfe*, Ann. Scuola Norm. Sup. Pisa (2) **2** (1933), 381–399.

[3] *Sulle proprietà caratteristiche delle indicatrici di crescenza delle transcendenti intere d'ordine finito*, Reale Accad. Ital. Mem. Cl. Sci. Fis. Mat. Nat. **7** (1936), 131–189.

BOAS, JR., R. P.

[1] *Entire functions*, Academic Press, New York, 1954.

[2] *Asymptotic properties of functions of exponential type*, Duke Math. J. **20** (1953), 433–448.

[3] *The growth of analytic functions*, Duke Math. J. **13** (1946), 471–481.

BOAS, JR., R. P., SCHAEFFER, A. C.

[1] *A theorem of Cartwright*, Duke Math. J. **9** (1942), 879–883.

BOHR, H.

[1] *Fastperiodische Funktionen*, Springer, Berlin, 1932; English transl., Chelsea, New York, 1947.

[2] *Ueber fastperiodische ebene Bewegungen*, Comment. Math. Helv. **4** (1932), 51–64.

BORG, G.

[1] *Eine Umkehrung der Sturm-Liouvilleschen Eigenwertaufgabe. Bestimmung der Differentialgleichung durch die Eigenwerte*, Acta Math. **78** (1946), 1–96.

BRATIŠČEV, A. V., KOROBEĬNIK, JU. F.

[1] *On some characteristics of the growth of subharmonic functions*, Mat. Sb. **106 (148)** (1978), 44–65; English transl. in Math. USSR Sb. **35** (1978).

BUCK, R. C.

[1] *On the distribution of the zeros of an entire function*, J. Indian Math. Soc. **16** (1952), 147–149.

CARLEMAN, T.

[1] *Über die Approximation analytischer Funktionen durch lineare Aggregate von vorgegebenen Potenzen*, Ark. Mat. Astr. Fys. **17** (1922), no. 9.

CARLSON, F.

[1] *Sur une classe de séries de Taylor*, Thesis, Upsala, 1914.

[2] *Über ganzwertige Funktionen*, Math. Z. **11** (1921), 1–23.

CARTAN, H.

[1] *Sur les systèmes de fonctions holomorphes à variétés linéaires lacunaires et leurs applications*, Ann. Sci. École Norm. Sup. (3) **45** (1928), 255–346.

CARTWRIGHT, M. L.

[1] *On functions which are regular and of finite order in an angle*, Proc. London Math. Soc. (2) **38** (1934/35), 158–179.

[2] *On certain integral functions of order one*, Quart. J. Math. Oxford Ser. **7** (1936), 46–55.

[3] *Integral functions*, Cambridge Univ. Press, New York, 1956.

ČEBOTAREV, N. G.

[1] *On some modifications of Hurwitz's problem*, C. R. (Dokl.) Acad. Sci. URSS **35** (1942), 223–226.

[2] *On a particular type of transcendental equations*, C. R. (Dokl.) Acad. Sci. URSS **34** (1942), 38–41.

[3] *On entire functions with real interlacing roots*, C. R. (Dokl.) Acad. Sci. URSS **35** (1942), 195–197.

[4] *On the methods of Sturm and Fourier for transcendent functions*, C. R. (Dokl.) Acad. Sci. URSS **34** (1942), 2–4.

[5] *The Hurwitz problem for transcendental functions*, Trudy LVVAKA, Moscow, 1945. (Russian)

[6] *Über die Realität von Nullstellen ganzer transzendenter Funktionen*, Math. Ann. **99** (1928), 660–686.

ČEBOTAREV, N. G., MEĬMAN, N. N.

[1] *The Routh-Hurwitz problem for polynomials and entire functions*, Trudy Mat. Inst. Steklov. **26** (1949). (Russian)

ČERNOLJAS, V. I.

[1] *On the distribution of zeros of entire functions of generalized completely regular growth of variable order*, Dokl. Akad. Nauk SSSR **224** (1975), 549–552; English transl. in Soviet Math. Dokl. **16** (1975).

[2] *On entire functions of generalized completely regular growth of arbitrary upper and lower order*, Dokl. Akad. Nauk SSSR **234** (1977), 772–775; English transl. in Soviet Math. Dokl. **18** (1977).

DELSARTE, J.

[1] *Sur une extension de la formule de Taylor*, J. Math. Pures Appl. (9) **17** (1938), 213–231.

DOOB, J. L.

[1] *Boundary limit theorems for a half-space*, J. Math. Pures Appl. (9) **37** (1958), 385–392.

DUFFIN, R. J., SCHAEFFER, A. C.

[1] *Power series with bounded coefficients*, Amer. J. Math. **67** (1945), 141–154.

ESSÉN, M. R.

[1] *The* cos $\pi\lambda$ *theorem*, Lecture Notes in Math., Vol. 467, Springer-Verlag, Berlin and New York, 1975.

EVGRAFOV, M. A.

[1] *Asymptotic estimates and entire functions*, 2nd rev. ed., Fizmatgiz, Moscow, 1962; English transl. of 1st ed., Gordon and Breach, New York, 1961.

FAĬNBERG, E. D.

[1] *Estimates of the indicators of functions that are analytic and have nonintegral finite order in a half plane*. I, Teor. Funkciĭ, Funkcional. Anal. i Priložen. Vyp. 21 (1974), 13–30. (Russian)

[2] *Estimates of the indicators of functions that are analytic and have nonintegral finite order in a half plane*. I, Dopovīdī Akad. Nauk Ukraïn. RSR Ser. A **1976**, 24–27. (Ukrainian)

FIRSAKOVA, O. S.

[1] *Some problems concerned with interpolation by means of integral functions*, Dokl. Akad. Nauk SSSR **120** (1958), 477–480. (Russian)

FRIDMAN, G. A.

[1] *On the asymptotic behavior of functions which are dual in the sense of Young*, Mat. Sb. **70** (**112**) (1966), 518–525. (Russian)

[2] *Slowly increasing functions and their application*, Sibirsk. Mat. Z. **7** (1966), 1139–1160; English transl. in Siberian Math. J. **7** (1966).

FUCHS, W. H. J.

[1] *On the growth of functions of mean type*, Proc. Edinburgh Math. Soc. (2) **9** (1953/58), 53–70.

FUJIWARA, M.

[1] *Über die Nullstellen der ganzen Funktionen von Geschlecht Null und Eins*, Tôhoku Mat. J. **25** (1925), 27–35.

GANAPATHY IYER, V.

[1] *On the order and type of integral functions bounded at a sequence of points*, Ann. of Math. (2) **38** (1937), 311–320.

GEL'FAND, I. M., LEVITAN, B. M.

[1] *On the determination of a differential equation from its spectral function*, Izv. Akad. Nauk SSSR Ser. Mat. **15** (1951), 309–350; English transl. in Amer. Math. Soc. Transl. (2) **1** (1955).

GEL'FAND, I. M., RAĬKOV, D. A., ŠILOV, G. E.

[1] *Commutative normed rings*, Fizmatgiz, Moscow, 1960; English transl., Chelsea, New York, 1964.

GEL'FOND, A. O.

[1] *The calculus of finite differences*, GITTL, Moscow, 1952; German transl., VEB Deutscher Verlag, Berlin, 1958; French transl., Dunod, Paris, 1963.

[2] *The problem of the representation and uniqueness of an entire analytic function of first order*, Uspehi Mat. Nauk **3** (1937), 144–174. (Russian)

[3] *Sur un théorème de M. M. Weigert-Leau*, Mat. Sb. **36** (1929), 99–101.

[4] *Linear differential equations of infinite order with constant coefficients and asymptotic periods of entire functions*, Trudy Mat. Inst. Steklov. **38** (1951), 42–67; English transl. in Amer. Math. Soc. Transl. (1) **2** (1962).

[5] *Sur les systèmes complets de fonctions analytiques*, Mat. Sb. (N.S.) **4** (**46**) (1938), 149–156.

GEL'FOND, A. O., LEONT'EV, A. F.

[1] *On a generalization of Fourier series*, Mat. Sb. (N.S.) **29** (**71**) (1951), 477–500. (Russian)

GOL'DBERG, A. A.

[1] *The extremal indicator for an entire function with positive zeros*, Sibirsk. Mat. Ž. **3** (1962), 170–177. (Russian)

[2–5] *An integral with respect to a semiadditive measure and its application to the theory of entire functions*. I, II, III, IV, Mat. Sb. **58** (**100**) (1962), 289–334; **61** (**103**) (1963), 334–349; **65** (**107**) (1964), 414–453; **66** (**108**) (1965), 411–457; English transls. in Amer. Math. Soc. Transl. (2) **88** (1969).

[6] *Indicators of entire functions and an integral with respect to a nonadditive measure*, Contemporary Problems Theory Analytic Functions (Proc. Internat. Conf., Erevan, 1965), "Nauka", Moscow, 1966, pp. 88–93. (Russian)

GOL'DBERG, A. A., KORENKOV, N. E.

[1] *The asymptotic behavior of the logarithmic derivative of an entire function of completely regular growth*, Ukrain. Mat. Ž. **30** (1978), 25–32; English transl. in Ukrainian Math. J. **30** (1978).

[2] *Asymptotic behavior of the logarithmic derivative of an entire function of order zero*, Ukrain. Mat. Ž. **30** (1978), 291–298; English transl. in Ukrainian Math. J. **30** (1978).

GOL'DBERG, A. A., OSTROVSKIĬ, I. V.

[1] *Distribution of values of meromorphic functions*, "Nauka", Moscow, 1970. (Russian)

[2] *The derivatives and primitives of entire functions of completely regular growth*, Teor. Funkciĭ, Funkcional. Anal. i Priložen. Vyp. 18 (1973), 70–81. (Russian)

[3] *New investigations on the growth and distribution of values of entire and meromorphic functions of genus zero*, Uspehi Mat. Nauk **16** (1961), no. 4 (100), 51–62; English transl. in Russian Math. Surveys **16** (1961).

GONČAROV, V. L.

[1] *Interpolation processes and entire functions*, Uspehi Mat. Nauk **3** (1937), 113–143. (Russian)

GOVOROV, N. V.

[1] *On the indicator of functions of nonintegral order that are analytic and of completely regular growth in a half plane*, Dokl. Akad. Nauk SSSR **162** (1965), 495–498; English transl. in Soviet Math. Dokl. **6** (1965).

[2] *On the indicator of functions of integral order that are analytic and of completely regular growth in a half plane*, Dokl. Akad. Nauk SSSR **172** (1967), 763–766; English transl. in Soviet Math. Dokl. **8** (1967).

[3] *Extremal indicator of an entire function with positive zeros of a given upper and lower density*, Depovīdī Akad. Nauk Ukraïn. RSR **1966**, 148–150. (Ukrainian)

[4] *The estimation from below of a function that is subharmonic in a disk*, Teor. Funkciĭ, Funkcional. Anal. i Priložen. Vyp. 6 (1968), 130–150. (Russian)

GRIŠIN, A. F.

[1] *Functions holomorphic and of order zero in the interior of an angle*, Teor. Funkciĭ, Funkcional. Anal. i Priložen. Vyp. 1 (1965), 41–56. (Russian)

[2–4] *Regularity of the growth of subharmonic functions*. I–II, III, IV, Teor. Funkciĭ, Funkcional. Anal. i Priložen. Vyp. 6 (1968), 3–29; Vyp. 7 (1968), 59–84; Vyp. 8 (1969), 126–135. (Russian)

GROMMER, J.

[1] *Ganze transzendente Funktionen mit lauter reellen Nullstellen*, J. Reine Angew. Math. **144** (1914), 114–166.

GRUMAN, L.

[1] *The growth of entire solutions of differential equations of finite and infinite order*, Ann. Inst. Fourier (Grenoble) **22** (1972), fasc. 1, 211–238.

HAVINSON, S. JA.

[1] *The theory of extremal problems for bounded analytic functions satisfying additional conditions inside the domain*, Uspehi Mat. Nauk **18** (1963), no. 2 (110), §12, pp. 90–95; English transl. in Russian Math. Surveys **18** (1963).

HAYMAN, W. K.

[1] *Questions of regularity connected with the Phragmén-Lindelöf principle*, J. Math. Pures Appl. (9) **35** (1956), 115–126.

HEĬFIC, A. I.

[1] *A characterization of the zeros of certain special classes of entire functions of finite degree*, Teor. Funkciĭ, Funkcional. Anal. i Priložen. Vyp. 9 (1969), 3–13. (Russian)

HELLERSTEIN, S., WILLIAMSON, J.

[1] *Derivatives of entire functions and a question of Pólya*. I, II, Trans. Amer. Math. Soc. **227** (1977), 227–249; **234** (1977), 497–503.

HERMITE, C.

[1] *Extract d'une lettre de M. Ch. Hermite de Paris à Mr. Borchardt de Berlin sur le nombre des racines d'une équation algébrique comprises entre des limites données*, J. Reine Angew. Math. **52** (1856), 39–51; reprinted in his *Oeuvres*, Vol. I, Gauthier-Villars, Paris, 1905, pp. 397–414.

HIRSCHMAN, JR., I. I., JENKINS, J. A.

[1] *Note on a result of Levine and Lifschitz*, Proc. Amer. Math. Soc. **1** (1950), 390–393.

IBRAGIMOV, I., KELDYŠ, M.

[1] *Sur l'interpolation des fonctions entières*, Mat. Sb. **20 (62)** (1947), 283–291. (Russian; French summary)

JENSEN, J. L. W. V.

[1] *Recherches sur la théorie des équations*, Acta Math. **36** (1912/13), 181–195.

JESSEN, B., TORNEHAVE, H.

[1] *Mean motions and zeros of almost periodic functions*, Acta Math. **77** (1945), 137–279.

JUROV, P. G.

[1] *Asymptotic estimates of entire functions that are given by canonical products*, Mat. Zametki **10** (1971), 641–648; English transl. in Math. Notes **10** (1971).

KAC'NELSON, V. Ė.

[1] *Generalization of the Paley-Wiener theorem on the representation of entire functions of finite degree*, Teor. Funkciĭ, Funkcional. Anal. i Priložen. Vyp. 1 (1965), 99–110. (Russian)

KAHANE, J.-P.

[1] *Quasi analyticité des fonctions moyenne-périodiques*, C. R. Acad. Sci. Paris **236** (1953), 569–571.

[2] *Sur quelques problèmes d'unicité et de prolongement, relatifs aux fonctions approchables par des sommes d'exponentielles*, Ann. Inst. Fourier (Grenoble) **5** (1953/54), 39–130.

KARASEVA, T. M.

[1] *On the inverse Sturm Liouville problem for a non-Hermitian operator*, Mat. Sb. **32 (74)** (1953), 477–484. (Russian)

KOLOMIĬCEVA, T. A.

[1] *The asymptotic behavior of an entire function with regular distribution of zeros*, Teor. Funkciĭ, Funkcional. Anal. i Priložen. Vyp. 15 (1972), 35–43. (Russian)

KONDRATJUK, A. A.

[1] *Entire functions with a finite maximal density of zeros*. I, II, Teor. Funkciĭ, Funkcional. Anal. i Priložen. Vyp. 10 (1970), 57–70; Vyp. 11 (1970), 35–40. (Russian)

[2] *The extremal indicator for entire functions with positive zeros*. I, II, Litovsk. Mat. Sb. **7** (1967), 79–117; **8** (1968), 65–85.

[3] *A Fourier series method for entire and meromorphic functions of completely regular growth*, Mat. Sb. **106 (148)** (1978), 386–408; English transl. in Math. USSR Sb. **35** (1978).

KONDRATJUK, A. A., FRIDMAN, A. N.

[1] *The lower indicator of an entire function of zero genus with positive zeros*, Ukrain. Mat. Ž. **24** (1972), 106–109; English transl. in Ukrainian Math. J. **24** (1962).

[2] *The limit value of the lower indicator and estimates from below for entire functions with positive zeros*, Ukrain. Mat. Ž. **24** (1972), 488–494; English transl. in Ukrainian Math. J. **24** (1972).

KOREVAAR, J.

[1] *The zeros of approximating polynomials and the canonical representation of an entire function*, Duke Math. J. **18** (1951), 573–592.

[2] *Approximation and interpolation applied to entire functions*, Thesis, Univ. of Leiden, 1949.

KRASIČKOV [-TERNOVSKIĬ], I. F.

[1] *Comparison of entire functions of finite order by means of the distribution of their zeros*, Mat. Sb. **70 (112)** (1966), 198–230. (Russian)

[2] *Estimates from below for entire functions of finite order*, Sibirsk. Mat. Ž. **6** (1965), 840–861; English transl. in Amer. Math. Soc. Transl. (2) **83** (1969).

[3] *Comparison of entire functions of integral order by means of the distribution of their zeros*, Mat. Sb. **71 (113)** (1966), 405–419. (Russian)

KREĬN, M. G.

[1] *Concerning a special class of entire and meromorphic functions*, Article VI in Ahiezer and Kreĭn [1].

[2] *A contribution to the theory of entire functions of exponential type*, Izv. Akad. Nauk SSSR Ser. Mat. **11** (1947), 309–326. (Russian; English summary)

[3] *The fundamental propositions of the representation theory of Hermitian operators with deficiency index* (m, m), Ukrain. Mat. Ž. **1** (1949), no. 2, 3–66; English transl. in Amer. Math. Soc. Transl. (2) **97** (1971).

[4] *On the indeterminate case in the Sturm-Liouville boundary value problem in the interval* $(0, \infty)$, Izv. Akad. Nauk SSSR Ser. Mat. **16** (1952), 293–324. (Russian)

[5] *On the representation of functions by Fourier-Stieltjes integrals*, Učen. Zap. Kuĭbyšev. Gos. Ped. Inst. **7** (1943), 123–148. (Russian)

[6] *Solution of the inverse Sturm-Liouville problem*, Dokl. Akad. Nauk SSSR **76** (1951), 21–24. (Russian)

[7] *Determination of the density of a nonhomogeneous symmetric cord by its frequency spectrum*, Dokl. Akad. Nauk SSSR **76** (1951), 345–348. (Russian)

KREĬN, M. G., LEVIN, B. JA.

[1] *On entire almost periodic functions of exponential type*, Dokl. Akad. Nauk SSSR **64** (1949), 285–287. (Russian)

KUDINA, L. S.

[1] *Estimates for functions that are representable in the form of a difference of functions subharmonic in the ball*, Teor. Funkciĭ, Funkcional. Anal. i Priložen. Vyp. 14 (1971), 58–67. (Russian)

LAGUERRE, E.

[1] *Sur les fonctions du genre zéro et du genre un*, C. R. Acad. Sci. Paris **95** (1882), 828–831.

[2] *Oeuvres*, Vol. I, Gauthier-Villars, Paris, 1898.

LANDKOF, N. S.

[1] *Foundations of modern potential theory*, "Nauka", Moscow, 1966; English transl., Springer-Verlag, Berlin and New York, 1972.

LAPIN, G. P.

[1] *On interpolation in the class of entire functions of finite order and finite type*, Mat. Sb. **29 (71)** (1951), 565–580. (Russian)

[2] *Interpolation in the class of entire functions of finite order*, Izv. Vysš. Učebn. Zaved. Matematika **1959**, no. 5 (12), 146–153. (Russian)

LEAU, L.

[1] *Recherche des singularités d'une fonction définie par un développement de Taylor*, J. Math. Pures Appl. (5) **5** (1899), 365–425.

LEONT'EV, A. F.

[1] *Series of Dirichlet polynomials and their generalizations*, Trudy Mat. Inst. Steklov. **39** (1951). (Russian)

[2] *On interpolation in the class of entire functions of finite order and normal type*, Dokl. Akad. Nauk SSSR **66** (1949), 153–156. (Russian)

[3] *On an interpolation problem*, Dokl. Akad. Nauk SSSR **66** (1949), 331–334. (Russian)

[4] *On completeness of a system of exponential functions in a curvilinear strip*, Mat. Sb. (N.S.) **36 (78)** (1955), 555–568. (Russian)

[5] *On a question of interpolation in the class of entire functions of finite order*, Mat. Sb. (N.S.) **41 (83)** (1957), 81–96. (Russian)

[6] *Values of an entire function of finite order at given points*, Izv. Akad. Nauk SSSR Ser. Mat. **22** (1958), 387–394. (Russian)

[7] *On interpolation in the class of entire functions of finite order*, Proc. Third All-Union Math. Congr. (Moscow, 1956), Vol. 1, Izdat. Akad. Nauk SSSR, Moscow, 1956, pp. 86–87. (Russian)

[8] *On the analyticity of the kernel of an integral transformation*, Mat. Sb. **62 (104)** (1963), 31–38. (Russian)

[9] *On the convergence of a sequence of Dirichlet polynomials*, Dokl. Akad. Nauk SSSR **108** (1956), 23–26. (Russian)

LEVIN, B. JA.

[1] *Sur la croissance d'une fonction entière suivant un rayon et la distribution de ses zéros suivant leurs arguments*, Mat. Sb. **2 (44)** (1937), 1097–1142. (Russian; French summary)

[2] *Sur certaines applications de la série d'interpolation de Lagrange dans la théorie des fonctions entières*, Mat. Sb. **8 (50)** (1940), 437–454. (Russian; French summary)

[3] *On entire functions of finite degree which are of regular growth*, Dokl. Akad. Nauk SSSR **71** (1950), 601–604. (Russian)

[4] *Sur la constant séculaire d'une fonction holomorphe presque périodique*, C. R. (Dokl.) Acad. Sci. URSS **33** (1941), 182–185.

[5] *Hermite's criterion for integral functions of exponential type*. I, II, C. R. (Dokl.) Acad. Sci. URSS **41** (1943), 47–50, 99–100.

[6] *On some extremal properties of entire functions of finite degree*, Dokl. Akad. Nauk SSSR **65** (1949), 605–608. (Russian)

[7] *On a special class of entire functions and on related extremal properties of entire functions of finite degree*, Izv. Akad. Nauk SSSR Ser. Mat. **14** (1950), 45–84. (Russian)

[8] *Some extremal properties of entire functions of several variables*, Dokl. Akad. Nauk SSSR **78** (1951), 861–864. (Russian)

[9] *On a class of entire functions*, Dokl. Akad. Nauk SSSR **78** (1951), 1085–1088. (Russian)

[10] *The general form of special operators on entire functions of finite degree*, Dokl. Akad. Nauk SSSR **79** (1951), 397–400. (Russian)

[11] *On functions determined by their values on a certain interval*, Dokl. Akad. Nauk SSSR **70** (1950), 949–952. (Russian)

[13] *On a generalization of the Fejér-Riesz theorem*, C. R. (Dokl.) Acad. Sci. URSS **52** (1946), 291–294.

[14] *On functions holomorphic in a half-plane*, Trudy Odessa Ped. Inst. **3** (1941), 5–14. (Russian)

[15] *Operators preserving inequalities between entire functions*, Dokl. Akad. Nauk SSSR **89** (1953), 605–608. (Russian)

[16] *Generalization of a theorem of Cartwright concerning an entire function of finite degree bounded on a sequence of points*, Izv. Akad. Nauk SSSR Ser. Mat. **21** (1957), 549–558. (Russian)

[17] *On the convergence of sequences of H-polynomials*, Mat. Sb. **66 (108)** (1965), 384–397. (Russian)

[18] *Entire functions*, Izdat. Moskov. Gos. Univ., Moscow, 1971. (Russian)

LEVIN, B. JA., LIVŠIC, M.

[1] *Quasi-analytic functions represented by Fourier series*, Mat. Sb. **9 (51)** (1941), 693–711. (Russian)

LEVIN, B. JA., OSTROVSKIĬ, I. V.

[1] *On the dependence of the growth of an entire function on the distribution of the zeros of its derivatives*, Sibirsk. Mat. Z. **1** (1960), 427–455; English transl. in Amer. Math. Soc. Transl. (2) **32** (1963).

LEVINSON, N.

[1] *Gap and density theorems*, Amer. Math. Soc., Providence, R. I., 1940.

LEVITAN, B. M.

[1] *Almost periodic functions*, GITTL, Moscow, 1953. (Russian)

[2] *Über eine Verallgemeinerung der Ungleichungen von S. Bernstein and H. Bohr*, C. R. (Dokl.) Acad. Sci. URSS **15** (1937), 169–172.

[3] *The application of generalized shift operators to linear differential equations of the second order*, Uspehi Mat. Nauk **4** (1949), no. 1 (29), 3–112; English transl. in Amer. Math. Soc. Transl. (1) **10** (1962).

LIHT, M. K.

[1] *A remark on the Paley-Wiener theorem on entire functions of finite degree*, Uspehi Mat. Nauk **19** (1964), no. 1 (115), 169–171. (Russian)

LINDELÖF, E.

[1] *Sur les fonctions entières d'ordre entier*, Ann. Sci. École Norm. Sup. (3) **22** (1905), 369–395.

[2] *Le calcul des résidues et ses applications à la théorie des fonctions*, Gauthier-Villars, Paris, 1905.

[3] *Mémoire sur la théorie des fonctions entières de genre fini*, Acta Soc. Sci. Fenn. **31** (1903), no. 1.

LINDWART, E., PÓLYA, G.

[1] *Über einen Zusammenhang zwischen der Konvergenz von Polynomfolgen und der Verteilung ihrer Wurzeln*, Rend. Circ. Mat. Palermo **37** (1914), 297–304.

LJUBARSKIĬ, JU. I.

[1] *On the approximation of an entire function of exponential type by trigonometric polynomials*, Math. Phys. and Functional Anal. Vyp. 3, Fiz.-Tehn. Inst. Nizkih Temperatur Akad. Nauk Ukrain. SSR, Kharkov, 1972, pp. 56–67. (Russian)

LOGVINENKO, V. N.

[1] *Construction of an entire function with a prescribed indicator in the case of a prescribed integral proximate order*, Funktional. Anal. i Priložen. **6** (1972), no. 4, 87–88; English transl. in Functional Anal. Appl. **6** (1972).

[2] *Binomial asymptotics for a class of entire functions*, Dokl. Akad. Nauk SSSR **205** (1972), 1037–1039; English transl. in Soviet Math. Dokl. **13** (1972).

[3] *On a theorem of Valiron and Titchmarsh*, Vestnik Har'kov. Univ. no. 83, Ser. Meh.-Mat. Vyp. 37 (1972), 3–13. (Russian)

LUNC, G. L.

[1] *On a theorem connected with the growth of entire functions of integer order*, Izv. Akad. Nauk Armjan. SSR Ser. Mat. **5** (1970), 358–370. (Russian)

MACAEV, V. I.

[1] *On the growth of entire functions that admit a certain estimate from below*, Dokl. Akad. Nauk SSSR **132** (1960), 283–286; English transl. in Soviet Math. Dokl. **1** (1960).

MACINTYRE, A. J.

[1] *Laplace's transformation and integral functions*, Proc. London Math. Soc. (2) **45** (1938), 1–20.

MANDELBROJT, S.

[1] *Séries de Fourier et classes quasi-analytiques de fonctions*, Gauthier-Villars, Paris, 1935.

MARČENKO, V. A.

[1] *Some questions of the theory of one-dimensional linear differential operators of the second order*, I, II, Trudy Moskov. Mat. Obšč. **1** (1952), 327–420; **2** (1953), 3–83; English transl. in Amer. Math. Soc. Transl. (2) **101** (1973).

[2] *On some questions of the approximation of continuous functions on the whole real axis*. III, Har'kov, Gos. Univ. Učen. Zap. **34** = Zap. Mat. Otd. Fiz.-Mat. Fak. i Har'kov. Mat. Obšč. (4) **22** (1950), 115–125. (Russian)

MARKUŠEVIČ, A. I.

[1] *Theory of analytic functions*, GITTL, Moscow, 1950; English transl., Vols. I, II, III, Prentice-Hall, Englewood Cliffs, N. J., 1965, 1967.

[2] *Sur les bases dans l'espace des fonctions analytiques*, Mat. Sb. **17** (**59**) (1945), 211–252; English transl. in Amer. Math. Soc. Transl. (2) **22** (1962).

MATISON, H.

[1] *Certain integral functions related to exponential sums*, Duke Math. J. **4** (1938), 9–29.

MEĬMAN, N. N.

[1] *On the distribution of the zeros of an integer function*, C. R. (Dokl.) Acad. Sci. URSS **40** (1943), 179–181.

[2] *On conditions under which the derivative of a majorant function is a majorant for the derivative of the function*, Dokl. Akad. Nauk SSSR **71** (1950), 609–612. (Russian)

[3] See N. G. Čebotarev and N. N. Meĭman [**1**].

NEĬMARK, JU. I.

[1] *Stability of linearized systems*, LKVVIA, Leningrad, 1949. (Russian)

NEVANLINNA, R.

[1] *Eindeutige analytische Funktionen*, Springer, Berlin, 1936; rev. English transl., 1970.

[2] *Über die Eigenschaften meromorpher Funktionen in einem Winkelraum*, Acta Soc. Sci. Fenn. **50** (1920/26), no. 12 (1925).

NOBLE, M. E.

[1] *Extensions and applications of a Tauberian theorem due to Valiron*, Proc. Cambridge Philos. Soc. **47** (1951), 22–37.

OBRECHKOFF, N. [OBREŠKOV, N.]

[1] *Sur les fonctions entières limites de polynomes dont les zéros sont réels et entrelacés*, Mat. Sb. **9 (51)** (1941), 421–428.

OSTROVSKIĬ, I. V.

[1] *Generalization of a theorem of M. G. Kreĭn*, Dokl. Akad. Nauk SSSR **116** (1957), 742–745. (Russian)

[2] *The connection between the growth of a meromorphic function and the distribution of the arguments of its values*, Izv. Akad. Nauk SSSR Ser. Mat. **25** (1961), 277–328. (Russian)

[3] *The connection between the growth of a meromorphic function and the distribution of the arguments of its values*, Dokl. Akad. Nauk SSSR **132** (1960), 48–51; English transl. in Soviet Math. Dokl. **1** (1960).

[4] *Some asymptotic properties of entire functions with real negative zeros*, Har′kov. Gos. Univ. Učen. Zap. **120** = Zap. Meh.-Mat. Fak. i Har′kov. Mat. Obšč. (4) **28** (1961), 23–32. (Russian)

PALEY, R. E. A. C., WIENER, N.

[1] *Fourier transforms in the complex domain*, Amer. Math. Soc., Providence, R. I., 1934.

PFLUGER, A.

[1] *Die Wertverteilung und das Verhalten von Betrag und Argument einer speziellen Klasse analytischer Funktionen*. I, II, Comment. Math. Helv. **11** (1938/39), 180–214; **12** (1939/40), 25–65.

[2] *Über Interpolation ganzer Funktionen*, Comment. Math. Helv. **14** (1941/42), 314–349.

[3] *Über das Anwachsen von Funktionen, die in einem Winkelraum regulär und vom Exponentialtypus sind*, Compositio Math. **4** (1937), 367–372.

[4] *Über ganze Funktionen ganzer Ordnung*, Comment. Math. Helv. **18** (1945/46), 177–203.

[5] *Über gewisse ganze Funktionen vom Exponentialtypus*, Comment. Math. Helv. **16** (1943/44), 1–18.

[6] *On analytic functions bounded at the lattice points*, Proc. London Math. Soc. (2) **42** (1936/37), 305–315.

PHRAGMÉN, E., LINDELÖF, E.

[1] *Sur une extension d'un principe classique de l'analyse et sur quelques propriétés des fonctions monogènes dans le voisinage d'un point singulier*, Acta Math. **31** (1908), 381–406.

PLANCHEREL, M., PÓLYA, G.

[1] *Fonctions entières et intégrales de Fourier multiples*, Comment. Math. Helv. **9** (1936/37), 224–248.

PÓLYA, G.

[1] *Untersuchungen über Lücken und Singularitäten von Potenzreihen*, Math. Z. **29** (1929/30), 549–640.

[2] *Über Annäherung durch Polynome mit lauter reellen Wurzeln*, Rend. Circ. Mat. Palermo **36** (1913), 279–295.

[3] *Algebraische Untersuchungen über ganze Funktionen vom Geschlechte Null und Eins*, J. Reine Angew. Math. **145** (1915), 224–249.

[4] *Sur une question concernant les fonctions entières*, C. R. Acad. Sci. Paris **158** (1914), 330–333.

PÓLYA, G., SCHUR, J.

[1] *Über zwei Arten von Faktorenfolgen in der Theorie der algebraischen Gleichungen*, J. Reine Angew. Math. **144** (1914), 89–113.

PONTRJAGIN, L. S.

[1] *On the zeros of some elementary transcendental functions*, Izv. Akad. Nauk SSSR Ser. Mat. **6** (1942), 115–134; English transl. in Amer. Math. Soc. Transl. (2) **1** (1955).

POTAPOV, V. P.

[1] *On the divisors of an almost periodic polynomial*, Sb. Trudov Inst. Mat. Akad. Nauk Ukrain. SSR **12** (1949), 36–81. (Russian)

POVZNER, A. JA.

[1] *On differential equations of Sturm-Liouville type on a half-axis*, Mat. Sb. (N.S.) **23 (65)** (1948), 3–52; English transl. in Amer. Math. Soc. Transl. (1) **4** (1962).

REGENSBURGER, M.

[1] *Asymptotische Darstellung und Lage der Nullstellen spezieller ganzer Funktionen (Exponentialsummen)*, Math. Ann. **111** (1935), 505–540.

RONKIN, L. I.

[1] *On approximation of entire functions by trigonometric polynomials*, Dokl. Akad. Nauk SSSR **92** (1953), 887–890. (Russian)

RUBEL, L. A., TAYLOR, B. A.

[1] *A Fourier series method for meromorphic and entire functions*, Bull. Soc. Math. France **96** (1968), 53–96.

ŠAGINJAN, A. L.

[1] *A fundamental inequality in function theory, and its applications*, Izv. Akad. Nauk Armjan. SSR Ser. Fiz.-Mat. Nauk **12** (1959), 3–25. (Russian)

[2] *On some inequalities and their applications in function theory*, Dokl. Akad. Nauk SSSR **129** (1959), 284–287. (Russian)

SANSONE, G.

[1] *Equazioni differenziali nel campo reale*. I, 2nd ed., Zanichelli, Bologna, 1948.

SCHUR, J.

[1] *Zwei Sätze über algebraische Gleichungen mit lauter reellen Wurzeln*, J. Reine Angew. Math. **144** (1914), 75–88.

ŠEREMETA, M. N.

[1] *On the connection between the growth of the maximum of the modulus of an entire function and the moduli of the coefficients of its power series expansion*, Izv. Vysš. Učebn. Zaved. Matematika **1967**, no. 2 (57), 100–108; English transl. in Amer. Math. Soc. Transl. (2) **88** (1970).

[2] *The connection between the growth of functions of order zero which are entire or analytic in a disk and their power series coefficients*, Izv. Vysš. Učebn. Zaved. Matematika **1968**, no. 6 (73), 115–121. (Russian)

SERGIENKO, E. N.

[1] *The growth of entire functions that admit a certain estimation from below*, Teor. Funkciĭ, Funkcional. Anal. i Priložen. Vyp. 15 (1972), 78–96. (Russian)

[2] *On the growth of meromorphic functions that admit a certain estimate from below*, Teor. Funkciĭ, Funkcional. Anal. i Priložen. Vyp. 21 (1974), 83–104. (Russian)

SUBHANKULOV, M. L.

[1] *Tauberian theorems with remainder*, "Nauka", Moscow, 1976. (Russian)

SUBHANKULOV, M. L., TOLBAEV, B.

[1] *Distribution of the zeros of some classes of entire functions*, Dokl. Akad. Nauk Tadžik. SSR **7** (1964), no. 11, 3–5. (Russian)

TITCHMARSH, E. C.

[1] *The theory of functions*, 2nd ed., Clarendon Press, Oxford, 1939.

[2] *Introduction to the theory of Fourier integrals*, Clarendon Press, Oxford, 1937.

TJAN, M. M.

[1] *On an application of the Carleman-Subhankulov Tauberian theorem*, Izv. Akad. Nauk UzSSR Ser. Fiz.-Mat. Nauk **1963**, no. 3, 18–20. (Russian)

TOLBAEV, B.

[1] *On the zeros of an entire function of finite degree*, Dokl. Akad. Nauk Tadžik. SSR **6** (1963), no. 7, 9–12. (Russian)

[2] *A sharpening and generalization of a theorem of Titchmarsh on the zeros of entire functions*, Izv. Akad. Nauk Tadžik. SSR Otdel. Fiz.-Tehn. i Him. Nauk **1965**, No. 1, 37–42. (Russian)

TROŠIN, G. D.

[1] *On the interpolation of functions analytic in a half plane*, Mat. Sb. **36 (78)** (1955), 39–56. (Russian)

[2] *On the interpolation of functions analytic in an angle*, Mat. Sb. **39 (81)** (1956), 239–252. (Russian)

[3] *On the problem of interpolating functions analytic in an angle*, Izv. Vysš. Učebn. Zaved. Matematika **1963**, no. 1 (32), 150–157. (Russian)

UŠAKOVA, I. V.

[1] *Some uniqueness theorems for functions subharmonic and meromorphic in the unit circle*, Dokl. Akad. Nauk SSSR **137** (1961), 1319–1322; English transl. in Soviet Math. Dokl. **2** (1961).

[2] *Some estimates for functions subharmonic in a disc*, Har'kov. Gos. Univ. Učen. Zap. 135 = Zap. Meh.-Mat. Fak. i Har'kov. Mat. Obšč. (4) **29** (1963), 53–66. (Russian)

[3] *Some estimates for functions subharmonic in a ball*, Teor. Funkciĭ, Funkcional. Anal. i Priložen. Vyp. 1 (1965), 219–223. (Russian)

VALIRON, G.

[1] *Lectures on the general theory of integral functions*, Privat, Toulouse, 1923.

[2] *Sur la formule d'interpolation de Lagrange*, Bull. Sci. Math. (2) **49** (1925), 181–192, 203–224.

[3] *Sur les fonctions entières d'ordre nul et d'ordre fini, et en particulier les fonctions à correspondance régulière*, Ann. Fac. Sci. Univ. Toulouse (3) **5** (1913), 117–257.

VIDENSKIĬ, V. S.

[1] *Entire transcendental functions and their applications to the study of N-functions*, Mat. Sb. **62 (104)** (1963), 121–139. (Russian)

WEYL, H.

[1] *Mean motion*. I, II, Amer. J. Math. **60** (1938), 889–896; **61** (1939), 143–148.

WIENER, N.

[1] *Tauberian theorems*, Ann. of Math. (2) **33** (1932), 1–100.

[2] *The Fourier integral and certain of its applications*, Cambridge Univ. Press, Cambridge, 1933; reprint, Dover, New York, 1959.

WIENER, N., PITT, H. R.

[1] *On absolutely convergent Fourier-Stieltjes transforms*, Duke Math. J. **4** (1938), 420–436.

WIGERT, S.

[1] *Sur les fonctions entières*, Öfversigt Kongl. Vetenskaps Akad. Förhandlingar **57** (1900), 1001–1011 (1901).

ZYGMUND, A.

[1] *Trigonometrical series*, PWN, Warsaw, 1935.